# Spread-Spectrum Communications

Advanced Microprocessors, *Edited by A. Gupta and H. D. Toong*
Biological Effects of Electromagnetic Radiation, *Edited by J. M. Osepchuk*
Electronic Switching: Digital Central Office Systems of the World, *Edited by A. E. Joel, Jr.*
A Guide for Writing Better Technical Papers, *Edited by C. Harkins and D. L. Plung*
Compendium of Communication and Broadcast Satellites: 1958 to 1980, *Edited by M. P. Brown*
Digital MOS Integrated Circuits, Edited by *M. I. Elmasry*
Modern Active Filter Design, Edited by *R. Schaumann, M. A. Soderstrand, and K. R. Laker*
Optical Fiber Technology, II, *Edited by C. K. Kao*
Analog MOS Integrated Circuits, *Edited by P. R. Gray, D. A. Hodges, and R. W. Brodersen*
Interference Analysis of Communication Systems, *Edited by P. Stavroulakis*
Data Conversion Integrated Circuits, *Edited by D. J. Dooley*
Satellite Communications, *Edited by H. L. Van Trees*
Programs for Digital Signal Processing, *Edited by the Digital Signal Processing Committee, IEEE*
Automatic Speech & Speaker Recognition, *Edited by N. R. Dixon and T. B. Martin*
Speech Analysis, *Edited by R. W. Schafer and J. D. Markel*
The Engineer in Transition to Management, *Edited by I. Gray*
Modern Spectrum Analysis, *Edited by D. G. Childers*
Phase-Locked Loops & Their Application, *Edited by W. C. Lindsey and M. K. Simon*
Spread Spectrum Techniques, *Edited by R. C. Dixon*
Literature Survey of Communications Satellite Systems and Technology, *Edited by J. H. W. Unger*
Communications Satellite Systems: An Overview of the Technology, *Edited by R. G. Gould and Y. F. Lum*
Waveform Quantization and Coding, *Edited by N. S. Jayant*
Computer Networking, *Edited by R. P. Blanc and I. W. Cotton*
Communications Channels: Characterization and Behavior, *Edited by B. Goldberg*
Optical Fiber Technology, *Edited by D. Gloge*
Selected Papers in Digital Signal Processing, II, *Edited by the Digital Signal Processing Committee, IEEE*
A Guide for Better Technical Presentations, *Edited by R. M. Woelfle*
Data Communications Via Fading Channels, *Edited by K. Brayer*
Computer Communications, *Edited by P. E. Green, Jr. and R. W. Lucky*

# Spread-Spectrum Communications

**Edited by**

## Charles E. Cook
## Fred W. Ellersick

Communications Division
MITRE Corporation

## Laurence B. Milstein
Department of Electrical Engineering and
Computer Sciences
University of California, San Diego

## Donald L. Schilling
Department of Electrical Engineering
City College of New York

Published for the IEEE Communications
Society by the IEEE PRESS.

**IEEE
PRESS**

The Institute of Electrical and Electronics Engineers, Inc., New York

Copyright © 1983 by

THE INSTITUTE OF ELECTRICAL AND ELECTRONICS ENGINEERS, INC.

345 East 47th Street, New York, NY 10017

*All rights reserved.*

PRINTED IN THE UNITED STATES OF AMERICA

Sole Worldwide Distributor (Exclusive of the IEEE)

JOHN WILEY & SONS, INC.

605 Third Ave.

New York, NY 10158

Wiley Order Number: 471-87886-3

IEEE Order Number: PC01636

**Library of Congress Cataloging in Publication Data**

Main entry under title:

Spread-spectrum communications.

Includes indexes.
1. Spread spectrum communications.   I. Cook,
Charles E. (Charles Emerson), 1926-    .  II. IEEE
Communications Society.
TK5102.5.S666   1983        621.38'043        83-12665
ISBN 0-87942-170-3

iv

# Contents

# Spread-Spectrum Communications

# Part I
# Perspectives

SPREAD-SPECTRUM communications systems—surely one of the most important and dynamic areas of communications today—have many applications. The most important application, and the one most frequently discussed, is the use of spectrum spreading to communicate in the presence of intentional interference (or "jamming"). Other important applications are to reject *unintentional* interference, to lower the probability of a "friendly" transmission being detected by an "enemy" receiver, to combat multipath problems, and to provide multiple access to a communications system shared by a number of users. This book addresses all of these applications of spread-spectrum communications, and considers both military and commercial situations.

Many military situations present a severe challenge to the designer or operator of a communications system. This challenge is often twofold. Can a communications signal be designed in such a way as to make it extremely difficult for the enemy to detect? Or, if the signal *is* detected and the enemy attempts to jam the intended receiver, can spectrum spreading be employed to overcome the effects of the jammer? Depending upon the nature of the jammer's tactics, the signal designer may decide to use either linear or nonlinear spreading sequences, possibly in conjunction with error correction coding and interleaving; a classic game theory situation results from this action-reaction interplay.

In commercial situations, more efficient spectrum utilization can sometimes be achieved by the use of spectrum spreading. The U.S. Federal Communications Commission is currently investigating commercial usage of spread-spectrum communications systems in order to establish, modify, or expand the rules and regulations concerning such usage.[1]

In a third arena—academia—research in spread spectrum is becoming increasingly common, although the number of courses devoted to spread spectrum is still relatively small. Such courses, however, are probably not long in coming, since they can make a student collect all of his knowledge concerning error-correction coding, modulation, channel characteristics, effects of Gaussian and non-Gaussian noise, and synchronization, and then apply this cumulative knowledge at the system level to solve real-world problems. The papers collected in this book address the issues identified above, and provide essential background material for courses in spread-spectrum communications.

The first paper deals with the early history of spread-spectrum communications; the next three papers provide additional historical anecdotes. The fifth paper is a tutorial addressing the general concepts of spread-spectrum communications. The sixth is concerned with techniques for detecting an unknown spread-spectrum signal in noise. Other areas covered in papers in succeeding parts are interference rejection, multiple access, synchronization, and practical hardware problems.

The early history of spread-spectrum communications is presented in a paper by Scholtz ("The Origins of Spread-Spectrum Communications") that is truly archival in nature. It provides many fascinating details and interesting photographs, and represents the combined efforts of the author and many other contributors. Additional historical anecdotes are presented in the next three papers: Scholtz ("Notes on Spread-Spectrum History"), Price ("Further Notes and Anecdotes on Spread-Spectrum Origins"), and Bennett ("Secret Telephony as a Historical Example of Spread-Spectrum Communication").

The fifth paper is an extensive tutorial by Pickholtz, Schilling, and Milstein ("Theory of Spread Spectrum Communications—A Tutorial"). It includes discussions on antijamming considerations, code-division multiple access, multipath, properties of common spreading techniques, and techniques for acquisition and tracking.

Additional perspective and tutorial information is provided in the paper by Krasner ("Optimal Detection of Digitally Modulated Signals"). It includes a discussion of those parts of detection theory necessary to understand the low probability of intercept (LPI) aspects of a variety of spectrum-spreading techniques. The emphasis is on cases wherein the signal-to-noise ratio per symbol is small compared to unity.

[1] "Authorization of spread spectrum and other wideband emissions not presently provided for in the FCC rules and regulations," FCC Docket 81-413.

# The Origins of Spread-Spectrum Communications

ROBERT A. SCHOLTZ, FELLOW, IEEE

*Abstract*—This monograph reviews events, circa 1920–1960, leading to the development of spread-spectrum communication systems. The WHYN, Hush-Up, BLADES, F9C-A/Rake, CODORAC, and ARC-50 systems are featured, along with a description of the prior art in secure communications, and introductions to other early spread-spectrum communication efforts. References to the available literature from this period are included.

## I. INTRODUCTION

*"Whuh? Oh,"* said the missile expert. *"I guess I was off base about the jamming. Suddenly it seems to me that's so obvious, it must have been tried and it doesn't work."*

*"Right, it doesn't. That's because the frequency and amplitude of the control pulses make like purest noise— they're genuinely random. So trying to jam them is like trying to jam FM with an AM signal. You hit it so seldom, you might as well not try."*

*"What do you mean, random? You can't control anything with random noise."*

The captain thumbed over his shoulder at the Luanae Galaxy. *"They can. There's a synchronous generator in the missiles that reproduces the* same *random noise, peak by pulse. Once you do that, modulation's no problem. I don't know* how *they do it. They just do. The Luanae can't explain it; the planetoid developed it."*

England put his head down almost to the table. *"The same random,"* he whispered *from the very edge of sanity.*

−from "The Pod in the Barrier" by Theodore Sturgeon, in *Galaxy*, Sept. 1957; reprinted in *A Touch of Strange* (Doubleday, 1958).

LED by the Global Positioning System (GPS) and the Joint Tactical Information Distribution System (JTIDS), the spread-spectrum (SS) concept has emerged from its cloak of secrecy. And yet the history of this robust military communication technique remains largely unknown to the modern communication engineer. Was it a spark of genius or the orderly evolution of a family of electronic communication systems that gave birth to the spread-spectrum technique? Was it, as Frank Lehan said, an idea whose time had come? Was the spread-spectrum technique practiced in World War II, as Eugene Fubini declares? Was it invented in the 1920's as the U.S. Patent Office records suggest? Was Theodore Sturgeon's lucid description of a jam-proof guidance system precognition, extrasensory perception, or a security leak? Let's examine the evidence.

Manuscript received August 26, 1981; revised January 29, 1982. This work was supported in part by the Army Research Office under Grant DAAG-29-79-C-0054.

The author is with the Department of Electrical Engineering, University of Southern California, Los Angeles, CA 90007.

The basic signal characteristics of modern spread-spectrum systems are as follows.

1) The carrier is an unpredictable, or pseudorandom, wide-band signal.

2) The bandwidth of the carrier is much wider than the bandwidth of the data modulation.

3) Reception is accomplished by cross correlation of the received wide-band signal with a synchronously generated replica of the wide-band carrier.

The term "pseudorandom" is used specifically to mean random in appearance but reproducible by deterministic means. A key parameter of SS systems is the number of essentially orthogonal signaling formats which could be used to communicate a data symbol. Here two signaling formats are orthogonal in the sense that the signals employed in one format for communication would not be detected by a processor for the other format, and vice versa. We shall call the number of possible orthogonal signaling formats the multiplicity factor of the communication link.

While conventional communication systems other than wide-band frequency modulation (FM) have a multiplicity factor near unity, SS systems typically have multiplicity factors in the thousands. Thus, a well-designed SS system forces a jammer to guess which of a multiplicity of orthogonal signaling formats is being used, or to reduce significantly his power per format by jamming all possibilities. The receiver is not confronted with a similar problem since it is privy to the pseudorandom sequence of signaling formats which the transmitter will use for communication. Excluding the notion 2) that the multiplicity factor be large, all of these characteristics are apparent in Sturgeon's story.

The multiplicity factor is the nominal value of the more widely used term, processing gain. In terms of signal-to-interference power ratios (SIR's), the processing gain of an SS system is the factor by which the receiver's input SIR is multiplied to yield the SIR at the output of the receiver's correlation detector. The input SIR can be interpreted as a computation over the ensemble of possible orthogonal signaling formats, while the output SIR pertains only to the system selected by the transmitter and receiver for communication.

Spread-spectrum systems, because of the nature of their signal characteristics, have at least five important performance attributes.

1) Low probability of intercept (LPI) can be achieved with high processing gain and unpredictable carrier signals when power is spread thinly and uniformly in the frequency domain, making detection against noise by a surveillance receiver difficult. A low probability of position fix (LPPF) attribute goes one step further in including both intercept and direction finding (DFing) in its evaluation. Low probability of signal exploita-

Reprinted from *IEEE Trans. Commun.*, vol. COM-30, pp. 822–854, May 1982.

tion (LPSE) may include additional effects, e.g., source identification, in addition to intercept and DFing.

2) Antijam (AJ) capability can be secured with an unpredictable carrier signal. The jammer cannot use signal observations to improve its performance in this case, and must rely on jamming techniques which are independent of the signal to be jammed.

3) High time resolution is attained by the correlation detection of wide-band signals. Differences in the time of arrival (TOA) of the wide-band signal, on the order of the reciprocal of the signal bandwidth, are detectable. This property can be used to suppress multipath and, by the same token, to render repeater jammers ineffective.

4) Transmitter–receiver pairs using independent random carriers can operate in the same bandwidth with minimal cochannel interference. These systems are called spread-spectrum code-division multiple-access (CDMA) systems.

5) Cryptographic capabilities result when the data modulation cannot be distinguished from the carrier modulation, and the carrier modulation is effectively random to an unwanted observer. In this case the SS carrier modulation takes on the role of a key in a cipher system. A system using indistinguishable data and SS carrier modulations is a form of privacy system.

We will see how the search for a system with one or more of these features led to independent discoveries of the spread-spectrum concept.

Three basic system configurations for accomplishing the reception of a wide-band, seemingly unpredictable carrier have been pioneered:

1) Transmitted reference (TR) systems accomplish detection of the unpredictable wide-band carrier by transmitting two versions of the carrier, one modulated by data and the other unmodulated. These versions, being separately recoverable by the receiver (e.g., they may be spaced apart in frequency), are the inputs to a correlation detector which extracts the data (see Fig. 1).

2) Stored reference (SR) systems require independent generation at transmitter and receiver of pseudorandom wideband waveforms which are identical in their essential characteristics. The receiver's SS carrier generator is adjusted automatically to keep its output in close synchronism with the arriving SS carrier. Detection then proceeds in a manner similar to the TR system (see Fig. 2).

3) Filter systems generate a wide-band transmitted signal by pulsing a matched filter (MF) having a long, wide-band, pseudorandomly controlled impulse response. Signal detection at the receiver is accomplished by an identically pseudorandom, synchronously controlled matched filter which performs the correlation computation (see Fig. 3). Rapid pseudorandom variation of the transmitter's impulse response ensures the unpredictability of the wide-band carrier.

Theodore Sturgeon's missile guidance system was an SR-SS system, the configuration which is prevalent today.

Spread-spectrum systems are also classified[1] by the tech-

---

[1] In the sequel, the uncapitalized word "classified" will usually be a formal security designation; likewise for the words "secret" and "confidential."

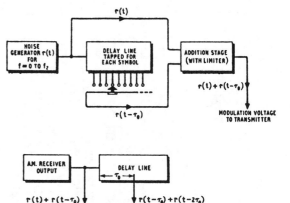

Fig. 1. Block diagram of a basic transmitted reference system using pure noise as a carrier and time-shift keying for data modulation. No effort has been made to separate the data and reference channels in this system proposed by the East German Professor F. H. Lange. His configuration is nearly identical to that suggested in Project Hartwell a decade earlier. (Diagram from [11].)

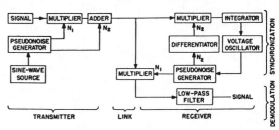

Fig. 2. JPL's first attempt at a stored reference spread-spectrum design is shown here. This particular system uses one unmodulated noise signal $N_2$ for synchronizing the receiver's pseudonoise generator and another $N_1$ for carrying data. Most SR-SS systems do not use a separate signal for SS signal synchronization. (Diagram from [132].)

nique which they employ to achieve the wide-band carrier signal. Here are some digital system examples.

1) Pure noise was sometimes used as a carrier in early experimental systems, giving ideal randomness properties. However, pure noise is useful only in a TR system. If a jammer for some reason cannot use the reference channel signal to jam the data channel signal, then the multiplicity factor for a system using antipodal modulation of binary data on the noise carrier is

multiplicity factor = 2(data bit time)(carrier bandwidth).

When the jammer can gain access to both channels, the multiplicity factor reduces to unity, i.e., there is no AJ advantage.

2) Direct sequence (DS) systems employ pseudorandom sequences, phase-shift-keyed (PSK) onto the carrier, for spreading. The time spent in transmitting a single carrier symbol from this sequence is called the chip time of the system. With

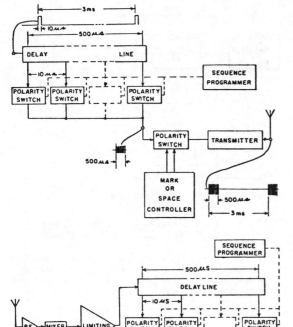

Fig. 3. Costas and Widmann's Phantom system employs a pulsed delay line with pseudorandomly controlled taps summed to provide an SS signal for modulation. An identically structured system with a synchronous replica of the tap controller is used to construct a matched pseudorandom filter for data detection at the receiver. (Diagrams modified from [157].)

binary PSK data antipodally modulated on this SS carrier, the resultant system's multiplicity factor is given by

$$\text{multiplicity factor} = (\text{data bit time})/(\text{chip time}).$$

Direct sequence systems possess excellent TOA resolution and are efficient in power amplifier operation.

3) Frequency modulation with frequency wobbled over a wide bandwidth is a carryover from early radar technology. Some FM-SS systems may have more predictable carrier modulation formats (e.g., linear FM, chirp) and, hence, may be more susceptible to jamming. If the jammer does not use the modulation structure to its advantage then the multiplicity factor for an FM system is approximately

$$\text{multiplicity factor} = (\text{data bit time})(\text{FM carrier bandwidth}).$$

4) Frequency hopping (FH) systems achieve carrier spreading by driving a frequency synthesizer with a pseudorandom sequence of numbers spanning the range of the synthesizer. In the pure form of this system, data is usually frequency-shift-keyed (FSK) onto the spread carrier. With binary FSK modulation at one data bit per carrier hop, the multiplicity factor is

given by

$$\text{multiplicity factor} = (\text{hop time}) (\text{frequency range})$$

assuming the frequencies used are packed as tightly as orthogonality permits. Typically the new carrier phase cannot be predicted when a frequency hop occurs. However, fully coherent FH is possible, e.g., with a minimum-shift-keying (MSK) format, which is virtually indistinguishable from DS operation. Present technology achieves the highest multiplicity factor using frequency hopping, provided that a sufficient bandwidth can be allocated.

5) Time hopping (TH) to spread the carrier is achieved by randomly spacing narrow transmitted pulses. In TH systems, the reciprocal of the average duty factor is a measure of the multiplicity factor. That is,

$$\text{multiplicity factor} = (\text{average pulse spacing})/(\text{pulse width}).$$

Time hopping is useful as a form of random time multiplexing allowing both transmitter and receiver use of the same antenna.

Some systems are hybridized from the above to achieve the advantages of several different techniques. For example, JTIDS uses TH, FH, and DS modulation simultaneously for carrier spreading.

Analog (e.g., voice) modulated SS systems have been developed, with the multiplicity factor for a well-designed system given approximately by

$$\text{multiplicity factor} = (\text{carrier bandwidth})/(\text{output}$$

$$\text{bandwidth})$$

the output bandwidth being the bandwidth of the receiver correlator's output signal.

A historical look at the development of spread-spectrum systems will not only shed light on their origins, but will also provide an interesting case history of the interaction between basic research and the evolution of technology.

## II. PRIOR KNOWLEDGE

Before we can assess the ingenuity which went into the development of the first spread-spectrum systems, we must examine the state of the art in communication theory and technology in the 1940's. Here are capsule summaries of technical events in the prehistory of SS communications.

### Radar Innovations

From the 1920's through World War II, many systems incorporating some of the characteristics of spread-spectrum systems were studied. The birth of RADAR, i.e., RAdio Detection And Ranging, occurred in the mid-1920's when scientists used echo sounding to prove the existence of an ionized gas layer in the upper atmosphere. British scientists E. V. Appleton and M. A. F. Barnett performed this feat by transmitting a frequency modulated wave upward and listening for the return echo [1]. Applications of this concept to aircraft instrumentation were obvious and FM altimetry became a reality in the 1930's, with all major combatants in World War II making use of this technology [2]. Typically, linear-sawtooth or sinusoidal modulations were used in these early systems. The frequency modulation generally serves two purposes, 1) it ameliorates the problem of interference due to leakage of the transmitted signal directly into the receiver, and 2) it makes possible the measurement of propagation delay and, hence, range.

Historically, the development of pulsed radars has received more attention than that of continuous wave (CW) radars, since isolation of the transmitting and receiving systems is a lesser problem in this case. By the end of World War II, the Germans were developing a linear FM pulse compression (chirp) system called Kugelschale, and a pulse-to-pulse frequency-hopping radar called Reisslaus [3]. In 1940 Prof. E. Huttman was issued a German patent on a chirp pulse radar, while U.S. patents on this type of system were first filed by R. H. Dicke in 1945 and by S. Darlington in 1949 [4]. The mid-1940's also saw the formulation of the matched filter concept for maximum signal-to-noise ratio (SNR) pulse detection by North [5] and Van Vleck and Middleton [6]. This development indicated that the performance of optimum signal detection procedures in the presence of white noise depends only on the ratio of signal energy to noise power spectral density, thus leaving the choice of waveform open to satisfy other design criteria (e.g., LPI or AJ). Resolution, accuracy, and ambiguity properties of pulse waveforms finally were placed on a sound theoretical basis by P. M. Woodward [7] in the early 1950's and excellent treatises on this subject are now available [8], [9].

Spectrum spreading was a natural result of the Second World War battle for electronic supremacy, a war waged with jamming and antijamming tactics. On the Allied side by the end of the war, every heavy bomber, excluding Pathfinders, on the German front was equipped with at least two jammers developed by the Radio Research Laboratory (RRL) at Harvard [10]. The use of chaff was prevalent, the Allies consuming 2000 tons per month near the end. On the German side, it is estimated that at one time as many as 90 percent of all available electronic engineers were involved in some way in a tremendous, but unsuccessful, AJ program. Undoubtedly Kugelschale and Reisslaus were products of this effort.

In a postwar RRL report [10], the following comment on AJ design is notable:

"In the end, it can be stated that the best anti-jamming is simply good engineering design and the spreading of the operating frequencies."

Certainly, spectrum spreading for jamming avoidance (AJ) and resolution, be it for location accuracy or signal discrimination (AJ), was a concept familiar to radar engineers by the end of the war.

• • •

In the late 1950's and early 1960's the East German scientist F. H. Lange toured Europe and the United States collecting (unclassified) material for a book on correlation techniques. Published first in 1959 with its third edition being translated into English [11] a few years later, Lange's book contains some references all but unnoticed by researchers on this side of the Atlantic. The most intriguing of these is to the work of Gustav Guanella of Brown, Boveri, and Company in Switzerland. Among Guanella's approximately 100 patents is one [12] filed in 1938, containing all the technical characteristics of an SR-SS radar! The radiated signal in Guanella's CW radar is "composed of a multiplicity of different frequencies the energies of which are small compared with the total energy" of the signal. His prime examples of such signals are acoustic and electrical noise, and an oscillator whose frequency is "wobbled at a high rate between a lower and upper limit."

Ranging is accomplished by adjusting an internal signal delay mechanism to match the external propagation delay experienced by the transmitted signal. Delay matching errors are detected by cross correlating the internally delayed signal with a 90 degree phase-shifted (across the whole transmission band) version of the received signal. Thus, if the transmitted signal is of the form

$$\sum_n a_n \cos(\omega_n t + \phi_n),$$

the propagation delay is $\tau_p$, and the internal delay is $\tau_i$, then the measured error is proportional to

$$\sum_n a_n^2 \sin[\omega_n(\tau_p - \tau_i)].$$

This ensemble of phase-locked loops, all rolled up into one neat package, possesses a tracking loop S-curve which looks like the Hilbert transform of the transmitted signal's autocorrelation function. Undoubtedly, Guanella's patent contains possibly the earliest description of a delay-locked loop. In addition to accurate range measurement, the patent further indicates improved performance against interference.

Guanella used the same type of error-sensing concept in an earlier patent filed in 1936 [13]. Many of his inventions are cited as prior art in later patents. For a modern treatment of delay-locked loops see [14], [15].

### Developments in Communication Theory

Probabilistic modeling of information flow in communication and control systems was the brainchild of the preeminent mathematician Norbert Wiener of the Massachusetts Institute of Technology (M.I.T.). In 1930 Wiener published his celebrated paper "Generalized Harmonic Analysis" [16] developing the theory of spectral analysis for nonperiodic infinite-duration functions. When World War II began, Wiener was asked by the National Defense Research Committee (NDRC) to produce a theory for the optimal design of servomechanisms. Potential military applications for this theory existed in many gunfire control problems [17]. The resultant work [18], published initially in 1942 as a classified report and often referred to as the "Yellow Peril," laid the groundwork for modern continuous-parameter estimation theory. By 1947 Wiener's filter design techniques were in the open literature [19].

• • •

In 1915 E. T. Whittaker concluded his search for a distinctive function among the set of functions, all of which take on the same specified values at regularly spaced points along the real line. This "function of royal blood whose distinguished properties set it apart from its bourgeois brethren" is given by

$$x(t) = \sum_n x(n/2W) \sin \left[\pi(2Wt - n)\right] / \left[\pi(2Wt - n)\right]$$

where $x(n/2W)$ represents the specified values and $x(t)$ is the cardinal function of the specified values, a function whose Fourier transform is strictly band limited in the frequency domain [20]-[23]. Based on this result, the sampling theory used in a communication context by Hartley [24], Nyquist [25], Kotelnikov [26], and Shannon [27] states that a function band limited to $W$ Hz can be represented without loss of information by samples spaced $1/(2W)$ seconds apart. Generalizations [28], [29] of this result indicate that a set of approximately $2TW$ orthogonal functions of $T$ seconds duration and occupying $W$ Hz can be constructed. In SS theory, this provides the connection between the number of possible orthogonal signaling formats and system bandwidth. Although earlier Nyquist [25] and later Gabor [30] both had argued using Fourier series that $2TW$ samples should be sufficient to represent a $T$-second segment of such a band-limited signal, it was Shannon who made full use of this classical tool.

• • •

Claude E. Shannon, who had known Wiener while a graduate student at M.I.T., joined the Bell Telephone Laboratories (BTL) in 1941, where he began to establish a fundamental theory of communication within a statistical framework. Much of his work, motivated in good part by the urge to find basic cryptographic and cryptanalytic design principles [31], was classified well past the end of the Second World War. In a paper [27] first presented in 1947, Shannon invoked the cardinal expansion in formulating a capacity for delivering information (negentropy [31]) over channels perturbed solely by additive Gaussian noise. He showed that this channel capacity was maximized by selectively spreading the signaling spectrum so that wherever deployed within designated bandwidth confines—but only there—the sum of its power spectral density plus that of the independent noise should lie as uniformly low as possible, yet utilize all the average transmitter power available. Moreover, this capacity was met by sending a set of noise-like waveforms and distinguishing between them at the receiver via a minimum-distance criterion akin to correlation-testing the observed signal against locally stored waveform replicas. Even though Shannon's theory did not apply directly to many interference/jamming situations, his remarkable concepts and results [32] profoundly influenced communication engineers' thinking.

Driven by the intense interest in the theories of Wiener and Shannon, the Institute of Radio Engineers (IRE) formed the Professional Group on Information Theory, which commenced publishing in 1953 [33]. The first three chairmen of this Group were, in order, Nathan Marchand, W. G. Tuller, and Louis deRosa. Marchand and deRosa, close friends, were at that time playing key roles in the development of SS systems; Tuller had independently but rather heuristically arrived at one of Shannon's capacity formulas.

*Correlator Mechanization*

One of the difficult problems which Guanella faced (by his account without any knowledge of Wiener's work) was to fabricate a device which will perform a weighted correlation computation on two inputs. Specifically, a means was needed for taking two inputs $x_1(t)$ and $x_2(t)$ and computing

$$y(t) = \int^t x_1(u)x_2(u)w(t - u)\, du$$

where $y(t)$ is the device output and $w(t)$ is the weighting function. The difficulty here is not with the weighting (i.e., filtering) operation, but with the prior multiplication of $x_1(t)$ by $x_2(t)$, and in particular with the range of inputs over which accurate multiplication can be accomplished. As shall be seen later, the ability to mechanize the correlation operation precisely is essential in building high-performance SS systems.

• • •

In 1942 Nathan Marchand, then a 26-year-old engineer working for ITT's Federal Telephone and Radio Corporation in New York, discussed his radio receiver invention with ITT engineer and patent attorney Paul Adams. Marchand had developed a converter for demodulating a received FM signal of known frequency wobbulation by mixing it with a time-aligned, heterodyned replica of the wobbulated signal to produce a signal of constant intermediate frequency (IF) which could then be narrow-band filtered. The receiver's antimultipath attributes designed by Marchand and additional anti-interference features suggested by Adams appear in a 1947 patent [34]. Later during World War II, after studying Wiener's "Yellow Peril," Marchand was able to dub his converter a bandpass correlator.

• • •

At M.I.T. in 1947, Prof. Yuk Wing Lee commenced research into the implications of Wiener's theories and the new directions they inspired for engineering science. Soon thereafter Lee was joined by Jerome Wiesner and Thomas Cheatham, and their collective efforts led to the development of the first high-performance electronic correlators. In August, 1949, they applied for a patent [35] and in October they reported applications of correlation techniques to detection problems [36]. Continuing this work, Henry Singleton proceeded to innovate an all-digital correlator [37].

*Protected Communications*

The earliest patent [38] presently construed by the U.S. Patent Office as being spread spectrum in nature was filed in 1924 by Alfred N. Goldsmith, one of the three founders of the IRE. Goldsmith proposed to counteract the fading effects encountered in short wave communication, due to multipath, by

"radiating a certain range of wave frequencies which are modulated in accordance with the signal and actuating a receiver by means of energy collected on all the frequencies, preferably utilizing a wave which is continuously varied in wave frequency over a certain range of cycles recurring in a certain period."

Certainly, we can identify this as a form of FM-SS transmission. However, the envisioned data modulation was by amplitude (AM) with reception by a broadly tuned AM receiver. Hence, the correlation detector necessary to achieve the full benefits of SS operation was not inherent in Goldsmith's disclosure. For a World War II disclosure on an FM-SS chirp communication system with a more sophisticated receiver, claiming a primitive form of diversity reception for multipath signals and a capability against narrow-band interference, see [39].

• • •

In 1935 Telefunken engineers Paul Kotowski and Kurt Dannehl applied for a German patent on a device for masking voice signals by combining them with an equally broad-band noise signal produced by a rotating generator [40]. The receiver in their system had a duplicate rotating generator, properly synchronized so that its locally produced noise replica could be used to uncover the voice signal. The U.S. version of this patent was issued in 1940, and was considered prior art in a later patent [48] on DS-SS communication systems. Certainly, the Kotowski–Dannehl patent exemplifies the transition from the use of key-stream generators for discrete data encryption [41] to pseudorandom signal storage for voice or continuous signal encryption. Several elements of the SS concept are present in this patent, the obvious missing notion being that of bandwidth expansion.

The Germans used Kotowski's concept as the starting point for developing a more sophisticated capability that was urgently needed in the early years of World War II. Gottfried Vogt, a Telefunken engineer under Kotowski, remembers testing a system for analog speech encryption in 1939. This employed a pair of irregularly slotted or sawtoothed disks turning at different speeds, for generating a noise-like signal at the transmitter, to be modulated/multiplied by the voice signal. The receiver's matching disks were synchronized by means of two transmitted tones, one above and one below the encrypted voice band. This system was used on a wire link from Germany, through Yugoslavia and Greece, to a very- and/or ultrahigh frequency (VHF/UHF) link across the Mediterranean to Rommel's forces in Derna, Libya.

Bell Telephone Laboratories improved on Telefunken's original scheme and applied for patents on their telephony apparatus in 1941 [42], [43]. BTL's disclosures and applications were placed under secrecy order since their system was being depended on by Roosevelt, Churchill, and other Allied leaders during World War II [44]. This system, officially called the X System and nicknamed the Green Hornet, changed its prerecorded keys daily for security. BTL continued its work on key-stream generation and in the mid-1940's filed for patents on all-electronic key generators which combined several short keys of relatively prime lengths to produce key streams possessing long periods [45], [46]. Such schemes also had been studied by Shannon [31] at BTL, but his comments on these were deleted before republication of his declassified report on secrecy systems in the *Bell System Technical Journal*. All of these BTL

patent filings remained under secrecy order until the 1970's when the orders were rescinded and the patents issued.

• • •

One can view the advanced Telefunken system as an avatar of a TR system since specialized signals are transmitted to solve the disk synchronization problem. Another novel variation of TR voice communication was conceived in the U.S. during the war years by W. W. Hansen. This Sperry/M.I.T. Radiation Laboratory scientist is noted for his invention of the microwave cavity resonator and for his joint effort with the Varian brothers in originating the Klystron. In a 1943 patent application [47], Hansen describes a two-channel system with the reference channel used solely for the transmission of noise, and the intelligence channel bearing the following signal (in complex notation):

$$\exp\left\{ j\int^t [\omega_1 + An(t')]\, dt' \right\} \cdot \exp\left\{ j\int^t [\omega_2 + Bv(t')]\, dt' \right\}$$

where $n(t)$ is a filtered version of the noise communicated via the reference channel, $v(t)$ is the voice signal, and assuming $n(t)$ and $v(t)$ are at comparable levels, $A \gg B$. The intelligence signal is the result of combining a wide-swing noise-modulated FM waveform with a narrow-swing voice-modulated FM waveform in a device "similar in principle of operation to the mixers used in superheterodyne receivers."

At the receiver, the reference channel signal is used to reconstruct the first of the above factors, and that in turn is mixed with the received intelligence signal to recover the voice-modulated waveform represented by the second factor. This receiver mixer appears to be similar in many respects to Marchand's bandpass correlator.

To overcome some of the fundamental weaknesses of TR systems (more on this later), Hansen threw in an additional twist: the filtering of the reference channel signal, used to generate $n(t)$, was made time dependent, with transmitter and receiver filters required to change structure in virtual synchronism under the control of a chronometer. This structural change could not be detected in any way by observing the reference channel.

When presenting his design along the TR-FM-SS lines, Hansen notes that the intelligence signal cannot be heard by unauthorized narrow-band receivers because "such wide-swing modulations in effect tune the transmitted wave outside the frequency band of the unauthorized listener's receiver for the greater portion of the time and thus make such a receiver inoperative." Concerned about the fact that a wide-band FM receiver might conceivably recover the signal $An(t) + Bv(t)$, he also concludes that "if therefore the noise [$n(t)$] has important components throughout the range of signal frequencies and if the swing due to the noise is large compared to the swing due to the signal [$v(t)$], deciphering is impossible."

Curiously enough, due to the use of an exponential form of modulation, Hansen's design is constructed as a TR-FM-SS communication system at radio frequency (RF), but equivalently at demodulated baseband, it is simply a "typical noise masking" add/subtract TR system. (This latter appraisal of

[47] is from the case file—open to the public as for any issued patent—in Crystal City on an SR-SS invention [48] of major importance to a later period in this history.) Moreover, except for its TR vulnerabilities, Hansen's system is good AJ design, and as he points out, a large amount of additional noise can be injected at the RF output of the transmitter's intelligence channel for further masking without seriously degrading system performance.

Surprisingly, without the spectral spreading and chronometer-controlled reference signal filters, Hansen's system would bear a strong resemblance to a TR-FM system described in 1922 by Chaffee and Purington [49]. Hence, the concept of transmitting a reference signal to aid in the demodulation of a disguised information transmission is at least 60 years old!

• • •

Dr. Richard Gunther, an employee of the German company Siemens and Halske during World War II, recalls another speech encryption system involving bandwidth expansion and noise injection. In a fashion similar to the Western Electric B1 Privacy System, the voice subbands were pseudorandomly frequency scrambled to span 9 kHz and pure noise was added to fill in the gaps. The noise was later eliminated by receiver filtering in the speech restoration process. Tunis was the terminus of a link operated at 800 MHz and protected by this system.

• • •

With a German invasion threatening, Henri Busignies of ITT's Paris laboratories made an unprecedented visit to the French Patent Office to remove all vestiges of material on his latest inventions. He then headed across the Atlantic, joined ITT's Federal Telephone and Radio Corporation, and quickly filed a landmark patent on a radar moving-target indicator. Busignies, a remarkably prolific inventor who over his lifetime was granted about 140 patents, soon collaborated with Edmond Deloraine and Louis deRosa in applying for a patent [50] on a facsimile communication system with intriguing antijam possibilities here set forth:

> The system uses a transmitter which sends each character "a plurality of times in succession," and a receiver in which the character signals are visually reproduced, "one on top another . . . to provide a cumulative effect." If "the interference signals are not transmitted to provide such a cumulative effect, the interference will form only a bright background but will not prevent the signals from being read."

From a jamming viewpoint, the real novelty in the disclosure is in the fact that the mechanisms which read the characters at the transmitter and write the characters at the receiver synchronously vary in rate of operation. Thus, attempts to jam the system with periodic signals which might achieve the "cumulative effect" at the receiver output will be unsuccessful.

In a sequel patent filed six weeks later [51], it is specified that the facsimile pulse modulation should have a low average duty cycle, be characterized by steep wavefronts, and have high peak-to-average power, in order to attain superior protec-

tion. This time-wobbling system is obviously an early relative of modern TH-SS systems. Concurrently with these efforts, deRosa covered similar applications in the field of radar by filing what may be the first patent on random jittering of pulse repetition frequencies [52].

Test results of the facsimile system are mentioned briefly in a 1946 NDRC Division 15 report [53] which also points out in a radar context that

> "There is factual evidence that tunability is foremost as an AJ measure. Frequency spread of radars, which serves the same function, is a corollary and equally important." With regard to communications, "RF carrier frequency scrambling and time modulation of pulses with time scrambling" are possible communication antijam measures.

The report's final recommendations state that "any peacetime program to achieve protection against jamming should not be concerned with the type of equipment already in service, but should be permitted an unrestricted field of development." This was sensible advice to follow, when practical, in the postwar years.

• • •

Another study of protected communications was launched when ITT submitted Proposal 158A to the NDRC for consideration. Although the original proposal only suggested the use of redundancy in time or frequency as a possible AJ measure, a 1944 report [54] stated with regard to jamming that

> "The enemy can be forced to maintain a wide bandwidth if we use a coded frequency shifting of our narrower printer bandwidth so that it might at any time occupy any portion of a wider band."

This clear suggestion of FH-SS signaling was not explored further in the last year of the contract. Several different tone signaling arrangements were considered for communication to a printer at rates on the order of one character per second. Synchronization of these digital signaling formats was accomplished in open-loop fashion using precision tuning forks as reference clocks.

> "These forks are temperature compensated over a wide range and are mounted in a partial vacuum, so that their rate is not affected by the low barometric pressures encountered at high altitudes. Their accuracy is of the order of one part in a million, so that once the receiving distributor was phased with the transmitted signal, it remained within operable limits for two hours or more. A differential gear mechanism, operated by a crank handle on the front panel, was provided for rephasing the receiving distributor whenever this became necessary."

The receiving distributor controlled the reinitializing of $L$-$C$ tank circuits tuned to detect transmitted tones. Due to their high $Q$, these circuits performed an integrate-and-dump operation during each distributor cycle. This detector was a significant improvement over the prior art, a fact indeed recognized intuitively by ITT, rather than derived from correlation principles.

ITT's printer communication system was tested at Rye Lake Airport on February 21, 1945. The printer performed well in the presence of jamming 11 dB stronger than the desired signal, and under conditions where voice on the same channel was not intelligible [55]. The interference in this test consisted of an AM radio station.

• • •

As far as technology is concerned, all of the above communication systems share a common propensity for the use of electromechanical devices, especially where signal storage and synchronization are required. Undoubtedly in the 1940's the barriers to be overcome in the development of SS communications were as much technological as they were conceptual. The final 1940's state-of-the-art vignette to follow is in an area whose need for lightweight, rugged systems did much to drive communication technology toward all-electronic and eventually all-solid-state systems.

*Missile Guidance*

During World War II the NDRC entered the realm of guided missiles with a variety of projects [56] including the radio control in azimuth only (AZON) of conventionally dropped bombs (VB's) which trailed flares for visibility, radar-controlled glide bombs (GB's) such as the Pelican and the Bat, and the remotely controlled ROC VB-10 using a television link. Now documented mostly through oral history and innocuous circuit patents, one of several secure radio guidance efforts took place at Colonial Radio, predecessor of the Sylvania division at Buffalo, NY. This project was under the direction of Madison Nicholson, with the help of Robert Carlson, Alden Packard, Maxwell Scott, and Ernest Burlingame. The secret communications system concept was stimulated, so Carlson thinks, by talks with Navy people who wanted a system like the "Flash" system which the Germans used for U-boat transmissions. However, it wasn't until the Army Air Force at Wright Field posed the following problem that the Colonial Radio effort began seriously.

The airfoil surfaces of the glide bombs were radio controlled by a mother plane some distance away, sometimes with television display (by RCA) relayed back to the plane so that closed-loop guidance could be performed. It was feared that soon the Germans would become adept at jamming the control. To solve this problem Colonial Radio developed a secure guidance system based on a pulsed waveform which hopped over two diverse frequency bands. This dual band operation led to the system's nickname, Janus, after the Roman god possessing two faces looking in opposite directions. Low duty cycle transmission was used, and although the radio link was designed to be covert, the system could withstand jamming in one of its two frequency bands of operation and still maintain command control.

The Colonial Radio design's transmitter for the mother aircraft was designated the AN/ARW-4, and the corresponding glide bomb receiver was the AN/CRW-8. Testing of the radio

guidance system took place at Wright Field in 1943, under the direction of Lt. Leonhard Katz, Capts. Walter Brown and Theodore Manley, and Project Engineer Jack Bacon. The contract, including procurement of two transmitters and seven receivers, was completed by June 1944 [57].

ITT also participated in these World War II guidance programs, notably with a system called Rex [58]. One patent, evidently resulting from this work and filed in 1943 by Emile Labin and Donald Grieg [59], is interesting because it suggests CDMA operation in pulse code modulation (PCM) systems by slight changes in the pulse repetition frequency. In addition, the patent notes the jammer's inherent problem of trying to deliver its interference to the victim receiver in synchronism with the transmitted pulse train. However, the notion of multiplicity factor or spectrum spreading is not mentioned.

A third guidance system for the control of VB's and GB's was proposed by the Hammond Laboratory, a privately organized research group with a history in radio guidance dating back to 1910 [60], [61]. The Hammond system used a complicated modulation format which included a carrier wobbled over 20 kHz to protect against tone interference, and FM control signals amplitude modulated onto this frequency-modulated carrier [58]. More notable in this history than the system itself is the fact that Ellison Purington of the Hammond Laboratory in 1948 came close to describing a TH and FH carrier for a radio control system in a patent application [62]. The actual details describe a TH-SS system with control signals coded into the transmission using frequency patterns. Magnetic or optical recording "on a rotating member driven by a constant speed motor" was one suggestion for the storage of different time hopping patterns, while another possibility mentioned involves delay line generation of pulse train patterns. Control keys are hidden in the way that the patterns are mapped onto different frequencies to create "radiations . . . randomly distributed in time and in frequency."

Other salient patents, based on World War II AJ and command/control efforts, include those of Hoeppner [63] and Krause and Cleeton [64].

## III. EARLY SPREAD-SPECTRUM SYSTEMS

The following accounts of early SS developments are given to some extent as system genealogies. As we shall see, however, the blood lines of these system families are not pure, there being a great deal of information exchange at the conceptual level despite the secrecy under which these systems were developed. Approximate SS system time lines for several of the research groups tracked here are shown in Fig. 4. Since the SS concept was developed gradually during the same period that Shannon's work on information theory became appreciated, J. R. Pierce's commentary [65] on the times should be borne in mind:

"It is hard to picture the world before Shannon as it seemed to those who lived in it. In the face of publications now known and what we now read into them, it is difficult to recover innocence, ignorance, and lack of understanding. It is easy to read into earlier work a generality that came only later."

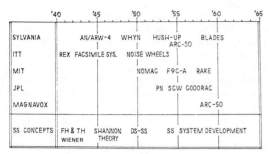

Fig. 4.  Approximate time lines for the systems and concepts featured in this history.

## WHYN

Many of the roots of SS system work in the U.S.A. can be traced back to the pioneering of FM radar by Major Edwin Armstrong during the early phases of World War II. The Armstrong technique involved transmitting a sinusoidally modulated wide-band FM signal, and then heterodyne-mixing the return from the target with a frequency offset replica whose identical sinusoidal modulation could be phase-shifted manually. When properly adjusted, the output of the mixer was very narrowband and the phase difference between the transmitted modulation and that of the replica then gave a measure of the two-way propagation delay to the target. Certainly, this created a bandwidth expansion and compression methodology, primitive though it was since the FM wobbulation was simply a sine wave.

Sylvania's Bayside Laboratories on Long Island received the contract in World War II to continue development of the Armstrong radar, and Bayside engineers started considering more exotic modulation signals to improve its ranging characteristics. This led, in 1946, to a Sylvania subcontract from Republic Aviation under Army Air Force Project MX-773, to develop a guidance system for a 500–1500 mile surface-to-surface missile. Although celestial and inertial navigation were possibilities, it was decided that a radio-controlled system using FM ranging would be the most easily realized. Two navigation systems were studied, the first being a circular-navigation, two-ground-station system in which the range to each station was determined separately using the FM radar technique. For each range measurement a pair of ultrastable oscillators would be used, one in the ground station and one in the missile. After oscillator initialization at launch, the phase difference between the received signal modulation and the replica modulation would be proportional to range.

The second system was designed to overcome location errors that would occur in the first system due to drift between the oscillators. A third ground station was introduced for transmitting a reference signal to which the missile and ground station oscillators were locked. Then the difference between the ranges to the three ground stations could be measured at the missile, the intersection of the corresponding hyperbolic loci indicating its location. The acronym WHYN, standing for Wobbulated HYperbolic Navigation, was the descriptor coined by Norman Harvey for this system. From the receiver's point of view, the circular navigation system was a primitive SR-FM-SS system while the WHYN system was TR-FM-SS.

Accurate high-frequency (HF) ranging requires that the receiver extract the ground wave propagation and ignore the potentially strong skywave multipath, as well as atmospheric noise and jamming. The MX-773 subcontract specifications called for satisfactory discrimination against interferences of the following types:

1) Skywave, identical in modulation to the ground wave guidance signal, but 40 times greater in amplitude and delayed 100–250 $\mu$s.

2) Other guidance signals identical in modulation, but 15 times greater in amplitude and differing in arrival time by 50–2000 $\mu$s.

3) Unmodulated, pulse, or noise-modulated interference up to 20 times the guidance signal in amplitude.

The Bayside engineering team, headed by Norman Harvey, Walter Serniuk, and Meyer Leifer, and joined in 1947 by Nathan Marchand, felt that an FM signal with a more complex modulation than Armstrong's would satisfy requirements. The concept was bench tested via analog simulation with perfect guidance signal synchronization being wired in. Using multiple tone modulation under a maximum frequency deviation constraint of 10 kHz, no simple multitone FM modulation satisfying the contractual constraints could be found. However, low-frequency noise modulation was shown on the bench test to give "an excellent discrimination function with no secondary peaks."

The Sylvania team recognized that noise modulation was "very appealing from the anti-jamming and security aspects," but its utility in WHYN was questionable since the recording and reproduction requirements in the actual system would be severe. Accordingly, electronic generation of a reproducible multitone modulation function remained the preferred approach. Although the above are quoted from [66], these revealing results were in classified print by October, 1948 [67], simultaneously with Shannon's open publication of pseudorandom signaling.

When Republic Aviation's missile development was discontinued, Sylvania work proceeded on WHYN under the auspices of the Air Force's Watson Laboratories [later to become the Rome Air Development Center (RADC)] with this support spanning the 1948–1952 time frame. Noise modulation never made it into the WHYN system but correlation detection certainly did. In fact, it was noted [68] in 1950 that "Had the full significance of cross-correlation been realized [at the beginning], it is probable that the name [WHYN] would be different." Advocacy of correlation detection reached an artistic peak when the following classified Sylvania jingle was heard at a 1950 autumn meeting in Washington.

> "Correlation is the best,
>   It outdoes all the rest,
> Use it in your guided missile
>   And all they'll hear will be a whistle.
>   Whistle, whistle, whistle . . ."

Sung to the tune of a popular Pepsi-Cola commercial, this bit of creativity may have been inspired by the arrival, at Sylvania's helm, of Pepsi's chief executive.

The earliest public disclosure of the concepts which had evolved in the first WHYN study appears circumspectly in the last paragraph of an October, 1950, article by Leifer and Marchand in the *Sylvania Technologist* [69]:

"... The factors determining signal bandwidth and receiver noise bandwidth are entirely different; in the former it is resolution and in the latter, rate of flow of information. A signal that provides good resolution and, hence, has fairly large bandwidth, should be made more complex in nature within this bandwidth for anti-jamming characteristics. Finally, it is important to note that nowhere has the type of modulation of the signals been specified; the conclusions apply equally to pulse-, frequency-, and phase-modulated signals."

Ideas and analyses which were prompted by the Sylvania Bayside work appeared in the literature [70]-[73], in two Harvey patents, the first on WHYN [74] and the second on a collision warning radar [75] which could employ noise modulation, and in another patent [76] on spectrum shaping for improving TOA measurement accuracy in correlation detectors. With continued study, the need for bandwidth expansion to improve system performance became even more apparent, and it was declared that [77]

"Jamming signals which are noise modulated or nonsynchronous cw or modulated signals are rejected to the same extent that general noise is rejected, the improvement in signal over interference in terms of power being equivalent to the ratio of the transmission bandwidth to the receiver bandwidth."

This improvement property of SS systems is usually referred to as processing gain, which nominally equals the multiplicity factor of the system. By suitably setting these bandwidth parameters, acceptable receiver operation from 40 to 60 dB interference-to-signal ratio was reported in laboratory tests, and navigation receivers operating at $-25$ dB SNR were predicted [78].

*A Note on CYTAC*

WHYN was one of the competitors in the development of LORAN (LOng RAnge Navigation), a competition which was eventually won by Sperry Gyroscope Company's CYTAC [79]. Developed in the early 1950's, the CYTAC system and its CYCLAN predecessor had many of the attributes of WHYN, but signal-wise, CYTAC was different in two regards. First, pulse modulation was used so that earliest arriving skywaves could be rejected by gating, and second, phase coding of the pulses was innovated to reject multihop skywaves. These same properties, designed into the system and later patented by Robert Frank and Solomon Zadoff [80], were also used to discriminate between signals from different LORAN stations. The polyphase codes originally designed for CYTAC's pulse modulation were patented separately by Frank [81], but were eventually replaced in LORAN-C by biphase codes to reduce complexity [82]. A certain degree of receiver mismatching also was employed for enhancing time resolution, a similar stratagem having been used for the WHYN system [76].

Fig. 5. One of the last snapshots of Madison "Mad" Nicholson, at age 51, on a cold Easter day in 1958. As a tribute to this dedicated scientist who died suddenly in mid-January, 1959, the library at Sylvania's Amherst Laboratory was named the Madison G. Nicholson, Jr., Memorial Library. (Photo courtesy of Dana Cole.)

Since narrow-band interference was a potential problem in LORAN, the anti-interference capabilities of this pulse-compression type of signaling were appreciated and reported in 1951 [83]. To further improve performance against in-band CW interference, manually tuned notch filters were added to CYTAC in 1955 and automatic anti-CW notch filters [84], [85] were added to LORAN-C in 1964. To indicate progress, Frank notes that LORAN receivers with four automatically tunable notch filters are now on the market, some for under $1600.

*Hush-Up*

In the summer of 1951 Madison Nicholson (see Fig. 5) of Sylvania Buffalo headed a proposal effort for the study of a communication system which he called "Hush-Up." Undoubtedly, the SS ideas therein were distilled versions of those brought to Colonial Radio from the WHYN project by Norman Harvey shortly before that subsidiary lost its identity and was absorbed by Sylvania in February, 1950. Nicholson coaxed his old colleague, Robert M. Brown, who had worked at Bayside on the Armstrong radar in World War II while Nicholson had led the AN/ARW-4 team at Colonial, back to Sylvania to work with him and Allen Norris for the duration of the proposal effort. Harvey, by then chiefly responsible for commercial television work, left the realm of military communications research and development. In due course Wright Air Development Center (WADC) gave Sylvania a contract beginning in May, 1952, and Nicholson's team went "behind closed doors" to begin work.

Having boned up on Sylvania Bayside's WHYN reports, the engineers at Buffalo set out to verify that a noise-like signal could be used as a carrier, and received coherently, without causing insoluble technical problems. Independently adopting a pattern of experimentation which was being pursued secretly by other researchers at the time, detector operation was initially examined in the laboratory using a broad-band carrier whose source was thermal noise generated in a 1500 Ω resistor. This wide-band carrier signal was wired directly to the receiver as one input of the correlation detector, thereby temporarily bypassing the remaining major technical problem, the generation of a noise-like carrier at the transmitter and the internal production of an identical, synchronous copy of the same noise-like carrier at the receiver.

In 1953, as the follow-on contract for Hush-Up commenced, James H. Green was hired specifically to develop digital techniques for producing noise-like carriers. John Raney, a Wright Field Project Engineer who had worked on WHYN, also joined Nicholson as System Engineer in early 1953. Nicholson and Raney almost certainly deserve the credit for coining the now universally recognized descriptor "spread spectrum," which Sylvania termed their Hush-Up system as early as 1954.

During the second contractual period, which lasted into 1957, Green and Nicholson settled on the form of noise-like carrier which Hush-Up would employ in place of WHYN's FM, namely, a pseudorandomly generated binary sequence PSK-modulated (0 or 180 degrees ) onto an RF sinusoid. Such binary sequences with two-level periodic correlation were called "perfect words" by Nicholson. In the end a variety of perfect word known as an $m$-sequence was advocated for implementation (more on $m$-sequences later). Synchronization of the DS-SS signal was accomplished by an early–late gate, dithered tau tracker ($\tau$ = delay). Nicholson and Green's tau tracker invention has been, until recently, under patent secrecy order.

As development progressed, the system evolving from the Hush-Up effort was officially designated the ARC-50. Sylvania engineer Everard Book fabricated the ARC-50/XA-2 "flying breadboards." In 1956, flight testing began at Wright-Patterson Air Force Base (WPAFB) with WADC Project Engineers Lloyd Higginbotham and Charles Arnold at the ground end of the ARC-50/XA-2 test link and Capt. Harold K. Christian in the air. The assigned carrier frequency for the tests was the WPAFB tower frequency; the ground terminal of the ARC-50 was about 100 yards from the tower antenna, and communication with the airborne terminal was acceptable at ranges up to 100 miles. Vincent Oxley recalls that tower personnel, and the aircraft with whom they were conducting normal business, were never aware of ARC-50 transmissions. While the tests were successful, it must have been disheartening to Buffalo engineers when Sylvania failed to win the production contract for the ARC-50.

## BLADES[2]

In the mid-1950's, Madison Nicholson spent part of his considerable creative energies in the development of methods for generating signals having selectable frequency deviation from

a reference frequency. Nicholson achieved this goal with notable accuracy by creating an artificial Doppler effect using a tapped delay line. Even though patent searches uncovered similar frequency-synthesis claims by the Hammond Organ Company, the resulting inventions [86], [87] were a breakthrough for Sylvania engineers working on SS systems.

In addition to being used to slew the time base in the Hush-Up receiver, Nicholson's "linear modulator" (or "cycle adder") was an essential part of another system which Jim Green named the Buffalo Laboratories Application of Digitally Exact Spectra, or BLADES for short. Initiated with company funds in 1955 and headed by Green and Nicholson, the BLADES effort was originally intended to fill Admiral Raeburn's Polaris submarine communications requirements.

Perhaps due to concern for the serious distortions that multipath could cause in long-range HF communications, the ARC-50 DS configuration was abandoned in favor of an FH-SS system. In 1957 a demonstration of the breadboard system, operating between Buffalo, NY, and Mountain View, CA, was given to a multiservice group of communications users. Vincent Oxley was system engineer on this development, as well as for the follow-on effort in 1958 to produce a packaged prototype.

The original breadboard contained only an FH-SS/FSK anti-jam mode. The system achieved its protection ratio (Sylvania's then current name for processing gain) by using the code generator to select two new frequencies for each baud, the final choice of frequency being dictated by the data bit to be transmitted. To be effective, a jammer would have to place its power on the other (unused) frequency, or as an alternative, to place its power uniformly over all potentially usable frequencies. Because of the possibility that a jammer might put significant power at the unused frequency, or that the selected channel frequency might be in a fade, a (15, 5) error-correcting code was developed and implemented for the prototype, and was available as an optional mode with a penalty of reducing the information transmission rate to one-third.

While apparently no unclassified descriptions of BLADES are available, glimpses of the system can be seen in several "sanitized" papers and patents produced by Sylvania engineers. Using the results of Pierce [88], Jim Green, David Leichtman, Leon Lewandowski, and Robert Malm [89] analyzed the performance improvements attainable through the diversity achieved by FH combined with coding for error correction. Sylvania's expertise in coding at that time is exemplified by Green and San Soucie's [90] and Fryer's [91] descriptions of a triple-error-correcting (15, 5) code, Nicholson and Smith's patent on a binary error-correction system [92], and Green and Gordon's patent [93] on a selective calling system. All are based on properties of the aforementioned perfect words called $m$-sequences, which were investigated in Sylvania Buffalo's Hush-Up studies. Also involved in BLADES development were R. T. Barnes, David Blair, Ronald Hileman, Stephen Hmelar, James Lindholm, and Jack Wittman at Sylvania, and Project Engineers Richard Newman and Charles Steck at the Navy's Bureau of Ships.

The prototype design effort was aimed at equipment optimization. Extremely stable, single quartz crystal, integrate-

---

[2] It is convenient to recount this Sylvania system next, even though chronologically it would belong toward the end of the monograph.

and-dump filters were developed. Based on their success, a bank of 32 "channel" filters was implemented for an $M$-ary FSK optional mode to transmit a full character (5 bits) per baud. Loss of a single baud in this case meant loss of a full character because the (15, 5) decoder could only correct 3 bit errors per codeword. A "noodle slicer" was implemented to avoid this problem by interleaving five different codewords, so that each baud carried one bit from each word. This interleaving technique was the subject of a patent filed in 1962 by Sylvania engineers Vincent Oxley and William De Lisle [94]. Noodle slicing was never employed in the FH binary FSK mode.

BLADES occupied nearly 13 kHz of bandwidth in its highest protection mode. In addition to being a practical AJ system, Vincent Oxley recalls that during initial breadboard on-air tests, the system also served very well as an unintentional jammer, efficiently clearing all other users from the assigned frequency band.

After considerable in-house and on-air testing between the Amherst Laboratories at Williamsville, NY, and San Juan, PR, the packaged prototype was finally delivered for shipboard testing in 1962. Such a system was evidently carried into the blockade associated with the Cuban missile crisis but was not tested there due to a radio silence order. In 1963 BLADES was installed on the command flagship Mt. McKinley for operational development tests. Successful full-duplex field trials over intercontinental distances were observed by Sylvania engineer Gerry Meiler, who disembarked at Rota, Spain, leaving the system in the hands of Navy personnel. Further into the Mediterranean, intentional jamming was encountered, and BLADES provided the only useful communication link for the McKinley. Thus, BLADES was quite likely the earliest FH-SS communication system to reach an operational state.

*Noise Wheels*

At the end of World War II, ITT reorganized and constructed a new facility at Nutley, NJ, incorporated as Federal Telecommunication Laboratories (FTL), with Henri Busignies as Technical Director. There in 1946 a group of engineers in Paul Adams's R-16 Laboratory began working on long-range navigation and communication techniques to meet the requirements of the expanding intercontinental air traffic industry. In the available frequency bands, it was expected that multipath generated by signal ducting between the ionosphere and the earth would cause significant distortion, while the prime source of independent interference at the receiver would consist of atmospheric noise generated for the most part by lightning storms in the tropics. A major effort was initiated to study the statistical properties of the interference and to learn how to design high performance detectors for signals competing with this interference.

This was the situation in 1948 when Shannon's communication philosophy, embracing the idea that noise-like signals could be used as bearers of information, made a distinct impression on FTL engineers. Mortimer Rogoff, one of the engineers in R-16 at the time, was an avid photographic hobbyist. He conceived of a novel experimental program using photographic techniques for storing a noise-like signal and for build-

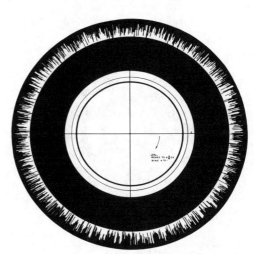

Fig. 6. Only two copies of Rogoff's secret noise wheel, shown here, were made to support ITT's early research on spread-spectrum systems. The noise wheel concept was revived briefly in 1963 when two more wheels were produced and tested in a system at ITT. (Photo from [96], courtesy of ITT.)

ing an ideal cross correlator. Supported by ITT funds and doing some work in a makeshift home lab, Rogoff prepared a 4 in × 5 in sheet of film whose transmissivity varied linearly in both directions, thus creating a mask whose transmission characteristic at every point $(X, Y)$ was proportional to the product $XY$. Two signals then were correlated by using them as the $X$ and $Y$ inputs to the oscilloscope, reading the light emitted from the masked oscilloscope face with a photomultiplier, and low-pass filtering the resultant output.

Rogoff's noise-like carrier came straight from the Manhattan telephone directory. Selecting at random 1440 numbers not ending in 00, he radially plotted the middle two of the last four digits so that the radius every fourth of a degree represented a new random number (see Fig. 6). This drawing was transferred to film which, in turn, when rotated past a slit of light, intensity-modulated a light beam, providing a stored noise-like signal to be sensed by a photocell.

In initial experiments Rogoff mounted two identical noise wheels on a single axle driven by a Diehl 900 rpm synchronous motor (see Fig. 7). Designed and assembled by Rogoff and his colleague, Robert Whittle, separate photocell pickups were placed on each wheel, one stationary and one on an alidade, so that the relative phase between the two signals could be varied for test purposes. Using time shift keying (an extra pickup required) to generate MARK or SPACE, one noise wheel's signal was modulated and then combined with interference to provide one correlator input, while the other input came directly from the second noise wheel. These baseband experiments, with data rates on the order of a bit per second and, hence, a multiplicity factor of well over 40 dB, indicated that a noise-like signal hidden in ambient thermal noise could still accurately convey information.

In another part of FTL, highly compartmentalized for security purposes, Louis deRosa headed the R-14 Electronic Warfare Group. DeRosa, who earlier had collaborated with

Fig. 7. ITT's equipment constructed for bench-testing a communication system based on noise-like carriers stored on wheels. (Photo from [96], courtesy of ITT.)

Busignies and Deloraine, and who had exchanged many friendly arguments with Nathan Marchand concerning the merits of IF correlation (à la Marchand [34]) versus baseband correlation via homodyning (deRosa's favorite), held an umbrella contract through Dr. George Rappaport, Chief of the Electronic Warfare Branch at WADC, to pursue a variety of electronic countermeasures and counter-countermeasures. The contract, codenamed Project Della Rosa, spanned the 1947–1951 time frame and, hence, was concurrent with Rogoff's work.

The first written indication of deRosa's visualization of an SR-SS technique occurs in one of this prolific inventor's patents, filed in January, 1950, with L. G. Fischer and M. J. DiToro [95], and kept under secrecy order for some time. The fine print of this patent calls out the possibility of using an arbitrarily coded waveform generated at the transmitter and an identical, synchronous, locally generated waveform at the receiver to provide a reference for a correlation detector, to reliably recover signals well below the noise level.

On August 1, 1950, deRosa gave a laboratory demonstration of Rogoff's noise wheels to visiting U.S. Air Force personnel, with the system extracting signals 35 dB below the interfering noise. Later the same month deRosa and Rogoff produced a secret proposal [96] outlining Rogoff's work and proposing several refinements including PSK data modulation, wider bandwidth carrier generation (either by scaling Rogoff's original system or by introducing flying spot scanners reading a pseudorandom image), and quicker-response drives for the receiver's noise wheel synchronizing servo.

Whittle recalls that in mid-1951 the wheels were separated by about 200 yards in the first test of his synchronization system for the noise wheel drives. During these tests Bing Crosby's crooning on radio station WOR provided the jamming as Morse code was successfully transmitted at −30 dB SNR. Tapes of the test were made and taken on unsuccessful Washington, DC, marketing trips, where there was considerable interest but evidently the government could not grasp the full significance of the results.

In 1952 an FTL Vice President, retired General Peter C. Sandretto, established relations between deRosa and Eugene Price, then Vice President of Mackay Marine, whereby Mackay facilities in Palo Alto, CA, were made available for transcontinental tests of the FTL equipment. Testing began in late November and ended before Christmas, 1952, with Whittle and Frank Lundburg operating an ARC-3 Collins transmitter at the Mackay installation, and deRosa and Frank Bucher manning the receiver at Telegraph Hill, NJ.

Coordination of these field trials was done by telephone using a codeword jargon, with

"crank it" = bring up transmitter power,

"take your foot off" = reduce transmitter power,

"ring it up" = advance the sync search phase,

"the tide is running" = severe fading is being encountered,

"go north" = increase transmission speed.

Initial synchronization adjustments typically took 3–5 min. Matched tuning forks, ringing at a multiple of 60 Hz, provided stable frequency sources for the drives with the receiver synchronizer employing a war-surplus Bendix size 10 selsyn resolver for phase shifting purposes. Rogoff's original noise wheels were retained for the transcontinental tests, as was his photo-optical multiplier, although the multiplier was improved to handle both positive and negative inputs. Using ionospheric prediction charts, transmission was near the maximum usable frequency (where multipath is least), in the 12–20 MHz range, without FCC license. The system bandwidth was fixed at 8 kHz, the data rate varied down to a few bits per second, and the transmitter power was adjustable between 12 and 25 W.

Although documentation of the test results has not yet been made available, Whittle recalls that during a magnetic storm that happened to occur, a 50 kW Mackay transmitter could not communicate with the East Coast using its conventional modulation, while FTL's test system operated successfully on 25 W. Often the noise-wheel system communicated reliably, even while interference in the same frequency band was provided by the high-power Mackay transmitter. Air Force observer Thomas Lawrence, Project Engineer on Della Rosa and Chief of the Deceptive Countermeasures Section at WADC (another WADC team member was Frank Catanzarite), also recalls witnessing these capabilities.

However, some problems were encountered. In addition to the FTL system once being detected out-of-band (probably in the vicinity of the transmitter), propagation effects apparently caused trouble at times. The signals received at Telegraph Hill were preserved by a speed-lock tape recorder which had been built from scratch to have adequate stability. In the months following the transcontinental tests, John Groce performed correlation recovery experiments on the taped signals, experiencing considerable difficulty with multipath.

Due to government-decreed project isolation, Rogoff was not told about the above tests of his noise wheels. In fact, Rogoff could not follow developments after 1950, except to participate in a patent application with deRosa, for which [96] served as the disclosure. With the help of patent attorney Percy Lantzy, the application, which described a full-fledged SR-SS single-sideband communication system based on Rogoff's noise wheels, was filed in March, 1953. (Incidentally, the original patent claims placed few restrictions on the DS modulation technique to be employed, but subsequently these were struck out in favor of single-sideband specification.)

In June, 1953, the Bureau of Ships placed a secrecy order against the application, which stood until July, 1966, when the Navy recommended recision of the order and issuance of the patent. Technically, this was accomplished in November, 1966, but before the printing presses in the U.S. Patent Office had begun to roll, a civil servant at the National Security Agency (NSA) noted the invention and was able to get secrecy reimposed. This order stood until 1978 when NSA permitted wholesale recision on scores of patents including at least a dozen on SS techniques. The deRosa–Rogoff patent [48] was finally awarded in November, 1979, nearly thirty years after the invention's conception.

The emphasis in both invention and early experimental work at FTL was on covert communication and on suppressing atmospheric noise. It is impossible to determine exactly when FTL engineers appreciated the fully robust AJ capabilities of their system. In 1950 they suspected that broad-band noise jamming would be the best attack against the receiver's signal processor, while the receiver itself might be disabled by any strong signal if it did not possess sufficient dynamic range. The deRosa–Rogoff patent, although using the phrase "secrecy and security" several times, never specifically claims AJ capabilities. However, during the course of their work, FTL engineers coined the term "chip" to denote an elementary pulse which is modulated by a single random or pseudorandom variable, and they realized that high performance against atmospheric noise. or when hiding beneath a strong signal like radio station WOR, required many chips per data bit of transmission.

For unknown reasons, FTL was unable to capitalize significantly on this early entrance into the SS field. When in June, 1970, as an Assistant Secretary of Defense, Louis deRosa (see Fig. 8) was asked about later developments involving the FTL system, he mentioned only Project Dog, a U.S. Navy covert communications operation in the North Korean theater.

*The Hartwell Connection*

In January, 1950, the Committee on Undersea Warfare of the National Research Council addressed a letter to Admiral C. B. Monsen, Assistant Chief of Naval Operations, in which the commmittee urged the determination of a long-range program against submarines [97]. This was the beginning of a sequence of events which led to the formation of a classified study program known as Project Hartwell, held at M.I.T. in June through August, 1950. Under the direction of Prof. Jerrold Zacharias, the study brought together highly qualified experts from the military, industry, and universities, to find new ways to protect overseas transportation.

A subsequent history [98] of the Research Laboratory of Electronics (RLE) at M.I.T. indicates that Hartwell was pos-

Fig. 8. Louis deRosa remained with ITT until 1966 when he joined the Philco-Ford Corporation as Director of Engineering and Research. In 1970 he left a Corporate Vice President position at Philco-Ford to be sworn in (above) by Melvin Laird as Assistant Secretary of Defense for Telecommunications, the first holder of that office. He died unexpectedly in 1971 after a long workout on the tennis court. (Photo courtesy of Mrs. Louis deRosa, standing next to Secretary Laird.)

sibly the most successful of M.I.T.'s summer study projects, motivating the development of "the Mariner class of merchant vessels; the SOSUS submarine detection system; the atomic depth charge; a whole new look at radar, sonar, and magnetic detection; and a good deal of research on oceanography." This 1966 history omitted (perhaps due to classification) the fact that transfer of an important concept in modern military communications took place at Hartwell.

One of the many ideas considered was the possibility of hiding fleet communication transmissions so that enemy submarines could not utilize them for direction finding. Appendix G of the secret final report on Project Hartwell suggested that a transmitter modulated by a wide band of noise be employed, reducing the energy density of the transmitted signal "to an arbitrarily small value." If at the same time the actual intelligence bandwidth were kept small, covert communications should be possible in certain situations.

Three systems for accomplishing covert communications were described in the report. One, acknowledged to be the suggestion of FTL's Adams and deRosa (Adams alone was an attendee), was an SR-SS system. A second system, attributed to J. R. Pierce of BTL, used very narrow pulses to achieve frequency spreading, pulse pair spacing to carry intelligence, and coincidence detection at the receiver. It was noted that if synchronized (random) pulse sources were available at transmitter and receiver, then cryptographic-like effects were possible, presumably by transmitting only the second of each pulse pair.

A third system, with no proponent cited, is the only one described by a block diagram in the final report (see Fig. 1). To avoid the synchronization problems inherent in stored reference systems, it was proposed that the noise-like carrier alone be transmitted on one channel, and that an information-bearing delay-modulated replica of the carrier also be transmitted at either the same frequency or at an offset frequency. A cross-correlation receiver still would be employed in this TR-SS system, but the carrier storage and synchronization problems of an SR-SS system would be traded for the headaches of a second channel.

The Hartwell report noted that the SR system was cryptographically more secure than the TR system, which transmitted a copy of the wide-band carrier in the clear. Furthermore, it would be improper to transmit the intelligence-free wide-band carrier on the same channel as the intelligence-modulated carrier with a fixed delay $\tau$ between them, since this delay-line addition would impose a characteristic $\cos(\pi f \tau)$ periodic ripple on the power spectral density of the transmitted signal. This ripple might be detectable on a panoramic receiver, compromising the covertness of the transmission. Although not mentioned in the report, it was realized at about the same time that multipath could produce a similar delay-line effect with similar results on any wide-band signal, including SR-SS transmissions.

To close this revealing discussion on noise modulation, the Hartwell report suggested that several of these kinds of systems, using different wide-band carriers, could operate simultaneously in the same band with little effect on each other. This concept, which, it is noted, would eliminate the cooperative synchronization required in time-division multiple-access (TDMA) systems, is one of the earliest references to CDMA operation.

Among the attendees at Project Hartwell was Jerome B. Wiesner, then Professor of Electrical Engineering at M.I.T. and Associate Director of RLE. Concerning Wiesner's place in the development of modern communications, it was later said by an M.I.T. professor [99], "Perhaps one might put it that Wiener preached the gospel and Wiesner organized the church. Jerry's real strength . . . lies in his ability to spot the potential importance of an idea long before others do."

Certainly Wiesner appreciated the possibilities of the wide-band communication systems discussed at Hartwell. Shortly after Hartwell, Wiesner met Robert Fano in a hallway near the Building 20 bridge entrance to the RLE secret research area and told Fano of a "Navy study idea" for using a noise-modulated carrier to provide secure military communications. Even though Fano was familiar with Shannon's precepts and had been an early contributor to the new field of information theory, this made a profound impression on him. He in turn discussed the concept with Wilbur Davenport, a then recent recipient of the Sc.D. degree from M.I.T. They decided to split the research possibilities, with Fano studying radar applications and Davenport developing the communication applications. This was a fortunate juxtaposition with radar work alongside communications since covertness could not be maintained in radar applications and jamming was always a possibility. The AJ potential of SS systems was appreciated immediately and reported in a series of RLE secret Quarterly Progress Reports.

The year 1951 saw another secret summer study, known as Project Charles, in action at M.I.T. Under the direction of F. W. Loomis of the University of Illinois, Project Charles investigated air defense problems, including electronic warfare. Appendix IV-1 of the Charles Report [100], written by Harry Nyquist of BTL, suggests that carrier frequencies be changed in accordance with a predetermined random sequence, and that by using this FH pattern over a wide band, the effects of jamming could be minimized. (Nyquist's experience as an NSA consultant may have played a role here.) In the next section of Appendix IV, the Charles Report proposes that a ground wave radar use a noise-modulated CW carrier to achieve security against countermeasures, and indicates that M.I.T. is investigating this technique (over a decade after Guanella's original conception).

## NOMAC

Correlation methodology is so basic to modern communications that it may be difficult to imagine a time when the technique was not widely accepted. Fano, commenting on that era, has said, "There was a heck of a skepticism at the time about crosscorrelation . . . it was so bad that in my own talks I stopped using the word crosscorrelation. Instead I would say, 'You detect by multiplying the signals together and integrating.'" Nevertheless, by 1950 M.I.T. researchers were leading proponents of correlation techniques, and were finding more and more problems which correlation might help solve.

It was into this climate that Wiesner brought the noise-like wide-band carrier concept from Project Hartwell to M.I.T. researchers. Within a year of this event Lincoln Laboratory received its organizational charter and commenced operation, its main purpose being the development of the SAGE (Semi-Automatic Ground Environment) air defense system defined by Project Charles. Soon thereafter, the classified work at RLE was transferred to Lincoln Laboratory and became Division 3 under the direction of William Radford. There, fundamental SS research was performed, to a significant extent by M.I.T. graduate students, guided by Group Leaders Fano and Davenport. The acronym NOMAC, classified confidential at the time and standing for "NOise Modulation And Correlation," was coined by one of these students, Bennett Basore, to describe the SS techniques under study. The term "spread spectrum" was never heard at M.I.T. in those days.

Basore's secret Sc.D. thesis [101], the first on NOMAC systems, was completed under Fano, Davenport, and Wiesner in 1952. It consisted of a comparison of the performances of transmitted- and stored-reference systems operating in the presence of broad-band Gaussian noise. An RF simulation of a NOMAC system with multiplicity factors up to 45 dB was used to back up theoretical analyses. As in Nicholson's and Rogoff's initial experiments, the synchronization problem of the SR system was bypassed in the experimental setup. The carrier was obtained by amplifying thermal and tube noise, while the interfering noise was produced by some old radar RF strips made originally for M.I.T.'s Radiation Laboratory. Data were on-off keyed. A bandpass correlator was employed in which two inputs at offset frequencies were inserted into an appropriate nonlinearity, the output signal then observed at the difference frequency through a narrow bandpass integrating filter, and the result envelope-detected to recover correlation magnitude. Basore's conclusion was that the effect of noise in the reference channel was to reduce the receiver's output SNR by the ratio of the signal power level to the signal-plus-noise power level in the reference channel.

The main advantages of TR-SS systems are

1) no SS carrier synchronization problems at the receiver

2) no SS carrier storage or generation required at receiver.

On the other hand, there are apparent disadvantages to the TR system:

1) relatively poor performance at low SNR's in the signal and reference channels

2) extra bandwidth may be required for reference channel

3) no privacy feature when the clear SS carrier is available to all listeners

4) difficulties in matching reference and data channel characteristics, e.g., group delays

5) easily jammed when the difference between the reference and data channel center frequencies is known

6) no multipath rejection capability.

While the advantages of TR systems have since dwindled due to the development of synchronization techniques for the SR system, the disadvantages of TR systems are to a great extent fundamental. Considerable experimental work on TR-NOMAC systems was performed at M.I.T. in the 1950-1952 time frame. Davenport's Group 34 at Lincoln Laboratory developed several TR-SS systems, including one called the P9D. An HF version of the P9D was tested between Lincoln Laboratory and a Signal Corps site in New Jersey and, according to Davenport, worked "reasonably well." This led to the development of a VHF version intended for an ionospheric scatter channel to a Distant Early Warning (DEW) radar complex near Point Barrow, AK. Since the need for LPI and AJ was marginal, SS modulation was not considered necessary and the DEW-Line link was eventually served by more conventional equipment.

A TR system study also was carried out by U.S. Army Signal Corps Capt. Bernard Pankowski in a secret Master's degree thesis [102], under the direction of Davenport. Published at the same time as Basore's thesis, Pankowski's work details several ideas concerning jamming, multiplexing, and CDMA operation of TR-NOMAC systems. In particular, it noted that jamming a TR system is accomplished simply by supplying the receiver with acceptable alternative reference and data signals, e.g., a pair of sine waves in the receiver's passbands at the appropriate frequency separation.

Bernie Pankowski offered three possible solutions to the jamming problem, namely, going to the MF or SR systems which others were studying at the time, or developing a hybrid pure noise-TR, FH-SR system with one of the two channels frequency hopped to deny offset frequency knowledge to the jammer. Similarly, CDMA operation was achieved by assigning each transmitter–receiver pair a different frequency offset between their data and reference channels. Laboratory experiments on various single-link TR system configurations with two multiplexed circuits sharing the same reference channel were carried out for a channel bandwidth of 3000 Hz and a data bandwidth of 50 Hz.

There were several exchanges of ideas with other research groups during the time period following Basore's and Pankowski's theses. For example, at Lincoln Laboratory on October 2, 1952, Sylvania, Lincoln, and Air Force personnel participated in discussions led by Meyer Leifer and Wilbur Davenport on the subject of secure communications [103]. In February, 1953, Sylvania, Lincoln, and Jet Propulsion Laboratory researchers attended the (Classified) RDB Symposium on the Information Theory Applications to Guided Missile Problems at the California Institute of Technology [78], [104]. Detailed records of these kinds of exchanges appear to be virtually nonexistent. (RDB: the Pentagon's Research and Development Board.)

As Group 34 studied the TR approach, it became apparent that the SR approach had advantages that could not be overlooked. The task of solving the key generation and synchronization problems for an SR system was given to another of Davenport's Sc.D. candidates, Paul Green. Green's secret thesis [105] is a clearly written comparison of several NOMAC system configurations, the aim of which is to determine a feasible SR design. Comparisons are based on the relationship between input and output signal-to-noise (or jamming) ratios for the receiver's signal processor, and the degradations in this relationship due to synchronization error and multipath. Green deduced that correlation at baseband would require a phase-locked carrier for good correlator performance, while correlation at IF à la Basore, with the correlator output being the envelope of the bandpass-filtered IF signal, would require SS carrier sync error to be bounded by the reciprocal of the SS carrier bandwidth.

Green then designed and built (see Fig. 9) a digitally controlled SS carrier generator in which five stagger-tuned resonant circuits were shock-excited by pseudorandom impulse sequences which in turn were generated from 15 stored binary sequences of lengths 127, 128, and 129 (see Fig. 10). The resultant signal had a long period and noise-like qualities in both the time and frequency domains, yet was storable and reproducible at an electronically controlled rate at both ends of a communication link. The proposed SS carrier synchronization procedure at the receiver was quite similar to then contemporary tracking-radar practice, progressing through search, acquisition, and track modes with no change in signal structure. Tracking error was sensed by differencing correlator outputs for slightly different values of clock oscillator phase. Based on Green's results which indicated that an SR system was feasible, and on jamming tests which confirmed TR system vulnerability [106], Group 34's resources were turned toward prototyping an SR system. This marked the end of TR system research at Lincoln Laboratory.

*F9C-A/Rake*

The prototype SR-NOMAC system developed for the Army Signal Corps by Lincoln Laboratory was called the F9C. Its evolution to a final deployed configuration, which spanned the 1953-1959 time frame, was carried out in cooperation with the Coles Signal Laboratory at Ft. Monmouth, in particular with the aid of Harold F. Meyer, Chief of the Long Range Radio Branch, and Bernard Goldberg, Chief of the Advanced Development Section, and also Lloyd Manamon and Capt. H. A. ("Judd") Schulke, all of that Laboratory. This effort had the

Fig. 9. These two racks of equipment constitute the transmitter and receiver used to carry out the experimental portion of Paul Green's secret Sc.D. dissertation. The SS carrier generators occupy the upper half of each rack, with the plug boards allowing the operator to change the structure of the 15 stored binary sequences. Later in the F9C system, these plug boards were replaced by punched card readers. (Photo from [105], courtesy of M.I.T. Lincoln Laboratory.)

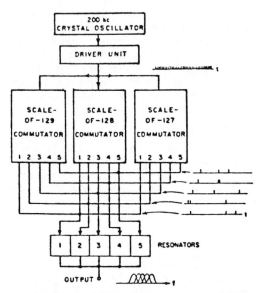

Fig. 10. The boxes in the above diagram of Paul Green's SS signal generator are located so that they correspond to the physical layout in the equipment racks of Fig. 9. An SS signal generator similar to the one shown here, combining waveforms of relatively prime periods, was chosen for the F9C system. (Diagram from [105].)

wholehearted support of Lt. Genl. James D. O'Connell, then Chief Signal Officer of the U.S. Army.

Paul Green remained at Lincoln Lab after completing his thesis, and was placed in charge of building and testing F9C equipment. Included in the group of engineers contributing to the development of the F9C were Bob Berg, Bill Bergman, John Craig, Ben Eisenstadt, Phil Fleck, Bill McLaughlin, Bob Price, Bill Smith, George Turin, and Charles Wagner (originator of the Wagner "code," the simplest version of the Viterbi algorithm).

The F9C system [107] occupied 10 kHz of bandwidth and originally employed frequency-shift data modulation at a rate of approximately 22 ms/bit. This resulted in a multiplicity factor greater than 200. The F9C radioteletype system was intended for long-range fixed-plant usage over HF channels. Initially, SS signal generation was accomplished by combining the 28 outputs of four 7-stage counters (fixed to have periods 117, 121, 125, and 128) using an array of "and" gates, and driving a bandpass filter with the resultant pulse train. For security against jamming, the gate array connections were controlled by changeable punched cards and this served the role of a key for the system. At the time there were discussions concerning the possibility of making the SS signal provide cryptographic security as well, but this idea was eventually dropped in favor of conventional data encryption before modulation.

Both the SS signal generator and data modulation technique were later modified to improve spectral and correlation characteristics and change the SS signal period, thereby increasing AJ and privacy capabilities [108]. (For a discussion of the effects encountered in combining sequences of different periods, see [109]–[111]. Also, in a March, 1955, secret report, Price proposed improving DS-SS by resorting to error correction coding in combination with soft or hard decisions against CW or pulse jammers; but this AJ strategy was not implemented in the F9C system.)

At the suggestion of Signal Corps Capt. John Wozencraft, the bandpass filter in Basore's bandpass correlator was replaced by an active filter [112] employing a diode-quenched high-$Q$ $L$-$C$ tank circuit, thereby attaining true IF integrate-and-dump correlation operation. A different circuit achieving the same matched-filter-type improvement on sinusoids was developed independently by M. L. Doelz for the Collins Radio Company [113].

Synchronization of the SS signal was accomplished initially by sending a tone burst at a preagreed frequency to start the four 7-stage counters in near synchronism. A fine search then began to bring the receiver's SS modulation clock into precise alignment with the received modulation. When synchronism was achieved, a tracking loop was closed to maintain sync. The fine search was conducted at a rate of 1000 s for each second of relative delay being swept. The frequency standards used in the system were stable enough that even with propagation variations, a disablement of the tracking loop for a day would cause a desynchronization of at most 10 ms. Eventually it was demonstrated [114] that the tone burst was not necessary and the four 7-stage clocks were approximately aligned by time of day at 5 min intervals in initial search situations.

```
RGR
PHIL MOVED YESTERDAY MOST OF HIS JUNK THAT IS THOUGH HE IS STILL AT
ROBINSON RD
THERE IS NOT TOO MUCH DOING AT THE LAB AT THE MOMENT
WE ARE TRYING TO FIND ERUIPMENT FOR THE ATLATIC PULSE TESTS
THE IPSWICH SITE IS COMING ALONG SLOWLY BECAUSE OF THE LONG DELAY IN
STARTING
MARIE WILL BE FLYING OUT TO SFO ON MONDAY
SHE SAYS R  THAT SHE WILL GIVE YOU A CALL WHEN SHE GETS TR  THERE
  I DONT KNOW IF SHE HAS A JOB YET

RGR I WAS WONDERING WHEN SHE WOULD ARRIVE OUT HERE AND ALSO IF I COULD
BE OF ANY ASSISTANCE TO HERE  I WILL WAIT FOR HER CALL AT EDL ON MONDAY
IN REGARD TO MY OWN PLANS I PLAN TO STAY OUT HERE UNTIL THE FIRST EQUIP
AT EDL IS SHOWN TO BE OK AND THE IF TIME PERMITS WILL GO BACK TO THE

LAB FOR A SHORT STAY  IT WILL BE SHORT SINCE I AM GOING UP TO TACOME
TO BE MARRIED ON THE 25 OF AUGUST AND THEN I PLAN TO GO TO SWEDEN ON
THE 28 OF SEPTEMBER FOR AT LEST A MONTH
SO I WILL BE ON VACATION FOR ABOUT THREE MONTHS  GA

OH REALLY ??
YES
WE ARE ALL BREAKING OUR NECKS TRYING TO GT  GET AT THIS

MACHINE TO SAY CONGRATULATIONS!

AND HER I EXPECTLD PITY  HHHHHH

WELL THIS IS RATHER SUDDEN AS THE GIRLS SAY

YEP

HELLO BOB THIS IS P GREEN     I KNEW IT WOULD COME TO THIS SOME DAY

CHEER UP THE SORST IS YET TO COME     CONGRATULZYIONS

CI
    YES I UNDERST,$ 5-5
IHATISHEEDSHYTHOUMSHOULD HOPE FOR THE BEST AND BE SATE  SATISFIR SATISH
H
AND BE SATISFIED WITH THE WORST  GA
```

Fig. 11.   This duplex teletype output, made during coast-to-coast tests of the F9C system, includes undoubtedly the first wedding announcement afforded the security of spread-spectrum communications. (Copy courtesy of the announcer at the West Coast station, Robert Berg.)

Transcontinental field trials of the F9C system commenced in August, 1954 [108]. The transmitter was located in Davis, CA, and the receiver in Deal, NJ, to provide an eastbound HF link for F9C tests. A conventional teletype link was supplied for westbound communication (see Fig. 11). Initial tests verified what many suspected, namely, that multipath could severely reduce the effectiveness of SS systems. While at low data rates an ordinary FSK receiver would operate based on the energy received over all propagation paths, the high time resolution inherent in an SS receiver would force the receiver to select a single path for communication, resulting in a considerable loss in signal level. Based on these early trials, several of the previously mentioned modifications were made, and in addition it was decided to add diversity to the system to combat multipath. Two receivers with antennas displaced by 550 feet were used for space diversity tests, and two correlators were employed to select signals propagated by different paths, in tau-diversity (time delay) tests.

A second set of field trials began in February, 1955, to determine the effects of these changes on performance of the transcontinental link. Results showed that an ordinary FSK system with space diversity and integrate-and-dump reception still significantly outperformed the F9C, with tau-diversity showing some hope of improving F9C performance. Both local and remote jamming tests were conducted in this second series, the interfering signal being an in-band FSK signal with MARK and SPACE frequency spacing identical to that of the F9C data modulation. The remote jammers were located at Army Communication Station ABA in Honolulu, HI, and at the Collins Radio Company in Cedar Rapids, IA. With tau-diversity, the F9C achieved a rough average of 17 dB improvement over FSK against jamming in the presence of multipath, justifying transition to an F9C-A production phase.

While initially the F9C MARK-SPACE modulation was FSK, this was eventually changed to another, equally phase-insensitive form of orthogonal signaling called the "mod-clock" approach. The mod-clock format, conceived by Neal Zierler and Bill Davenport, consisted of either transmitting the SS code in its original form (SPACE), or transmitting it with every other pulse from the SS code generator inverted (MARK).

Perhaps it was a case of serendipity that several years earlier Fano had suggested communication-through-multipath as an Sc.D. thesis topic to Bob Price. In any event, after a particularly frustrating day of field tests in which they encountered highly variable F9C performance, Price and Green got together in their Asbury Park boarding house to discuss multipath problems. Price already knew the optimal answer to some questions that were to come up that evening. Since receiving his doctorate and having been rehired by Davenport after trying his hand at radio astronomy in Australia, he had been polishing his dissertation with "lapidary zeal" (Green's witticism). Price had in fact statistically synthesized a signal processing technique for minimum-error-probability reception of signals sent over a channel disturbed by time-varying multipath as well as noise [115].

Green separately had been trying to determine how to weight the outputs of a time-staggered bank of correlators in order to improve F9C performance, and, acting on Jack Wozencraft's suggestion, had decided to choose weights to maximize the resultant transversal filter's output signal-to-noise ratio. Of course, the TOA resolution capability of the F9C was sufficient to guarantee that the outputs of different correlators in the bank represented signals arriving via different paths. Thus, the problem was one of efficiently recombining these signals. It took little time for Price and Green to realize that the results of their two approaches were nearly identical, and from that evening onward, the "Rake" (coined by Green) estimator–correlator became part of their plans for the F9C. Price took charge of building the Rake prototype, with the assistance of John Craig and Robert Lerner.

Related to Wiesner and Lee's work on system function measurements using cross correlation [116], Brennan's work on signal-combining techniques [117], and Turin's multipath studies [118], the Rake receiver could in turn be viewed as a predecessor of adaptive equalizers [119]. The Rake processor [120]-[122] (patented at Davenport's prompting) is adaptive in the sense that the weight on each MARK–SPACE tap pair is determined by the outputs of that MARK–SPACE tap pair, averaged over a multipath stability time constant. (See Fig. 12.) In its ultimate form, the magnetostrictive tapped delay line (patented by Nicholson [123]), around which the processor was built, contained 50 taps spanning 4.9 ms, the spacing being the reciprocal of the NOMAC signal bandwidth.

In addition to solving the multipath dilemma and thereby securing the full 23 dB of potential processing gain, Rake also allowed the sync search rate to be increased so that only 25 s were necessary to view one second of delay uncertainty [114]. (Readers of this early literature should note that to prevent disclosure of the actual F9C SS signal structure, all unclassified discussions of Rake, e.g., [121], invoked $m$-

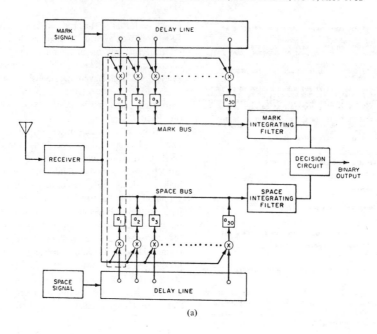

(a)

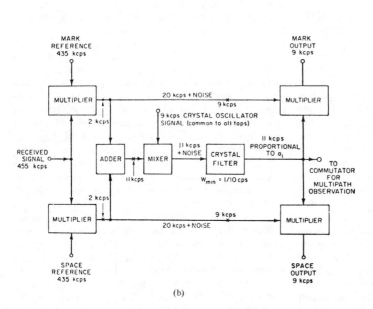

(b)

(c)

Fig. 12. (a) This two-delay-line version of Rake shows how signals arriving via different path delays are recombined for MARK and SPACE correlation detection. In practice, a single delay line configuration was adopted. (b) The tap unit diagrammed here includes a long-time-constant crystal filter whose output signal envelope is proportional to the combining weight ($a_i$). This processing corresponds to that shown in the dashed box in (a). Rejection traps to eliminate undesirable cross products are shown by X's. (c) This Rake rack contains 30 tap units, two helical magnetostrictive delay lines, and a commutator chassis. (Diagrams taken from [120], photo courtesy of M.I.T. Lincoln Laboratory.)

sequences for signal spreading. In addition, mod-clock MARK–SPACE modulation was never mentioned in this open literature.)

The F9C-A production contract was let to Sylvania Electronic Defense Laboratory (EDL) at Mountain View, CA, in 1955, with Judd Schulke acting as Project Engineer for the Signal Corps, and Bob Berg as Lincoln Lab's representative, resident at EDL. By December, 1956, the first training manuals had been published [124]. Originally 16 F9C-A transmitter-receiver pairs were scheduled to be made, but funds ran out after production of only six pairs. The first installation was made for Washington, DC, near Woodbridge, VA/La Plata, MD. Worldwide strategic deployment commenced with the installation in Hawaii in January, 1958, and was followed by installations in Germany (Pirmasens/Kaiserslautern, February, 1958), Japan, and the Philippines. With the threat of a blockade of Berlin, the equipment assigned to Clark Field in the Philippines was moved in crates of Philippine mahogany to Berlin in the spring of 1959.

Rake appliques for the F9C-A receivers were fabricated later by the National Company of Malden, MA. These were produced with an improved yet simplified circuit configuration, invented by General Atronics [125], which employed tap units having a full 10 kHz of internal bandwidth instead of being structured as in Fig. 12(b). Additionally, the F9C-A/Rake appliques introduced a novel method of ionospheric multipath display, in which the multipath-matched tap-combining weights were successively sensed by a short pulse traveling along the magnetostrictive delay line, the pulse duty cycle being low enough to have negligible effect on the Rake signal processing. Bernie Goldberg was the Project Director for this effort and Robert L. Heyd served as the Project Engineer. Together they also developed Goldberg's innovative "stored ionosphere" concept [126] in which the F9C-A/Rake's multipath measurement function was used to record ionospheric channel fluctuations for their later re-creation in testing shortwave apparatus. This measurement capability was also employed to assess multipath effects, between Hawaii and Tokyo, of a high altitude nuclear detonation in the Pacific in July, 1962.

The F9C-A/Rake is no longer on-site, operational, or supported by the Army.

## A Note on PPM

As the Hartwell report indicated, J. R. Pierce of BTL had suggested that covertness be achieved by using extremely narrow pulses for communication, thereby spreading the transmission spectrum. This idea was undoubtedly based on BTL's postwar work on pulse position modulation (PPM) [65]. After discussing the CDMA idea generally in a 1952 paper [127], Pierce and Hopper make the following observations:

"There are a number of ways in which this sort of performance can be achieved. One has been mentioned: the use of random or noise waveforms as carriers. This necessitates the transmission to or reproduction at the receiver of the carrier required for demodulation. Besides this, the signal-to-noise ratio in such a system is relatively poor even in the absence of interference unless the bandwidth used is many times the channel bandwidth. . . . In the system discussed here, the signal to be sent is sampled at somewhat irregular intervals, the irregularity being introduced by means of a statistical or 'random' source. The amplitude of each of the samples is conveyed by a group of pulses, which also carries information as to which transmitter sent the group of pulses. A receiver can be adjusted to respond to pulse groups from one transmitter but to reject pulse groups from other transmitters."

This early unclassified reference not only mentions the disadvantages of certain SS systems, but also indicates a PPM technique for achieving the CDMA property of an SS system. PPM systems evidently remained of interest to BTL engineers for some time (e.g., see [128]), and also formed the basis for some Martin Company designs [129], [130].

## CODORAC

In 1952 the Jet Propulsion Laboratory (JPL) of the California Institute of Technology was attempting to construct a radio command link for the purpose of demonstrating remote control of the Corporal rocket. The two groups most closely connected with the formulation of a system for accomplishing this task were the Telemetry and Control Section under Frank Lehan and the Guidance and Control Section under Robert Parks, both reporting to William Pickering.

One novel concept was formulated by Eberhardt Rechtin, a recent Cal Tech Ph.D. under Parks, who decided that the current radio design approach, calling for the IF bandwidth to match the Doppler spread of the signal, could be improved dramatically. Rechtin's solution was to adjust the receiver's local oscillator automatically to eliminate Doppler variations, thereby significantly reducing the receiver's noise bandwidth. This automated system used a correlator as its error detector, with the correlator inputs consisting of the received signal and the derivative of the estimate of the received signal. The resultant device, called a phase-locked loop (PLL), with its characteristics optimized for both transient and steady-state performance [131], was a key ingredient of all later JPL guidance and communication systems. Surprisingly, when attempts were made to patent an advanced form of PLL, the prior claim which precluded the award did not come from television, which also had synchronization problems, but came instead from a 1905 patent on feedback control. In retrospect, Eb Rechtin feels that perhaps his greatest contribution in this area consisted of "translating Wiener's 'Yellow Peril' into English," and converting these elegant results into practice.

In struggling with blind-range problems occurring in the integration of a tracking range radar into the Corporal guidance system, Frank Lehan realized that his problems were due to the shape of the radar signal's autocorrelation function. The thought that the autocorrelation function of broad-band noise would be ideal led Lehan to formulate the concept of an elementary TR-SS communication system using a pure noise carrier. In May, 1952, Lehan briefly documented his partially developed ideas and their potential for LPI and AJ in a memo to Bill Pickering. Lincoln Laboratory's NOMAC work was quickly discovered, both JPL's and Lincoln's being sponsored by the Army, and the wealth of information contained in Lincoln's detailed reports was made available to JPL researchers.

By the spring of 1953 JPL had decided upon a DS-SS configuration for the Corporal guidance link, and Rechtin, noting applications for his tracking loop theory in SS code synchronization, transferred to Lehan's section to head a group studying this problem. Seeing the value of the M.I.T. documentation, JPL began a series of bimonthly progress reports in February, 1953, these later being combined and annotated for historical purposes in 1958 [132].

The term "pseudonoise" with its abbreviation "PN" was used consistently from 1953 onward in JPL reports to denote the matched SS signal generators used in a DS system. Two PN generators initially were under consideration (see Fig. 13), the first being a product of 12 digitally generated (±1) square waves having relative periods corresponding to the first 11 primes. This type-I generator was eventually dropped due to its excessive size and weight. The type-II PN generator was

" . . . based on the equation

$$x(t + m) = x(t)x(t + n)$$

where $t$ represents time, $m$ and $n$ are integers ($m$ represents a time displacement greater than $n$), and the functions $x(t + m)$, $x(t)$, and $x(t + n)$ may equal ±1 only. . . . If the correct values of $m$ and $n$ are chosen, the period before repeat is $2^m - 1$. . . . The correlation function of all type-II PN generators consists of a triangle of height unity and of a width equal to twice the shift time standing on a block of height $(2^m - 1)^{-1}$."

This origination of an almost perfect spike-like autocorrelation function, accompanied by descriptions of shift register hardware, positive results of baseband synchronization experiments at −20 dB SNR's, and a table of suitable values of $m$ and $n$ for values of $m$ up to 20, was reported as progress through August, 1953 [132], [133]. In later works by other researchers, these PN sequences were called shift-register sequences or linear-recurring sequences due to their particularly convenient method of generation, and were also termed $m$-sequences since their period is maximal.

On January 18, 1954, a JPL PN radio system was operated over a 100 yard link and two independent commands were communicated. Initial synchronization was achieved with the aid of a land line which was disconnected after sync acquisition. The system was able to withstand jammer-to-signal power ratios of 15–20 dB before losing lock, against a wide variety of jamming threats. This test was the assurance that JPL engineers needed regarding the practicality of SR-SS communications.

At this point work on the command system was temporarily dropped and a major effort was begun to optimize a pure ranging system, called the Scrambled Continuous Wave (SCW) system, which consisted of a "very narrow-band CW system scrambled externally by a PN sequence." On July 27, 1954, Corporal round 1276-83 carrying an SCW transponder was launched at White Sands Proving Ground. The transponder operated successfully from takeoff to near impact 70 miles away, providing range and range rate without loss of lock in

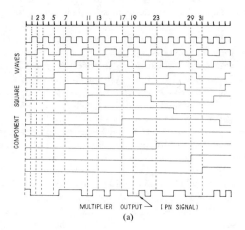

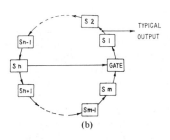

| Time Displacement | | Length of |
|---|---|---|
| Value of $m$ | Value of $n$ or $m-n$ | Sequence |
| 2 | 1 | 3 |
| 3 | 1 | 7 |
| 4 | 1 | 15 |
| 5 | 2 | 31 |
| 6 | 1 | 63 |
| 7 | 1 | 127 |
| 7 | 3 | 127 |
| 9 | 4 | 511 |
| 10 | 3 | 1023 |
| 11 | 2 | 2047 |
| 15 | 1 | 32767 |
| 15 | 4 | 32767 |
| 15 | 7 | 32767 |
| 17 | 3 | 131071 |
| 17 | 5 | 131071 |
| 18 | 7 | 262143 |
| 20 | 3 | 1048575 |

(c)

Fig. 13. (a) The type-I PN generator uses a multiplier to combine the outputs of binary (+1 or −1) signal shapers which in turn are driven by the outputs of relatively prime frequency dividers operating on the same sinusoid. The component square waves and the resultant PN product signal are shown here. (b) JPL's type-II generator was an $m$-stage linear-feedback shift register which produced binary (0 or 1) sequences of maximal period. The output of the $m$th and $n$th stages are added modulo 2 to produce the input of the first stage S1. (c) This first list of connections for the type-II generator was produced at JPL by hand and computer search. (Diagrams and table redrawn from [132].)

the synchronization circuitry. Rechtin, engineer Walter Victor, and Lehan (who left JPL in 1954) later filed an invention disclosure based on the SCW system results from this test, and called the system a COded DOppler RAdar Command (CODORAC) system. This acronym was used to describe the radio guidance systems developed for the Sergeant and later the Jupiter missiles in the 1954–1958 time frame.

Throughout this period one of the major problems in establishing one-way communication to a missile was to make the PN generator tough enough to withstand high temperatures and vibrations as well as small and light enough to fit into the missile design. A variety of devices (e.g., subminiature hearing aid tubes) and potting compounds were tested. In 1954 Signal Corps liaison official G. D. Bagley was able to obtain approximately 100 of the Western Electric type 1760 transistors (the first available outside BTL) for use by JPL engineer Bill Sampson in the construction of a PN generator. The resulting circuitry was an interesting combination of distributed-constant delay lines and transistor amplifiers and logic, chosen because it minimized the number of active elements required [134]. This general method of construction remained the norm at JPL through 1958.

Late in 1954 a separate group under Sampson was formed for the purpose of investigating possible countermeasures against the SCW system equipment designed by a group headed by Walt Victor. Designed to make this phase of the program as objective as possible, this organization brought forth a thoroughly designed system with high countermeasures immunity. Here are three issues on which significant progress was made.

1) It was hoped that repeater jamming would be ineffective due to the high TOA resolution capability of SS and to the excess propagation delay incurred by the repeater. The period of the PN sequence was made longer than the missile flight time so that it would be impossible for a repeater to store a PN coded signal for nearly a full period and deliver it to the victim receiver in synchronism with the directly transmitted PN sequence one period later. A weakness in this regard still existed in a simple *m*-sequence generator based on a linear recursion. Specifically, these sequences possessed a "cycle-and-add" property (for example, see [135]) by which the modulo 2 sum of a sequence and a delayed version of that sequence results in the production of the same sequence at still another delay. The equivalent "shift-and-multiply" property for the ±1 version of these *m*-sequences, satisfying the equation quoted earlier in this subsection, conceivably could be used by a jammer to produce an advance copy of the sequence without waiting a full period. In an effort to completely rule out this possibility, Cal Tech graduate student Lloyd Welch was hired in 1955 to study the generation of sequences which avoid the cycle-and-add property by resorting to nonlinear recursions [136]. Although laboratory system work continued to use linearly generated PN sequence for test purposes, final designs were to be based on nonlinear generators.

2) Initial jamming tests revealed weaknesses in the SCW system when confronted by certain narrow-band jammers. Most of these were due to problems in the mechanization of the multiplications required in the PN scrambler and corre-

lator descrambler. For example, if the descrambler effectively mechanizes a multiplication of the jamming signal by a constant plus the receiver's PN sequence replica (the constant representing a bias or imbalance in the multiplication/modulation process), then the multiplier output will contain an unmodulated replica of the jamming signal which has a free ride into the narrow-band circuitry following the descrambler. The sure cure for this problem is to construct better balanced multipliers/modulators, since the processing gain achievable in an SS system is limited by the "feedthrough" (or bias) in its SS multipliers. In the mid-1950's JPL was able to build balanced modulators which would support systems with processing gains near 40 dB. For a recent discussion of this problem area, see [137].

3) Another major concern of system designers was the decibel range and rates of variation of signal strength, due to missile motion and to pulsed or intermittent jamming. At the circuit level the two approaches to controlling signal levels in critical subassemblies were automatic gain control (AGC) and limiting. The AGC approach suffers from the possibility that its dynamics may make it susceptible to pulse jamming, while limiters, although instantaneous in nature, generate harmonics which might be exploited by a jammer. The eventual design rule-of-thumb was that limiters could be used when necessary on narrow-band signals (e.g., prior to PLL phase detectors), and that AGC techniques should be used in the wide-band portions of the system. Analytical support for this work came from JPL's own studies of AGC circuits [138], [139], and from Davenport's classic paper on limiters [140]. It was not realized until much later that in some instances the limiter theory was not appropriate for coherent signal processing analyses [141].

Many of these kinds of problems remain with the designer today, the differences being in the technology available to solve them.

Both the Sergeant and Jupiter guidance programs were terminated when decisions were made to choose all-inertial jam-proof guidance designs as the baseline for those missile systems. However, CODORAC technology survived in the JPL integrated telemetry, command, tracking, and ranging system for the Deep Space Program, and in the later projects of subcontractors who had worked for JPL in the Jupiter program. A modified version of CODORAC became the Space Ground Link Subsystem (SGLS) now used routinely in U.S. Department of Defense missile and space instrumentation.

### *m-Sequence Genesis*

The multiplicative PN recursion given in [132] and its linear recursion counterpart in modulo 2 arithmetic, namely

$$y(t + m) = y(t) \oplus y(t + n)$$

were among those under study by 1954 at several widely separated locations within the United States. Lehan recalls that the idea of generating a binary sequence recursively came out of a discussion which he had with Morgan Ward, Professor of Mathematics at Cal Tech, who had suggested a similar decimal arithmetic recursion for random number generation.

Fig. 14.  Cut from a 1953 photograph of a summer session class on mathematical problems of communication theory, this picture salutes (from left to right) Yuk Wing Lee, Norbert Wiener, Claude Shannon, and Robert Fano. It is ironic that Wiener could not prevent the transfer of his theories, through meetings like the one at which this picture was taken, to the military research which he refused to support after World War II. (Photo courtesy of M.I.T.)

It is hard to determine if this idea was mentioned at the (Classified) RDB Symposium held at Cal Tech in February, 1953. Lincoln Laboratory's Bill Davenport remembers that the first time that he had seen a PN generator based on the above recursion was in Lehan's office on one of his trips west. This generator, built to Rechtin's specifications, was used to extend Rechtin's hand-calculated table of $m$-sequence generators from a shift register length of at most 7, to lengths up to 20 (see Fig. 13).

Sol Golomb, then a summer hire at the Glenn L. Martin Company in Baltimore, MD, first heard of shift register-generated sequences from his supervisor, Tom Wedge, who in turn had run across them at a 1953 M.I.T. summer session course on the mathematical problems of communication theory. (This meeting included an elite group of the founding fathers of information theory and statistical communications. See Fig. 14.)

On the other hand, Neal Zierler, who joined Lincoln Laboratory in 1953, recalls discovering shift register generation of PN sequences while looking for ways to simplify the SS signal generators used for the F9C system. Golomb's [135], [142], [143] and Zierler's [144]-[146] work established them as leading theorists in the area of pseudonoise generation. However, Zierler's shift register-generated sequences were never used in the F9C-A system due to their cryptanalytic weaknesses. Golomb's work gained further recognition after he joined JPL in August, 1956.

Madison Nicholson's early attempts at PN sequence design date back to 1952 [103]. Nicholson's first exposure to the pseudorandomness properties of linearly recurring sequences probably came from Allen Norris, who remembers relating to Nicholson ideas developed from lectures by the noted mathematician, A. A. Albert, of the University of Chicago. Coworkers recollect that Nicholson used paper-and-pencil

methods for finding shift register logics which generated $m$-sequences. Jim Green in due course joined in this exploration, and built demonstration hardware. Bob Hunting was assigned to investigate the generation of long $m$-sequences and spent a considerable amount of time exercising Sylvania's then-new UNIVAC 1 in the Corporate Computer Center at Camillus, NY, and eventually produced an extensive list of "perfect word" generators. R. L. San Soucie and R. E. Malm developed nonlinear sequence-combining techniques for the BLADES prototype, the result being an SS carrier with a period of about 8000 centuries. Oliver Selfridge of Lincoln Laboratory's Group 34 became the government representative whose approval was required on Sylvania's SS code designs for Air Force contracts, but was not involved with the Navy's BLADES effort.

Early work by others on linear-feedback shift registers includes that of Gilbert [147], Huffman [148], and Birdsall and Ristenbatt [149]. Additional insights were available from the prewar mathematical literature, especially from Ward [150], Hall [151], and Singer [152], [153]. Of course, in the top secret world of cryptography, key-stream generation by linear recursions very well may have been known earlier, particularly since Prof. Albert and others of similar stature were consultants to NSA. But it is doubtful that any of these had a direct impact on the pioneering applications to SS communication in 1953-1954.

### ARC-50

In 1953 a group of scientists interested in the design of computers left the University of California at Los Angeles and formed a research laboratory under an agreement with the Magnavox Corporation. Their first contact with SS systems came when JPL approached them with the problem of building DS-SS code generators for the Jupiter missile's proposed radio navigation link. This exposure to JPL's work on PN sequences and their application to radio guidance paid dividends when Lloyd Higginbotham at WADC became interested in getting high-speed, long-period generators for the ARC-50 system which was emerging from the Hush-Up study at Sylvania Buffalo. At Sylvania, Hush-Up had started out under the premise of radio silence, and was aimed for an application to the then-new air-to-air refueling capability developed by the Strategic Air Command (SAC). After a demonstration of the wired system at Sylvania, a SAC representative made the "obvious" statement, "When you are in radar range, who needs radio silence?" From that time onward, the design was based on AJ considerations.

The AJ push resulted in NSA being brought into the program for their coding expertise. However, due to the nature of NSA, they passed technical judgment rather than provided any concrete guidance. The NSA view was that the SS codes had to be cryptographically secure to guarantee AJ capability, and Lincoln Laboratory had established that the proposed ARC-50 SS PN code was easily breakable. On this point Lloyd Higginbotham says, "At that time we felt we were being treated unfairly because the system was still better than anything else then in existence."

By 1958 Magnavox had parlayed their knowledge of high-speed PN generators into a development contract for the ARC-50 system, won in competition with Sylvania. Magnavox Research Laboratories operated out of a garage on Pico Boulevard in Santa Monica in those early days, with Jack Slattery as General Manager and Ragnar Thorensen as Technical Director. From their few dozen employees a team was organized to design the code generators and modem, while RF equipment was built at Magnavox's Fort Wayne facility. Shortly thereafter, Magnavox Research Laboratories moved to Torrance, CA, into a new facility sometimes referred to as "the house the ARC-50 built." Harry Posthumus came from Fort Wayne as Program Manager and teamed with system designers Tom Levesque, Bob Grady, and Gene Hoyt, system integrator Bob Dixon, and Bill Judge, Bragi Freymodsson, and Bob Gold.

Although retaining the spirit of the DS-SS system developed at Sylvania, technologically the design evolved through several more phases at Magnavox. Nowhere was this more obvious than in the design of the SS code generators, the heart of the system. The earliest Magnavox code generators were built using a pair of lumped constant delay lines, run in syncopated fashion to achieve a rate of 5 Mchips/s. This technology was expensive with a code generator costing about $5000, and was not completely satisfactory technically. The first improvement in this design came when the delay lines were transistorized, and a viable solution was finally achieved when 100 of the first batch of high-$\beta$, gold-doped, fast rise-time 2N753 transistors made by Texas Instruments were received and used to build a single-register code generator operating at 5 Mchips/s.

Originally to facilitate SS code synchronization, the system employed a synchronization preamble of 1023 chips followed by an $m$-sequence produced by a 31 stage shift register. Register length 31 was chosen because the period of the resultant $m$-sequence, namely, 2, 147, 483, 653, is prime, and it seemed unlikely that there would exist some periodic substructure useful to a jammer. Lacking knowledge of the proper connections for the shift register, a special machine was built which carried out a continuing search for long $m$-sequences. Problems were encountered involving false locks on correlation sidelobe peaks in the sync preamble (sometimes it seemed that a certain level of noise was necessary to make the system work properly), and concerning interference between different ARC-50 links due to poor cross-correlation properties between SS codes.

The ARC-50 was configured as a fully coherent system in which the SS code was first acquired, and the sinusoidal carrier was then synchronized using PLL techniques. Because of apprehension that jamming techniques might take advantage of coupling between the RF oscillator and the code chip clock, these two signals were generated independently in the transmitter. The receiver's PLL bandwidth was constrained by the fact that no frequency search was scheduled in the synchronization procedure, the assumption being that the pull-in range of the PLL was adequate to overcome both oscillator drifts and Doppler effects. Being a push-to-talk voice system which could operate either as a conventional AM radio or in an SS mode, a 5 s sync delay was encountered each time the SS modem was activated. Ranging up to 300 miles was possible with the measurement time taking about 40 s. To retain LPI

capability in this AJ system, transmitter power was adjustable from minute fractions of a watt up to 100 W.

Testing of the Magnavox ARC-50 began in 1959. Bob Dixon, joined by John G. Smith and Larry Murphy of Fort Wayne, put the ARC-50 through preliminary trials at WPAFB, and later moved on to the Verona site at RADC. One radio was installed in a C131 aircraft and the other end of the link resided in a ground station along with a 10 kW, CW jammer (the FRT-49). Testing consisted of flying the aircraft in the beam of the jammer's 18 dB antenna while operating the ARC-50. Limited results in this partially controlled environment indicated that the receiver could synchronize at jammer-to-noise ratios near those predicted by theory.

Shortly after these flight tests, an upgraded version of the ARC-50 was developed with significantly improved characteristics. To alleviate SS-code correlation problems, a new design was adopted, including an $m$-sequence combining procedure developed by Bob Gold [154], [155] which guaranteed low SS-code cross correlations for CDMA operation. The SS sync delay was reduced to one second and an improved ranging system yielded measurements in two seconds.

Even though the ARC-50 possessed obvious advantages over existing radios such as the ARC-27 or ARC-34, including a hundredfold improvement in mean time between failures, there was Air Force opposition to installing ARC-50's in the smaller fighter aircraft. The problem revolved around the fact that pilots were accustomed to having two radios, one being a backup for the other, and size-wise a single ARC-50 would displace both of the prior sets.

Certainly, the ARC-50 was a success, and Magnavox became an acknowledged leader in SS technology. Among the descendants of the ARC-50 are the GRC-116 troposcatter system which was designed free of a sync preamble, and the URC-55 and URC-61 ground-station modems for satellite channels. An applique, the MX-118, for the Army's VRC-12 family of VHF-FM radios was developed, but never was procured due in part to inadequate bandwidth in the original radios (see Fig. 15).

## IV. BRANCHES ON THE SS TREE

Many designs of the 1940's and 1950's have not yet been mentioned, but those described thus far seem in retrospect to have been exemplary pioneering efforts. It is time now to take notice of several SS systems left out of the previous accounting, some of which were never even prototyped.

### Spread-Spectrum Radar

With the exception of the 1940's state-of-the-art descriptions of technology, we have made a distinction between the use of SS designs for communication and their use for detection and ranging on noncooperative targets, and have omitted a discussion of the latter. The signal strength advantage which the target holds over the radar receiver in looking for the radar's transmission versus its echo means that LPI is very difficult to achieve. Moreover, the fact that an adversary target knows *a priori* its relative location means that even with pure noise transmission the radar is vulnerable to false echo creation by a delaying repeater on the target.

(a)

(b)

Fig. 15. Examples of early- and mid-1960's technology. (a) SS code generator portion of a TH system developed by Brown, Boveri, and Company for surface-to-air missile guidance. (Photo courtesy of I. Wigdorovits of Brown, Boveri, and Co.) (b) 1965 picture of the MX-118 applique, a member of the ARC-50 radio family and the first to use Gold codes. (Photo courtesy of Robert Dixon.)

Nonetheless, SS signaling has some advantages over conventional low time–bandwidth-product radar signaling: in better range (TOA) resolution for a peak-power-limited transmitter (via pulse compression techniques), in range ambiguity removal, in greater resistance to some nonrepeater jammers [4], and in a CDMA-like capability for sharing the transmission band with similar systems. Modern uses of SS radars include fusing (for a patent under wraps for 24 years, see [156]) and pulse compression, the latter's applications extending to high-resolution synthetic-aperture ground mapping.

*Other Early Spread-Spectrum Communication Systems*

Despite the security which once surrounded all of the advances described in previous sections, the SS system concept could not be limited indefinitely to a few companies and research institutions. The following notes describe several other early SS design efforts.

*Phantom:* An MF-SS system developed by General Electric (GE) for the Air Force, this system was built around tapped delay line filters. As shown in Costas and Widmann's patent [157], the tap weights were designed to be varied pseudorandomly for the purpose of defeating repeater jammers (see Fig. 3). Constructed in the late 1950's, the Phantom spread its signal over 100 kHz. As with the F9C-A, this system was eventually used also to measure long-haul HF channel properties. For a description of other SS-related work performed at GE, this in the 1951–1954 time frame and under the direction of Richard Shuey, see [158].

*WOOF:* This Sylvania Buffalo system hid an SS signal by placing within its transmission bandwidth a high-power, friendly, and overt transmitter. Thereby the SS transmission would be masked by the friendly transmitter, either completely escaping notice or at least compounding the difficulties encountered by a reconnaissance receiver trying to detect it.

*RACEP:* Standing for Random Access and Correlation for Extended Performance, RACEP was the name chosen by the Martin Company to describe their asynchronous discrete address system that provided voice service for up to 700 mobile users [129]. In this system, the voice signal was first converted to pulse position modulation, and then each pulse in the resultant signal was in turn converted to a distinctive pattern of three narrow pulses and transmitted at one of a possible set of carrier frequencies. With the patterns serving also as addresses, this low duty cycle format possessed some of the advantages of SS systems.

*Cherokee:* Also by the Martin Company, this was a PN system with a transmission bandwidth of nearly a megahertz and a processing gain of about 16 dB [129]. Both RACEP and Cherokee were on display at the 15th Annual Convention of the Armed Forces Communications and Electronics Association in June, 1961.

*MUTNS:* Motorola's Multiple User Tactical Navigation System was a low frequency, hyperbolic navigation system employing PN signaling. Navigation was based on stable ground wave propagation with the SS modulation used to discriminate against the skywave, as it was in Sylvania's WHYN. Motorola, a subcontractor to JPL on the Jupiter CODORAC link, began Army-supported work on MUTNS in 1958. The first complete system flight test occurred on January 23, 1961 [159],[160].

*RADA:* RADA(S) is a general acronym for Random Access Discrete Address (System). Wide-band RADA systems developed prior to 1964 include Motorola's RADEM (Random Access DElta Modulation) and Bendix's CAPRI (Coded Address Private Radio Intercom) system, in addition to RACEP [161].

*WICS:* Jack Wozencraft, while on duty at the Signal Corps Engineering Laboratory, conceived WICS, Wozencraft's Iterated Coding System. This teletype system was an SR-FH-SS system employing 155 different tones in a 10 kHz band to communicate at 50 words/min. Each bit was represented by two successively transmitted tones generated by either the MARK or the SPACE pseudorandomly driven frequency programmer. Bit decisions were made on detecting at least one of the two transmitted frequencies in receiver correlators, and parity checking provided further error correction capability. The subsequent WICS development effort by Melpar in the mid-1950's contemplated its tactical usage as an applique to radios then in inventory [114]. However, just as in ITT's early system concepts, the intended generation of pseudorandom signals via recording [162] did not result in a feasible production design.

*Melpar Matched Filter System:* A more successful mid-1950's development, this MF-SS design was largely conceived by Arthur Kohlenberg, Steve Sussman, David Van Meter, and Tom Cheatham. To transmit a MARK in this teletype system an impulse is applied to a filter composed of a pseudorandomly selected, cascaded subset of the several hundred sections of an all-pass lumped-constant linear-phase delay line. The receiver's MARK matched filter is synchronously composed of the remaining sections of the delay line. The same technique was used to transmit SPACE [114]. Patents [163], [164] filed on the system and its clever filter design, the latter invented by Prof. Ernst Guillemin who was a Melpar consultant, were held under secrecy along with the WICS patent until the mid-1970's. An unclassified discussion of an MF-SS system for use against multipath is given in [165].

*Kathryn:* Named after the daughter of the inventor, William Ehrich, and developed by General Atronics, Kathryn's novel signal processing effected the transmission of the Fourier transform of a time-multiplexed set of channel outputs combined with a PN signal. Upon reception, the inverse transform yielded the original PN × multiplexed-signal product, now multiplied by the propagation medium's system function, thereby providing good or bad channels in accordance with that function. When jamming is present, the data rate is reduced by entering the same data bit into several or all data channels. In this case a Rake-like combiner is used to remerge these channels at the output of the receiver's inverse Fourier transformer [114], [166]. The modern SS enhancement technique of adaptive spectral nulling against nonwhite jamming was at least implicitly available in this system.

*Lockheed Transmitted Reference System:* Of the several TR-SS systems patented, this one designed by Jim Spilker made it into production in time to meet a crisis in Berlin, despite the inherent weaknesses of TR systems [167]. The interesting question here is, "What circumstances would cause someone to use a TR system?" Evidently, extremely high chip rates are part of the answer. For an earlier TR patent that spent almost a quarter-century under secrecy order, see [168].

*NOMAC Encrypted-Voice Communications:* In 1952 at the suggestion of Bob Fano, Bennett Basore, with the help of Bill McLaughlin and Bob Price, constructed and briefly tested an IF model of a NOMAC-TR-FM voice system. At first surprised by the clarity of communication and lack of the self-noise which typifies NOMAC-AM systems, Basore soon realized that SS-carrier phase noise was eliminated in the bandpass correlation process and that SS-carrier amplitude noise was re-

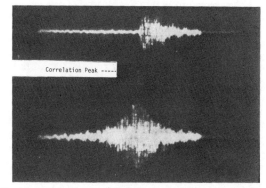

Fig. 16. Fano's elegant matched filter experiment consisted of transmitting an acoustic pulse into a chamber containing many reflectors. The upper signal shown here represents the sound sensed by a microphone in the room, and tape recorded. The tape was then reversed (not rewound) and replayed into the chamber, the microphone this time sensing the lower of the above two signals, specifically intended as the autocorrelation of the upper signal. Fano recalls being startled by his inability to see at first the extremely narrow peak of the autocorrelation function on the oscilloscope screen. The peak was soon discovered when the display intensity was increased. (Photo courtesy of Robert Fano.)

moved by the limiting frequency-discriminator. Little more was done until years later when, in 1959, John Craig of Lincoln Laboratory designed an experimental SR-SS system based on low-deviation phase modulation of a voice signal onto an F9C-like noise carrier. The system provided fair quality voice with negligible distortion and an output SNR of about 15 dB, the ever-present noise deriving from system flaws. Simulated multipath caused problems in this low-processing-gain system, and it was postulated that Rake technology might alleviate the problem [169], [170], but the work was abandoned.

*NOMAC Matched Filter System:* Based on Fano's research into high time–bandwidth-product matched filters (see Fig. 16), an MF-SS teletype communication system was suggested in 1952 [171]. Research at Lincoln Laboratory on this SS communication system type was confined to exploring a viable filter realization. This communication approach apparently was dropped when full scale work began on the F9C system. Fano later patented [172] the wide-band matched filter system concept, claiming improved performance in the presence of multipath.

*Spread Eagle:* This matched filter system was pursued by Philco in the late 1950's.

*SECRAL:* This ITT missile guidance system development of the late 1950's was a DS-SS design.

*Longarm and Quicksilver:* These are both early FH antimultipath systems built by Hughes Aircraft Company, under the leadership of Samuel Lutz and Burton Miller, and sponsored by Edwin McCoppin of WPAFB.

*Spread-Spectrum Developments Outside the United States*

This historical review has concentrated on SS developments in the United States for several reasons.

1) The theories of Wiener and especially Shannon, which propounded the properties of and motivated the use of random and pseudorandom signals, were available in the U.S. before such basics were appreciated elsewhere (with the exception of Guanella). This gave U.S. researchers a significant lead time, an important factor near the outset of the Cold War when the Voice of America was being jammed intensively. Additional impetus for SS development came in urgent response to the threats posed by the onslaught of the Korean War and the tense confrontations over Berlin.

2) SS development occurred just after the Second World War, at a point in time when many of the world's technological leaders had suffered tremendous losses in both manpower and facilities, and additionally in Germany's case, political self-control. Research and industry in the U.S., on the other hand, were unscathed and the U.S. became the home for many leading European scientists, e.g., Henri Busignies and Wernher von Braun, to name two among many.

3) The unclassified literature available to this author (virtually all the references in this history are now unclassified)[3] points to the earliest SS developments having arisen in the United States.

We will now look at evidence of some SS beginnings outside the U.S.A.

Bill Davenport remembers a secret interchange with a visiting British delegation in which pre-Rake NOMAC concepts were discussed. Later he was informed that the British had not pursued that approach to secure, long-range communication because they envisioned major problems from multipath [173]. Frank Lehan recalls a discussion with a British scientist who told him that the British had studied PN sequences several years before JPL developed the idea. Bob Dixon dates Canada's experimental Rampage system to the early 1950's, with no further details yet available [174]. So it seems that the closest friends of the U.S. were at least cognizant of the SS concept, knew something of PN generation, and to some extent had experimented with the idea. Further information on these early efforts has not been uncovered.

In neutral Switzerland, Brown, Boveri and Company developed, starting in the late 1950's, an SS guidance system (see Fig. 15). This was no doubt stimulated by the pioneering inventor of noise-modulated radar [12] (and of encryption schemes which the NDRC had sought to decipher during World War II), Gustav Guanella. He quickly appreciated, and may well have seen the true significance of, the Rake concept upon its publication. Now, an intriguing question is, "When did the Soviet bloc become privy to the SS concept and realize its potential?"

In the mid-1950's some members of a high-level task force were convinced that the Russians knew about SS techniques and in fact might be using it themselves. For example, Eugene Fubini personally searched the U.S. Patent Office open files to see what a foreign country might be able to learn there of this new art; nomenclature was a problem and he had to look under "pulse communications" as well as many other patent classi-

3 There is available from the author upon request an extensive bibliography compiled by J. M. Smith, which focuses on the prime documentation for those Sylvania, FTL (ITT), M.I.T., and JPL spread-spectrum developments that are of major significance.

fications. (This difficulty was eased recently when the Patent Office created a special subclass 375-1 entitled Spread Spectrum.) Also curious about this issue, Paul Green determined to try to find out for himself the status of Russian knowledge about NOMAC techniques. After studying the language he examined the Russian technical literature, surveying their work in information theory and attempting to uncover clues that might lie there to noise modulation concepts. Green came to believe that there was no plausible reason to suspect that the Soviets were then developing spread-spectrum systems, due in part to lack of technology and possibly to no perceived need for AJ communications capability.

Later Paul Green visited the Soviet Union and gave a talk in Russian on the use of Rake to measure properties of the ionosphere, which seemingly was accepted at face value. Because of this contact and his literature scrutiny, in the mid-1960's Green decided to postpone his plans to write an unclassified account of Lincoln Laboratory's NOMAC work, toward which full military clearances had already been granted.

The earliest Soviet reference (as cited in, e.g., [175]) proposing noise-like, intelligence-bearing signals is a 1957 publication by Kharkevich [176] on amplitude or frequency modulation of pure noise. Like Goldsmith's [38], Kharkevich's work is missing a key ingredient, namely, the attainment of synchronous detection via correlation with a stored or transmitted reference. Within a few months of the approved 1958 publication of the Rake concept for using wide-band signals ostensibly to counter multipath, that paper was translated into Russian, and hardly a year later an exposition of Rake appeared in Lange's first book *Korrelationselektronik* [177]. Thus began the revelation of the SS concept in the U.S. literature from scientific journals and conference proceedings to magazines such as *Electronics*, *Electronic Design*, and *Aviation Week*. Here is a small sample of U.S. open papers referenced in the Soviet literature:

a) March 1958. Rake remedy for multipath, using wide-band signals [121].

b) December 1959. Use of wideband noise-like signals, CDMA, and jamming [178], [179].

c) Fall 1960. PN-controlled TH-SS command link for missile guidance [130].

d) January 1961. Analysis of a pure noise (TR) communication system [180].

e) March 1961. Discussion of RADA systems [181].

f) 1963. 200 Mcps PN generator construction [182].

g) December 1963. Wideband communication systems including Rake, RACEP, and RADEM [161].

It is clear from these citations and other evidence that the Russians were studying PN sequences no later than 1963 [183], and by 1965 had carefully searched and reported [184] on the U.S. open literature discussing Rake, Phantom, and the various RADA systems. Between 1965 and 1971 the Soviets published several books [175], [185]–[188] concerned with SS principles and their applications to secure communication, command, and control.

## V. A VIEWPOINT

One can paint the following picture of the development of spread-spectrum communications. During World War II the

Allies and the Axis powers were in a desperate technological race on many fronts, one being secure communications. Jamming of communication and navigation systems was attempted by both sides and the need for reliable communication and accurate navigation in the face of this threat was real. One major AJ tactic of the war was to change carrier frequency often and force the jammer to keep looking for the right narrow band to jam. While this was possible to automate in the case of radar, communication frequency hopping was carried out by radio operators, in view of the major technological problem of providing an accurate synchronous frequency reference at the receiver to match the transmitter. Thus, at least frequency hopping and, to a similar extent, time hopping were recognized AJ concepts during the early 1940's.

Many of the early "secure" or "secret" non-SS communication systems seem to have been attempts to build analog equivalents of cryptographic machines and lacked the notion of bandwidth expansion (e.g., the Green Hornet, the Telefunken dual wheels system). The initial motivation for direct sequence systems appears, on the other hand, to have come from the need for accurate and unambiguous time-of-arrival measurements in navigation systems (e.g., WHYN and CODORAC), and from the desire to test or extend Shannon's random-signaling concept and thus communicate covertly (e.g., Rogoff's noise wheels experiment). The DS concept followed the FH and TH concepts by several years partly because the necessary correlation detection schemes were just emerging in the late 1940's.

Who first took these diverse system ideas and recognized the unifying essential requirements of a spread-spectrum system (e.g., high carrier-to-data bandwidth ratio, an unpredictable carrier, and some form of correlation detection)? From the available evidence it appears that Shannon certainly had the insight to do it but never put it in print, and that two close friends, Nathan Marchand and Louis deRosa, both key figures in the formation of the IRE's Group on Information Theory, led Sylvania Bayside and FTL, respectively, toward a unified SS viewpoint. It seems that Sylvania Bayside had all the ingredients of the direct sequence concept as early as 1948, but did not have the technology to solve some of the signal processing problems. It remained for Mortimer Rogoff to provide a method for storing pseudonoise (a technique reminiscent of Telefunken's wheels), giving ITT the complete system assembled and tested under the Della Rosa contract and documented to a government agency.

Meanwhile the idea either was propagated to or was independently conceived by several research and design groups, notably at M.I.T. in 1950 and at JPL in 1952. Group 34 at M.I.T. Lincoln Laboratory, sparked by Bill Davenport, Paul Green, and Bob Price, is generally credited with building the first successful SS communication system for several reasons.

1) The Rake system was the first wide-band pseudorandom-reference system to send messages reliably over the long-range HF multipath channel.

2) The F9C-A system, soon followed by the Rake applique, was probably the first deployed (nonexperimental) broadband communication system which differed in its essentials from wide-deviation FM, PPM, or PCM.

Fig. 17. VIP's at the IEEE NAECON '81 included Robert Larson, Wilbur Davenport, Paul Green, B. Richard Climie, Mrs. Mortimer Rogoff, Mortimer Rogoff, Mrs. Louis deRosa, and Robert Price. Featured at this meeting was the presentation of the Pioneer Award to deRosa (posthumously), Rogoff, Green, and Davenport for their ground-breaking work in the development of spread-spectrum communications. (Photo courtesy of W. Donald Dodd.)

3) The Rake system was the first such SR communication system to be discussed in the open literature, other than information theoretic designs.

JPL's radio control work, in competition with inertial guidance systems, did not reach a deployment stage until suitable applications appeared in the Space Program. In addition to opening new vistas in the development of PN generation techniques, JPL's contribution to SS technology has been the innovation of tracking loop designs which allow high-performance SS systems to be placed on high-speed vehicles with results comparable to those of stationary systems. Both the M.I.T. and JPL programs have left a legacy of excellent documentation on spread-spectrum signal processing, spectral analysis, and synchronization, and have provided some of the finest modern textbooks on communications.

A very successful long-term SS system development began at Sylvania Buffalo under Madison Nicholson and later Jim Green, and ended up merging with some JPL-based experience at Magnavox in the production of the ARC-50 family of systems. The ARC-50 was the first deployed SS system with any of the following characteristics:

1) avionics packaging,
2) fully coherent reception (including carrier tracking),
3) a several megahertz chip rate, and
4) voice capability.

Although losing the ARC-50 final design and production contract to Magnavox, Sylvania continued on to develop BLADES, the earliest FH-SS communication system used operationally. Moreover, BLADES represented, by publication (e.g., [90]) and actual hardware, the start of real-world application of shift-register sequences to error correction coding, an algebraic specialty that would flourish in coming years.

Since the 1950's when the SS concept began to mature, the major advances in SS have been for the most part technological, with improvements in hardware and expansion in scope of application continuing to the present day. Now with the veil of secrecy being lifted, the contributions of some of the earliest pioneers of SS communications are being recognized (see Fig. 17). We hope that this historical review has also served that

## ACKNOWLEDGMENT

This history is the result of the efforts of a great number of dedicated people. Paul E. Green gave this work its initial impetus by making his own archive of declassified Lincoln Laboratory documents available to the author and by backing them up with his personal commentary. The breadth of this account is due to a significant extent to the untiring work of Robert Price, who nearly single-handedly has supplied the author with a file drawer full of material culled from publications worldwide and numerous interviews, and whose comments have appreciably aided the production of this monograph.

Unreferenced as well as referenced items for this history have been collected over a period of almost three years by the author, Paul, and Bob, from many sources and in many forms. We would like to thank the following for their first-hand recollections: Paul R. Adams (deceased), Bennett Basore, Robert S. Berg, Nelson Blachman, Everard Book, Robert M. Brown, William Budd, Henri Busignies (deceased), Robert Carlson, Dana Cole, John Costas, Wilbur Davenport, Donald Disinger, Robert Dixon, Robert Fano, Robert Frank, Edgar Gilbert, Bernard Goldberg, Solomon Golomb, Harold Graves, James H. Green, John Groce, Richard Gunther, Hunter Harris, Norman Harvey, Lloyd Higginbotham, Leonhard Katz, Leon Kraft, Ernst Krause, Frank Lehan, David Leichtman, Leon Lewandowski, James Lindholm, Nathan Marchand, Charles J. Marshall, Harold F. Meyer, Allen Norris, Vincent Oxley, Bernard Pankowski, Ellison Purington, John Raney, George Rappaport, Eberhardt Rechtin, Mortimer Rogoff, Donald Rothschild, William Sampson, Peter Sandretto, Robert San Soucie, Herbert Schulke, Oliver Selfridge, Walter Serniuk, Robert A. Smith, William B. Smith, James Spilker, Seymour Stein, Steven Sussman, Ragnar Thorensen, John Travis, Richard Turyn, Walter Victor, Gottfried Vogt, Lloyd Welch, Robert Whittle, John Wozencraft, and Neal Zierler.

The author additionally wishes to thank the following people for their support in various ways: J. Meredith Smith for his diligent searching of sources (including a 1970 interview with Louis deRosa), Robert Kulinyi for uncovering the early German work, Mrs. Louis deRosa for information on her husband's career, and John R. Pierce (pseudonym J. J. Coupling) for aid in locating Theodore Sturgeon.

The following librarians, archivists, historians, and document control officers have also been very helpful: John Hewitt of the M.I.T. Research Laboratory of Electronics, H. Alan Steeves of the Sperry Research Center, Mary Granese and Robert S. Clarke (deceased) of the M.I.T. Lincoln Laboratory, Kay Haines of the Jet Propulsion Laboratory, R. Dean Porter of the Wright-Patterson Air Force Base, Charles Dewing and Franklin Burch of the U.S. National Archives, Karl Wildes of M.I.T., and Stephen L. Johnston.

Among the many other people whose ready assistance, encouragement, or significant interest have made this research a memorable experience, are: James W. Allen, J. Neil Birch, Ian Blake, Robert Bowie, Walter Braun, Charles R. Cahn, John W. Craig, Michael DiToro (deceased), Paul Drouilhet, Benjamin Eisenstadt, Laurin Fischer, Eugene Fubini, Irving Gabelman, Gustav Guanella, George Harvey, Leon Himmel, Conrad Hoeppner, George Hulst, George Jacoby, Thomas Lawrence, Yuk Wing Lee, Meyer Leifer, Arnold Levine, William C. Lindsey, Robert Malm, Theodore Manley, James Massey, Walter Morrow, David Nicholson, J. Francis Reintjes, R. Kenneth Ricker, James Roche, Theodore Sturgeon, Charles Stutt, George Valley, Richard Waer, Jerome Wiesner, I. Wigdorovits, and Henry Zimmermann.

We only regret that we have not been able to express here our sincere appreciation to everyone contacted in this fascinating pursuit.

## REFERENCES

[1] E. V. Appleton and M. A. F. Barnett, "On some direct evidence for downward atmospheric reflection of electric rays," *Proc. Roy. Soc., Ser. A*, vol. 109, pp. 621–641, Dec. 1, 1925.

[2] D. G. C. Luck, *Frequency Modulated Radar*. New York: McGraw-Hill, 1949.

[3] S. L. Johnston, "Radar ECCM history," in *Proc. NAECON*, May 1980, pp. 1210–1214.

[4] G. R. Johnson, "Jamming low power spread spectrum radar," *Electron. Warfare*, pp. 103–112, Sept.–Oct. 1977.

[5] D. O. North, "An analysis of the factors which determine signal/noise discrimination in pulsed carrier systems," RCA Lab., Princeton, NJ, Rep. PTR-6C, June 25, 1943; see also *Proc. IEEE*, vol. 51, pp. 1015–1028, July 1963.

[6] J. H. Van Vleck and D. Middleton, "A theoretical comparison of the visual, aural, and meter reception of pulsed signals in the presence of noise," *J. Appl. Phys.*, vol. 17, pp. 940–971, Nov. 1946.

[7] P. M. Woodward, *Probability and Information Theory, with Applications to Radar*. New York: Pergamon, 1953.

[8] C. E. Cook and M. Bernfeld, *Radar Signals, An Introduction to Theory and Application*. New York: Academic, 1967.

[9] A. W. Rihaczek, *Principles of High Resolution Radar*. New York: McGraw-Hill, 1969.

[10] F. E. Terman, "Administrative history of the Radio Research Laboratory," Radio Res. Lab., Harvard Univ., Cambridge, MA, Rep. 411-299, Mar. 21, 1946.

[11] F. H. Lange, *Correlation Techniques*. Princeton, NJ: Van Nostrand, 1966.

[12] G. Guanella, "Distance determining system," U. S. Patent 2 253 975, Aug. 26, 1941 (filed in U. S. May 27, 1939; in Switzerland Sept. 26, 1938).

[13] ——, "Direction finding system," U.S. Patent 2 166 991, July 25, 1939 (filed in U.S. Nov. 24, 1937; in Switzerland Dec. 1, 1936).

[14] J. J. Spilker, Jr. and D. T. Magill, "The delay-lock discriminator—An optimum tracking device," *Proc. IRE*, vol. 49, pp. 1403–1416, Sept. 1961.

[15] J. J. Spilker, Jr., "Delay-lock tracking of binary signals," *IRE Trans. Space Electron. Telem.*, vol. SET-9, pp. 1–8, Mar. 1963.

[16] N. Wiener, "Generalized harmonic analysis," *Acta Math.*, vol. 55, pp. 117–258, May 9, 1930; reprinted in *Generalized Harmonic Analysis and Tauberian Theory (Norbert Wiener: Collected Work*, Vol. 2), P. Masani, Ed. Cambridge, MA: M.I.T. Press, 1979.

[17] "Gunfire control," Nat. Defense Res. Committee, Office Sci. Res. Develop., Washington, DC, Summary Tech. Rep., Div. 7, 1946 (AD 200795).

[18] N. Wiener, *Extrapolation, Interpolation, and Smoothing of Stationary Time Series with Engineering Applications*. Cambridge, MA: M.I.T. Press, 1949.

[19] N. Levinson, "The Wiener rms (root mean square) error criterion in filter design and prediction," *J. Math. Phys.*, vol. 25, pp. 261–278, Jan. 1947.

[20] J. M. Whittaker, *Interpolatory Function Theory* (Cambridge Tracts in Mathematics and Mathematical Physics, no. 33). New York: Cambridge Univ. Press, 1935.

[21] E. T. Whittaker, "On the functions which are represented by the

expansions of the interpolation theory,'' *Proc. Roy. Soc. Edinburgh*, vol. 35, pp. 191–194, 1915.

[22] J. McNamee, F. Stenger, and E. L. Whitney, ''Whittaker's cardinal function in retrospect,'' *Math. Comput.*, vol. 25, pp. 141–154, Jan. 1971.

[23] A. J. Jerri, ''The Shannon sampling theorem—Its various extensions and applications: A tutorial review,'' *Proc. IEEE*, vol. 65, pp. 1565–1596, Nov. 1977.

[24] R. V. L. Hartley, ''The transmission of information,'' *Bell Syst. Tech. J.*, vol. 7, pp. 535–560, 1928.

[25] H. Nyquist, ''Certain topics in telegraph transmission theory,'' *AIEE Trans.*, vol. 47, pp. 617–644, Apr. 1928.

[26] V. A. Kotelnikov, ''Carrying capacity of 'ether' and wire in electrical communications'' (in Russian), *Papers on Radio Communications, 1st All-Union Conv. Questions of Technical Reconstruction of Communications*, All-Union Energetics Committee, USSR, 1933, pp. 1–19.

[27] C. E. Shannon, ''Communication in the presence of noise,'' *Proc. IRE*, vol. 37, pp. 10–21, Jan. 1949.

[28] D. Slepian, ''On bandwidth,'' *Proc. IEEE*, vol. 64, pp. 292–300, Mar. 1976.

[29] A. H. Nuttall and F. Amoroso, ''Minimum Gabor bandwidth of $M$ orthogonal signals,'' *IEEE Trans. Inform. Theory*, vol. IT-11, pp. 440–444, July 1965.

[30] D. Gabor, ''Theory of communication,'' *J. Inst. Elec. Eng. (London)*, vol. 93, part 3, pp. 429–457, Nov. 1946.

[31] C. E. Shannon, ''A mathematical theory of cryptography,'' Bell Tel. Lab., memo., Sept. 1, 1945; later published in expurgated form as ''Communication theory of secrecy systems,'' *Bell Syst. Tech. J.*, vol. 28, pp. 656–715, Oct. 1949.

[32] ——, ''A mathematical theory of communication,'' *Bell Syst. Tech. J.*, vol. 27, pp. 379–423, July, and 623–656, Oct. 1948.

[33] *Report of Proceedings, Symp. Inform. Theory*, London, England, Sept. 26–29, 1950; reprinted in *Trans. IRE Professional Group Inform. Theory*, vol. PGIT-1, Feb. 1953.

[34] N. Marchand, ''Radio receiver,'' U.S. Patent 2 416 336, Feb. 25, 1947 (filed May 21, 1942).

[35] Y. W. Lee, J. B. Wiesner, and T. P. Cheatham, Jr., ''Apparatus for computing correlation functions,'' U.S. Patent 2 643 819, June 30, 1953 (filed Aug. 11, 1949).

[36] Y. W. Lee, T. P. Cheatham, Jr., and J. B. Wiesner, ''The application of correlation functions in the detection of small signals in noise,'' M.I.T. Res. Lab. Electron., Tech. Rep. 141, Oct. 13, 1949 (ATI 066538, PB 102361).

[37] H. E. Singleton, ''A digital electronic correlator,'' M.I.T. Res. Lab. Electron., Tech. Rep. 152, Feb. 21, 1950.

[38] A. N. Goldsmith, ''Radio signalling system,'' U.S. Patent 1 761 118, June 3, 1930 (filed Nov. 6, 1924).

[39] C. B. H. Feldman, ''Wobbled radio carrier communication system,'' U.S. Patent 2 422 664, June 24, 1947 (filed July 12, 1944).

[40] P. Kotowski and K. Dannehl, ''Method of transmitting secret messages,'' U.S. Patent 2 211 132, Aug. 13, 1940 (filed in U.S. May 6, 1936; in Germany May 9, 1935).

[41] D. Kahn, *The Codebreakers*. New York: Macmillan, 1967.

[42] R. C. Mathes, ''Secret telephony,'' U.S. Patent 3 967 066, June 29, 1976 (filed Sept. 24, 1941).

[43] R. K. Potter, ''Secret telephony,'' U.S. Patent 3 967 067, June 29, 1976 (filed Sept. 24, 1941).

[44] S. V. Jones, ''After 35 years, secrecy lifted on encoded calls,'' *New York Times*, p. 27, July 3, 1976.

[45] A. J. Busch, ''Signalling circuit,'' U.S. Patent 3 968 454, July 6, 1976 (filed Sept. 27, 1944).

[46] A. E. Joel, Jr., ''Pulse producing system for secrecy transmissions,'' U.S. Patent 4 156 108, May 22, 1979 (filed Jan. 21, 1947).

[47] W. W. Hansen, ''Secret communication,'' U.S. Patent 2 418 119, Apr. 1, 1947 (filed Apr. 10, 1943).

[48] L. A. deRosa and M. Rogoff, ''Secure single sideband communication system using modulated noise subcarrier,'' U.S. Patent 4 176 316, Nov. 27, 1979 (filed Mar. 20, 1953); reissue appl. filed Sept. 4, 1981, Re. Ser. No. 299 469.

[49] E. Chaffee and E. Purington, ''Method and means for secret radiosignalling,'' U.S. Patent 1 690 719, Nov. 6, 1928 (filed Mar. 13, 1922).

[50] E. M. Deloraine, H. G. Busignies, and L. A. deRosa, ''Facsimile system,'' U.S. Patent 2 406 811, Sept. 3, 1946 (filed Dec. 15, 1942).

[51] ——, ''Facsimile system and method,'' U.S. Patent 2 406 812, Sept. 3, 1946 (filed Jan. 30, 1943).

[52] L. A. deRosa, ''Random impulse system,'' U.S. Patent 2 671 896, Mar. 9, 1954 (filed Dec. 13, 1942).

[53] ''Radio countermeasures,'' Nat. Defense Res. Committee, Office Sci. Res. Develop., Washington, DC, Summary Tech. Rep. Div. 15, vol. I, 1946 (AD 221601).

[54] E. M. Deloraine, ''Protected communication system,'' Fed. Radio Tel. Lab., New York, NY, Rep. 937-2, Apr. 28, 1944 (from the National Archives, Record Group 227; this report was written to Division 15 of the National Defense Research Committee, Office of Scientific Research and Development, on Project RP-124).

[55] H. Busignies, S. H. Dodington, J. A. Herbst, and G. R. Clark, ''Radio communication system protected against interference,'' Fed. Tel. Radio Corp., New York, NY, Final Rep. 937-3, July 12, 1945 (same source as [54]).

[56] B. Gunston, *Rockets and Missiles*. New York: Crescent, 1979.

[57] Air Force Corresp. File on Contr. W535-sc-707 with Colonial Radio Corp., May 1943–June 1944.

[58] ''Guided missiles and techniques,'' Nat. Defense Res. Committee, Office Sci. Res. Develop., Washington, DC, Summary Tech. Rep. Div. 5, vol. I, 1946 (AD 200781).

[59] E. Labin and D. D. Grieg, ''Method and means of communication,'' U.S. Patent 2 410 350, Oct. 29, 1946 (filed Feb. 6, 1943).

[60] J. H. Hammond, Jr. and E. S. Purington, ''A history of some foundations of modern radio-electronic technology,'' *Proc. IRE*, vol. 45, pp. 1191–1208, Sept. 1957.

[61] L. Espenschied *et al.*, ''Discussion of 'A history of some foundations of modern radio-electronic technology','' *Proc. IRE*, vol. 47, pp. 1253–1268, July 1959.

[62] E. S. Purington, ''Radio selective control system,'' U.S. Patent 2 635 228, Apr. 14, 1953 (filed June 2, 1948).

[63] C. H. Hoeppner, ''Pulse communication system,'' U.S. Patent 2 999 128, Sept. 5, 1961 (filed Nov. 4, 1945).

[64] E. H. Krause and C. E. Cleeton, ''Pulse signalling system,'' U.S. Patent 4 005 818, Feb. 1, 1977 (filed May 11, 1945).

[65] J. R. Pierce, ''The early days of information theory,'' *IEEE Trans. Inform. Theory*, vol. IT-19, pp. 3–8, Jan. 1973.

[66] ''The WHYN guidance system,'' Phys. Lab., Sylvania Elec. Products, Bayside, NY, Final Eng. Rep., Modulation Wave Form Study & F-M Exciter Develop., Contr. W28-099ac465, June 1949 (AD895816).

[67] ''The WHYN guidance system,'' Phys. Lab., Sylvania Elec. Products, Flushing, NY, Interim Eng. Rep. 5, Contr. W28-099ac465, Oct. 1948 (ATI 44524).

[68] ''The WHYN guidance system,'' Phys. Lab., Sylvania Elec. Products, Bayside, NY, Final Eng. Rep., Equipment Develop. & East Coast Field Test, Contr. W28-099ac465, June 1950 (AD 895815).

[69] M. Leifer and N. Marchand, ''The design of periodic radio systems,'' *Sylvania Technologist*, vol. 3, pp. 18–21, Oct. 1950.

[70] N. Marchand and H. R. Holloway, ''Multiplexing by orthogonal functions,'' presented at the IRE Conf. Airborne Electron., Dayton, OH, May 23–25, 1951.

[71] N. Marchand and M. Leifer, ''Cross-correlation in periodic radio systems,'' presented at the IRE Conf. Airborne Electron., Dayton, OH, May 23–25, 1951.

[72] W. E. Budd, ''Analysis of correlation distortion,'' M.E.E. thesis, Polytech. Inst. Brooklyn, Brooklyn, NY, May 1955.

[73] N. L. Harvey, M. Leifer, and N. Marchand, ''The component theory of calculating radio spectra with special reference to frequency modulation,'' *Proc. IRE*, vol. 39, pp. 648–652, June 1951.

[74] N. L. Harvey, ''Radio navigation system,'' U.S. Patent 2 690 558, Sept. 28, 1954 (filed Feb. 4, 1950).

[75] ——, ''Collision warning radar,'' U.S. Patent 2 842 764, July 8, 1958 (filed Feb. 21, 1951).

[76] H. C. Harris, Jr., M. Leifer, and D. W. Cawood, ''Modified cross-correlation radio system and method,'' U.S. Patent 2 941 202, June 14, 1960 (filed Aug. 4, 1951).

[77] ''The WHYN guidance system,'' Phys. Lab., Sylvania Elec. Products, Bayside, NY, Final Eng. Rep., Equipment, Syst. Lab. Tests & Anal., Contr. W28-099ac465, June 1953 (AD 024044).

[78] M. Leifer and W. Serniuk, ''Long range high accuracy guidance system,'' presented at the RDB Symp. Inform. Theory Appl. Guided Missile Problems, California Inst. Technol., Pasadena, Feb. 2–3, 1953.

[79] W. P. Frantz, W. N. Dean, and R. L. Frank, ''A precision

multi-purpose radio navigation system," in *IRE Nat. Conv. Rec.*, New York, NY, Mar. 18–21, 1957, part 8, pp. 79–98.

[80] R. L. Frank and S. Zadoff, "Phase-coded hyperbolic navigation system," U.S. Patent 3 099 835, July 30, 1963 (filed May 31, 1956).

[81] R. L. Frank, "Phase-coded communication system," U.S. Patent 3 099 795, July 30, 1963 (filed Apr. 3, 1957).

[82] W. Palmer and R. L. Frank, in "1971 Pioneer Award," *IEEE Trans. Aerosp. Electron. Syst.*, vol. AES-7, pp. 1015–1021, Sept. 1971.

[83] [R. L. Frank and S. A. Zadoff], "Study and field tests of improved methods for pulse signal detection," Sperry Gyroscope, Great Neck, NY, Final Eng. Rep., Contr. AF28(099)-333, Sperry Eng. Rep. 5223-1245, June 1951 (ATI 150834).

[84] E. J. Baghdady, "New developments in FM reception and their application to the realization of a system of 'power-division' multiplexing," *IRE Trans. Commun. Syst.*, vol. CS-7, pp. 147–161, Sept. 1959.

[85] R. L. Frank, "Tunable narrowband rejection filter employing coherent demodulation," U.S. Patent 3 403 345, Sept. 24, 1968 (filed July 19, 1965).

[86] M. G. Nicholson, Jr., "Apparatus for generating signals having selectable frequency deviation from a reference frequency," U.S. Patent 2 972 109, Feb. 14, 1961 (filed June 11, 1956).

[87] ——, "Generator of frequency increments," U.S. Patent 2 923 891, Feb. 2, 1960 (filed July 6, 1956).

[88] J. N. Pierce, "Theoretical diversity improvement in frequency shift keying," *Proc. IRE*, vol. 46, pp. 903–910, May 1958.

[89] J. H. Green, Jr., D. K. Leichtman, L. M. Lewandowski, and R. E. Malm, "Improvement in performance of radio-teletype systems through use of element-to-element frequency diversity and redundant coding," presented at the 4th Annu. Symp. Global Commun., Washington, DC, Aug. 1–3, 1960.

[90] J. H. Green, Jr. and R. L. San Soucie, "An error correcting encoder and decoder of high efficiency," *Proc. IRE*, vol. 46, pp. 1741–1744, Oct. 1958.

[91] R. G. Fryer, "Analytical development and implementation of an optimum error-correcting code," *Sylvania Technologist*, vol. 13, pp. 101–110, July 1960.

[92] M. G. Nicholson, Jr. and R.A. Smith, "Data transmission systems," U.S. Patent 3 093 707, June 11,1963 (filed Sept. 24, 1959).

[93] J. H. Green, Jr. and J. Gordon, "Selective calling system," U.S. Patent 3 069 657, Dec. 18, 1962 (filed June 11, 1958).

[94] V. C. Oxley and W. E. De Lisle, "Communications and data processing equipment," U.S. Patent 3 235 661, Feb. 15, 1966 (filed July 11, 1962).

[95] L. A. deRosa, M. J. DiToro, and L. G. Fischer, "Signal correlation radio receiver," U.S. Patent 2 718 638, Sept. 20, 1955 (filed Jan. 20, 1950).

[96] L. A. deRosa and M. Rogoff, Sect. I (Communications) of "Application of statistical methods to secrecy communication systems," Proposal 946, Fed. Telecommun. Lab., Nutley, NJ, Aug. 28, 1950.

[97] "A report on security of overseas transport," Project Hartwell, M.I.T., Cambridge, MA, Sept. 21, 1950 (ATI 205035, ATI 205036; not available from M.I.T.).

[98] "R.L.E.: 1946 + 20," M.I.T. Res. Lab. Electron., Cambridge, MA, May 1966.

[99] D. Lang, *An Inquiry into Enoughness*. New York: McGraw-Hill, 1965.

[100] "Problems of air defense," Project Charles, M.I.T., Cambridge, MA, Final Rep., Aug. 1, 1951 (not available from M.I.T.).

[101] B. L. Basore, "Noise-like signals and their detection by correlation," M.I.T. Res. Lab. Electron. and Lincoln Lab., Tech. Rep. 7, May 26, 1952 (AD 004641).

[102] B. J. Pankowski, "Multiplexing a radio teletype system using a random carrier and correlation detection," M.I.T. Res. Lab. Electron. and Lincoln Lab., Tech. Rep. 5, May 16, 1952 (ATI 168857; not available from M.I.T.).

[103] "Engineering study and experimental investigation of secure directive radio communication systems," Sylvania Elec. Products, Buffalo, NY, Interim Eng. Rep., Contr. AF-33(616)-167, Aug. 5–Nov. 5, 1952 (AD 005243).

[104] W. B. Davenport, Jr., "NOMAC data transmission systems," presented at the RDB Symp. Inform. Theory Appl. Guided Missile Problems, California Inst. Technol., Pasadena, Feb. 2–3, 1953.

[105] P. E. Green, Jr., "Correlation detection using stored signals," M.I.T. Lincoln Lab., Tech. Rep. 33, Aug. 4, 1953 (AD 020524).

[106] B. M. Eisenstadt, P. L. Fleck, Jr., O. G. Selfridge, and C.A. Wagner, "Jamming tests on NOMAC systems," M.I.T. Lincoln Lab., Tech. Rep. 41, Sept. 25, 1953 (AD 020419).

[107] P. E. Green, Jr., "The Lincoln F9C radioteletype system," M.I.T. Lincoln Lab., Tech. Memo. 61, May 14, 1954 (not available from M.I.T.).

[108] P. E. Green, Jr., R. S. Berg, C. W. Bergman, and W. B. Smith, "Performance of the Lincoln F9C radioteletype system," M.I.T. Lincoln Lab., Tech. Rep. 88, Oct. 28, 1955 (AD 080345).

[109] N. Zierler, "Inverting the sum generator," M.I.T. Lincoln Lab., Group Rep. 34-48, Feb. 13, 1956 (AD 310397).

[110] B. M. Eisenstadt and B. Gold, "Autocorrelations for Boolean functions of noiselike periodic sequences," *IRE Trans. Electron. Comput.*, vol. EC-10, pp. 383–388, Sept. 1961.

[111] R. C. Titsworth, "Correlation properties of cyclic sequences," Ph.D. dissertation, California Inst. Technol., Pasadena, 1962.

[112] J. M. Wozencraft, "Active filters," U.S. Patent 2 880 316, Mar. 31, 1959 (filed Mar. 21, 1955).

[113] M. L. Doelz and E. T. Heald, "A predicted wave radio teletype system," in *IRE Conv. Rec.*, New York, NY, Mar. 22–25, 1954, part 8, pp. 63–69.

[114] B. Goldberg, "Applications of statistical communications theory," presented at the Army Sci. Conf., West Point, NY, June 20–22, 1962 (AD 332048); republished in *IEEE Commun. Mag.*, vol. 19, pp. 26–33, July 1981.

[115] R. Price, "Notes on ideal receivers for scatter multipath," M.I.T. Lincoln Lab., Group Rep. 34-39, May 12, 1955 (AD 224559).

[116] J. B. Wiesner and Y.W. Lee, "Experimental determination of system functions by the method of correlation," *Proc. IRE*, vol. 38, p. 205, Feb. 1950 (abstr.).

[117] D. G. Brennan, "On the maximum signal-to-noise ratio realizable from several noisy signals," *Proc. IRE*, vol. 43, p. 1530, Oct. 1955.

[118] G. L. Turin, "Communication through noisy, random-multipath channels," in *IRE Conv. Rec.*, New York, NY, Mar. 19–22, 1956, part 4, pp. 154–166.

[119] P. Monsen, "Fading channel communications," *IEEE Commun. Mag.*, vol. 18, pp. 16–25, Jan. 1980.

[120] R. Price and P. E. Green, Jr., "An anti-multipath communication system," M.I.T. Lincoln Lab., Tech. Memo. 65, Nov. 9, 1956 (not available from M.I.T.).

[121] ——, "A communication technique for multipath channels," *Proc. IRE*, vol. 46, pp. 555–570, Mar. 1958.

[122] ——, "Anti-multipath receiving system," U.S. Patent 2 982 853, May 2, 1961 (filed July 2, 1956).

[123] M. G. Nicholson, Jr., "Time delay apparatus," U.S. Patent 2 401 094, May 28, 1946 (filed June 23, 1944).

[124] "Lincoln F9C-A radio teletype system," Sylvania Electron. Defense Lab., Mountain View, CA, Instruction Manual EDL-B8, Dec. 21, 1956.

[125] D. E. Sunstein and B. Steinberg, "Communication technique for multipath distortion," U.S. Patent 3 168 699, Feb. 2, 1965 (filed June 10, 1959).

[126] B. Goldberg, R. L. Heyd, and D. Pochmerski, "Stored ionosphere," in *Proc. IEEE Int. Conf. Commun.*, Boulder, CO, June 1965, pp. 619–622.

[127] J. R. Pierce and A. L. Hopper, "Nonsynchronous time division with holding and with random sampling," *Proc. IRE*, vol. 40, pp. 1079–1088, Sept. 1952.

[128] A. R. Eckler, "The construction of missile guidance codes resistant to random interference," *Bell Syst. Tech. J.*, vol. 39, pp. 973–994, July 1960.

[129] A. Corneretto, "Spread spectrum com system uses modified PPM," *Electron. Design*, June 21, 1961.

[130] R. Lowrie, "A secure digital command link," *IRE Trans. Space Electron. Telem.*, vol. SET-6, pp. 103–114, Sept.–Dec. 1960.

[131] R. M. Jaffe and E. Rechtin, "Design and performance of phase-lock loops capable of near optimum performance over a wide range of input signal and noise levels," Jet Propulsion Lab., Pasadena, CA, Progress Rep. 20-243, Dec. 1954; see also *IRE Trans. Inform. Theory*, vol. IT-1, pp. 103–114, Mar. 1955.

[132] E. Rechtin, "An annotated history of CODORAC: 1953–1958," Jet Propulsion Lab., Pasadena, CA, Rep. 20-120, Contr. DA-04-495-Ord 18, Aug. 4, 1958 (AD 301248).

[133] Corporal Bimonthly Summary Rep. 37a (July 1–Sept. 1, 1953), Jet Propulsion Lab., Pasadena, CA, Oct. 1, 1953.

[134] W. F. Sampson, "Transistor pseudonoise generator," Jet Propulsion Lab., Pasadena, CA, Memo. 20-100, Dec. 7, 1954 (AD 056175).

[135] S. W. Golomb, *Shift Register Sequences*. San Francisco, CA: Holden-Day, 1967.

[136] B. L. Scott and L. R. Welch, "An investigation of iterative Boolean sequences," Jet Propulsion Lab., Pasadena, CA, Sect. Rep. 8-543, Nov. 1, 1955.

[137] R. C. Dixon, *Spread Spectrum Systems*. New York: Wiley, 1976.

[138] Jupiter Bimonthly Summary No. 6 (Mar. 15–May 15, 1957), Jet Propulsion Lab., Pasadena, CA, June 1, 1957.

[139] W. K. Victor and M. H. Brockman, "The application of linear servo theory to the design of AGC loops," *Proc. IRE*, vol. 48, pp. 234–238, Feb. 1960.

[140] W. B. Davenport, Jr., "Signal-to-noise ratios in bandpass limiters," *J. Appl. Phys.*, vol. 24, pp. 720–727, June 1953.

[141] J. C. Springett and M. K. Simon, "An analysis of the phase coherent–incoherent output of the bandpass limiter," *IEEE Trans. Commun. Technol.*, vol. CT-19, pp. 42–49, Feb. 1971.

[142] S. W. Golomb, "Sequences with randomness properties," Glenn L. Martin Co., Baltimore, MD, Terminal Progress Rep., Contract Req. No. 639498, June 1955.

[143] ——, "Remarks on orthogonal sequences," Glenn L. Martin Co., Baltimore, MD, Interdepartment communication, July 28, 1954.

[144] N. Zierler, "Two pseudo-random digit generators," M.I.T. Lincoln Lab., Group Rep. 34-24, July 27, 1954.

[145] ——, "Several binary sequence generators," M.I.T. Lincoln Lab., Tech. Rep. 95, Sept. 12, 1955 (AD 089135).

[146] ——, "Linear recurring sequences," *J. SIAM*, vol. 7, pp. 31–48, Mar. 1959.

[147] E. N. Gilbert, "Quasi-random binary sequences," Bell Tel. Lab., unpublished memo., Nov. 27, 1953.

[148] D. A. Huffman, "Synthesis of linear sequential coding networks," presented at the 3rd London Symp. Inform. Theory, Sept. 12–16, 1955; published in *Information Theory*, C. Cherry, Ed. New York: Academic, 1956.

[149] T. G. Birdsall and M. P. Ristenbatt, "Introduction to linear shift-register generated sequences," Univ. Michigan Res. Inst., Ann Arbor, Tech. Rep. 90, Oct. 1958 (AD 225380).

[150] M. Ward, "The arithmetical theory of linear recurring sequences," *Trans. Amer. Math. Soc.*, vol. 35, pp. 600–628, July 1933.

[151] M. Hall, "An isomorphism between linear recurring sequences and algebraic rings," *Trans. Amer. Math. Soc.*, vol. 44, pp. 196–218, Sept. 1938.

[152] J. Singer, "A theorem in finite projective geometry and some applications to number theory," *Trans. Amer. Math. Soc.*, vol. 43, pp. 377–385, May 1938.

[153] R. J. Turyn, "On Singer's parametrization and related matters," Appl. Res. Lab., Sylvania Electron. Syst., Waltham, MA, Eng. Note 197, Nov. 10, 1960.

[154] R. Gold, "Optimal binary sequences for spread spectrum multiplexing," *IEEE Trans. Inform. Theory*, vol. IT-13, pp. 619–621, Oct. 1967.

[155] ——, "Maximal recursive sequences with 3-valued cross-correlation functions," *IEEE Trans. Inform. Theory*, vol. IT-14, pp. 154–156, Jan. 1968.

[156] T. B. Whiteley and D. J. Adrian, "Random FM autocorrelation fuze system," U. S. Patent 4 220 952, Sept. 2, 1980 (filed Feb. 17, 1956).

[157] J. P. Costas and L. C. Widmann, "Data transmission system," U. S. Patent 3 337 803, Aug. 22, 1967 (filed Jan. 9, 1962).

[158] "Reliable tactical communications," General Electric Res. Lab., Schenectady, NY, Final Rep., Contr. DA-36-039sc-42693, Mar. 2, 1954 (AD 30344).

[159] E. J. Groth, "Notes on MUTNS, a hybrid navigation system," Conf. 6733 on Modern Navigation Systems, Univ. Michigan, Ann Arbor, Summer 1967 (Univ. Michigan Library Call No. VK 145.M624, 1967).

[160] E. J. Groth *et al.*, "Navigation, guidance, and control system for drone aircraft," Motorola, Final Eng. Rep., Contr. DA36-039sc78020, June 6, 1961 (AD 329101, AD 329102, AD 329103).

[161] L. S. Schwartz, "Wide-bandwidth communications," *Space/Aeronautics*, pp. 84–89, Dec. 1963.

[162] J. M. Wozencraft, "Reliable radio teletype coding," U.S. Patent 3 896 381, July 22, 1975 (filed Nov. 2, 1960).

[163] A. Kohlenberg, S. M. Sussman, and D. Van Meter, "Matched filter communication systems," U.S. Patent 3 876 941, Apr. 8, 1975 (filed June 23, 1961).

[164] E. Guillemin, "Matched filter communication systems," U.S. Patent 3 936 749, Feb. 3, 1976 (filed June 23, 1961).

[165] S. M. Sussman, "A matched filter communication system for multipath channels," *IRE Trans. Inform. Theory*, vol. IT-6, pp. 367–373, June 1960.

[166] W. G. Ehrich, "Common channel multipath receiver," U.S. Patent 3 293 551, Dec. 20, 1966 (filed Dec. 24, 1963).

[167] J. J. Spilker, Jr., "Nonperiodic energy communication system capable of operating at low signal-to-noise ratios," U.S. Patent 3 638 121, Jan. 25, 1972 (filed Dec. 20, 1960).

[168] H. G. Lindner, "Communication security method and system," U.S. Patent 4 184 117, Jan. 25, 1980 (filed Apr. 16, 1956).

[169] J. W. Craig and R. Price, "A secure voice communication system," *Trans. Electron. Warfare Symp.*, 1959.

[170] J. W. Craig, Jr., "An experimental NOMAC voice communication system," M.I.T. Lincoln Lab., Rep. 34G-0007, Aug. 29, 1960 (AD 319610).

[171] D. J. Gray, "A new method of teletype modulation," M.I.T. Lincoln Lab., Tech. Rep. 9, Sept. 22, 1952 (AD 000928).

[172] R. M. Fano, "Anti-multipath communication system," U.S. Patent 2 982 852, May 2, 1961 (filed Nov. 21, 1956).

[173] L. A. deRosa, M. Rogoff, W. B. Davenport, Jr., and P. E. Green, Jr., in "1981 Pioneer Award," *IEEE Trans. Aerosp. Electron. Syst.*, vol. AES-18, pp. 153–160, Jan. 1982.

[174] R. C. Dixon, Ed., *Spread Spectrum Techniques*. New York: IEEE Press, 1976.

[175] A. M. Semenov and A. A. Sikarev, *Shirokopolosnaya Radiosvyazy* [Wideband Radio Communications]. Moscow, USSR: Voyenizdat, 1970.

[176] A. A. Kharkevich, "The transmission of signals by modulated noise," *Telecommunications (USSR)*, vol. 11, no. 11, pp. 43–47, 1957.

[177] F. H. Lange, *Korrelationselektronik* [Correlation Electronics]. Berlin, Germany: VEB Verlag Technik, 1959.

[178] J. P. Costas, "Poisson, Shannon, and the radio amateur," *Proc. IRE*, vol. 47, pp. 2058–2068, Dec. 1959.

[179] ——, "Author's comment," *Proc. IRE*, vol. 48, p. 1911, Nov. 1960 (see also "Information capacity of fading channels under conditions of intense interference," *Proc. IEEE*, vol. 51, pp. 451–461, Mar. 1963).

[180] P. Bello, "Demodulation of a phase-modulated noise carrier," *IRE Trans. Inform. Theory*, vol. IT-7, pp. 19–27, Jan. 1961.

[181] H. Magnuski, "Wideband channel for emergency communication," in *IRE Int. Conv. Rec.*, New York, NY, Mar. 20–23, 1961, part 8, pp. 80–84.

[182] R. A. Marolf, "200 Mbit/s pseudo random sequence generator for very wide band secure communication systems," in *Proc. NEC*, Chicago, IL, 1969, vol. 19, pp. 183–187.

[183] A. I. Alekseyev, "Optimum noise immunity of noise like signals" (in Russian), presented at the 19th All-Union Conf. Popov Society, May 1963; trans. in *Telecommun. Radio Eng. (USSR)*, pt. 2, vol. 19, pp. 79–83, Aug. 1965.

[184] M. K. Razmakhnin, "Wideband communication systems" (in Russian), *Zarubezhnaya Radioelektronika* [Foreign Radio Electron.], no. 8, pp. 3–29, 1965.

[185] N. T. Petrovich and M. K. Razmakhnin, *Sistemy Svyazi s Shumopodobnymi Signalami* [Communication Systems with Noise-Like Signals]. Moscow: Sovetskoye Radio, 1969.

[186] A. I. Alekseyev, A. G. Sheremet'yev, G. I. Tuzov, and B. I. Glazov, *The Theory and Application of Pseudorandom Signals* (in Russian). Moscow: 1969.

[187] Yu. B. Okunev and L. A. Yakovlev, *Shirokopolosnye Sistemy Svyazi a Sostavnymi Signalami* [Wideband Systems of Communication with Composite Signals]. Moscow: Svyaz, 1968.

[188] L. S. Gutkin, V. B. Pestryakov, and V. N. Tipugin, *Radioupravleniye* [Radio Control]. Moscow: 1970.

33

# Notes on Spread-Spectrum History

ROBERT A. SCHOLTZ, FELLOW, IEEE

*Abstract*—This paper presents additional historical information relating to the spread-spectrum history described in [1]. Included here are ties with speech scrambling technology, wide-band matched filter design, stimulus of the WHYN system design by British research, ITT's efforts to protect vital information during World War II, and other miscellaneous notes and references.

## ADDITIONAL NOTES ON SPREAD-SPECTRUM HISTORY

IN reviewing the efforts which went into the writing of [1], it is obvious to the author that the publication of this history was greatly simplified by two modern technological wonders, the copy machine and the word processor, to facilitate information transfer and reorganization, respectively. I hope that readers will word-process the following comments into the original history. (Braces { } will be used to enclose reference numbers from [1].)

•  •  •

Gustav Guanella (see Fig. 1), the Swiss inventor who held pre-World War II patents on radars and direction finders employing noise-like signals {12}, {13}, died in January, 1982. Although he evidently did not pursue spread-spectrum techniques for many years thereafter, Guanella applied his innovative talents in the related field of speech scrambling systems. A 1946 NDRC report [2] in fact discusses his designs and the level of security which they provide.

The NDRC report also describes a speech privacy system in which the voice waveform is multiplied by a code signal before transmission, and, after reception, is further multiplied by a synchronous replica of the inverse of the code signal to recover the voice waveform. Hence, this early-1940's speech privacy concept possesses many of the ingredients of a spread-spectrum design, but does not mention bandwidth expansion, and may be appropriately dubbed a continuous-wave cryptographic system. It was further noted that this multiplicative system was quite vulnerable to "cryptanalysis" when the code signal possessed a short period. Several long-period code generators were discussed in the report.

The historically useful reports [2], {17}, {53}, {58} of the NDRC (National Defense Research Committee) were undoubtedly precursors of the postwar demobilization of that civilian organization. Military officials, recognizing the major contributions of the academic world during the war just ended, and realizing the need for a continued strong research base, initiated a program of support for advanced research in the nation's

Manuscript received July 1, 1982; revised November 12, 1982. This work was supported in part by the Army Research Office under Grants DAAG 29-79-C-0054 and DAAG 29-82-K-0142.

The author is with the Department of Electrical Engineering, University of Southern California, Los Angeles, CA 90089.

Fig. 1. Gustav Guanella, Swiss pioneer of noise-modulated radar and speech privacy systems.

universities [3, p. 98]. First installed at M. I. T., Harvard, Columbia, and Stanford, this Joint Services Electronics Program now sponsors research at many major U. S. institutions.

•  •  •

Louis Chereau, Director of Patents and Information under Maurice Deloraine at ITT's Paris Laboratories in 1940, has provided the author with excerpts from a history [4] of the early years at ITT. Written by Deloraine, one segment of the history recounts the efforts and risks undergone in France to protect valuable information, and to eventually transport it to the United States. Chereau indicates that he, not Busignies, visited the French Patent Office in Vichy just before the Germans crossed the demarcation line, for the purpose of removing Busignies' MTI patent application on "Elimination of fixed echoes," filed in Lyon on May 27, 1942, and also Ferdinand Bac's LORAN application. The MTI application was based on a Busignies memorandum, dated October 24, 1940, and the memorandum in turn was motivated by a "brainstorming session" initiated by Chereau who had put forth the idea of echo cancellation. The memorandum was written shortly before Deloraine and Busignies, along with Emile Labin and Georges Chevigny and their families, escaped via Marseilles, Algiers, Casablanca, Tangier, and Lisbon to New York, arriving on New Year's Eve, the last day of 1940. Busignies filed his application in the United States on March 5, 1941, and was granted U.S. Patent 2 570 203.

•  •  •

Reprinted from *IEEE Trans. Commun.*, vol. COM-31, pp. 82–84, Jan. 1983.

34

Fig. 2. Nathan Marchand, organizer and first chairman of the IRE Information Theory Group, early practitioner of correlation techniques, and codesigner of the WHYN system. He was the lyricist of the classified jingle [1, p. 830] presented at a Washington meeting where radar correlation–detection methodology was discussed [15].

With regard to Robert Fano's acoustic pulse experiment [1, Fig. 16], it should be noted that the peak of the remarkably compressed pulse became visible on the oscilloscope screen when the lights were turned off, not just by intensity adjustment as stated in [1]. (Unfortunately, turning the lights out will not improve peak visibility in Fig. 16.) This pseudorandom, high time–bandwidth product, matched-filter experiment, possibly the first not based on linear FM modulation, can be dated to approximately October, 1951, when Fano disclosed a multiple matched-filter communication system [5] to his colleagues.

While Fano's invention, which originally suggested a reverse-driven magnetic-drum recording for signal generation, basically employed analog signals, another then contemporary matched-filter invention by Ronald Barker [6] definitely used digital signals. Barker's design employs the binary patterns, which now bear his name, as frame sync markers in digital data streams. While this application is not inherently bandwidth expanding, the waveform correlation design objectives in frame sync applications are quite similar to those for SS-MF communication applications, as well as to those for radar pulse-compression.

Fano, who had expounded basic correlation and matched filter concepts [7], eventually learned of Guanella's pioneering noise-correlation radar, probably through patent searches carried out for his antimultipath communication invention {172}. Similarly in the early 1940's while processing the patent application on his correlator design {34}, Nathan Marchand (see Fig. 2) became aware of Ellison Purington's wobbled FM, stored-reference, secrecy invention which used a difference-frequency correlator in its receiver (cited in [8]). These connections illustrate the often-untapped source of information available through the U.S. Patent Office. As a further example, among the 30 patents referenced in the reissue application of the

deRosa–Rogoff patent {48}, Price [8] has noticed two early-1940's frequency-hopping patents which are surprising, both in their technical content and in the backgrounds of two of the inventors.

•••

Prompted by a recent concern of Nelson Blachman, H.A. ("Judd") Schulke, who was the Signal Corps Project Engineer for the Lincoln F9C-A system development and production, has acted to permanently preserve F9C-A documentation {124}. Pending its ultimate declassification, deposit of this historical material into the personal archives of (retired) Major General Schulke in the Special Collections Section of the U.S. Military Academy Library, West Point, should now assure its security against routine or inadvertent destruction. Such safeguarding is certainly merited for these paper remains of the first operationally deployed communication system to use spread-spectrum modulation as an electronic counter-countermeasure.

Proper archival retention is particularly appropriate in light of the presumed loss, during a disastrous fire at the St. Louis, MO, federal repository, of vital records of the pioneering Colonial/Sylvania-Buffalo spread-spectrum ECCM systems inspired by Madison Nicholson. Substantially all technical information on the Colonial Radio "Janus," an antijamming missile guidance system of World War II, is thus missing, there surviving merely its contract-support file {57} and five innocuous thyratron-circuit designs in U.S. Patents 2 407 399, 2 428 126, 2 438 962, 2 462 945, and 2 483 620.

Nicholson's postwar proposal, from which the Sylvania Hush-Up program grew, was lost to history in the same conflagration, along with Sylvania monthly reports (of February, 1954 through December, 1955) written during the follow-on ARC-50 development phase. The latter might have shed light on the source of the $m$-sequence concept at Buffalo, and documented in detail the voice-FM technique known to have been used in this early design. Another unfortunate casualty of the repository fire seems to be the operating manual [9] for the Sylvania ARC-50 "flying breadboards."

•••

Norman Harvey, the leader in the development of the WHYN concept, comments that phase-matching was accomplished in the Armstrong FM radar receiver by adjusting the frequency of the modulation (not by phase-shifting as in [1]), so that the round-trip propagation delay corresponded to one modulation period. Concerning the oral history of the origins of WHYN, Harvey states:

"The MX-773 project started out in a strictly study phase, with no special linkage to the FM radar project. It was the specification of 0.5 mile accuracy at 1500 miles that initially gave us the greatest concern. At the time, DECCA (England) was operating a pure cw, cycle-matching navigation system that attracted our interest because it promised that kind of accuracy under ideal conditions. Bob Bowie, John Wilmarth of Republic Aviation, and I went to England in the spring of 1946 to attend the Provisional International Conference on Air Organization (PICAO), to look particularly at DECCA, but also to see if there were other systems that might be adapted to meet our requirements.

Not long after that, the idea of combining a DECCA-type cycle matching system (for accuracy) with FM modulation techniques (for resolution), occurred to me, and the WHYN concept was born."

Harvey recalls that the multitone waveform studies did not begin until after the advent of the WHYN effort. Harvey's October 1950 paper [10], a companion to ₁69 , gives his views on the implications of waveform research studies carried out on the WHYN program.

• • •

As indicated in [1], it is quite likely that $m$-sequences had been studied as potential key-streams for cryptographic purposes before they ever were considered as carrier-spreading modulations for communications. Additional evidence of this possibility is a reference in a revised version [11] of {135}, to an NSA study [12] of irreducible polynomials under a cubic transformation. Dated July 7, 1953, the report on this useful tool for finding (among other things) $m$-sequence generator connections apparently was not soon available to spread-spectrum researchers.

Also concerning $m$-sequences, in March, 1955, David Huffman introduced shift register generators as potential error-correcting code hardware [13], a half year before he presented their pseudorandom and autocorrelation properties in {148}. This contribution set the stage for the pioneering BLADES error-correction hardware referred to in {90}.

• • •

I would like to call the reader's attention to a 1959 survey by B. J. DuWaldt [14] which describes several early spread-spectrum systems, and in particular, contains information about Philco's Spread Eagle.

The following documentation numbers may be useful in securing certain references from [1]:

AD 224557 for {115} (misprint in [1]).

ATI 014050 for {54}.

ATI 139962 for all three volumes of {100}.

The patent for the tau-tracker invention of James Green and Madison Nicholson (see [1, p. 832]), filed on June 17, 1958, was finally issued November 30, 1982, after 24 years of secrecy, as U.S. Patent 4 361 890.

Finally, please note the following corrections to [1]. The material abstracted from Pierce and Hopper {127} in [1, p.

841] is partly paraphrased, and not a perfect quote. The NDRC Division 15 Report {53} mentions test results for ITT's printer system {54}, {55}, not ITT's facsimile system {50}, {51}, as stated in [1]. Although the mod-clock jargon is an appropriate descriptor for the final F9C-A orthogonal modulation technique, that particular terminology was not used at M. I. T. during the 1950's (see [1, p. 839]).

## ACKNOWLEDGMENT

The author is indebted to Louis Chereau, Robert Frank, Norman Harvey, and Robert Price for certain materials contained in this commentary, and to J. Meredith Smith for drawing the author's attention to the DuWaldt survey.

## REFERENCES

[1] R. A. Scholtz, "The origins of spread-spectrum communications," IEEE Trans. Commun., vol. COM-30, pp. 822–854, May 1982.

[2] "Speech and facsimile scrambling and decoding," Nat. Defense Res. Committee, Off. Sci. Res. Develop., Washington, DC, Summary Tech. Rep. Div. 13, vol. 3, 1946 (AD 221609); reprinted by Aegean Park Press, Laguna Hills, CA.

[3] H. A. Zahl, Electrons Away, or Tales of a Government Scientist. New York: Vantage, 1968.

[4] M. Deloraine, When Telecom and ITT Were Young. New York: Lehigh, 1976; first published in French as Des Ondes et des Hommes: Jeunesse des Telecommunications et de l'ITT. Paris, France: Flammarion, 1974.

[5] R. M. Fano, "Patent disclosure," Nov. 14, 1951, disclosed orally to R. F. Schreitmueller, P. E. Green, Jr., and W. B. Davenport, Jr., Oct. 8, 1951.

[6] R. H. Barker, "Synchronising arrangements for pulse code systems," U.S. Patent 2 721 318, Oct. 18, 1955 (filed in U.S. Feb. 16, 1953; in Great Britain Feb. 25, 1952).

[7] R. M. Fano, "Signal-to-noise ratio in correlation detectors," M.I.T. Res. Lab. Electron., Tech. Rep. 186, Feb. 19, 1951 (PB 110543).

[8] R. Price, "Further notes and anecdotes on spread-spectrum origins," this issue, pp. 85–97.

[9] "Operation, alignment, and adjustment of radio set AN/ARC-50(XA-2)," Sylvania Elec. Products, Buffalo, NY, Contr. AF33 (600)-26734, Nov. 1, 1955.

[10] N. L. Harvey, "A new basis for the analysis of radio navigation and detection systems," Sylvania Technologist, vol. 3, pp. 15–18, Oct. 1950.

[11] S. W. Golomb, Shift Register Sequences, rev. ed. Laguna Hills, CA: Aegean Park Press, 1982.

[12] R. W. Sloan and R. W. Marsh, "The structure of irreducible polynomials mod 2 under a cubic transformation," NSA Sect. 314 Rep., July 7, 1953.

[13] D. A. Huffman, "A study of the memory requirements of sequential switching circuits," M.I.T. Res. Lab. Electron., Tech. Rep. 293, Mar. 14, 1955.

[14] B. J. DuWaldt, "Survey of radio communications securing techniques," Space Technol. Lab., Los Angeles, CA, Tech. Rep. TR-59-0000-00789, Aug. 31, 1959 (AD 358618).

[15] "Cross-correlation radar," Phys. Lab., Sylvania Elec. Products, Bayside, NY, Rep. YD-51-5, Feb. 1951.

# Further Notes and Anecdotes on Spread-Spectrum Origins

ROBERT PRICE, FELLOW, IEEE

*Abstract*—This sequel to R. A. Scholtz's monograph of May 1982, in this TRANSACTIONS, adds to the early history of spread-spectrum communications from circa 1900 until about 15 years ago. Relevant to the origins of this field of radio signaling, the following are included in a number of items of quite recent appreciation: the 1930 invention of secret communications featuring spectrum despreading via a locally generated reference at the receiver; a stratagem of electronic warfare during the Battle of the Bulge, in World War II, that cleverly made use of Armstrong's frequency modulation; and the role of a star actress from the motion picture field, in origination of frequency hopping for antijamming radio control of missiles.

Some further, and first-hand, reminiscences are given of germinal M.I.T. spread-spectrum contributions, and of similar noise-correlation radar art. Extensive footnotes cast sidelights on the beginnings of information theory, and also, among other things, on the keeping of military high-technology secrets over decades, even between security-cleared family members. A variety of references is appended.

## I. INTRODUCTION

THIS addendum to Robert A. Scholtz's very comprehensive, and highly reliable, history of early spread-spectrum (SS) communications [1], and to his sequel [2], summarizes a number of supplementary records that relate to the beginnings of SS-like methodologies. Some more informal remembrances are also offered below.

A good deal of the following material has only recently become known. Portions of it are interesting not only technologically, but also in view of who several of those persons were, in particular, among the many investigators involved in SS origination.

The next section is a revisit to the narrative [1, pp. 835–836] that has recalled a certain occurrence of intercorporation idea flow during 1950. The succeeding sections of the present sequel are generally in chronological order, from the turn of the twentieth century up to the late 1960's.

It should be understood that—with the conspicuous exception of most of the next section—much of what is sketched here neither bears directly on any of the systems already recounted in [1], nor is it in itself closely interrelated. Robert C. Dixon {174, pp. 6–7}[1] has similarly implied a need for caution against overdiligently seeking family ties between the various ancestors of SS communications technology.

## II. THE HARTWELL CONNECTION AND RECONNECTION

The phone-switchboard similes chosen for the headings here and in [1, p. 835], as jointly linked, allude to the back-and-forth transfers and exchanges of information on prospective telecommunication advances that happened in the

Manuscript received September 22, 1982; revised October 18, 1982.
The author is at 80 Hill St., Lexington, MA 02173. This paper does not necessarily represent any specific views or undertakings of the Sperry Corporation, which assumes no responsibility for its contents.

[1] As in [2], a number enclosed in a "{ }" denotes the correspondingly "[ ]"-enclosed reference-citation number from [1].

Hartwell project of 1950, and afterwards. These insights circulated between the Massachusetts Institute of Technology (M.I.T.) and two of the country's foremost telephone-and-telegraph research organizations.[2] Appendix G of the originally secret Hartwell report {97} summarizes four distinctly special techniques for secure communications envisioned there, one of these being attributed to John R. Pierce of the Bell Telephone Laboratories (BTL), and one to Paul R. Adams and Louis A. deRosa at the Federal Telecommunication Laboratories (FTL) of the International Telephone and Telegraph Corporation (ITT).

Additionally noteworthy is that of these four SS-relevant inventions, the three in which correlation methods are employed for reception are referred to, jointly, as arising from "concepts of information theory." This observation is made in Part II of the Hartwell report's Appendix D, thus recognizing them as embodiments of "the next order of subtlety" and "the next logical step" beyond the German "Flash" ("Kurier") system of World War II to which "...the U.S.S.R. has had access...to say nothing of its designers...." (One of the Kurier designers, Gottfried Vogt, fortuitously was later active in the SS deployment, by the U.S. Army Signal Corps, of the Lincoln F9C-A system which, as reviewed in [1], ultimately resulted from Project Hartwell.)

Of further interest is that this Appendix D, "Direction Finding" section of the Hartwell report made a prediction of when SS communication technology might have become "available against us": "1955+." This was, in fact, about the year that the F9C-A system went operational during the successful "positive program of our own" advocated by this same appendix. (Both Appendixes D and G had been noted, in the latter, to suggest ways of "technological feasibility" for meeting requirements in "greater resistance to jamming" as well as "greater security from interception.")

Returning to Appendix G of the Hartwell report, it is distinguished both in its originative telecommunications proposals, and in having two of its four Principal Contributors, Jerome B. Wiesner and Edward E. David, go on from M.I.T. to be later appointed the Science Advisers to Presidents Kennedy and Nixon, respectively. The other contributors (now deceased), Harald T. Friis[3] and Ralph K. Potter, were researchers

[2] A third such telephone company, the General Telephone and Electronics Corporation, can be credited (per [1, p. 832] but after the fact, through its acquisition of Sylvania Electric Products, Inc.) with originating the descriptor "spread spectrum." This term appeared in notes made by John J. Raney for a Sylvania marketing presentation of December, 1954. When inspected by the author on October 3, 1980, that record (later lost through burglary) contained as a section subtitle: "What We Mean by Spread Spectrum."
[3] Friis recounts [3]: "Dr. Wiesner presented my analyses of radio-wave propagation in and above water and Prof. Zacharias presented our general results at a special final meeting before the admirals..." (in Washington [4]).

Reprinted from *IEEE Trans. Commun.*, vol. COM-31, pp. 85–97, Jan. 1983.

known for their major accomplishments at BTL, the former in antennas and the latter in information transmission systems. Potter had been a prime innovator of the remarkably successful "X-System" (dubbed by its sponsor and deployer, the Signal Corps: "Sigsaly"). This was the pioneering, ultra-secret, and strikingly sophisticated ciphony equipment [1, p. 827], {43}, {44}, [5] that[4] during the Second World War protected vital transoceanic radiotelephone consultations among high leaders of the Allied forces, and very possibly between Roosevelt and Churchill in particular. (Wiesner, Friis, and Potter were Principal Contributors also to Appendix D, along with P. R. Adams of FTL.)

Up until the present, the editorship of Appendix G has not been adequately established from among its four Principal Contributors. Customarily, such identification would point to who conceived the two important, yet-unattributed telecommunications advances introduced there: transmitted-reference (TR) spread spectrum, and code-division multiple-access (CDMA). Robert M. Fano recently saw this Appendix for the first time and now has the "...impression...that it was written by Jerry Wiesner." Subsequently, Edward David upon similar query has replied that he feels it to be logical, from the information available to him (going back even to before Hartwell, when he was a graduate student of Wiesner's), to conclude directly that Jerome Wiesner is the independent inventor of both techniques in question.

These concurring responses appear to corroborate sufficiently a recollection made in 1970 by the now deceased SS pioneer, Louis deRosa, that, according to John M. Smith, "J. B. Wiesner read the reports of Federal Telecommunication Laboratories before suggesting similar techniques in connection with Project Hartwell." Smith, acknowledged in [1] as deRosa's interviewer, wrote this down in paraphrase shortly after phoning him. Smith has later added:

"Should this be interpreted to mean that deRosa was claiming that, until reading the FTL reports, Wiesner had no previously formulated ideas about signaling with noise-like waveforms? ... I do not believe that to have been deRosa's intent . . . I didn't get the impression he was trying to take anything away from Wiesner or M.I.T."

It has been reliably reported that Wiesner himself, because of the loss or, perhaps, purloining of his secret laboratory notebooks around the time that he became President Kennedy's Special Assistant for Science and Technology, does not want to rely solely on his own memory about these long-ago events. Some months before Project Hartwell, however, Wiesner had introduced a novel technique into the field of statistical communications which is of obvious relevance here. That [7, p. 341], {116} was his noise-stimulus cross-correlation sensor concept which (independently of {12} and {67}, and as witnessed by Leon G. "Jake" Kraft) significantly generalized Norbert Wiener's {18, p. 50} probing of filters via noise-

autocorrelations. This noise-signaling innovation, when taken together with the Fano–David corroborations, now unquestionably establishes Wiesner as an entirely original inventor of the TR type of SS communications. (William W. Hansen's prior TR-SS patent {47} seems to have gone unnoticed by inventors other than those of {48}.)

Of far more consequence is that Wiesner went on to conceive and document, in the Hartwell report, a completely new kind of multiplexing which imaginatively extended his TR-SS invention quite outside covertness applications. He argued there that such a format, using a multiplicity of different noise-carriers in combination with correlation reception, would offer freedom from the tight frequency-allocation and/or time-synchronization constraints previously required in multiuser communications. This CDMA invention of Wiesner's has matured over the past three decades into a well-utilized modulation method for the random-access sharing of radio channels. Now much in demand by the military—but usually using the stored-reference (SR) carrier technique à la FTL (as so credited in the Hartwell report and there, too, multiplex-extended by Wiesner)—that signaling innovation also holds promise for contemporary commercial and civil communications. (In the latter context, CDMA has been under consideration through FCC Docket 81-414 as cited in [8].)

In responding to the Hartwell appendix-editorship question, Edward David happened to recall that upon joining BTL in 1950, he brought with him the Appendix G noise-carrier concepts, which in turn stimulated considerable BTL discussion. This may have been additionally prompted by Claude E. Shannon's private suggestion of the CDMA possibility, made at BTL well before Hartwell. That Shannon notion is credited at the outset of {127} in having been conceived "some years ago" as of September, 1952, and recently has been dated[5] to

---

[4] The Sigsaly digital-voice system accelerated the transition in telecommunications technology from analog to digital. Quoting from [6, p. 958] (written just before this X-System was declassified): "It was Potter who was to take the lead, in the 1940s, in exploring and promoting this radical departure, which ultimately would have profound consequences in the transmission plant of the Bell System."

[5] This dating comes from recollections of a fundamental paper on information theory published by Shannon at that time, in which he found the minimax solution {27, pp. 19–20} to what might be construed as a game-theory problem in jamming/antijamming. When queried about this by the author on July 28, 1982, Claude Shannon replied that there well could have been such a military application in the back of his mind. Similar mathematics were later presented by William L. Root [9], [10] and Nelson M. Blachman [11], [12]—the former being directly motivated as a member of the Lincoln Laboratory NOMAC (NOise Modulation And Correlation) project, while the latter was influenced both by {27} and through his association with Sylvania's development of the operational Lincoln F9C-A system. (NOMAC/F9C-A efforts are described in [1, pp. 836–841].)

Should Shannon's minimax model indeed have been militarily intended in {27}, he would have been in a sense coming full circle—from his initial research in cryptography {31} and X-System speech scrambling [5, ref. 2, p. 317], onward to the consequent fulfillment of his aims for "information theory" (indeed, so quoted from the 1945 BTL memorandum cited in {31})—and from establishing that field on the firmest of foundations, back again to the advancement of basic principles for securing communications against threats from adversaries.

Shannon actually had embarked on a searching study of the communication process even prior to his World War II military work, as evident from a letter that he wrote on February 16, 1939, to Vannevar Bush [13, pp. 504–505]. Coincidentally, Alan M. Turing in England had identified by 1940 an entropy-measure for making evaluations in cryptanalysis, where with extreme secrecy he was speeding the decipherment of the German Enigma-machine codes [14], [15]. A very oblique reference [16, p. 125] was made to this in 1960, years before the "Ultra Secret" could be divulged, of the British breakthrough in reading Enigma traffic throughout the Second World War.

Someone looking into the origins of information theory—and now

early 1949 by Brockway McMillan and John R. Pierce. Apparently, Shannon himself never put his SR-SS multiplexing idea into print, nor did it leave BTL before 1952. In any event, nothing more on CDMA or closely similar SS formats seems to have been done at BTL up to 1961, notwithstanding[6] a news item to the contrary [18]. (It should be noted that the 1950 FTL disclosure {96} of SR-SS does describe a type of noise-carrier multiplexing, but one needing cooperative time synchronization, whose elimination is the essential contribution of both the Wiesner and Shannon concepts.)

Also in respect to the preceding, Yuk Wing Lee (joint contributor with SS pioneer Wilbur B. Davenport to an early, secret progress report from the soon-to-be Lincoln Laboratory Group 34) has written that the filing of a secret patent application was once contemplated among Wiesner, Davenport, and Fano. Whether this may have been to claim the CDMA or TR-SS inventions, or some other variety, Lee did not know. (Fig. 1 presents an early photograph of the four principals in M.I.T.'s beginnings of SS communications.)

In a book [2, ref. 3] authored in 1968 (with a foreword by Wiesner, shown then as M.I.T. Provost in a photograph opposite p. 63), Harold A. Zahl, Director of Research at the Ft. Monmouth, NJ, laboratories of the Army Signal Corps, compliments M.I.T.'s SS program implicitly. On p. 118 of this memoir, he refers to the M.I.T. Research Laboratory of Electronics as having received an "...added $600,000 for

Fig. 1. Jerome Wiesner (second from left, later President of M.I.T.) with his colleagues Yuk Wing Lee (far left), Robert Fano, and Wilbur Davenport (far right), shown at an M.I.T. session on modern communications in the summer of 1951. That year, Wiesner made the intriguing suggestion {105, p. 41} of using radio-astronomical emissions to provide at the transmitter and receiver the correlated noise references needed for "pseudonoise" spread-spectrum communications. (Photo courtesy of Karl Wildes, M.I.T.; unfortunately, space limitations have prevented larger presentation, for clarity, of this group photograph.)

Korean-born classified research...." This "...had blossomed into a lusty laboratory [on the top floor of Building 22 there, a Radiation Laboratory relic of World War II now long since demolished] producing ideas toward the solution of many military problems."

Finally concerning the "Hartwell connections," it should be noted that—although Jerome Wiesner is certainly the prime inventor of CDMA as published through the distribution of some 500 copies (per {98, p. 9}) of the Hartwell report {97}— the Lincoln Laboratory SS system successes for which he was the original motivator did not directly demonstrate a full-fledged CDMA capability.[7] That is, albeit Bernard J. Pankowski indeed created for his innovative thesis {102} the first noise-carrier multiplexing system, it was neither stored-reference, nor did it achieve a CDMA-like quality other than by frequency-offset [1, p. 837]. Nonetheless, this M.I.T. secret thesis of 1952 is significant not only for its early exploration of new approaches to multiplexing—but furthermore for its remarks that: "The idea of using both a noise carrier and a continuous correlation detector in a communication system occurred independently to Wiesner and DeRosa" (p. 2), and (p. 10) that[8] the "...'stored' noise system" was "suggested by Adams and DeRosa."

knowing of 1) Turing's additional ". . . work on an entirely new method of achieving security for voice transmissions, much needed at that time because voice 'scramblers' attached to telephone circuits were too easily broken" [14, p. 176]; 2) his trip to the U.S. for classified conferences midway in the war [15, p. 36]; 3) his verification there, during a visit to BTL, of the X-System secrecy as cited in [5]; and 4) Shannon's X-System role as footnoted here (which, unknown to him, was a counterpart to Turing's)—might all too easily jump to a conclusion, *wrongly,* that Turing introduced Shannon to informational negentropy.

While it is true, according to the author's July 28, 1982, interview with Claude Shannon, that he and Turing did confer during the war, they talked only about such then-academic topics as artificial intelligence. Shannon sensed that Turing was at BTL to discuss speech scrambling, but it being "wartime," he "knew better than to ask" about that. Although he had U.S. military clearance for radar research, and so was privy early to Wiener's "Yellow Peril" {18} of 1942 (which he helped to expound in {17, Part II} and in [17]), Shannon had no access whatever to Turing's secret endeavors, nor even to official American cryptology work other than that for his own evaluation of the X-System. The author has recently been advised of these security barriers by William R. Bennett in connection with [5], and by John W. Tukey (the coiner of "bit"). (Additional items of X-System history are given in Footnote 14.)

[6] Not long ago, Brockway McMillan reminisced that, inspired by the CDMA notion due (at BTL) to Shannon—which the latter in his July 28, 1982, interview with the author said that he had proposed as an appealingly "democratic" way for users to share the time-frequency dimensionality resources of a communication channel available to all—he immediately thought of SR-SS military applications. McMillan remarked that at first he saw the implications for antijamming communications, and then later for covert signaling, but that nothing came of these perceptions. It now appears that the report in [18] about BTL having ". . . built and tested various types of wide-band equipment with complex modulation schemes" may have been an SS-relevant reference to the World War II digitally secure voice system described in [5], or to its postwar speech-scrambling descendants, or possibly to "chirp" radar (per the Darlington patent {4, p. 103}).

Incidentally, McMillan has also related [13, p. 431] that Shannon once pointed out to him the use of entropy-measure for evaluating "goodness of fit" in cryptography (a topic on which Footnote 5 touches).

[7] "Definitionsmanship" might well arise in trying to determine when CDMA was first installed operationally. This probably occurred no later than 1965, the year that the Magnavox MX-118 spread-spectrum applique became available with its Gold codes [1, pp. 845-846].

[8] This observation paraphrases the footnote in {97, p. G-10}, which credits "Adams and DeRose [sic] of Federal Telecommunications [sic] Laboratories" for SR-SS with its "marked advantages." It is now appreciated that the attribution would have been better made, additionally to deRosa, to the SS pioneer Mortimer Rogoff instead of to his repre-

## III. VARIOUS ANTIJAMMING PURSUITS PRIOR TO THE SECOND WORLD WAR

Now turning back the pages of radio history by virtually a half-century more, it has been found {3}, [20, p. 1] that secrecy, jamming, and antijamming had each begun to stir attention as early as 1901. (Note: For further details see [81, pp. 38–39], where also a recommendation is mentioned that "the homing-pigeon service should be discontinued as soon as some system of wireless telegraphy is adopted.") The first two of these three differing adversary-recourses were astutely exercised together in reporting, by "wireless," the America's Cup yacht race of that year. They were invoked by a thereby countermeasures-pioneering engineer who was competing against both Guglielmo Marconi and Lee De Forest for live coverage of this nautical event. (Another such double stratagem, which simultaneously deployed the latter pair from the above trio of tactics, was played some 60 years later while BLADES transmission tests were first underway for the U.S. Navy, as recalled in [1, p. 833].)

Even before, in 1899, Marconi had experimented with frequency-selective reception in response to worries about radio interference {3}, [20, p. 1]. The Navy, whose due concern[9] had prompted this most basic advance in the radio art, reemphasized its apprehensions by warnings "...during 1904 on the dangers of relying on wireless in war" that "raised the questions of enemy intercept and interference which were to remain and grow into the major research and operational fields of radio intelligence and radio countermeasures" [22, p. 1243].[10]

Next on the scene seems to have been the Hammond Radio Research Laboratory [1, p. 829]. John Hays Hammond, Jr., an inventor whose "...development of radio remote control served as the basis for modern missile guidance systems" also "...developed techniques for preventing enemy jamming of remote control and invented a radio-controlled torpedo..." [23]. By 1914, he was actively exploring means to "...greatly minimize the possibility of an enemy determining the wave lengths used in the control of the craft and thereupon interfering with the control thereof" [24]. In the radio guidance of remote vehicles, Hammond was following in the footsteps of Adm. Bradley A. Fiske and of Nikola Tesla [25, p. iii]. (Note: See "Remote Radio Control" [81, ch. XXIX] for more background on Hammond, Fiske, and Tesla.) The efforts against jamming were, initially, pursued by Benjamin F. Miessner, Hammond's sole coworker during 1912, who had invented a primitive form of SS signaling [25, p. 32] which was quite likely the earliest to be thus "...transmitted in a peculiar way" [25, p. 65].

The Miessner–Hammond carrier broadening was incorporated into a military transmitter, which with its associated receiver (that additionally introduced amplitude-limiting action) was delivered to the U.S. Army in France just before World War I ended. There, Maj. Edwin H. Armstrong verified the Hammond system's ability to communicate in the face of powerful enemy interference, a challenge officially recognized as "...one of the most important matters connected with the war" {60, footnote 48}.

From the First World War on into World War II, the furtherance of SS techniques continued at the Hammond Laboratory, with concentration on frequency-wobbling methods to send secret signals which sounded like "...some new kind of man-made static..." {60, p. 1203}. The antijamming attribute of spectrum spreading became obvious as the frequency deviation of the wobbling increased so that its "...wider swing reduced the amount of time that a given narrow-band disturbance could affect the intermediate-frequency circuit of the receiver..." {60, p. 1202}. Contributions of this kind, originating in that era from the Hammond group (additionally to { 49}) are documented in patents on transmitted-reference—and on stored-reference—frequency wobbling which were filed, respectively, by Emory L. Chaffee [26] in 1922 and Ellison S. Purington [27] in 1930.

The Chaffee patent claims that his invention, involving "rapid and erratic" wobbling, is useful for "secret radio telephony" and for "...telegraphic signals or radio-dynamic control." With respect to the Purington patent [27] (which seems somehow to have been overlooked in the chronicles {60} and {61}), that SR invention for obtaining "...secrecy in which a rapid variation rate is necessary..." for the wobbling was a sequel to Hammond/Navy experiments[11] of 1921–1922 which "...established points of interest regarding information theory when the interference greatly exceeds the signal" {60, p. 1202}. This latter, 1957 statement was naturally made from hindsight, since the field of information theory did

---

sentative at the Hartwell summer-study, Paul Adams. Another SS pioneer, Paul E. Green, later wrote {105 p. 6} like Pankowski: "The idea of using noise-like signals in a correlation type communication system seems to have originated in 'Project Hartwell,' a study project conducted at M.I.T. in 1950, at which time the transmitted-reference type was suggested by Wiesner and the stored-reference type by DeRosa and Adams."

Green's statement augmented the following introductory paragraph from the final report [19] of the secret M.I.T. Research Laboratory of Electronics group, led by Davenport, that was undertaking the NOMAC effort: "The program stems generally from the theoretical work of N. Wiener of M.I.T. and C. E. Shannon of the Bell Laboratories, and from the unclassified program of theoretical and experimental research carried out at the Research Laboratory of Electronics, M.I.T. The particular systems being studied are based specifically on the suggestion made in Project Hartwell by J. B. Wiesner of M.I.T. that a random (noise-like) carrier and a correlation detection system be used in order to obtain a considerably greater freedom than usual from enemy intercept."

[9] The U.S. Navy, too, was the first to respond to the radar prophecy made by Marconi in 1922, when he received the IRE Medal of Honor [21].

[10] As noted in [1, p. 838], the lead author of [22], Genl. James D. O'Connell, had keenly supported the development of the Lincoln F9C-A system. Procurement of that system was effected by his deputy, Genl. W. Preston Corderman, who earlier had been cognizant of the X-System [5] and had worked with BTL in its deployment (per Footnote 14)—and who, as a successor to the eminent cryptologist William F. Friedman, headed the Army's Signal Intelligence Service (often called "Arlington Hall") from 1943 until the end of World War II.

In the article [22] cited here, Genl. O'Connell may have been alluding to this secretly successful M.I.T. innovation in remarking that: "... most important, reliability in military communications..." together with the "... recognition of ... the vulnerability of the Army command and communication network ..." had (within the first two decades of the post-World War II period) "... led to major action. Programs were carried out for new antenna–transmitter–receiver design and installation and for new modulation techniques." (When being interviewed per Footnote 17, Genl. O'Connell informed the author that all of the quotations from [22] which appear on this page were written by him.)

[11] Also, circa 1920, the Hammond company had contracts from the U.S. Army to show that "noninterferable characteristics" could be secured for the radio control of aircraft {61, p. 1262}.

not exist in a statistical formulation prior to the Second World War. The acknowledgment that, postwar, an SR-SS communication innovation—employing pseudonoise for the carrier instead of the rapid frequency wobbling disclosed in [27]—had arisen on concepts drawn in part from that Shannon-established discipline, was observed in the Hartwell report. This 1950 credit, noted before in the "Hartwell connection and reconnection" section of the present sequel, has been reconfirmed by SS inventor Mortimer Rogoff {173, p. 154}. (Moreover, it now appears established by [27] that Purington is the first to have conceived difference-frequency cross correlation.[12])

## IV. SHORTLY BEFORE PEARL HARBOR:
## THE LAMARR-ANTHEIL
## FREQUENCY-HOPPING INVENTION

Moving further forward in time, there has recently been made available a list of patents (in [28] with brief commentary, and extending as far back as Goldsmith's 1924 filing {38}) relevant to prior art in SS communications. (These findings, referred to in [2], are no doubt independent of the effort by Eugene G. Fubini mentioned in [1, p. 848].)

Included in the 30 patents cited by [28] are two of World War II vintage, which are distinctive both for their conceptual advancement of the SS field, and because of the recognition that all three of the inventors involved have received for their significant contributions elsewhere. That pair of patents covers particular stored-reference frequency-hopping (FH) apparatus for accomplishing secret communications (where the broad FH method had been invented [29] as early as 1929). The remainder of this section deals with that invention whose patent was the earlier one, of those two applied for during the war.

In mid-1941 an application for the FH patent [30] was filed by Hedy K. Markey (see Fig. 2) and George Antheil (of course, without their awareness of [29]). She being at that time a recent ex-spouse of Hollywood scriptwriter Gene Markey, the former-named inventor had been baptized Hedwig Eva Maria Kiesler. Growing up on Austria, this only child of a prominent Vienna banker had shown, at age 16, a flair for innovation by letting herself be filmed in total nudity while starring in the Czech-produced classic, *Esctasy* (the fifth of her many mo-

Fig. 2. Hedy Lamarr, inventress of the first frequency-hopping spread-spectrum technique explicitly conceived for antijamming communications. (Photo courtesy of Kenneth Galente, The Silver Screen, New York.)

tion pictures [31]). Several years later she was one of the small minority among non-Jewish Austrians who saw great danger to the world in Germany's early-1938 *Anschluss* of her homeland. That year, permanently leaving her country and her munitions-magnate husband, Friedrich A. "Fritz" Mandl, as washouts to Hitler, the actress came to the United States on a seven-year contract from Metro-Goldwyn-Mayer. There, she legalized her stage-renaming (by Louis B. Mayer) to Hedy Lamarr. She brought with her memories of company films, which the now-escaped wife had witnessed, of difficulties Mandl and his factory managers were encountering in getting their "aimlessly"[13] unguided torpedoes to hit evasive targets.

Once settled in southern California but still greatly concerned by the war then impending for the United States, Lamarr sought out the versatile and volatile symphony composer Antheil [33, ch. 32]. Quickly stimulating a new application of his creative talents, she led him to their joint conception of a radio-control scheme in which the transmitted carrier frequency would jump about via a prearranged, randomized, and nonrepeating FH code. A torpedo carrying a properly synchronized receiver could thereby be secretly guided from its launch site all the way to its target. Hedy Lamarr and George Antheil thought that such a stealthy "dirigible craft" capabil-

---

[12] It might be thought puzzling that this technique, so often relied on in the M.I.T. NOMAC/F9C research, was unknown to Davenport—whose wife was a niece of Purington and indeed was one of two such sisters who had married M.I.T. Lincoln Laboratory executives—to have originated with that inventor; and that despite their homes being near enough for occasional social calls, Purington—now seen through [27] as most likely the earliest inventor in SR-SS-style secret communications—himself did not learn of Davenport's having led the team effort that culminated in the first SR-SS antijamming communication system actually to be deployed.

It should be remembered, however, that in those years not long after World War II, and when Stalin was yet alive to foment the Korean War, military secrecy was scrupulously observed on all sides.

There exists another, equally striking example of reciprocally long-preserved confidentialities, again in this same SS technology of national-security import. Two good friends (who had first met while M.I.T. graduate students), Bennett L. Basore and John C. Groce, each never let the other know—for more than a quarter-century—of their secret studies in similar SR-SS, pseudonoise-carrier communications. Basore's early research had been performed at M.I.T. [1, pp. 836–837], while Groce's was done a couple of years later in deRosa's laboratory at FTL [1, p. 834].

[13] Quoted from a remark of H. Lamarr's made on August 5, 1982, during her interview by the author [32].

ity, for missiles as well as torpedoes, would soon be needed by Germany's opponents.

Drawing large diagrams while stretched out on Lamarr's carpeted floor, she and Antheil concentrated on them for weeks until they arrived at a secure and feasible FH-SS concept. The system design took special advantage of the composer's know-how, in their plan to synchronize the radio transmission and reception frequencies by means of twin, identically crypto-code slotted, paper music rolls like those used in player piano, audio-frequency (!) mechanisms. Indeed, Antheil had already achieved such synchronization precisely in his multi-player-piano opus of the 1920's, *Ballet Mécanique* [33, p. 185]. Their invention disclosure points out that an FH repertoire of 88 radio frequencies could readily be accommodated.

The Lamarr-Antheil invention promised to be "sturdy and foolproof," and well within the manufacturing capabilities of the 1940's—and its FH secrecy features were enhanced by the inventors' advocacy of short-pulse transmission to provide low detectability. But what seems, for that day, to be most perceptive in the initial installment [34] of the invention disclosure is presented quite boldly:

"... it is veritably impossible for an enemy vessel to 'jam' or in any way interfere with the radio-direction of such a previously synchronized torpedo because, simply, no ship may have enough sending stations aboard or nearby to 'jam' every air-wavelength possible or to otherwise succeed except by barest accident to deflect the course of the oncoming radio controlled torpedo—unless, of course, it happened to have the exact synchronization pattern between sender-ship and torpedo."

(Minor, mostly typographical improvements have been made here within this quote while dropping its originally all-capitals lettering; in the patent itself [30], "block control" is said rather than "jam.")

Lamarr next brought her and Antheil's secret system concept to the attention of the then newly government-established National Inventors Council, which soon (quoting from [36]) "...classed Miss Lamarr's invention as in the 'red hot' category. The only inkling of what it might be was the announcement that it was related to remote control of apparatus employed in warfare." That is how it was guardedly publicized by the U.S. Department of Commerce after being scrutinized by Charles F. Kettering, the noted General Motors inventor who was also a pioneer (with Elmer A. Sperry) in remotely piloted vehicles.

Despite this first reaction of enthusiasm mixed with caution, the Lamarr (Markey)-Antheil patent appears to have been routinely issued and published, curiously without imposition of a Secrecy Order. It may be that such potential restraint from the U.S. Patent Office was precluded by the fact that Lamarr's continuing (until 1953) Austrian citizenship rendered her an "enemy alien" during most of World War II. Although this personal circumstance is indeed reflected in the patent case file [35], it seems to have been a mere technicality, which did not impair her screen-actress career nor her image to the American public.

At one of the many war-bond rallies through which she expressed her loyalty, Hedy Lamarr told a crowd of 10 000 in Elizabeth, NJ, that "...she knew what Nazism would mean to

this country because she knew what it did to her native country, Austria. 'I'm giving all I can because I have found a home here and want to keep it'" [37]. She had certainly tried to contribute to the war effort, too, by her inventiveness.

Lamarr and Antheil seem, however, to have been more than a score of years ahead of their time, considering that FH-SS evidently was not used operationally against intentional jamming until the 1963 exercise, by the U.S. Navy, of the Sylvania BLADES system [1, p. 833]. It appears that no coded-paper-hole implementation ever resulted from their FH invention, and that in the intervening decades from the issuance of its patent until now, whatever scant notice it has received has been confined to the popular press [38].

## V. TO WAR'S END, AND SPREAD-SPECTRUM STATUS THEN

This section observes a few further occurrences in the communications field from the era of the Second World War, which relate to SS developments. The following items, added to the war-period findings recounted in [1], [2], and [5], still cannot be expected to complete the picture of World War II technology relevant to SS origins. Doubtless, more pertinent material will yet emerge from the archives of the principal combatants in that global ordeal.[14] (Note: There has just come

---

[14] A significant historical reference, relating to Footnote 5, is preserved at the Public Record Office of Great Britain, where in *CAB 79-25* on p. 299 there it has been located by David Kahn. He has kindly provided it to the author, as follows, in confirming the connection of Alan Turing with the BTL X-System of [5]. In a "Security" subitem under Item 6, "Establishment of Trans-Atlantic Radio Telephone Circuit—New Project," of the "Minutes of a [British] War Cabinet Chiefs of Staff Committee Meeting Held on Monday 15 February 1943 at 3:30 p.m.," is stated, "It was noted that the only Englishman who had yet been able to examine the apparatus [the Sigsaly equipment: see Footnote 4] was Dr. Turing of the Government Code and Cypher School [i.e., the "Bletchley Park" of Ultra/Enigma fame]."

Archival material generously made available from the George C. Marshall Research Foundation, Lexington, VA, now reveals (as indexed there under "Telephone") that Turing's desire to survey the X-System development at BTL became a matter of concern among high leaders of the Allied war effort. According to Field Marshal Sir John Dill, who was Chief of the British Joint Staff Mission, Turing was "absolutely reliable" and "in on every secret we [British] possess" about cryptanalytical devices. Yet, BTL clearance was granted (on January 9, 1943) to that noted mathematician only following exchange of considerable dialogue between Dill and Genl. George C. Marshall, the U.S. Army Chief of Staff (whose rank was equivalent to that of field marshal). Agreement hung on the understanding that each country reserved "the right to refuse to permit the 'exploitation' of these [Sigsaly encryption and Enigma decryption, both of ultra-level classification] secret devices by the other country."

Data about the inaugural Sigsaly transatlantic telephone call have also been found and thoughtfully supplied by D. Kahn. That milestone was met on July 15, 1943, as given in documents of the 805th Signal Service Company which are held at the U.S. National Archives (under SGCO/805/0.1, Record Group 407 AG). Present in London for the demonstration was Lieut. Genl. Jacob L. Devers, while those in attendance at the Pentagon included Lieut. Genl. Joseph T. McNarney, U.S. Army Deputy Chief of Staff; Lieut. Genl. Brehon Somervell, Commander of the Army Service Forces; Maj Genl. Harry C. Ingles, the Chief Signal Officer; and (the then) Col. W. Preston Corderman. The last-named officer has very recently recalled that although he found it "difficult" to identify who was speaking to him that day from overseas, the understandability of the voice was "good," once Corderman became psychologically adjusted to the apparently long reaction time (caused by delays inherent to the X-System's vocoder signal processing) of the person with whom he was conversing.

Kahn has additionally remarked to the author, in referring to p. 556

to hand a comprehensive bibliographic reference which lists a great many chronicles dealing with the American armed forces. In this compendium [79], the tenth chapter, "Science and Technology in the Twentieth Century," is especially of present interest. The citations there to [80] and [81], for example, have introduced the author only lately to informative histories of, respectively, the U.S. Army Signal Corps and the U.S. Navy, that span more than a century of communications and radar activity.

Early in January, 1943, a U.S. Army Signal Corps officer, Henry P. Hutchinson, applied for a patent on FH signaling for "maintaining secrecy of telephone conversations" or for "privately transmitting information" [39]. This was the later-filed application in the FH pair (taken from the survey given in [28]) introduced at the beginning of the preceding section, the earlier one being the Lamarr (Markey)–Antheil filing discussed there.

As with Hedy Lamarr's application, Capt. Hutchinson's received official evaluation [40] –but whereas the former came from an "enemy alien" and had no Secrecy Order placed against it, the latter was that of an American citizen and was held under U.S. Patent Office secrecy until 1950. Another contrast is that in Hutchinson's patent the pseudorandom FH pattern was supplied through the use of on-line cryptographic machines in place of the coded player-piano rolls advocated by Lamarr and Antheil. (Hutchinson has recently told the author that, although not apparent in his patent, at the time of its filing he was aware of the advantage his concept could have for avoiding interference.)

After the war, this inventor of secure-voice FH was a colleague of Harold Zahl at the Ft. Monmouth, NJ, laboratories and thus soon became familiar, through visits to the M.I.T. Lincoln Laboratory, with the Army-supported NOMAC/F9C research. Citations to Hutchinson's notable postwar inventions are included in a NASA (National Aeronautics and Space Administration) publication on pioneers in satellite and space communications.

Also in January, 1943, British troops captured, from Genl.

Erwin Rommel's forces in North Africa, a communication transceiver called the "Optiphone" (or "Photophone"). Developed in Germany in the mid-1930's (and covered by U.S. Patent No. 2 010 313 to the Zeiss works), this system could provide voice communications over a light-path up to four miles long under reasonably clear atmospheric conditions [41]. While it would be far-fetched to view such (prelaser) incoherent optical communications as SS in nature,[15] the interesting aspect here is that this apparatus seems to have been another example of Rommel's employment of relatively advanced technology in secure communications. As described in [1, p. 827], he had been using, as well, a trans-Mediterranean radiotelephone link which was protected with a stored-noise carrier-correlation scheme, patented {40} by Kotowski et al.,[16] that essentially provided analog encryption [1, p. 849].

In turning next to another battlefield of the Second World War, the following point of explanation is appropriate. The surveys of SS-communications origins, by [1] and its present sequels, generally do not include classical wide-band frequency modulation (FM) because most of the history of that now-conventional technique would be repetitious here. However, it seems not to be well known that FM fulfilled–SS-style–an urgent need for jam-resistant communications during a crisis in the war. This singular event took place at the late-1944 "Battle of the Bulge" which surged past the crossroads town of Bastogne, in the Belgian Ardennes, where extended American forces had been trapped and Genl. George S. Patton's Third Army was storming to their rescue. Quoting at some length from [46, pp. 163–164]:

"On 26 December Patton succeeded in forcing a narrow corridor through the German tanks ringing Bastogne. It was only three hundred yards wide, but it opened the way to the American troops cut off there and punctured the German bulge, which began slowly to deflate under combined British and American pressure.

"Now came the first and only battle test of Jackal, the high-powered airborne radio jammer AN/ART-3 developed by the Signal Corps for the AAF [Army Air Forces]. The First and Third U.S. Armies had been reluctant to try Jackal jamming because a portion of the frequency band used by their tank radios overlapped into the German band to be jammed. Earlier tests in

of his book {41}, that just two weeks after this Sigsaly preliminary test, a radiotelephone conversation between President Roosevelt and Prime Minister Churchill–concerning their intended actions in response to Mussolini's downfall–was intercepted by the Germans and quickly frequency-descrambled. Winston Churchill himself, in his memoirs of *The Second World War*, on p. 60 of the fifth volume: *Closing the Ring*, notes the mention of this phone call contained in a follow-up cable to Franklin Roosevelt. Kahn's account, in {41}, of the Nazi reaction to their intercept of July 29, 1943, together with Churchill's report (*loc.cit.*, p. 53) of Hitler's agitation upon learning that Benito Mussolini had been forced to resign his dictatorship four days earlier, suggest a possibility of serious repercussions (in German preparedness against the Allied invasion of Italy) having arisen from this top-echelon breach of Anglo-American communications security.

However, it was not much longer before there would be recommended to the White House (via the October 12, 1943, cautionary letter to Harry Hopkins cited in [5], and earlier on p. 959 of the 1948 edition of the biography by Robert E. Sherwood, *Roosevelt and Hopkins: An Intimate History*) the means for absolutely denying to Adolf Hitler all further such intelligence. A complete Sigsaly installation had by then become available to replace the relatively simple speech-privacy systems which had offered the only protection–dangerously weak (per [2, ref. 2, chs. 4, 5] which also, interestingly re [5], presents some *non-X-System* type privacy methods based on the vocoder, and its actual multi-channel output wave-traces, there in Table I and Fig. 16, respectively)–from transatlantic eavesdropping up to that time.

[15] Later, the idea of signaling via incoherent light was briefly discussed at the beginning of a memorandum [42] by Harry Nyquist and Stephen O. Rice. These investigators decided to examine the SS prospects offered at radio wavelengths, instead of optical, by such a noise-carrier scheme when applied to communication through multipath disturbances, a topic on which Kharkevich published nearly a decade later {176}. Harrison E. Rowe's papers [43], [44] are of collateral BTL relevance.

[16] According to a telephone conversation on May 20, 1982, between the author and G. Vogt, who was a designer of the "Kurier" system (as cited early in this sequel), Kotowski had additionally been an originator of that "Flash" innovation for strategic radio signaling. This World War II system development combined time compression of messages with subsequent short-burst transmission, and was intended to enable German U-boats to keep in touch, covertly, with their very distant naval bases.

Vogt added that Kotowski was interested, too, in jamming aspects concerning the Flash method of submarine communications. Hence, Kotowski apparently contributed to various kinds of secure communications throughout his career. Even at the height of the war, he mentioned secret codes ("Geheimkodes") in an open publication ([45], cited by [13]).

England had indicated somewhat inconclusively that little or no interference would be caused, since American radio for armored forces was FM while similar German sets and Jackal were AM [amplitude modulation]. Now that nearly the whole of the German *Sixth Panzer Army* [italics sic] was in the Ardennes fighting, it seemed a good opportunity to test Jackal.

"Accordingly, beginning on 29 December and continuing through 7 January [1945], Eighth Air Force B-24's, based in England and bearing the Jackal jammers blaring full blast, flew in relays over the battle area, coinciding with a Third Army counterthrust in the vicinity of Bastogne. The first results seemed inconclusive, but, according to later reports from German prisoners, Jackal effectively blanketed German armored communications during these crucial days. Nor were the American tankmen inconvenienced or made voiceless by the overlap in frequencies. The jammer effectively filled the German AM receivers with a meaningless blare, while the American FM sets heard nothing but the voices of their operators."

This intriguing FM episode[17] dramatized in real life (as twice again referred to in [46, p. 318, footnote 57; p. 324]) the simple FM/AM analogy later given in Theodore Sturgeon's science-fiction portrayal of SS communications [1, p. 822]. That FM radio (and FM radio-relay, too, each unique to the U.S. Army in this worldwide war, per [46, pp. 21, 107]) was even available, "beyond anything either the enemy or the other allied nations possessed" [46, p. 631], was due in large measure to three men. Through experiments conducted in the mid-1930's with the close cooperation of its inventor, Edwin Armstrong, (the then) Col. Roger B. Colton and Maj. James D. O'Connell (see Footnotes 10 and 17) recognized the clear superiority of this modulation method and pushed hard to have it ready for communicators on the battlefield. One

[17] In recalling electronic warfare events concerning the Battle of the Bulge, Lieut. Genl. James O'Connell (the retired Chief Signal Officer cited in Footnote 10) related to the author, on October 9, 1982, that on the day before this famous conflict erupted he made an inspection to the vicinity of Bastogne. There, a U.S. radio-intercept team "bitched" to him that their monitoring was being upset because a large number of the enemy's communication frequencies were "sidling up" very closely to those radiated by the Signal Corps. A few days after the assault, (the then) Col. O'Connell developed the conclusion that the German attack had been bolstered in its surprise elements by inducing the Americans to jam their own intelligence operations. In its exploitation of this novel tactic, the consequent interference that the other side had to accept was apparently quite low by virtue of the geographic dispositions of the opposing armies.

Such a stratagem, of voluntary self-jamming in order to gain advantage, appears thus to have been embraced by one foe just before that showdown, but then, during it, played by the other adversary in a different way. The latter incident, with its successful FM/AM tactic to which this footnote attaches, was coincidental retaliation in kind for the involuntary self-jamming cleverly inflicted, through the former one, against the American side during the mounting of Hitler's surprise offensive.

Genl. O'Connell has also told the author (in now recording for the first time these two anecdotes of his own about the Battle of the Bulge) how an important U.S. Army transmitter operation was ably sustained on a hilltop approximately 35 km northwest of Bastogne. Even before the encirclement of that town where the entrapped Genl. Anthony C. McAuliffe retorted "NUTS!" to German demand for surrender, this radio-relay station had become isolated. Remaining for some days undiscovered there, the responsible Signal Corps troops kept their motor/generator set running by stealing below each night and quietly filching from the gasoline supplies of surrounding Panzer tanks. (Note: More details of this exemplary resourcefulness may be found in "The Inconspicuous Relay" [80, pp. 183–187].)

colonel exclaimed: "I feel that every soldier that lived through the war with an Armored Unit owes a debt that he does not even realize to General Colton" [46, p. 631]. Armstrong patriotically donated to every military service of the United States, free of any royalty payments, license to the use of all his inventions. (Note: Additional information on Maj. Armstrong's contributions to the Signal Corps is given in [80]).

Except for special situations like those in the Battle of the Bulge, and in the earlier Battle of El Alamein where the British fought likewise by electronics as well as turret-guns against German tanks [47, p. 3],

"radio jamming, in World War II generally did not require, and did not receive, much attention. Everyone was far more concerned [for intelligence analysis] with listening in. However, when . . . inhuman radio-guided missiles put in their terrifying appearance, the electromagnetic frequencies employed by the new military engines became suddenly too dangerous to neglect."

These quoted remarks from [46, p. 301] are in a sense companion to the comment on p. 326 there, that improvement of "antijamming techniques...to protect friendly radio" communications was urged by Col. Robert W. Raynsford in an official Army letter written a half-year before the end of the war.

At the time he authored [47], Frederick E. Bond had been introduced to the secret NOMAC research being undertaken at M.I.T. "Recent advances in the art," and "interesting possibilities" for teletype transmission, thus included "noise modulation" [47, p. 14]. He further pointed out that, as has been recounted by Davenport {173, p. 158}, the wide scope and efficiency of Soviet jamming of international news broadcasts and other such operations might well foretell an ability to place the Army's global communications system in jeopardy.

This was a complete reversal, Bond emphasized, of the World War II doctrine, quoting from {53, p. 12}, that in the "attempt to jam long-distance point-to-point circuits, the success of such operations may be rendered unprofitable by the amount of power that may be needed, by the lack of suitable geographical locations for the jammer (which must blanket the terminal of the enemy channel with a frequency determined by the intended victim), by the large number of frequency channels that the enemy may have available for emergency use, and by the need for jamming all channels practically continuously if the 'blockade' is to be effective." All four of these points had been more or less nullified by 1952, the latter pair through postwar overcrowding of the short-wave, ionospheric-radio frequency spectrum. Bond considered (per p. 5 of the memorandum [47], which also cites a July, 1951, bibliography on antijamming that had been compiled by the Research and Development Board mentioned in [1, p. 837]) that:

"This change in thinking is an excellent example of how unpredictable ECM [Electronic Counter Measures] really is."

## VI. POSTWAR POTPOURRI

As a bridge into "modern" times, [48] might serve as a candidate, since it cites Sylvania's SS-progenitor WHYN system [1, pp. 830-831] as well as [47]. While referring to correlation methods on its p. 65, [48] hedges the promise of these techniques with its 1955 commentary, as follows, that a

> "number of organizations have put forth considerable effort in the design and development of secure communication systems which are radically different from those in general use. However, the available literature concerning these systems does not reveal that a satisfactory system has been attained as yet. The very fact that jamproofness is the most important feature of these systems makes them extremely complicated with the result that each of them has a number of problems which have not as yet a satisfactory solution."

Two decades later, R. C. Dixon {174, p. 7} observed that the Lincoln F9C-A/Rake system which resulted from the M.I.T. NOMAC program "consisted of a roomful of equipment." While this is indisputable, any conclusion that it was therefore never used operationally is, of course, historically incorrect per [1, p. 841] and {173, p. 159}. After all, even the hundreds of cubic feet taken up by each single-phone-circuit terminal of the X-System [5] had not precluded deployment of this memorable BTL innovation—to found the first truly secure, multicontinental voice network, whose vocoder signal processing was supported at every site, per encrypted conversation there (in full duplex), by five-ton air conditioning and 30 kW of emergency power—so urgent was the war-era demand for this inviolably secret speech system in which, e.g., all pauses between utterances, however long or short, were hidden perfectly.

The report [48] to the U.S. Air Force goes on to list several antijamming references from the first decade of the post-World War II period, including an Illinois Institute of Technology document (AD 015150 [see Footnote 24]) of 1953. The latter in turn contains a relevant bibliography which cites a 1949 antijamming study project, "Aunty J" [49]. Here, jamming is mitigated by biphase-switching of the transmitted radio frequency carrier to create a "subcarrier whose phase is varied in accordance with a predetermined pattern." At the receiver, the phase inversions are removed in usual SS-correlation fashion, where the needed replication of the pattern is accomplished through resort to a "frequency standard" that has been simultaneously "transmitted in such a form that it could not be detected by the enemy without the use of cryptographic means." (This scheme somewhat resembles the TR-SS invention of James J. Spilker {167}.) The project contractor, Reevesound Company, stated that the investigation was "utilizing principles and methods described in patents and patent applications" which included U.S. Patents 2 240 500 and 2 295 207 (whose French counterparts had previously been issued in 1936 and 1938, respectively). These were held by Leonide Gabrilovitch, who during the war had proposed for voice transmission security an apparently similar technique [2, ref. 2, p. 102].

Robert Fano's role in early SS communications research at the M.I.T. Lincoln Laboratory (which, as an Air Force contractor, must have been among the organizations referred to in the preceding quotation from [48]) is described in [1, pp. 836, 839, 847] regarding, respectively, the Wiesner–Fano stimulus from Project Hartwell, the inspiration given R. Price to undertake a thesis [50] on multipath problems, and the Fano SS matched-filter (MF) conception. In {173, p. 158}, Wilbur

Davenport has acknowledged the major influence exerted by Fano on the NOMAC project. This was a continuance, in part, of the latter professor's military electronics career, originally commenced with wartime radar development at the M.I.T. Radiation Laboratory. (Incidentally, Fano had suggested to the present author, in the same conversation that envisioned the topic which would lead to [50], a secret-dissertation alternative: research to improve the radar detection of submarine snorkels.) The Scholtz sequel [2] refers further to original SS-MF inventions, and also to noise-correlation radar [1, pp. 825, 845–846]. The present sequel offers opportunity to recount more details of Fano's ideas both in the genesis of SS communications and in correlation radar.

R. M. Fano recently advised the author that when the NOMAC communication effort began, he was already concerned about potential multipath impairments of the secret signaling scheme introduced to him by J. B. Wiesner. Fano felt that it would be preferable to perform the correlation operation at the receiver via a matched filter, since by pulse-compression action the transmitted signal could then be demodulated without having to synchronize to any particular propagation path delay. P. E. Green's dissertation {105} hence devotes a number of pages to this correlation approach. Later, Price and Green referred, in {121, pp. 556, 569}, respectively, to Fano's proposed MF methodology [51] along with his general advocacy of wide-band transmissions to combat multipath, and to his having "spent many hours and not a few graduate students...in pursuit of solutions to the multipath problem."

The first inkling of Fano's MF notion to appear publicly [52], abroad in England, involved "idealized" communication transmission by noise-wiggles à la Shannon but, purposely, with essentially no implication of possible MF reception. This 1952 item gave only passing mention to {6}, yet was largely based on very informal notes [53] of 1951 in which the MF correlation means had been made explicit through that same Van Vleck–Middleton reference. Yet even in [53], which according to Fano's recent recollection was written just after his successful MF experiment [1, Fig. 16], the pulse-compression feature was cautiously left unrevealed.

In discussion at London following his 1952 presentation (p. 530 of the book cited in [52]), Fano looked forward to theoretical effort against "non-additive random disturbances... such as those introduced by ionospheric phenomena" and remarked that "we need to generalize in such a direction the work on detection done by Woodward [as published, e.g., in {7} the following year] and Davies." It seems that Fano hoped Price's thesis could lead to some sensible application and particularly would impact the NOMAC endeavor. However, although [50] contained the germ of a statistical detection synthesis for combatting multipath, it lay academically dormant—by apparently yielding nothing much of practical significance (except, of course, a doctorate)—until the Rake concept[18] got under way {120}.

---

[18] Along the road to the production-model Rake receiver ultimately appliqued to the Lincoln F9C-A system [1, pp. 839-841], there evolved from [50] via {115} an estimator–correlator principle for the optimum detection of random signals in independently additive white Gaussian noise. (As noted by [2], there is an erroneous reference number in [1] for {115}: the correct citation is AD 224557.)

Not long ago, Robert Fano reminded the author that, just as his first publication on MF reception of noise-like signals was done secretly [59]—where the picturesque analogy is made: "Thus the waveforms recorded at the transmitter and the filters at the receiver act as a set of keys and the corresponding set of locks. The decoder observes which lock has been opened and thereby knows which key has been used"—both his patent {172} and that for the Rake receiver {122} were kept under Secrecy Order until their joint issuance on the same day in 1961. By 1956, however, security considerations had relaxed enough that Fano was able to draft the unclassified memorandum [51] (included among the references in [54]) elaborating his MF approach.

Willis W. Harman, citing [59] in 1954, observed that

> "The earliest suggestion for the use of these properties [pulse compression with processing gain in signal-to-noise ratio, via the matched-filtering of a received noise-like waveform having a large time–bandwidth product] in communication systems seems to be that of R. M. Fano" [60, p. 5].

Before pursuing the lead [59] into Fano's postwar radar innovations, reference should be made to a final—but really be-

---

The broad topic of estimator–correlator optimality soon attracted the attention of Thomas Kailath and his M.I.T. graduate school professor, John M. Wozencraft. Becoming an official consultant to the SS-researching Group 34 at Lincoln Laboratory while still a student, Kailath considerably extended the rules for optimally detecting random signals and presented the results in his doctoral dissertation.

Kailath published his contributions, first in Reference 21 of [54], even before receiving the doctorate. (The paper [54] by George L. Turin is a comprehensive MF elucidation and survey, including {165} and many other germane references, that introduced the special June, 1960, Matched Filter Issue of the IRE TRANSACTIONS ON INFORMATION THEORY admirably organized and edited by Paul Green.) Next he presented his work at London [55], where during the discussion that followed, a query was raised concerning "one of the remarkable features about the Rake system": that it required 10 kHz of bandwidth to support a single 60 word/min teletype channel. In responding to this prod on "bandwidth conservation in the reverse," Kailath did not reveal the true purpose of the system—although (as he has recently told the author) he had already surmised it from {178}, which "made a big impression" on him. Like Shannon vis-à-vis Turing (Footnote 5), the young engineer from India had no "need to know" in this secret realm, and also like Shannon in the X-System situation, Kailath "knew better than to ask" his Group 34 colleagues to confirm his guesses.

Thomas Kailath later went on to achieve a pleasingly surprising, universal generalization of estimator–correlator optimality [56]. Earlier in his master's thesis as well, Kailath had broken new ground. That research probably came about from an all-but-forgotten vignette recently recalled to the author by Group 34's W. Boyd Smith, remembrancer (with G. Vogt) of the Rommel noise-disks [1, p. 827]. Price and Green, not completely certain what should be the spacing between the taps on the delay line in the Rake receiver, consulted Fano in their concerns about this important bandpass-filtering question, and were quickly reassured that Nyquist's sampling theory indeed still prevailed. Thereby, neither undue expense nor loss of precious processing gain were suffered, at least with the quasi-static multipath faced by the F9C system.

But what if the channel dynamics had been much more rapid? Once again, Wozencraft had perceived the potentially deeper issues here, and in due course his (by then former) student presented [57] an additional paper abroad, that was largely based on the Kailath M.I.T. master's studies of time- and frequency-spread channels. John G. Proakis and Paul R. Drouilhet *et al.*, also from Lincoln Laboratory's Group 34, in another publication [58] motivated at least partially by Rake ideas, promoted the term "decision-directed."

---

ginning—contribution of his to SS communications. Back even before Project Hartwell, Bennett L. Basore, who later wrote the first thesis in the field {101}, had already received Fano's stimulation in composing a term paper which set forth some of the new prospects suggested by the Shannon information theory. Appropriately, this study happened to note [61, p. 23] that "'jamming' for political purposes" had become "a common practice."

Soon after the M.I.T. conception, by Jerome Wiesner (no doubt per [7, p. 341]), of modulating a radar transmitter by random noise and then statistically signal-correlating the returned echo against delayed replicas of the transmission, Charles A. Stutt and Jake Kraft began to investigate this noise-correlation proposition in a radar, of novel ground-wave type, also proposed to them by Wiesner. Robert Fano, too, wrote a memorandum [62] inspired by Wiesner's noise-probing ideas and assigned the topic to his then student, John Wozencraft, for a master's thesis [63]. Fano, who had eagerly formed his Lincoln Laboratory group upon such new radar concepts (together with techniques previously contemplated when he was in the Radiation Laboratory), followed up the memo and thesis by a perceptive secret report. Here he emphasized, among other factors, that with noise-modulated correlation-radar a jamming signal "cannot be effective unless its power is greater than that of the moving-target echo by the ratio of the receiver bandwidth to the Doppler filter bandwidth" ([64, p. 31], where, of course, Fano was assuming essentially MF reception).

Later leaders of the Fano group were Kraft and William M. Siebert, with Robert M. Lerner as a principal investigator. Important papers that they published include [65] and [66], both being cited in [54]. The work of this group culminated in the AN/FPS-17 radar, perhaps the first of its kind, which was reported to have been "capable of tracking rockets at a distance of 1600 km" [67, p. 149]. (The Czech book [67] is intriguing for its discussion of such matters as time-hopping, shift registers, and "more complex binary codes" that provide "increased transmission security even under conditions of jamming.")

Bernard Elspas [68], and George R. Cooper and Clare D. McGillem [69], have also notably investigated "spread-spectrum radar" (so termed in {4} and in [1, p. 845]). Numerous references to this field are given in [20]; of particular account are citations there on p. 153 to surveys by David K. Barton and to early research by R. Bourette, W. Fishbein, and O. N. Rittenbach. This annotated compilation [20] by Stephen L. Johnston further presents a reprint of a 1958 Sylvania paper, by Donald B. Brick and Janis Galejs, that openly described the Wiesner–Fano type of spread-spectrum, "noise-correlation" radar.

In the April 1960 issue of *Space/Aeronautics* (published by Conover-Mast, and additionally referenced in [20]), a special report about electronic countermeasures (which effectively broke a "barrier of silence" according to p. 6 there) highlighted the Brick–Galejs perspectives on its p. 129. Moreover, it divulged there on pp. 152-154, 131, and 132, respectively, some jamming ideas of Louis deRosa, antijamming precepts quoted from John P. Costas,[19] and the kind of "cuddle-up" stratagem surprisingly employed by the Germans in the Battle of the Bulge (see Footnote 17).

Finally of interest in this electronic-warfare vein of early postwar radar, an SS-type patent [71] was filed in 1951 by C. Chapin Cutler, originator of differential pulse-code modulation. Kept under Secrecy Order for some years, Cutler's invention employs precession of the radio-frequency phase shifts between successive radar-like pulses, which could, e.g., be used to obtain communications of a "both time division and frequency division multiplexing" type. Chapin Cutler has informed the author that Shannon's noise-carrier CDMA scheme, and Pierce and Hopper's {127} ideas too, influenced the suggestion of such applications in [71]. The patent further considers "different types of varying precession angles" to secure "'secrecy' systems" and "anti-jamming radar."

This sequel-chronicle closes at last by taking note of three foreign publications that themselves took note of the Lincoln Laboratory SS research quite soon after the Rake paper {121} appeared. One of these [72] is simply a four-year-earlier version of {185}. In addition to citing (in order) {121}, {178}, [55], and a 1961 paper by G. L. Turin, [72] references the interesting paper [73]. The latter, published in 1963 by the first person to patent "chirp" radar, the East German E. Huttmann {4, p. 103}, reviews both prewar and postwar SS-like developments, including {165}, a 1962 (second) edition of the Lange book {177}, and {121}. Huttmann at one point talks of "a series of new possibilities" being opened for the future in the field of his subject, while surely not imagining that less than 50 km from his office, the Lincoln F9C-A/Rake system is standing to attention in West Berlin (per Davenport's visit there at O'Connell's request, mentioned in {173, p. 159}).

At the finish, we return to Gustav Guanella, who figured so prominently in sophisticated signaling systems, before,[20] during, and after World War II. In the notes [75] from his 1960 lecture for the series "Krieg im Aether" ("War in the Ether"), prepared for a military audience, Guanella concisely covers a wide range of correlation methodology, including "shared spectrum" CDMA and the securing of secrecy by "non-periodic carriers of wide bandwidth." Included there

---

[19] Naturally, the Costas paper {178} was referenced in this 1960 ECM review, and later was cited also by Gordon Raisbeck [70, p. 71] in a discussion of jamming versus communication transmissions that employ noise-like waveforms of high time-bandwidth product. Further on in the 1960's, David Kahn mentioned such "new techniques of communication . . . that spread a transmission over so broad a frequency spectrum that anyone listening on one frequency band would hear only a faint crackle like static" {41, p. 708}.

[20] The late Gustav Guanella, whose work has been described in [1] and revisited in [2], stated to James L. Massey in 1980 that he had arrived at his noise-radar invention {12} of 1938 entirely on intuitive grounds, without any knowledge at that time of Wiener's papers. This is rather reminiscent of Michelson's having employed autocorrelation in optics long before Wiener's generalized harmonic analysis {16} became available [74]. On the other hand, it is evident that in {16} Wiener was well aware of optical analogies to his mathematics, e.g., when there on p. 129, in speaking of Lord Rayleigh's treatment of white light, he remarks that "one is led to admire the unfailing heuristic insight of the true physicist."

---

among the 51 references are {121} and numerous other American contributions to the field, dating all the way back to Wiener's[21] classic {16} of 1930.

## ACKNOWLEDGMENT

It has been an exceptionally enjoyable privilege for me to have worked closely with Robert Scholtz over the past several years, in efforts to preserve for posterity the various genealogies, whether latent or fruitful, of a vigorous and substantial subfield of the radio art. This sequel to [1] benefits much from Bob Scholtz's careful and constructive criticisms.

During, and after, the 1981 Pioneer Award activities of the IEEE Aerospace and Electronic Systems Society—that honored certain early and significant advances in spread-spectrum communications {173}—I have had the pleasure of friendly associations with Phyllis deRosa, Mortimer Rogoff, Paul Green, and Wilbur Davenport, which have helped to illuminate and validate our historical inquiry.

It is hoped that the information in this paper has been presented with acceptable accuracy and fairness. The author sincerely thanks the following contributors for their kind and generous assistance in the gathering of primary source material, including valuable personal recollections, some of which appear in the text and footnotes: Bennett Basore, William R. Bennett, Nelson Blachman, W. Preston Corderman, Chapin Cutler, Edward David, Robert Fano, Kenneth Galente, I. J. Good, John Groce, Henry Hutchinson, Stephen L. Johnston, David Kahn, Thomas Kailath, Leon Kraft, Solomon Kullback, Hedy Lamarr, Mary Price Lee,[22] Yuk Wing Lee, Robert Lerner, Anthony Loder, James Massey, John McConaghy, Brockway McMillan, James D. O'Connell, John R. Pierce, John Proakis, John Raney, J. Francis Reintjes, Stephen Rice, Henry Ross, M.D., Harrison Rowe, Herbert Schulke, Claude Shannon, John Meredith Smith,[23] William Boyd Smith, Charles Stutt, Mark Thompson, John Tukey, Gottfried Vogt, I. Wigdorovits, and Karl Wildes.

The author is particularly indebted to librarians H. Alan Steeves and Mary A. Granese for their dedicated efforts in continued support of both [1] and this sequel, and to Karl Green of the scientific division at the Library of Congress for his advice of relevant World War II material. The encouragement and cooperation of the Editors of this TRANSACTIONS is cordially acknowledged—with special appreciation to two Guest Editors of the May 1982 Spread-Spectrum Issue, Fred W. Ellersick and Charles E. Cook, and to two of the IEEE Publications Staff who have been so patiently helpful, Carolyne Elenowitz and Paul V. Olowacz.

---

[21] Wiener's pacifism after World War II, mentioned as ironic in the caption to [1, Fig. 14], and again ironic by his citation within these Guanella military notes [75], is a major theme of [76].

[22] The author's authoress-sister who, after [1] had reached final form for publication, ventured that she once had seen somewhere a publicity item about a Hedy Lamarr patent on some sort of secret communication scheme.

[23] The first, or certainly a peer among the first, of serious historians of spread-spectrum communications [77]—and a great-nephew of the great cryptologists, William F. and Elizebeth Smith Friedman.

Lastly but very gratefully, I thank my secretary, Susan D. Whyte-Lemke, for teaching me word processing—that marvelous aid indeed exemplary of a wish of Norbert Wiener's: that progressive enlightenment in *The Human Use of Human Beings* [78] may flow from the Second Industrial Revolution.

## REFERENCES

[1] R. A. Scholtz, "The origins of spread-spectrum communications," *IEEE Trans. Commun.*, vol. COM-30, pp. 822–854, May 1982 (Part I).

[2] ——, "Notes on spread-spectrum history," this issue, pp. 82–84.

[3] H. T. Friis, *Seventy-five Years in an Exciting World*. San Francisco, CA: San Francisco Press, 1971, p. 34.

[4] J. R. Zacharias, "The Hartwell project," presented at the 6th Undersea Symp., Washington, DC, May 9–10, 1951 (AD 103991).[24]

[5] W. R. Bennett, "Secret telephony as a historical example of spread-spectrum communication," this issue, pp. 98–104.

[6] M. D. Fagen, Ed., *A History of Engineering and Science in the Bell System: The Early Years (1875–1925)*, Bell Lab., Murray Hill, NJ, 1975.

[7] J. B. Wiesner, "Statistical theory of communication," in *Proc. Nat. Electron. Conf.*, Chicago, IL, Sept. 26–28, 1949, pp. 334–341.

[8] "Specialized communications systems," in *The Radio Amateur's Handbook*. Newington, CT: Amer. Radio Relay League, 1982, p. 14-35.

[9] W. L. Root, "Some notes on jamming, I," M.I.T. Lincoln Lab., Tech. Rep. 103, Jan. 3, 1956 (AD 090352; based on Lincoln Lab. Group Rep. 34-47, Dec. 15, 1955).

[10] ——, "Communications through unspecified additive noise," *Inform. Contr.*, vol. 4, pp. 15–29, 1961.

[11] N. M. Blachman, "Communication as a game," in *IRE Wescon Rec.*, San Francisco, CA, Aug. 20–23, 1957, part 2, pp. 61–66. (Note here a coincidental juxtaposition with a survey of U.S.S.R. literature, pp. 67–83—carried out *sub rosa* re SS by P. E. Green, Jr., per [1, p. 848].)

[12] ——, "On the capacity of a band-limited channel perturbed by statistically dependent interference," *IRE Trans. Inform. Theory*, vol. IT-8, pp. 48–55, Jan. 1962.

[13] F.-W. Hagemeyer, "Die Entstehung von Informationskonzepten in der Nachrichtentechnik" ["The origin of information theory concepts in communication technology"], Doctoral dissertation, Free Univ. Berlin, Berlin, Germany, Nov. 8, 1979.

[14] G. Welchman, *The Hut Six Story: Breaking the Enigma Codes*. New York: McGraw-Hill, 1982.

[15] I. J. Good, "Pioneering work on computers at Bletchley," in *A History of Computing in the Twentieth Century*, N. Metropolis *et al.*, Eds. New York: Academic, 1980, pp. 31–45; see also pp. 78–79.

[16] ——, "Weight of evidence, causality, and false-alarm probabilities," presented at the 4th London Symp. Inform. Theory, Aug. 29–Sept. 2, 1960; in *Information Theory*, C. Cherry, Ed. Washington, DC: Butterworths, 1961, pp. 125–136.

[17] H. W. Bode and C. E. Shannon, "A simplified derivation of linear least square smoothing and prediction theory," *Proc. IRE*, vol. 38, pp. 417–425, Apr. 1950.

[18] A. Corneretto, "Designers shifting to wide-band communications," *Electron. Design*, pp. 4–5, Apr. 26, 1961.

[19] Final Rep., Commun. and Components Div., M.I.T. Res. Lab. Electron. and Project Lincoln, Sept. 30, 1952, p. 137 (inside its red cover, the contents of this originally secret report are identical to those in ATI 170188).

[20] S. L. Johnston, *Radar Electronic Counter-Countermeasures*. Dedham, MA: Artech House, 1979.

[21] H. A. Zahl, *Radar Spelled Backwards*.[25] New York: Vantage Press, 1972, p. 54.

[24] AD (and ATI) numbers shown are those for documents available from the Defense Technical Information Center, Alexandria, VA 22314.

[25] Also in [21, p. viii], Donald G. Fink spotlights Zahl in his team-up with Emanuel R. Piore, John W. Marchetti, and John E. Keto to "develop the post-war concept of Joint-Service support of research in the universities." The many benefits that have come from this farsighted concept are acknowledged in [2], where another book by Zahl is cited (Reference 3 there). The present author, R. Price, during his graduate education owed a great deal to the early support thus granted the M.I.T. Research Laboratory of Electronics.

[22] J. D. O'Connell, A. L. Pachynski, and L. S. Howeth, "A summary of military communication in the United States—1860 to 1962," *Proc. IRE*, vol. 50, pp. 1241–1251, May 1962; see also [81].

[23] "John Hays Hammond, Jr.," in *Micropaedia*, vol. IV, *The New Encyclopaedia Britannica*. Chicago, IL: Encyclopaedia Britannica, 1975, pp. 877–878.

[24] J. H. Hammond, Jr., "System of aeroplane control," U.S. Patent 1 568 972, Jan. 12, 1926 (filed Mar. 7, 1914); see also Hammond's U.S. Patent 1 420 257 (filed 1910), and "Security of radio control" {60, pp. 1193–1197}.

[25] B. F. Miessner, *On the Early History of Radio Guidance*. San Francisco, CA: San Francisco Press, 1964.

[26] E. L. Chaffee, "System of radio communication," U.S. Patent 1 642 663, Sept. 13, 1927 (filed Aug. 11, 1922).

[27] E. S. Purington, "Single side band transmission and reception," U.S. Patent 1 992 441, Feb. 26, 1935 (filed Sept. 6, 1930).

[28] Reissue Appl. Serial No. 299 469 as cited in {48}.

[29] W. Broertjes, "Method of maintaining secrecy in the transmission of wireless telegraphic messages," U.S. Patent 1 869 659, Aug. 2, 1932 (filed Nov. 14, 1929; in Germany, Oct. 11, 1929).

[30] H. K. Markey and G. Antheil, "Secret communication system," U.S. Patent 2 292 387, Aug. 11, 1942 (filed June 10, 1941).

[31] C. Young, *The Films of Hedy Lamarr*. Secaucus, NJ: Citadel, 1978, pp. 92–97.

[32] R. Price, "An IEEE history leads to an interview with Hedy Lamarr," *'Round the Center*, Sperry Res. Center, Sudbury, MA, vol. 3, Sept. 1982.

[33] G. Antheil, *Bad Boy of Music*. Garden City, NY: Doubleday, 1945.

[34] "Designs and ideas for a radio controlled torpedo," Dec. 23, 1940, contained in [35].

[35] Case file for U.S. Patent Serial No. 397 412, supplied courtesy of J. D. McConaghy, Esq., Lyon and Lyon, Los Angeles, CA (per authorization of H. Lamarr—not publicly available).

[36] *New York Times*, p. 24, Oct. 1, 1941.

[37] *New York Times*, p. 32, Sept. 6, 1942.

[38] *Time*, vol. 103, p. 52, Feb. 18, 1974; *Parade* (e.g., in *Boston Sunday Globe*), p. 16, Aug. 1, 1982. (Both items indicate only that the patent is in military or secret communications; no mention is made of frequency hopping or antijamming.)

[39] H. P. Hutchinson, "Speech privacy apparatus," U.S. Patent 2 495 727, Jan. 31, 1950 (filed Jan. 7, 1943).

[40] D. O. Slater, "Speech privacy and synchronizing system devised by Captain Henry P. Hutchinson," Bell Tel. Lab., Rep. 20 under Project C-43 of NDRC (OSRD 4573B), July 31, 1943 (as cited on p. 126 in Ref. 2 of [2]; available from U.S. Nat. Archives, Washington, DC).

[41] L. Cranberg, "German Optiphone equipment 'Li-Spr-80,'" Memo., "Captured enemy equipment report no. 26," Signal Corps Ground Signal Agency, Bradley Beach, NJ, May 5, 1944 (PB 001531).

[42] H. Nyquist and S. O. Rice, "Modulation of random noise—Bandwidth requirements," Bell Tel. Lab., Memo. to File 36760-1, Feb. 19, 1948 (retained in archives of Bell Lab., Short Hills, NJ).

[43] H. E. Rowe, "Amplitude modulation with a noise carrier," *Proc. IEEE*, vol. 52, pp. 389–395, Apr. 1964.

[44] ——, "Frequency or phase modulation with a noise carrier," *Proc. IEEE*, vol. 52, pp. 396–408, Apr. 1964.

[45] P. Kotowski, "Bandbreitenaufwand und Sicherheit verschiedener drahtloser Telegraphieverfahren" ["Bandwidth requirement and reliability of various methods of wireless telegraphy"], *Elek. Nachrichtentech.*, vol. 20, pp. 270–276, Nov.–Dec. 1943.

[46] G. R. Thompson and D. R. Harris, *The Signal Corps: The Outcome (Mid-1943 Through 1945)* (*United States Army in World War II*, Vol. 6, Part 5: *The Technical Services*, vol. 3). Washington, DC: Off. Chief of Military History, U.S. Army, 1966.

[47] F. E. Bond, "Military radio communication systems development vs. vulnerability to electronic countermeasures," Coles Signal Lab., Ft. Monmouth, NJ, Tech. Memo. M-1458, Aug. 25, 1952 (ATI 175096).

[48] "Antijamming design practices; Part II. Communication and navigation," Gen. Electron. Lab., Contr. AF 33(616)-2914, WADC Tech. Rep. 55-59, Part II, July 1955 (AD 072875).

[49] "Practical method and means for protecting radio communications and remote radio control against enemy interference," Reevesound Co., New York, Interim Eng. Rep. I-1.01, Contr. AF 28(099)-78, Aug. 10, 1949 (ATI 119957).

[50] R. Price, "Statistical theory applied to communication through

multipath disturbances,'' M.I.T. Res. Lab. Electron. Tech. Rep. 266 and M.I.T. Lincoln Lab. Tech. Rep. 34, Sept. 3, 1953 (AD 028497).[26]

[51] R. M. Fano, ''On matched-filter detection in the presence of multipath propagation,'' Rough draft, M.I.T. Res. Lab. Electron., Aug. 3, 1956 (unpublished).

[52] ——, ''Communication in the presence of additive noise,'' presented at the Symp. Appl. Commun. Theory, London, England, Sept. 22–26, 1952; in *Communication Theory*, W. Jackson, Ed. New York: Academic, 1953, pp. 169–182.

[53] ——, ''Communication in the presence of noise,'' Lecture Notes, M.I.T. Res. Lab. Electron., 1951 (unpublished).

[54] G. L. Turin, ''An introduction to matched filters,'' *IRE Trans. Inform. Theory*, vol. IT-6, pp. 311–329, June 1960.

[55] T. Kailath, ''Optimum receivers for randomly varying channels,'' presented at the 4th London Symp. Inform. Theory, Aug. 29–Sept. 2, 1960; in *Information Theory*, C. Cherry, Ed. Washington, DC: Butterworths, 1961, pp. 109–122.

[56] ——, ''A general likelihood-ratio formula for random signals in Gaussian noise,'' *IEEE Trans. Inform. Theory*, vol. IT-15, pp. 350–361, May 1969; see also vol. IT-16, pp. 276–288, May, and pp. 393–396, July 1970.

[57] ——, ''Measurements on time-variant communication channels,'' presented at the Int. Symp. Inform. Theory, Brussels, Belgium, Sept. 3–7, 1962; in *IRE Trans. Inform. Theory*, vol. IT-8, pp. S229–S236, Sept. 1962.

[58] J. G. Proakis, P. R. Drouilhet, Jr., and R. Price, ''Performance of coherent detection systems using decision-directed channel measurement,'' *IEEE Trans. Commun. Syst.*, vol. CS-12, pp. 54–63, Mar. 1964.

[59] ''A scheme of communication using matched-filter detection,'' Sect. 1.05, Quart. Progr. Rep., M.I.T. Res. Lab. Electron. and Project Lincoln, Jan. 30, 1952 (ATI 133942).

[60] W. W. Harman, ''The feasibility of matched-filter radar,'' Electron. Res. Lab., Stanford Univ., Stanford, CA, Tech. Rep. 35, Nov. 30, 1954 (AD 053117).

[61] B. L. Basore, ''The concept of efficiency in the communication of information,'' Sem. Paper for Course 6.501, Dep. Elec. Eng., M.I.T., Cambridge, MA, Jan. 16, 1950.

[62] R. M. Fano, ''Memorandum on the use of correlation detection in radar,'' M.I.T. Res. Lab. Electron., Jan. 22, 1951.

[63] J. M. Wozencraft, ''Application of modern communication techniques to radar ranging,'' S.M. thesis, Dept. Elec. Eng., M.I.T., Cambridge, MA, Feb. 1951.

[64] R. M. Fano, ''A unified approach to MTI and Doppler radar,'' M.I.T. Res. Lab. Electron. and Lincoln Lab., Tech. Rep. 10, July 22, 1952 (AD 004905).

[65] W. M. Siebert, ''A radar detection philosophy,'' *IRE Trans. Inform. Theory*, vol. IT-2, pp. 204–221, Sept. 1956.

[66] R. M. Lerner, ''A matched filter detection system for complicated Doppler shifted signals,'' *IRE Trans. Inform. Theory*, vol. IT-6, pp. 373–385, June 1960.

[67] J. Jecmen, *Elektronika ve Vojenstvi [Electronics in Military Science]*. Prague, Czechoslovakia, 1964 (AD 826193).

[68] B. Elspas, ''A radar system based on statistical estimation and resolution considerations,'' Appl. Electron. Lab., Stanford Univ., Stanford, CA, Tech. Rep. 361-1, Aug. 1, 1955 (AD 207896).

[69] G. R. Cooper and C. D. McGillem, ''Research in random signal radar techniques as ECCM aids,'' School Elec. Eng., Purdue Univ., West Lafayette, IN, Rep. 1 on Contr. DAAB07-68-C-0343, Aug. 1969 (AD 857587). (See also *Proc. IEEE*, vol. 51, pp. 1060–1061, July 1963.)

[70] G. Raisbeck, *Information Theory*. Cambridge, MA: M.I.T. Press, 1963. (His note re Wiener and Shannon is of interest in *IEEE Trans. Inform. Theory*, vol. IT-19, p. 826, Nov. 1973.)

[71] C. C. Cutler, ''Heterodyne systems employing trains of pulses,'' U.S. Patent 2 956 128, Oct. 11, 1960 (filed Aug. 22, 1951).

[72] N. T. Petrovich and M. K. Razmakhnin, *Shirokopolosnye Kanaly Svyazi s Shumopodobnymi Signalami [Wideband Channels of Communication with Noise-Like Signals]*. Moscow: VZEIS Press, 1965.

[73] E. Huttmann, ''Moderne Breitbandsignal-Verfahren fur mehrdeutige Ubertragungswege'' [''Modern wide band signal methods for multiple transmission paths''], *Nachrichtentechnik*, vol. 13, pp. 153–159, Apr. 1963.

[74] P. E. Green, Jr., ''An historical note on the autocorrelation function,'' *Proc. IRE*, vol. 41, p. 1519, Oct. 1953.

[75] G. Guanella, ''Korrelationsmethoden in der Hochfrequenztechnik'' [''Correlation methods in high-frequency technology''], notes of lecture given for Dep. Military Sci., Swiss Fed. Inst. Technol., Zurich, Switzerland, Feb. 4, 1960.

[76] S. J. Heims, *John von Neumann and Norbert Wiener*. Cambridge, MA: M.I.T. Press, 1980.

[77] J. M. Smith, ''A brief history of correlation detection, matched filters, and signaling with noise-like waveforms,'' Appendix C in ''Efficient signals and receivers for a spread-spectrum, multiple access, binary communication system,'' Jet Propulsion Lab., California Inst. Technol., Pasadena, Rep. 900-396, Dec. 4, 1970.

[78] N. Wiener, *The Human Use of Human Beings: Cybernetics and Society*. New York: Avon, 1967 (orig. ed. 1950).

[79] R. Higham, Ed., *A Guide to the Sources of United States Military History*. Hamden, CT: Shoe String Press, 1975; see also Suppl. I, 1981.

[80] M. L. Marshall, Ed., *The Story of the U.S. Army Signal Corps*. New York: Franklin Watts, 1965.

[81] L. S. Howeth, *History of Communications-Electronics in the United States Navy*. Washington, DC: U.S. Gov. Print. Off., 1963.

[26] The research performed for this thesis began as secret, but declassification was permitted before its completion. The earlier theses, {101}, {102}, and {105}, whose secrecy was maintained by M.I.T., demonstrate that even under military regulations, academic work of high caliber can be accomplished so as to advance a discipline significantly.

# Secret Telephony as a Historical Example of Spread-Spectrum Communication

WILLIAM R. BENNETT, FELLOW, IEEE

*Abstract*—The spread-spectrum properties of the X-System for secret telephony developed by Bell Telephone Laboratories for use in World War II are examined. In this system, the bandwidth of the speech signal was reduced by a vocoder, the vocoder signals were sampled and quantized to base six, and a random, never-reused, six-valued key stream was added modulo six to obtain a public message which was undecipherable without the key. It is believed that this was the first practical example of digital speech transmission. Examples of its effectiveness are described, and a number of human-interest type anecdotes are related.

THE purpose of the present paper is to supplement Robert A. Scholtz's extensive historical review of spread-spectrum communication [1] by clarifying the position of the World War II X-System (nicknamed "Green Hornet") in the spread-spectrum family. The brief references to the X-System on pp. 827 and 849 of Scholtz's paper do not convey an adequate description and are, in fact, misleading in their implications that the notion of bandwidth expansion was lacking, and that there was a relation between the X-System and a Tele-funken method of multiplying analog voice signals by random noise. The confusion is understandable because of the extreme secrecy under which the X-System was developed and operated. This secrecy was not relaxed until 1976, when the basic patents were declassified. The review gives references to these patents but not to a comprehensive 1978 disclosure of the X-System and its history in a book published by Bell Telephone Laboratories [2]. As stated in [2, p. 296], the Signal Corps used the code name "Sigsaly" for the system. Other disclosures by the Bell Laboratories are cited in [18].

Since the BTL publication gives an excellent description, it is unnecessary here to discuss the technical features of the X-System in great detail. What we are concerned with primarily is the relation to the spread-spectrum concept. If we consider Scholtz's three basic signal characteristics enumerated on p. 822 of the review paper, i.e., 1) random or pseudorandom wide-band carrier, 2) carrier bandwidth wider than data bandwidth, and 3) detection by cross correlation, the X-System is not included. However, if we take the broader definition of spread spectrum as any transmission method utilizing a wider band than is occupied by the signal itself, the X-System belongs and is, in fact, one of the earliest successful applications. Armstrong's use of wide-band FM to improve signal-to-noise ratio was, of course, earlier. The spread-spectrum feature of the X-System is somewhat concealed because a preliminary bandwidth compression of the telephone signal by means of the vocoder takes place,

Manuscript received September 18, 1982; revised October 7, 1982.
The author is at 29 Tulip Lane, Colts Neck, NJ 07722.

and it is the vocoder signal on which the expansion of bandwidth is performed in order to transmit an enciphered digital representation with sufficient accuracy over a fading radio telephone channel.

The purpose of the X-System was to provide complete secrecy over available telephone channels. A known method of achieving unbreakable secret transmission at that time was the Vernam telegraph method of World War I [3], [4], in which a never-reused random binary key stream was added mod 2 to a binary signal at the transmitter and added mod 2 to the public message at the receiver. To apply the Vernam method to telephony required sampling, quantizing, and coding, a combination which was later called pulse code modulation (PCM). Since PCM expands bandwidth, the telephone channel could not carry the resulting enciphered signal. The availability of the vocoder [5], which provided intelligibility at lower quality by representing a voice wave by ten spectral amplitude channels and a pitch channel, each occupying a band from 0 to 25 Hz, enabled the problem to be solved. PCM could be applied to the vocoder channels without increasing their total bandwidth beyond that of the telephone channel.

It was determined experimentally that each spectral amplitude sample could be adequately represented by a choice from six discrete levels and that 6 × 6 = 36 levels were sufficient for samples of the pitch signal. Accordingly, the vocoder output was converted to ten one-digit and one two-digit hexary PCM signals. (Strict usage should perhaps replace the word "digit" by "higit.") In actual transmission, the quantized vocoder signals were treated as 12 trains of six-valued pulses, which were individually added mod 6 to 12 independent six-valued random key pulse trains. It was known [6] that the pulse rate should be twice the signal bandwidth, and hence there were 50 pulses per second in each of the 12 pulse trains. Because of the severe problem of selective fading in the transoceanic radio circuits which were used, the 12 enciphered pulse trains were sent by frequency modulating 12 carrier frequencies, which were uniformly spaced in the telephone band. Use of the entire telephone band enabled a wider spacing in frequency of the FSK values than would have been possible if the minimum vocoder bandwidth were used. Bandpass filters at the transmitting and receiving ends restricted the bands of the FM channels and enabled separation of them at the receiver. The complete designation of the vocoder transmission method is expressed by PCM-FM-FDM.

The encipherment and decipherment were accomplished by synchronized key sources at transmitter and receiver. The

Reprinted from *IEEE Trans. Commun.*, vol. COM-31, pp. 98–104, Jan. 1983.

key sources consisted of identical disk records to be used once and only once. The recorded wave was the sum of the 12 voiceband carriers, each of which was amplitude modulated by a train of random hexary pulses. The pulse amplitudes were obtained by six-level quantization of samples from the output of a gas tube noise generator. The distribution of the noise samples was made uniform for the six levels by a non-linear operation on the original Gaussian distribution. The keys for the individual vocoder pulses were obtained by separating the channels with bandpass filters, demodulating the AM on each channel independently, sampling at 20 ms intervals, and quantizing the samples at six levels. The key samples were added mod 6 to the signal samples at the transmitter and subtracted mod 6 from the enciphered message samples at the receiver.

A vital feature of the six-level quantization of the vocoder signals was that the levels were not equally spaced on an amplitude scale but on a logarithmic scale, 6 db apart, except for the zero level. It was determined experimentally that this was an optimum spacing to minimize the number of levels needed to achieve acceptable quality. If equiamplitude spacing were used, the number of necessary levels would have been increased to make an already marginal system unrealizable. The principle of "companding" amplitudes for PCM was to become crucially important in later PCM systems for telephony. It is stated in the BTL publication that it is not known just who first made the suggestion of unequal levels. It so happens that I know the answer to this question because I was there when this part of the problem was first discussed—the departure from equal spacing was proposed by Eugene Peterson on the very sensible grounds that weak sounds required more delicate treatment than strong. It was also consistent with the Weber–Fechner law governing physiological responses. The 36 levels representing the pitch channel were equally spaced.

As discussed in a recent conversation between Robert Price and myself, the X-System included a remarkable number of interesting "firsts." In addition to 1) the obvious first realization of unbreakably enciphered telephony, there were 2) the first quantized speech transmission, 3) the first transmission of speech by PCM, 4) the first use of companded PCM, 5) the first examples of multilevel FSK, 6) the first useful realization of speech bandwidth compression, 7) the use of FSK-FDM as a viable transmission method over a fading medium, and 8) use of a multilevel "eye pattern" to adjust sampling intervals.

We emphasize at this point that it was far from obvious at the beginning that the objectives of the X-System could be met. In many essential components of the system, the technical capabilities of that time were strained to the point of "just barely possible." In particular, the problems of interchannel and intersymbol interference arising from the use of the fading radio link were such that a veteran telegraph engineer, A. M. Curtis, said at one time: "What you are trying to do is to operate a telegraph system with zero margin." The size and complexity of the equipment were frightening; a naval barge filled to capacity was towed by the Pacific fleet to make the X-System available for its operations.

In spite of many pessimistic prognostications, the end result was surprisingly good. Some of the most feared obstacles turned out to be minor barriers. Synchronization of the key records at transmitter and receiver appeared difficult but, as described in the BTL book, the availability of sufficiently accurate frequency standards and the relatively long sampling interval of 20 ms simplified the problem. Testimony that the X-System actually worked and that high-level allied personnel used it effectively in World War II has been augmented by recent conversations with men who participated in the project. We include here some anecdotal material from these conversations to give recognition to the human side of the technical achievement.

As stated in the BTL history, terminals for the X-System were installed and used at various important centers including Washington, London, North Africa, Paris, Hawaii, Australia, and the Philippines. Although many of the participants in this operation are no longer among the living, a substantial number are still alive and have vivid memories of their experiences. Luther G. (Luke) Schimpf, who is still actively engaged in mobile radio work at Bell Laboratories, took a responsible part in the installation and operation of the X-terminal in Algiers for use during the Allied campaign in North Africa. The equipment was placed in a former wine cellar of an old hotel occupied by army headquarters. He recalls that to convince the military leaders that the system was useful it was arranged for General Eisenhower to talk to his wife, Mamie. This was in fact a severe test because the pitch measuring circuit was designed for male voices. On the infrequent occasions when women talked over the system, they were instructed to keep their voices to as low a pitch as possible. The commercial telephone transmitter did not have a good enough low-frequency response to obtain satisfactory pitch determination, and hence a special high-quality "eight-ball" moving-coil microphone was substituted in the telephone set.

The key records were delivered by courier and immediately placed in a safe with combination known only to two or three selected officers. The records, which ran 15 minutes, were destroyed immediately after use. Changing records during a conversation was no problem because two machines were used with provision for switching from one to the other as the end of a record was approached. Since two-way speech was provided by the system, a key source was also necessary for enciphering the outgoing signal. This had to be from an independent record because, apart from the problem of time delay, duplicate use of the same key for transmitting and receiving on separate radio channels would violate the essential security requirement of never reusing a key record. A self-destruct feature was provided to be operated by the last man leaving the terminal in case of imminent capture. A push button activated a thermite process which in effect welded together the critical parts. No such action was ever necessary. The system had no antijamming provisions, and for the most part, did not require any. Schimpf recalls that they were bothered at one time by a German CW radio station transmitting in one of their frequency bands. They were able to make out the call letters and turned the information over to the Air Force. Shortly after, the offending station disappeared from the air.

51

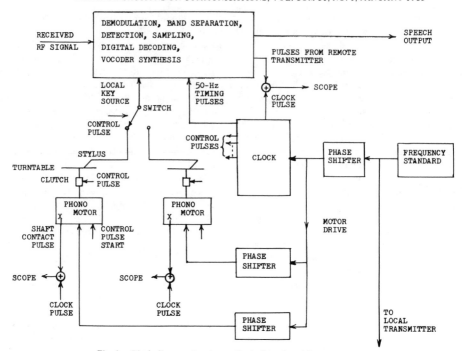

Fig. 1.   Block diagram showing method of synchronizing receiver with
transmitter.

At the request of the author, Luke Schimpf has reconstructed from memory the essential details of the synchronizing arrangements. Fig. 1 is a functional diagram showing how the receiver was synchronized with the remote transmitter. Duplicate apparatus was provided for the local transmitter except for the shared frequency standard. The latter was a highly stable 100 kHz quartz-crystal oscillator with appropriate frequency step-down circuits based on the art developed by Warren Marrison and his associates [12]. The accuracy of the frequency was within one part in ten million, which was more than sufficient to keep the tracking error within the required ±1 ms with very infrequent readjustments once synchronization had been established.

Initially, the clock time was set from Arlington radio signals with about one second accuracy. A scope displayed the sum of local clock pulses and timing pulses sent in one of the subcarrier channels from the remote terminal, in accordance with a prearranged pattern. Manual adjustment of the continuous phase shifter ahead of the clock input brought the clock pulse to synchronism with the remote transmitter. Control pulses from the clock were used to start the phonograph motors, which were then synchronized by adjustment of their individual phase shifters to match pulses derived from shaft contacts with pulses from the clock. The turntables were initially at rest with their styli carefully placed and indexed in the first grooves. When the motors reached synchronism, a control pulse from the clock actuated the clutch associated with one of the turntables. The clutch, which was designed by Ira Cole, made use of an ingenious adaptation of the telephone dial. Its task was to start the turntable smoothly without

making the stylus jump out of the groove. Readjusting the phase shifters while monitoring with the scope transformed the speech output gradually from random noise to silence as the sending and receiving keys were brought into synchronism and the decoding errors became less and less frequent. A similar adjustment was made at the same time on the receiver at the remote terminal. When both were completed, the system was ready for two-way conversation.

John M. Barstow, now living in retirement in Florida, cooperated with the 805th Signal Service Company to provide X-System facilities in London. The X-terminal itself was installed in a subbasement of Selfridge's department store, accessible from the Duke St. entrance. For use by Churchill and other high-echelon personnel, an OPEPS (off premise extension privacy system) link based on the A. G. Chapman patent described in [2, pp. 294-296] was installed between the X-terminal at Selfridge's and a deep basement at 10 Downing Street. Barstow recalls that when he was notified that the OPEPS equipment had been delivered, he went to a back entrance at 10 Downing Street and was stopped by two sentries with crossed bayonets. Eventually someone appeared at the door with authority for his entry. He went down four or five flights of stairs to a little room containing a bed, a table, and, in one corner, a cabinet of equipment with color-coded wires to be connected.

The OPEPS link operated by insertion of a hybrid coil (or bridge transformer) at the receiving end, with the telephone receiver and a noise source connected to conjugate arms of the hybrid. The line and its balancing network were connected to the other two arms. The noise was balanced out of the receiver but was added to the voice signal on the line. The noise was only 15 dB above the signal, but enclosure of the wires in a cable, and a noise alarm activated by disturbance of the balance if the wires were tapped, provided additional security. Available hybrid coils with 70 dB balance used with selected lines enabled a 40 dB balance to be maintained. OPEPS was also used in Washington and other places to obtain adequately secure extensions from the X-terminal, for example, from the Pentagon to the White House.

Some interesting material relevant to the X-System has been found by Dr. Price in an official history of the United States Army in World War II (reference 46 of [7]). On p. 145, dealing with the Allied invasion of Normandy in 1944, the following two paragraphs quoted in full give a flavor of the value furnished by a secret radio telephone channel between London and Washington.

"Replacement allowances for tank antennas had been based on training requirements and experiences in maneuvers in the United States. As with all replacement allowances, signal supply planners tried to anticipate the worst possible conditions that might be encountered and then added a generous allowance for error. Nevertheless, as soon as tanks were ashore in numbers in Normandy, requests for replacement antennas far in excess of anticipated losses poured in from every tank unit. The reason was that the low-branched trees and thick hedges of Normandy ripped off antennas with fearful regularity. No American tank in training or maneuvers had ever encountered such terrain—probably the worst "tank country" that could be found. Local reserves of antennas were quickly exhausted, and soon COMZ was shipping from the fast-disappearing theater reserve. The need for antennas became desperate.

"In the emergency, General Rumbough turned to the secret, direct radiotelephone channel to Washington. From London he talked directly to General Ingles, the Chief Signal Officer, urgently requesting the earliest possible shipment of more antennas. Within forty-eight hours the needed antennas were delivered by air."

Charles W. Vadersen, now living in retirement in South Carolina, was sent with X-equipment to install a terminal in Brisbane, Australia, the location of General MacArthur's headquarters. The Pacific part of the trip was made on a twin-Diesel, 20 000 ton ship, the *Bosch Fontein*, capable of 18 knots. There was no convoy, and attempt was made to make as much of the trip as possible during the dark of the moon. The ship left San Francisco in August, 1943, and took about 18 days to reach Australia. There was a 5 in gun on the stern, a 3 in gun on the forward deck, and two Bofors 20 mm antiaircraft guns on each side. Practice attempts to hit balloons and towed targets were uniformly unsuccessful and did little to give any assurance of safety. An aircraft sighted after four or five days turned out to be friendly. Eight days out of San Francisco, one of the two engines failed. The ship was forced to crawl along at 9 knots until the engine was restored to action about 36 hours later. Vadersen remembers crossing the International Date Line just in time to miss his birthday, which was September 2.

The original plan was to make the first landing at Brisbane, deliver the X-equipment with its assigned personnel, and then to proceed to New Guinea with the remainder of the cargo destined for the artillery engaged in operations there. To facilitate the program, the X-equipment had been loaded last in order to be unloaded first. However, a radio order was received en route to go to New Guinea as soon as possible because of the urgent need for the artillery supplies there. Accordingly, Vadersen and two of the military personnel were dropped off at Townsville on the east coast of Australia about 800 miles from Brisbane. Their instructions were to find their way to Brisbane in any way possible and to make advance preparations there for reception of the equipment when the ship arrived after the diversion to New Guinea. This necessitated an unloading and reloading of the X-equipment on the beach at Port Moresby, an incident referred to in the BTL history [2, p. 315]. According to Vadersen, the chief damage was severe bending of the steel panels provided for the power supply. He had to find an Australian blacksmith to put them back in shape. Getting from Townsville to Brisbane was an adventure in itself. The military personnel were able to get passage on official aircraft within a couple of days, but it was six days before Vadersen was able to "bum a ride" in a Sunderland flying boat. While waiting, he had some harrowing experiences with the local intelligence services which prompted him to ask to be put into some kind of uniform when he arrived in Brisbane. All the identification he had up to that time consisted of a passport and his orders. He was given a simulated rank of captain for the remainder of his stay.

The Brisbane terminal was put into operation on schedule in spite of the delay and damage in the delivery of equipment. Everyone worked straight through the last 36 hours to complete the job. General Kenney of the Air Force and Admiral Nimitz of the Navy were prominent consistent users of the facility. General MacArthur refused to talk over it because he was not completely convinced of the security. He preferred to communicate over the secure teletype channel. Some dignitaries found the quality of the restored speech unsatisfactory, but this was characteristically the result of a refusal to listen to the few simple instructions on how to talk over the system most effectively. Most users were willing to accept the price of some distortion in return for complete security. The intelligibility was good and improved with the experience of the listener. Recognition of voices depended more on cadence than on characteristic timbre. An outstanding demonstration of the superior performance of the system was a successful experimental call between Schimpf in Algiers and Vadersen in Brisbane over the transatlantic short-wave radio, across land lines from the Atlantic to the Pacific coasts, and transpacific radio to Australia. This was the event alluded to in [2, p. 298]. It is a tribute to the ruggedness of the system that it operated successfully over fading radio channels at least to the point at which commercial voice transmission failed.

Thomas W. Thatcher, Jr., still an active employee of BTL, recalls his experiences in charge of the X-equipment during American operations in the Philippines and later at Guam. Every Wednesday morning the Chief of Naval Operations talked over the system with the Commander of the Western Pacific Submarine Forces at Guam. The system was also used by the Air Force, notably by General Hap Arnold, now deceased. Thatcher suggested that one of the talkers at Guam might have been General Curtis LeMay. Dr. Price was able to call General LeMay by telephone. LeMay said that he used only the secure teletype but suggested others, still alive, who might have talked on the X-System. One of these, Lt. Genl. Barney M. Giles [8], [9], turned out to be a treasure of information.

Reached by telephone at his home in San Antonio, Texas, General Giles, who celebrated his 90th birthday on September 13, 1982, talked lucidly to Dr. Price about experiences with the secret telephone system. Giles said he had talked over it from Guam to Washington three or four times on important occasions, that he trusted its security, and was agreeably surprised by its good quality. He said that he received instructions on how to use it effectively from a communications expert who was probably a civilian. When asked if the telephone instrument was unusually heavy, he stated that it definitely was. There seems to be no doubt that it was the X-System he had used, although he did not know it by that name.

General Giles' association with secret telephony coincided with some momentous events which occurred at that time. From his position as Air Chief of Staff directly under General Arnold, he was made Deputy Commander of the 20th Bomber Command, succeeding General Millard Harmon, who was killed during a trip from Guam to Washington. Giles continued to report directly to Arnold. When President Roosevelt died, Giles was summoned back to Washington to stand in for Arnold, who was ill, at a high-level meeting in President Truman's office. The names of the others present included Marshall, Stimson, Forrestal, King, and Leahy. Their task was to tell Truman about the atomic bomb. Later, from Guam, Giles discussed deployment of the bomb with Arnold in Washington. He feels that the X-System was used at least once in discussing the logistics of the delivery of the atomic bombs at Hiroshima and Nagasaki.

There is no evidence that unauthorized listeners had any awareness of the nature and purpose of the system. Curiously enough, a 1941 article [10] in the *Brown Boveri Review*, a Swiss journal, described the possibility of obtaining secrecy with a device which was not given a name but, from its description, was unquestionably a vocoder. There is no record of any application. An NDRC 1946 report [11] not declassified until 1960 gives a review of various speech scrambling systems as of the end of World War II. The vocoder was discussed but no reference to the methods of the X-System were included. This is understandable because of the extreme sensitivity regarding even the existence of such a system.

Since there is no record of any realization by an interceptor as to what the nature and purpose of the "Green Hornet" signal was, the perfection of the key itself was, as far as is known, never subjected to the ultimate test of attempted recovery by a sophisticated enemy. N. D. Newby recalls submitting a sample of key values to B. D. Holbrook for an opinion as to its randomness. Holbrook performed some simple random tests as known at that time and asked J. Riordan for further

evaluation. Riordan stated that he could not prove that the values were random, nor could he prove that they were non-random.

It has been recounted that the opinions of various experts, e.g., Nyquist, Shannon, and Turing, were solicited at the beginning of the X-program as to whether or not the method was totally secure. The fact that none found any fault was reassuring, but elementary reasoning shows that there could be no valid objection to the concept if the technical requirements were met. If the key is completely random, then all decipherings of the public message are equally probable, and possession of the public message gives no information whatsoever. The only practical vulnerabilities would be disloyalty, indiscretion, and incompetence on the part of the operating personnel. The fact that no security leaks were reported testifies to the high qualities of the many people engaged in the program. The handling of the key material was the most sensitive part, since even if the enemy knew all about the method, he would still be helpless without the key.

R. C. Mathes, one of the coinventors of the system, and now at the age of 90 operating an avocado ranch in California, recalls the presence of an armed guard in the room where the master records were made. I. M. Kerney had the responsibility of delivering the masters to a record-pressing company in Manhattan and of making sure that the masters were destroyed after two and only two copies of each were made.

Although it had been assumed from the beginning that the X-System would be particularly helpful to President Roosevelt and Prime Minister Churchill, hard evidence of instances in which these two actually talked to each other over the system has not yet been found. A published reference to their telephone conversations exists in a book by James Leasor [13] based on the experiences of General Leslie Hollis, who had a responsible part in the development and use of a safe underground British Cabinet Headquarters at Storey's Gate, London, known as "The Hole in the Ground." On p. 38 of this book, there is the following statement:

> "When Mr. Churchill became Prime Minister in 1940, however, he arranged for a special line to be run to President Roosevelt at the White House, in Washington, and the complicated scrambler fitted to this to ensure complete security for their frequent calls was the size of a battleship's boiler. Even with one hundred and fifty rooms, and more than a mile of twisting corridors in the "Hole" by that time, it was impossible to find space for this gigantic scrambler. Eventually, engineers fitted it in the cellars of a department store in Oxford Street."

The description and location of the "gigantic scrambler" clearly identifies it as the X-System, but the 1940 date is far too early for it to have been put into use. Documents of the 805th Signal Service Company cited by Dr. Price in [7, p. 90] show that the inaugural call over the X-System between Washington and London took place on July 15, 1943. When reached recently by Bob Price by overseas telephone, Mr. Leasor agreed that the date of 1940 should not have been associated with the availability of the secret telephone system. The difficulty remains that the reader cannot be sure of what part of the telephone calls between Churchill and Roosevelt described in the book were actually made on the commercial transatlantic radiotelephone, which was equipped with so-called privacy arrangements providing no real security against a moderately competent interceptor.

A significant addition to the record has been obtained from Stephen M. Geis, who, as a lieutenant in the 805th Signal Service Company, had the responsibility of introducing Churchill to the use of Sigsaly late in April, 1944. The location was the little room in Churchill's underground headquarters, as described in Leasor's book [13], and as appearing in the photograph from *Life* magazine referred to in the BTL history [2, p. 315]. The Prime Minister accepted the instructions for use of the system most graciously. Geis then withdrew from the room, and Churchill talked on the telephone for about 10 minutes. Other details recalled by Geis were that Churchill wore his famous "siren suit" and that he graciously accepted a "stogie" offered to him but did not smoke it at the time.

Further light is thrown on the question by documents [14] unearthed by Dr. Price from the Franklin D. Roosevelt Library at Hyde Park, N.Y. In a memorandum for Admiral Brown on White House stationery dated 8 October 1943, Harry Hopkins wrote: "Captain Fenn of Naval Censorship called me up and recommended very strongly that the President, in calling the Prime Minister, call an agreed upon telephone number and the Prime Minister call a telephone number in this country, and that calls not be put in directly for the Prime Minister. . . . He also said we ought to give consideration to using the Army's scrambler system which I told him I understood was in existence."

From the same source, a memorandum for Mr. Hopkins on the stationery of the War Department, Office of the Chief of Staff, dated 12 October 1943, signed by Frank McCarthy, stated:

"A day or two ago one of our Signal Corps people came in to report that Censorship had recently listened to a commercial telephone conversation between you and the Prime Minister. He added that you had tactfully, but consistently, urged the Prime Minister to be careful of what he said, but that the Prime Minister cited names and places in such a way as to create possible danger for himself and others. In addition, this equipment furnishes a very low degree of security, and we know definitely that the enemy can break the system with almost no effort. We have in the War Department . . . equipment which guarantees almost complete security from the enemy . . .. The London terminal of this equipment is within one block of Number 10 Downing Street."

These examples show that Roosevelt and Churchill were not using the X-System as late as October, 1943. It seems certain that shortly after this time they did begin to talk to each other on it, but no specific instance has as yet been substantiated.

An interesting chapter in the Sigsaly saga was turned up by Price in a conversation with Margaret Truman's bodyguard, Frank Stoner. Stoner was formerly in the Secret Service and was assigned to code room duties at the White House. He was sent to Yalta for the Churchill–Roosevelt–Stalin conference there and recalls that Sigsaly equipment was carried on the U.S.S. *Catoctin* as far as Sevastopol. Telephone wires were strung on poles to cover the distance of about 50 miles from there to Yalta. The purpose was to provide secret telephone conversation between Washington and the top level participants. The site of the conference, which was an old palace formerly used by czars, had relatively primitive accommodations. There were very few telephones, and messages were delivered within the palace by runners. Stoner said that he received an official call from Washington on the "Green Hornet" from his father, General Frank E. Stoner, Chief of the Army Communications Service, to check on the operation of the system. He was surprised at the clarity of the reception at that great distance.

In the Pacific sector, General Dayton Eddy of the Signal Corps recalls that he "nursemaided" a Sigsaly installation in a "cement barge" on an island-hopping series of applications from Manila Bay to Tokyo Bay. He said that the equipment weighed "a couple of pounds less than a sawmill." A principal user of the system was General Spencer B. Akin, Chief Signal Officer to MacArthur.

A solid verification of use by Churchill of the X-System after Roosevelt's death was tracked down by Price in the diary of Fleet Admiral William D. Leahy [15]. Discovery of this source depended on a chain of helpful referrals starting with Marie-Louise Friendly, private secretary to Averell Harriman; Ronald D. Landa, historian for the State Department; Francis L. Loewenheim, coauthor of a book about Roosevelt and Churchill; George M. Elsey, president of the American Red Cross and formerly an expert in the Map Room of the White House, where he handled coded cables for Roosevelt; and Henry H. Adams, author of a biography of Hopkins and now working on a biography of Leahy.

Some background for the disclosures in the Leahy diary is furnished by a subsequently discovered entry in the private diary of Eben A. Ayers [16], Assistant Press Secretary to President Truman. Under the date of April 25, 1945, the entry is: "Around noon, the President went to the Pentagon without warning. The press got wind of it and were told it was an 'inspection.' Some learned that he went into the communication room. The fact was that he went over to talk over the European telephone, I believe, to Churchill." This is further substantiated by a statement in Leahy's published autobiography [17, p. 358], as follows: "At the direction of the President, I talked with him [Churchill] from the same highly guarded room in the Pentagon Building where Truman and the Prime Minister had, on April 25, discussed the Himmler peace offer." More details of the April 25 conversation are given on p. 354.

Leahy's diary [15] records that at 10:10 A.M., 7 May 1945, he was asked by President Truman to go to the Pentagon and make a call on the secret telephone to Churchill in reply to a telegram received from Churchill regarding reports that Germany had surrendered. The diary gives a verbatim record of the call, which was transcribed by Admiral Leahy's secretary. Following is the initial part of this transcript:

LEAHY: Admiral Leahy speaking.
CHURCHILL: It is me, the Prime Minister.
LEAHY: Colonel Warden, yes, sir.
CHURCHILL: You've got my telegram?
LEAHY: I have your telegram, sir. This is a message

which the other Admiral asked me to convey to Colonel Warden.

CHURCHILL: We are on the "secret" now, so we can talk quite freely. A message that he asked you to convey to me was what?

LEAHY: I convey the following message to you. In view of agreements already made, my Chief asks me to tell you that he cannot act without the approval of Uncle Joe. Did you understand, sir?

CHURCHILL: Will you let somebody with a younger ear listen to it? I am not quite sure I got it all down. I have got my secretary here. My ears are a bit deaf, you know. . . .

An entry in the diary for 11:10 A.M., the same day, reads: "Mr. Churchill called me from London on the open telephone telling me that crowds celebrating in the streets of London were beyond control and that he must make an announcement of the victory at noon."

It seems clear that the verdict of history is that the X-System successfully fulfilled its objectives and represents an achievement for which all participants can be proud.

## ACKNOWLEDGMENT

In conclusion, I wish to express my thanks to Robert Price for his tireless efforts in tracking down and collecting testimony from key witnesses. I wish to thank also my many friends of the X-System days who exchanged reminiscences with me in numerous telephone calls to reconstruct the remote past. In addition to the persons already named in the text, recognition is due to Alex Fowler, Doren Mitchell, Ralph Miller, George Stibitz, Amos Joel, Morton Fagan, and Lester Hochgraf. I also wish to thank my wife, Viola M. Bennett, for her invaluable editorial assistance.

## REFERENCES

[1] R. A. Scholtz, "The origins of spread-spectrum communications," *IEEE Trans. Commun.*, vol. COM-30, pp. 822–854, May 1982.

[2] M. D. Fagen, Ed., *A History of Engineering and Science in the Bell System. National Service in War and Peace (1925–1975)*, Bell Lab., Murray Hill, NJ, 1978, pp. 296–317.

[3] G. S. Vernam, "Cipher printing telegraph systems for secret wire and radio telegraphic communications," *AIEE J.* vol. 45, pp. 109–115, Feb. 1926; also *AIEE Trans.*, vol. 45, pp. 295–301, Feb. 1926.

[4] W. R. Bennett, *Introduction to Signal Transmission*. New York: McGraw-Hill, 1970, pp. 231–233.

[5] H. W. Dudley, "Remaking speech," *J. Acoust. Soc. Amer.*, vol. 11, pp. 169–177, Oct. 1939. See also [4, pp. 71–74].

[6] W. R. Bennett, "Time-division multiplex systems," *Bell Syst. Tech J.*, vol. 20, pp. 199–221, Apr. 1941.

[7] R. Price, "Further notes and anecdotes on spread-spectrum origins," this issue, pp. 85–97.

[8] "Barney McKinney Giles," *Current Biography: Who's News and Why, 1944*, pp. 233–235, 1945.

[9] H. S. Truman, *Memoirs of Harry S. Truman, Volume One: Year of Decisions*. Garden City, NY: Doubleday, 1955, p. 17.

[10] G. Guanella, "Methods for the automatic scrambling of speech," *Brown Boveri Rev.*, pp. 397–408, Dec. 1941.

[11] "Speech and facsimile scrambling and decoding," NDRC Div. 13, Rep. AD-221 609, Washington, DC, 1946.

[12] W. A. Marrison, "The evolution of the quartz crystal clock," *Bell Syst. Tech. J.*, vol. 27, pp. 510–588, July 1948.

[13] J. Leasor, *The Clock with Four Hands*. New York: Reynal, 1959. (Published in England under the title *War at the Top*.)

[14] H. Hopkins papers, Franklin D. Roosevelt Library, Hyde Park, NY.

[15] Diary of Adm. W. D. Leahy, Manuscript Div., Library of Congress, Washington, DC, pp. 68–78.

[16] Private Diary of E. A. Ayers, Harry S. Truman Library, Independence, MO.

[17] W. D. Leahy, *I Was There*. New York: McGraw-Hill, 1950.

[18] M. D. Fagen, Ed., *Impact, A Compilation of Bell System Innovations in Science and Engineering*. Bell Lab., 1st ed., 1971, p. 57; 2nd ed., 1981, pp. 33–34.

# Theory of Spread-Spectrum Communications—A Tutorial

RAYMOND L. PICKHOLTZ, FELLOW, IEEE, DONALD L. SCHILLING, FELLOW, IEEE,
AND LAURENCE B. MILSTEIN, SENIOR MEMBER, IEEE

*Abstract*—**Spread-spectrum communications, with its inherent interference attenuation capability, has over the years become an increasingly popular technique for use in many different systems. Applications range from antijam systems, to code division multiple access systems, to systems designed to combat multipath. It is the intention of this paper to provide a tutorial treatment of the theory of spread-spectrum communications, including a discussion on the applications referred to above, on the properties of common spreading sequences, and on techniques that can be used for acquisition and tracking.**

## I. INTRODUCTION

SPREAD-spectrum systems have been developed since about the mid-1950's. The initial applications have been to military antijamming tactical communications, to guidance systems, to experimental antimultipath systems, and to other applications [1]. A definition of spread spectrum that adequately reflects the characteristics of this technique is as follows:

"Spread spectrum is a means of transmission in which the signal occupies a bandwidth in excess of the minimum necessary to send the information; the band spread is accomplished by means of a code which is independent of the data, and a synchronized reception with the code at the receiver is used for despreading and subsequent data recovery."

Under this definition, standard modulation schemes such as FM and PCM which also spread the spectrum of an information signal do not qualify as spread spectrum.

There are many reasons for spreading the spectrum, and if done properly, a multiplicity of benefits can accrue simultaneously. Some of these are

- Antijamming
- Antiinterference
- Low probability of intercept
- Multiple user random access communications with selective addressing capability
- High resolution ranging
- Accurate universal timing.

Manuscript received December 22, 1981; revised February 16, 1982.
R. L. Pickholtz is with the Department of Electrical Engineering and Computer Science, George Washington University, Washington, DC 20052.
D. L. Schilling is with the Department of Electrical Engineering, City College of New York, New York, NY 10031.
L. B. Milstein is with the Department of Electrical Engineering and Computer Science, University of California at San Diego, La Jolla, CA 92093.

The means by which the spectrum is spread is crucial. Several of the techniques are "direct-sequence" modulation in which a fast pseudorandomly generated sequence causes phase transitions in the carrier containing data, "frequency hopping," in which the carrier is caused to shift frequency in a pseudorandom way, and "time hopping," wherein bursts of signal are initiated at pseudorandom times. Hybrid combinations of these techniques are frequently used.

Although the current applications for spread spectrum continue to be primarily for military communications, there is a growing interest in the use of this technique for mobile radio networks (radio telephony, packet radio, and amateur radio), timing and positioning systems, some specialized applications in satellites, etc. While the use of spread spectrum naturally means that each transmission utilizes a large amount of spectrum, this may be compensated for by the interference reduction capability inherent in the use of spread-spectrum techniques, so that a considerable number of users might share the same spectral band. There are no easy answers to the question of whether spread spectrum is better or worse than conventional methods for such multiuser channels. However, the one issue that is clear is that spread spectrum affords an opportunity to give a desired signal a power advantage over many types of interference, including most intentional interference (i.e., jamming). In this paper, we confine ourselves to principles related to the design and analysis of various important aspects of a spread-spectrum communications system. The emphasis will be on direct-sequence techniques and frequency-hopping techniques.

The major systems questions associated with the design of a spread-spectrum system are: How is performance measured? What kind of coded sequences are used and what are their properties? How much jamming/interference protection is achievable? What is the performance of any user pair in an environment where there are many spread spectrum users (code division multiple access)? To what extent does spread spectrum reduce the effects of multipath? How is the relative timing of the transmitter–receiver codes established (acquisition) and retained (tracking)?

It is the aim of this tutorial paper to answer some of these questions succinctly, and in the process, offer some insights into this important communications technique. A glossary of the symbols used is provided at the end of the paper.

## II. SPREADING AND DIMENSIONALITY— PROCESSING GAIN

A fundamental issue in spread spectrum is how this technique affords protection against interfering signals with

Reprinted from *IEEE Trans. Commun.*, vol. COM-30, pp. 855–884, May 1982.

finite power. The underlying principle is that of distributing a relatively low dimensional (defined below) data signal in a high dimensional environment so that a jammer with a fixed amount of total power (intent on maximum disruption of communications) is obliged to either spread that fixed power over all the coordinates, thereby inducing just a little interference in each coordinate, or else place all of the power into a small subspace, leaving the remainder of the space interference free.

A brief discussion of a classical problem of signal detection in noise should clarify the emphasis on finite interference power. The "standard" problem of digital transmission in the presence of thermal noise is one where both transmitter and receiver know the set of $M$ signaling waveforms $\{S_i(t), 0 \leq t \leq T; 1 \leq i \leq M\}$. The transmitter selects one of the waveforms every $T$ seconds to provide a data rate of $\log_2 M/T$ bits/s. If, for example, $S_j(t)$ is sent, the receiver observes $r(t) = S_j(t) + n_w(t)$ over $[0, T]$ where $n_w(t)$ is additive, white Gaussian noise (AWGN) with (two-sided) power spectral density $\eta_0/2$ W/Hz.

It is well known [3] that the signal set can be completely specified by a linear combination of no more than $D \leq M$ orthonormal basis functions (see below), and that although the white noise, similarly expanded, requires an infinite number of terms, only those within the signal space are "relevant" [3]. We say that the signal set defined above is $D$-dimensional if the minimum number of orthonormal basis functions required to define all the signals is $D$. $D$ can be shown to be [3] approximately $2B_D T$ where $B_D$ is the total (approximate) bandwidth occupancy of the signal set. The optimum (minimum probability of error) detector in AWGN consists of a bank of correlators or filters matched to each signal, and the decision as to which was the transmitted signal corresponds to the largest output of the correlators.

Given a specific signal design, the performance of such a system is well known to be a function only of the ratio of the energy per bit to the noise spectral density. Hence, against white noise (which has infinite power and constant energy in every direction), the use of spreading (large $2B_D T$) offers no help. The situation is quite different, however, when the "noise" is a jammer with a fixed *finite* power. In this case, the effect of spreading the signal bandwidth so that the jammer is uncertain as to where in the large space the components are is often to force the jammer to distribute its finite power over many different coordinates of the signal space.

Since the desired signal can be "collapsed" by correlating the signal at the receiver with the known code, the desired signal is protected against a jammer in the sense that it has an effective power advantage relative to the jammer. This power advantage is often proportional to the ratio of the dimensionality of the space of code sequences to that of the data signal. It is necessary, of course, to "hide" the pattern by which the data are spread. This is usually done with a pseudonoise (PN) sequence which has desired randomness properties and which is available to the cooperating transmitter and receiver, but denied to other undesirable users of the common spectrum.

A general model which conveys these ideas, but which uses random (rather than pseuodrandom) sequences, is as follows. Suppose we consider transmission by means of $D$ equiprobable and equienergy orthogonal signals imbedded in an $n$-dimensional space so that

$$S_i(t) = \sum_{k=1}^{n} S_{ik} \phi_k(t); \qquad 1 \leq i \leq D; \qquad 0 \leq t \leq T$$

where

$$S_{ik} = \int_0^T S_i(t) \phi_k(t) \, dt$$

and where $\{\phi_k(t); 1 \leq k \leq n\}$ is an orthonormal basis spanning the space, i.e.,

$$\int_0^T \phi_l(t) \phi_m(t) \, dt = \delta_{lm} \triangleq \begin{cases} 1 & l = m \\ 0 & l \neq m. \end{cases}$$

The average energy of each signal is

$$\int_0^T \overline{S_i^2(t)} \, dt = \sum_{k=1}^{n} \overline{S_{ik}^2} \triangleq E_s; \qquad 1 \leq i \leq D \qquad (1)$$

(the overbar is the expected value over the ensemble).

In order to hide this $D$-dimensional signal set in the larger $n$-dimensional space, choose the sequence of coefficients $S_{ik}$ independently (say, by flipping a fair coin if a binary alphabet is used) such that they have zero mean and correlation

$$\overline{S_{ik} S_{il}} \triangleq \frac{E_s}{n} \delta_{kl}; \qquad 1 \leq i \leq D. \qquad (2)$$

Thus, the signals, which are also assumed to be known to the receiver (i.e., we assume the receiver had been supplied the sequences $S_{ik}$ before transmission) but denied to the jammer, have their respective energies uniformly distributed over the $n$ basis directions as far as the jammer is concerned.

Consider next a jammer

$$J(t) = \sum_{k=1}^{n} J_k \phi_k(t); \qquad 0 \leq t \leq T \qquad (3)$$

with total energy

$$\int_0^T J^2(t) \, dt = \sum_{k=1}^{n} J_k^2 \triangleq E_J \qquad (4)$$

which is added to the signal with the intent to disrupt communications. Assume that the jammer's signal is independent of the desired signal. One of the jammer's objectives is to devise a strategy for selecting the components $J_k$ of his fixed total energy $E_J$ so as to minimize the postprocessing signal-to-noise ratio (SNR) at the receiver.

The received signal

$$r(t) = S_i(t) + J(t) \tag{5}$$

is correlated with the (known) signals so that the output of the $i$th correlator is

$$U_i \triangleq \int_0^T r(t)S_i(t)\, dt = \sum_{k=1}^n (S_{ik}^2 + J_k S_{ik}). \tag{6}$$

Hence,

$$E(U_i \mid S_i) = \sum_{k=1}^n \overline{S_{ik}^2} = E_s \tag{7}$$

since the second term averages to zero. Then, since the signals are equiprobable,

$$E(U_i) = \frac{E_s}{D}. \tag{8}$$

Similarly, using (1) and (2),

$$\mathrm{var}\,(U_i \mid S_i) = \sum_{k,l} J_k J_l \overline{S_{ik} S_{il}}$$

$$= \sum_{k=1}^n J_k^2 \overline{S_{ik}^2}$$

$$= \frac{E_s}{n} E_J \tag{9}$$

and

$$\mathrm{var}(U_i \mid S_m) = \frac{E_s}{n} E_j + \frac{E_s^2}{n}, \quad m \neq i. \tag{10}$$

A measure of performance is the signal-to-noise ratio defined as

$$\mathrm{SNR} \triangleq \sum_{m=1}^D \frac{E^2(U_i \mid S_m)}{\mathrm{var}\,(U_i \mid S_m)} P(S_m) = \frac{E_s}{E_J} \frac{n}{D}. \tag{11}$$

This result is *independent of the way that the jammer distributes his energy*, i.e., regardless of how $J_k$ is chosen subject to the constraint that $\Sigma_k J_k^2 = E_J$, the postprocessing SNR (11) gives the signal an advantage of $n/D$ over the jammer. This factor $n/D$ is the *processing gain* and it is exactly equal to the ratio of the dimensionality of the *possible* signal space (and therefore the space in which the jammer must seek to operate) to the dimensions needed to actually transmit the signals. Using the result that the (approximate) dimensionality of a signal of duration $T$ and of approximate bandwidth $B_D$ is $2B_D T$, we see the processing gain can be written as

$$G_P = \frac{n}{D} \cong \frac{2B_{ss}T}{2B_D T} = \frac{B_{ss}}{B_D} \tag{12}$$

where $B_{ss}$ is the bandwidth in hertz of the (spread-spectrum)

signals $S_i(t)$ and $B_D$ is the minimum bandwidth that would be required to send the information if we did not need to imbed it in the larger bandwidth for protection.

A simple illustration of these ideas using random binary sequences will be used to bring out some of these points. Consider the transmission of a single bit $\pm\sqrt{E_b/T}$ with energy $E_b$ of duration $T$ seconds. This signal is one-dimensional. As shown in Figs. 1 and 2, the transmitter multiplies the data bit $d(t)$ by a binary $\pm 1$ "chipping" sequence $p(t)$ chosen randomly at rate $f_c$ chips/s for a total of $f_c T$ chips/bit. The dimensionality of the signal $d(t)p(t)$ is then $n = f_c T$. The received signal is

$$r(t) = d(t)p(t) + J(t), \quad 0 \leqslant t \leqslant T, \tag{13}$$

ignoring, for the time being, thermal noise.

The receiver, as shown in Fig. 1, performs the correlation

$$U \triangleq \sqrt{\frac{E_b}{T}} \int_0^T r(t)p(t)\, dt \tag{14}$$

and makes a decision as to whether $\pm\sqrt{E_b/T}$ was sent depending upon $U \gtrless 0$. The integrand can be expanded as

$$r(t)p(t) = d(t)p^2(t) + J(t)p(t) = d(t) + J(t)p(t), \tag{15}$$

and hence the data bit appears in the presence of a code-modulated jammer.

If, for example, $J(t)$ is additive white Gaussian noise with power spectral density $\eta_{0J}/2$ (two-sided), so is $J(t)\, p(t)$, and $U$ is then a Gaussian random variable. Since $d(t) = \pm\sqrt{E_b/T}$, the conditional mean and variance of $U$, assuming that $\pm\sqrt{E_b/T}$ is transmitted, is given by $E_b$ and $E_b(\eta_{0J}/2)$, respectively, and the probability of error is [3] $Q(\sqrt{2E_b/\eta_{0J}})$ where $Q(x) \triangleq \int_x^\infty (1/\sqrt{2\pi})e^{-y^2/2}\, dy$. Against white noise of unlimited power, spread spectrum serves no useful purpose, and the probability of error is $Q(\sqrt{2E_b/\eta_{0J}})$ regardless of the modulation by the code sequence. White noise occupies all dimensions with power $\eta_{0J}/2$. The situation is different, however, if the jammer power is limited. Then, not having access to the random sequence $p(t)$, the jammer with available energy $E_J$ (power $E_J/T$) can do better than to apply this energy to one dimension. For example, if $J(t) = \sqrt{E_J/T}$, $0 \leqslant t \leqslant T$, then the receiver output is

$$U = E_b + \sqrt{E_b E_J}\, \frac{1}{n} \sum_{i=1}^n X_i \tag{16}$$

where the $X_i$'s are i.i.d.[1] random variables with $P(X_i = +1) = P(X_i = -1) = \frac{1}{2}$. The signal-to-noise ratio (SNR) is

$$\frac{E^2(U)}{\mathrm{var}\,(U)} = \frac{E_b}{E_J} n. \tag{17}$$

Thus, the SNR may be increased by increasing $n$, the process-

[1] Independent identically distributed.

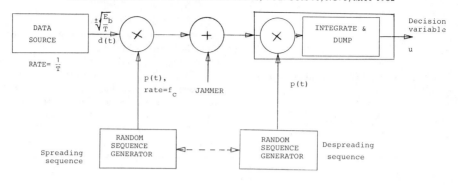

TRANSMITTER                          RECEIVER

Fig. 1.   Direct-sequence spread-spectrum system for transmitting a
single binary digit (baseband).

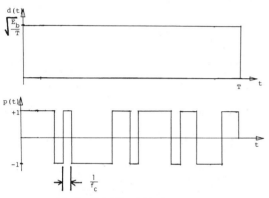

Fig. 2.   Data bit and chipping sequence.

ing gain, and it has the form of (11). As a further indication of this parameter, we may compute the probability $P_e$ that the bit is in error from (16). Assuming that a "minus" is transmitted, we have

$$P_e = P(U > 0)$$

$$= P(Z_n > \alpha n)$$

$$= \begin{cases} \dfrac{1}{2^n} \displaystyle\sum_{[\alpha n]}^{n} \binom{n}{k}; & \dfrac{E_b}{E_J} < 1 \\[4mm] 0; & \dfrac{E_b}{E_J} \geqslant 1 \end{cases} \qquad (18)$$

where

$Z_n \triangleq \dfrac{1}{2} \displaystyle\sum_{i=1}^{n} (1 + X_i)$ is a Bernoulli random variable with

mean $\dfrac{n}{2}$ and variance $\dfrac{n}{4}$,

$\alpha \triangleq \dfrac{1}{2}\left(1 + \sqrt{\dfrac{E_b}{E_J}}\right),$

and $[X]$ is defined as the integer portion of $X$. The partial binomial sum on the right-hand side of (18) may be upper bounded [2] by

$$P_e \leqslant \frac{1}{2^n}\left(\frac{1}{1-\alpha}\right)^n\left(\frac{1-\alpha}{\alpha}\right)^{\alpha n}; \qquad \tfrac{1}{2} < \alpha \leqslant 1$$

or

$$P_e \leqslant 2^{-n[1-H(\alpha)]}; \qquad \tfrac{1}{2} < \alpha \leqslant 1 \qquad (19)$$

where $H(\alpha) \triangleq -\alpha \log_2 \alpha - (1-\alpha)\log_2 (1-\alpha)$ is the binary entropy function. Therefore, for any $\alpha > \tfrac{1}{2}$ (or $E_b \neq 0$), $P_e$ may be made vanishingly small by increasing $n$, the processing gain. (The same result is valid even if the jammer uses a chip pattern other than the constant, all-ones used in the example above.) As an example, if $E_J = 9E_b$ (jammer energy 9.5 dB larger than that of the data), then $\alpha = 2/3$ and $P_e \leqslant 2^{-0.085n}$. If $n = 200$ (23 dB processing gain), $P_e < 7.6 \times 10^{-6}$.

An approximation to the same result may be obtained by utilizing a central limit type of argument that says, for large $n$, $U$ in (16) may be treated as if it were Gaussian. Then

$$P_e = P(U < 0) \cong Q\left(\sqrt{\frac{E_b}{E_J}}\, n\right) \qquad (20)$$

and, if $E_b/E_J = -9.5$ dB and $n = 200$ (23 dB), $P_e \cong Q(\sqrt{22}) \simeq 1.5 \times 10^{-6}$. The processing gain can be seen to be a multiplier of the "signal-to-jamming" ratio $E_b/E_J$.

A more traditional way of describing the processing gain, which brings in the relative bandwidth of the data signal and that of the spread-spectrum modulation, is to examine the power spectrum of an *infinite sequence* of data, modulated by the rapidly varying random sequence. The spectrum of the random data sequence with rate $R = 1/T$ bits/s is given by

$$S_D(f) = T\left(\frac{\sin \pi f T}{\pi f T}\right)^2 \qquad (21)$$

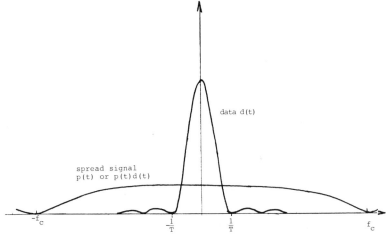

Fig. 3. Power spectrum of data and of spread signal.

and that of the spreading sequence [and also that of the product $d(t) p(t)$] is given by

$$S_{ss}(f) = \frac{1}{f_c} \left( \frac{\sin \pi f/f_c}{\pi f/f_c} \right)^2. \tag{22}$$

Both are sketched in Fig. 3. It is clear that if the receiver multiplies the received signal $d(t)p(t) + J(t)$ by $p(t)$ giving $d(t) + J(t)p(t)$, the first term may be extracted virtually intact with a filter of bandwidth $1/T \triangleq B_D$ Hz. The second term will be spread over at least $f_c$ Hz as shown in Fig. 3. The fraction of power due to the jammer which can pass through the filter is then roughly $1/f_cT$. Thus, the data have a power advantage of $n = f_cT$, the processing gain. As in (12), the processing gain is frequently expressed as the ratio of the bandwidth of the spread-spectrum waveform to that of the data, i.e.,

$$G_p \triangleq \frac{B_{ss}}{B_D} = f_c T = n. \tag{23}$$

The notion of processing gain as expressed in (23) is simply a power improvement factor which a receiver, possessing a replica of the spreading signal, can achieve by a correlation operation. It must not be automatically extrapolated to anything else. For example, if we use frequency hopping for spread spectrum employing one of $N$ frequencies every $T_H$ seconds, the total bandwidth must be approximately $N/T_H$ (since keeping the frequencies orthogonal requires frequency spacing $\approx 1/T_H$). Then, according to (12), $G_p = (N/T_H)/B_D$. Now if we transmit 1 bit/hop, $T_HB_D \approx 1$ and $G_P = N$, the number of frequencies used. If $N = 100$, $G_p = 20$ dB, which seems fairly good. But a single spot frequency jammer can cause an average error rate of about $10^{-2}$, which is not acceptable. (A more detailed analysis follows in Section IV below.) This effectiveness of "partial band jamming" can be reduced by the use of coding and interleaving. Coding typically precludes the possibility of a small number of fre-

quency slots (e.g., one slot) being jammed causing an unacceptable error rate (i.e., even if the jammer wipes out a few of the code symbols, depending upon the error-correction capability of the code, the data may still be recovered). Interleaving has the effect of randomizing the errors due to the jammer. Finally, an analogous situation occurs in direct sequence spreading when a pulse jammer is present.

In the design of a practical system, the processing gain $G_p$ is not, by itself, a measure of how well the system is capable of performing in a jamming environment. For this purpose, we usually introduce the *jamming margin* in decibels defined as

$$M_J = G_P - \left( \frac{E_b}{\eta_{0J}} \right)_{\min} - L. \tag{24}$$

This is the residual advantage that the system has against a jammer after we subtract both the minimum required energy/bit-to-jamming "noise" power spectral density ratio $(E_b/\eta_{0J})_{\min}$ and implementation and other losses $L$. The jamming margin can be increased by reducing the $(E_b/\eta_{0J})_{\min}$ through the use of coding gain.

We conclude this section by showing that regardless of the technique used, spectral spreading provides protection against a broad-band jammer with a finite power $P_J$. Consider a system that transmits $R_0$ bits/s designed to operate over a bandwidth $B_{ss}$ Hz in white noise with power density $\eta_0$ W/Hz. For any bit rate $R$,

$$\left( \frac{E_b}{\eta_0} \right)_{\text{actual}} = \frac{P_s}{\eta_0 R} = \frac{P_s}{P_N} \frac{B_{ss}}{R} \tag{25}$$

where

$$P_s \triangleq E_b R = \text{signal power}$$

$$P_N \triangleq \eta_0 B_{ss} = \text{noise power}.$$

Then for a specified $(E_b/\eta_0)_{\min}$ necessary to achieve mini-

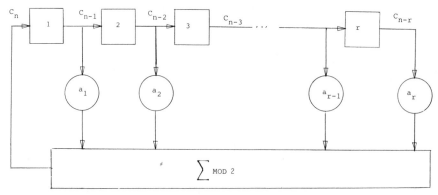

Fig. 4. Simple shift register generator (SSRG).

mum acceptable performance,

$$R \leqslant \frac{P_s}{P_N} \frac{B_{ss}}{(E_b/\eta_0)_{\min}} \triangleq R_0. \tag{26}$$

If a jammer with power $P_J$ now appears, and if we are already transmitting at the maximum rate $R_0$, then (25) becomes

$$\left(\frac{E_b}{\eta_0}\right)_{\text{actual}} = \frac{P_s}{P_N + P_J} \frac{B_{ss}}{R_0}$$

$$= \left(\frac{E_b}{\eta_0}\right)_{\min} \frac{P_N}{P_N + P_J}$$

or

$$\left(\frac{E_b}{\eta_0}\right)_{\text{actual}} = \left(\frac{E_b}{\eta_0}\right)_{\min} \frac{\eta_0}{\eta_0 + P_J/B_{ss}}. \tag{27}$$

Thus, if we wish to recover from the effects of the jammer, the right-hand side of (27) should be not much less than $(E_b/\eta_0)_{\min}$. This clearly requires that we increase $B_{ss}$, since for any finite $P_J$, it is then possible to make the factor $\eta_0/(\eta_0 + P_J/B_{ss})$ approach unity, and thereby retain the performance we had before the jammer appeared.

### III. PSEUDORANDOM SEQUENCE GENERATORS

In Section II, we examined how a purely random sequence can be used to spread the signal spectrum. Unfortunately, in order to despread the signal, the receiver needs a replica of the transmitted sequence (in almost perfect time synchronism). In practice, therefore, we generate pseudorandom or pseudonoise (PN) sequences so that the following properties are satisfied. They

1) are easy to generate
2) have randomness properties
3) have long periods
4) are difficult to reconstruct from a short segment.

Linear feedback shift register (LFSR) sequences [4] possess

properties 1) and 3), most of property 2), but not property 4). One canonical form of a binary LFSR known as a simple shift register generator (SSRG) is shown in Fig. 4. The shift register consists of binary storage elements (boxes) which transfer their contents to the right after each clock pulse (not shown). The contents of the register are linearly combined with the binary (0, 1) coefficients $a_k$ and are fed back to the first stage. The binary (code) sequence $C_n$ then clearly satisfies the recursion

$$C_n = \sum_{k=1}^{r} a_k C_{n-k} \ (\text{mod } 2); \qquad a_r = 1. \tag{28}$$

The periodic cycle of the states depends on the initial state and on the coefficients (feedback taps) $a_k$. For example, the four-stage LFSR generator shown in Fig. 5 has four possible cycles as shown. The all-zeros is always a cycle for any LFSR. For spread spectrum, we are looking for *maximal length* cycles, that is, cycles of period $2^r - 1$ (all binary $r$-tuples except all-zeros). An example is shown for a four-state register in Fig. 6. The sequence output is $1\,0\,0\,0\,1\,1\,1\,1\,0\,1\,0\,1\,1\,0\,0\,\cdots$ (period $2^4 - 1 = 15$) if the initial contents of the register (from right to left) are 1000. It is always possible to choose the feedback coefficients so as to achieve maximal length, as will be discussed below.

If we do have a maximal length sequence, then this sequence will have the following pseudorandomness properties [4].

1) There is an approximate balance of zeros and ones ($2^{r-1}$ ones and $2^{r-1} - 1$ zeros).

2) In any period, half of the runs of consecutive zeros or ones are of length one, one-fourth are of length two, one-eighth are of length three, etc.

3) If we define the $\pm1$ sequence $C_n' = 1 - 2C_n$, $C_n = 0$, 1, then the autocorrelation function $R_{C'}(\tau) \triangleq 1/L$ $\sum_{k=1}^{L} C_k' C_{k+\tau}'$ is given by

$$R_{C'}(\tau) = \begin{cases} 1, & \tau = 0, L, 2L \cdots \\ -\dfrac{1}{L}, & \text{otherwise} \end{cases} \tag{29}$$

where $L = 2^r - 1$. If the code *waveform* $p(t)$ is the "square-wave" equivalent of the sequences $C_n'$, if $L \gg 1$, and if we

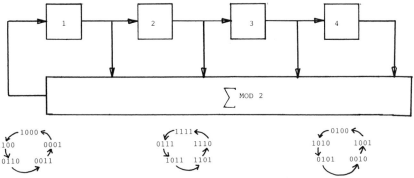

Fig. 5. Four-stage LFSR and its state cycles.

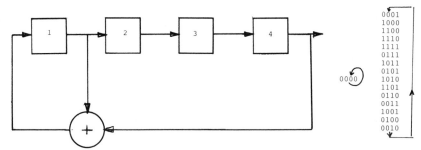

Fig. 6. Four-stage maximal length LFSR and its state cycles.

define

$$q(\tau) \triangleq \begin{cases} 1 - |\tau| f_c; & |\tau| \leqslant \dfrac{1}{f_c} \\ 0; & \text{otherwise} \end{cases}$$

then

$$R_p(\tau) \simeq \sum_i q\left(\tau - \frac{iL}{f_c}\right). \tag{30}$$

Equation (29), and therefore (30), follow directly from the "shift-and-add" property of maximal length (ML) LFSR sequences. This property is that the chip-by-chip sum of an MLLFSR sequence $C_k$ and any shift of itself $C_{k+\tau}$, $\tau \neq 0$ is the *same* sequence (except for some shift). This follows directly from (28), since

$$(C_n + C_{n+\tau}) = \sum_{k=1}^{L} a_k (C_{n-k} + C_{n+\tau-k}) \,(\text{mod } 2). \tag{31}$$

The shift-and-add sequence $C_n + C_{n+\tau}$ is seen to satisfy the same recursion as $C_n$, and if the coefficients $a_k$ yield maximal length, then it must be the same sequence regardless of the initial (nonzero) state. The autocorrelation property (29) then follows from the following isomorphism:

$$(\{0, 1\}, +) \leftrightarrow (\{1, -1\}, \times).$$

Therefore,

$$C_k + C_{k+\tau} \leftrightarrow C_k{}' C_{k+\tau}{}'$$

and if $C_k{}'$ is an MLLFSR $\pm 1$ sequence, so is $C_k{}' C_{k+\tau}{}', \tau \neq 0$. Thus, there are $2^{r-1}$ 1's and $(2^{r-1} - 1)$ −1's in the product and (29) follows. The autocorrelation function is shown in Fig. 7(a).

Property 3) is a most important one for spread spectrum since the autocorrelation function of the code sequence *waveform* $p(t)$ determines the spectrum. Note that because $p(t)$ is *pseudorandom*, it is periodic with period $(2^r - 1) \cdot 1/f_c$, and hence so is $R_p(\tau)$. The spectrum shown in Fig. 7(b) is therefore the line spectrum

$$S_p(f) = \left[ \sum_{\substack{m=-\infty \\ m \neq 0}}^{\infty} \delta(f - mf_0) \right] \frac{L+1}{L^2} \left( \frac{\sin \pi f / f_c}{\pi f / f_c} \right)^2$$

$$+ \frac{1}{L^2} \delta(f) \tag{32}$$

where

$$f_0 = \frac{f_c}{2^r - 1}.$$

If $L = 2^r - 1$ is very large, the spectral lines get closer together, and for practical purposes, the spectrum may be viewed as being continuous and similar to that of a purely random binary waveform as shown in Fig. 3. A different, but commonly used implementation of a linear feedback shift register is the modular shift register generator (MSRG) shown in Fig. 8. Additional details on the properties of linear feedback shift registers are provided in the Appendix.

63

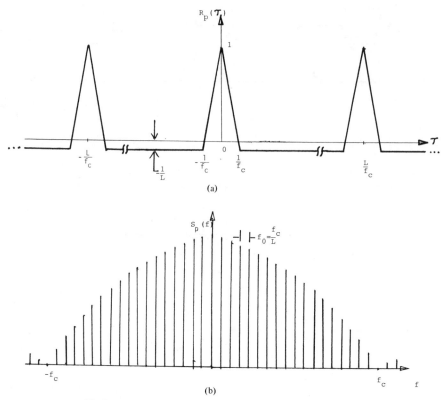

Fig. 7. Autocorrelation function $R_p(\tau)$ and power spectral density of MLLFSR sequence waveform $p(t)$. (a) Autocorrelation function of $p(t)$. (b) Power spectral density of $p(t)$.

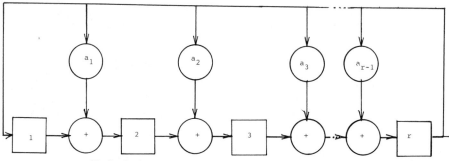

Fig. 8. Implementation as a modular shift register generator (MSRG).

For spread spectrum and other secure communications (cryptography) where one expects an adversary to attempt to recover the code in order to penetrate the system, property 4) cited in the beginning of this section is extremely important. Unfortunately, LFSR sequences do not possess that property. Indeed, using the recursion (28) or (A8) and observing only $2r - 1$ consecutive bits in the sequence $C_n$ allows us to solve for the $r - 1$ middle coefficients and the $r$ initial bits in the register by linear simultaneous equations. Thus, even if $r = 100$ so that the length of the sequence is $2^{100} - 1 \simeq 10^{30}$, we would be able to construct the entire sequence from 199 bits by solving 198 linear equations

(mod 2), which is neither difficult nor that time consuming for a large computer. Moreover, because the sequence $C_n$ satisfies a recursion, a very efficient algorithm is known [7], [8] which solves the equations or which equivalently synthesizes the shortest LFSR which generates a given sequence.

In order to avoid this pitfall, several modifications of the LFSR have been proposed. In Fig. 9(a) the feedback function is replaced by an arbitrary Boolean function of the contents of the register. The Boolean function may be implemented by ROM or random logic, and there are an enormous number of these functions ($2^{2^r}$). Unfortunately, very little is known [4] in the open literature about the properties of such non-

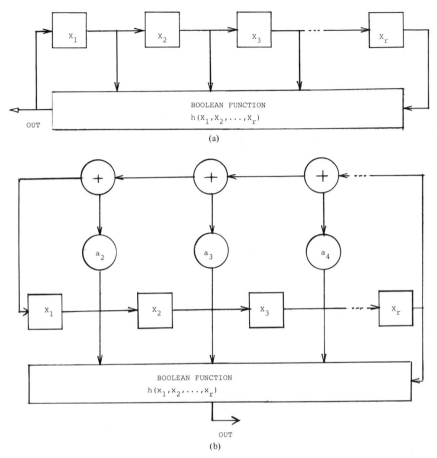

Fig. 9. Nonlinear feedback shift registers. (a) Nonlinear FDBK. Number of Boolean functions = $2^{2^r}$. (b) Linear FSR, nonlinear function of state, i.e., nonlinear output logic (NOL).

linear feedback shift registers. Furthermore, some nonlinear FSR's may have no cycles or length > 1 (e.g., they may have only a *transient* that "homes" towards the all-ones state after any initial state). Are there feedback functions that generate only *one* cycle of length $2^r$? The answer is yes, and there are exactly $2^{2^{r-1}-r}$ of them [9]. How do we find them? Better yet, how do we find a subset of them with all the "good" randomness properties? These are, and have been, good research problems for quite some time, and unfortunately no general theory on this topic currently exists.

A second, more manageable approach is to use an MLLFSR with nonlinear output logic (NOL) as shown in Fig. 9(b). Some clues about designing the NOL while still retaining "good" randomness properties are available [10]-[12], and a measure for judging how well condition 4) is fulfilled is to ask: What is the degree of the shortest LFSR that would generate the same sequence? A simple example of an LFSR with NOL having three stages is shown in Fig. 10(a). The shortest LFSR which generates the same sequence (of period 7) is shown in Fig. 10(b) and requires six stages.

When using PN sequences in spread-spectrum systems, several additonal requirements must be met.

1) The "partial correlation" of the sequence $C_n'$ over a window $w$ smaller than the full period should be as small as possible, i.e., if

$$\rho(w; j, \tau) \triangleq \sum_{n=j}^{j+w-1} C_n' C_{n+\tau}',$$

$$\rho(w) = \max_{j, \tau} |\rho(w; j, \tau)| \qquad (33)$$

should be $\ll L = 2^r - 1$.

2) Different code pairs should have uniformly low cross correlation, i.e.,

$$R_{C'C''}(\tau) \triangleq \frac{1}{L} \sum_{k=1}^{L} C_k' C_{k+\tau}'' \qquad (34)$$

should be $\ll 1$ for all values of $\tau$.

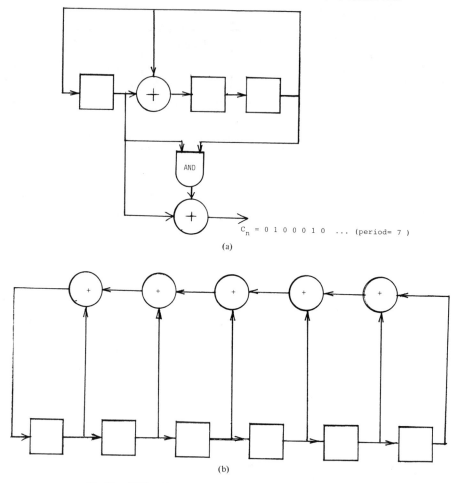

$C_n = 0\ 1\ 0\ 0\ 0\ 1\ 0\ \ldots$ (period= 7 )

(a)

(b)

Fig. 10.    LFSR with NOL and its shortest linear equivalent. (a) Three-stage LFSR with NOL. (b) LFSR with $f(x) = 1 + x + x^2 + x^3 + x^4 + x^5 + x^6$ which generates the same sequence as that of (a) under the initial state 1 0 0 0 1 0.

3)   Since the code sequences are periodic with period $L$, there are two correlation functions (depending on the relative polarity of one of the sequences in the transition over an initial point $\tau$ on the other). If we define the finite-cross-correlation function [13] as

$$f_{C'C''}(\tau) \triangleq \frac{1}{L} \sum_{k=1}^{\tau} C_k{}' C_{k+\tau}{}'', \tag{35}$$

then the so-called even and odd cross-correlation functions are, respectively,

$$R_{C'C''}{}^{(e)}(\tau) = f_{C'C''}(\tau) + f_{C'C''}(L - \tau)$$

and

$$R_{C'C''}{}^{(o)}(\tau) = f_{C'C''}(\tau) - f_{C'C''}(L - \tau)$$

and we want

$$\max_{\tau} |R_{C'C''}{}^{e}(\tau)| \quad \text{and} \quad \max_{\tau} |R_{C'C''}{}^{(o)}(\tau)|$$

to be $\ll 1$.

The reason for 1) is to keep the "self noise" of the system as low as possible since, in practice, the period is very long compared to the integration time per symbol and there will be fluctuation in the sum of any filtered (weighted) subsequence. This is especially worrisome during acquisition where these fluctuations can cause false locking. Bounds on $\rho(w)$ [14] and averages over $j$ of $\rho(w; j, \tau)$ are available in the literature.

Properties 2) and 3) are both of direct interest when using PN sequences for code division multiple access (CDMA) as will be discussed in Section V below. This is to ensure minimal cross interference between any pair of users of the common spectrum. The most commonly used collection of sequences which exhibit property 2) are the Gold codes [15]. These are sequences derived from an MLLFSR, but are *not* of maximal length. A detailed procedure for their construction is given in the Appendix.

Virtually all of the known results about the cross-correlation properties of useful PN sequences are summarized in [16].

As a final comment on the generation of PN sequences for spread spectrum, it is not at all necessary that feedback shift registers be used. *Any* technique which can generate "good" pseudorandom sequences will do. Other techniques are described in [4], [16], [17], for example. Indeed, the generation of good pseudorandom sequences is fundamental to other fields, and in particular, to cryptography [18]. A "good" cryptographic system can be used to generate "good" PN sequences, and vice versa. A possible problem is that the specific additional "good" properties required for an operational spread-spectrum system may not always match those required for secure cryptographic communications.

## IV. ANTIJAM CONSIDERATIONS

Probably the single most important application of spread-spectrum techniques is that of resistance to intentional interference or jamming. Both direct-sequence (DS) and frequency-hopping (FH) systems exhibit this tolerance to jamming, although one might perform better than the other given a specific type of jammer.

The two most common types of jamming signals analyzed are single frequency sine waves (tones) and broad-band noise. References [19] and [20] provide performance analyses of DS systems operating in the presence of both tone and noise interference, and [21]–[26] provide analogous results for FH systems.

The simplest case to analyze is that of broad-band noise jamming. If the jamming signal is modeled as a zero-mean wide sense stationary Gaussian noise process with a flat power spectral density over the bandwidth of interest, then for a given fixed power $P_J$ available to the jamming signal, the power spectral density of the jamming signal must be reduced as the bandwidth that the jammer occupies is increased.

For a DS system, if we assume that the jamming signal occupies the total RF bandwidth, typically taken to be twice the chip rate, then the despread jammer will occupy an even greater bandwidth and will appear to the final integrate-and-dump detection filter as approximately a white noise process. If, for example, binary PSK is used as the modulation format, then the average probability of error will be approximately given by

$$P_e = Q\left(\sqrt{\frac{2E_b}{\eta_0 + \eta_{0J}}}\right). \qquad (36)$$

Equation (36) is just the classical result for the performance of a coherent binary communication system in additive white Gaussian noise. It differs from the conventional result because an extra term in the denominator of the argument of the $Q(\cdot)$ function has been added to account for the jammer. If $P_e$ from (36) is plotted versus $E_b/\eta_0$ for a given value of $P_J/P_s$, where $P_s$ is the average signal power, curves such as the ones shown in Fig. 11 result.

Expressions similar to (36) are easily derived for other modulation formats (e.g., QPSK), and curves showing the performance for several different formats are presented, for example, in [19]. The interesting thing to note about Fig. 11 is that for a given $\eta_{0J}$, the curve "bottoms out" as $E_b/\eta_0$ gets larger and larger. That is, the presence of the jammer will cause an irreducible error rate for a given $P_J$ and a given $f_c$. Keeping $P_J$ fixed, the only way to reduce the error rate is to increase $f_c$ (i.e., increase the amount of spreading in the system). This was also noted at the end of Section II.

For FH systems, it is not always advantageous for a noise jammer to jam the entire RF bandwidth. That is, for a given $P_J$, the jammer can often increase its effectiveness by jamming only a fraction of the total bandwidth. This is termed *partial-band jamming*. If it is assumed that the jammer divides its power uniformly among $K$ slots, where a slot is the region in frequency that the FH signal occupies on one of its hops, and if there is a total of $N$ slots over which the signal can hop, we have the following possible situations. Assuming that the underlying modulation format is binary FSK (with noncoherent detection at the receiver), and using the terminology MARK and SPACE to represent the two binary data symbols, on any given hop, if

1) $K = 1$, the jammer might jam the MARK only, jam the SPACE only, or jam neither the MARK nor the SPACE;

2) $1 < K < N$, the jammer might jam the MARK only, jam the SPACE only, jam neither the MARK nor the SPACE, or jam both the MARK and the SPACE;

3) $K = N$, the jammer will always jam both the MARK and the SPACE.

To determine the average probability of error of this system, each of the possibilities alluded to above has to be accounted for. If it is assumed that the $N$ slots are disjoint in frequency and that the MARK and SPACE tones are orthogonal (i.e., if a MARK is transmitted, it produces no output from the SPACE bandpass filter (BPF) and vice versa), then the average probability of error of the system can be shown to be given by [23], [24]

$$P_e = \frac{(N-K)(N-K-1)}{N(N-1)} \frac{1}{2} \exp\left(-\tfrac{1}{2}\,\text{SNR}\right)$$

$$+ \frac{K(N-K)}{N(N-1)} \exp\left[-\frac{1}{\dfrac{2}{\text{SNR}} + \dfrac{1}{\text{SJR}}}\right] + \frac{K(K-1)}{N(N-1)} \frac{1}{2}$$

$$\cdot \exp\left[-\frac{1}{2}\frac{1}{\dfrac{1}{\text{SNR}} + \dfrac{1}{\text{SJR}}}\right] \qquad (37)$$

where SNR is the ratio of signal power to thermal noise power at the output of the MARK BPF (assuming that a MARK has been transmitted) and SJR is the ratio of signal

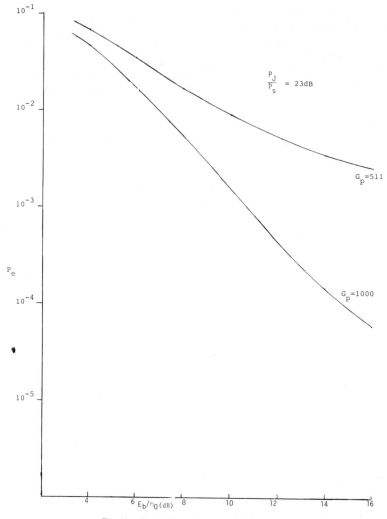

Fig. 11.   Probability of error versus $E_b/\eta_0$.

power to *jammer power per slot* at the output of the MARK BPF. By jammer power per slot, we mean the total jammer power divided by the number of slots being jammed (i.e., SJR $= P_s/(P_J/K)$).

The coefficients in front of the exponentials in (37) are the probabilities of jamming neither the MARK nor the SPACE, jamming only the MARK or only the SPACE, or jamming both the MARK and the SPACE. For example, the probability of jamming both the MARK and the SPACE is given by $K(K-1)/N(N-1)$. In Fig. 12, the $P_e$ predicted by (37) is plotted versus SNR for $K = 1$ and $K = 100$ for a $P_J/P_s$ of 10 dB. These two curves are labeled "uncoded" on the figure.

Often, a somewhat different model from that used in deriving (37) is considered. This latter model is used in [26], and effectively assumes that either MARK and SPACE are simultaneously jammed or that neither of the two is jammed. For this case, a parameter $\rho$, where $0 < \rho \leqslant 1$, representing the fraction of the band being jammed, is defined. The

resulting average probability of error is then maximized with respect to $\rho$ (i.e., the worst case $\rho$ is found), and it is shown in [26] that

$$P_{e\max} > \frac{e^{-1}}{E_b/\eta_0}$$

where $e$ is the base of the natural logarithm. It can be seen that partial band jamming affords the jammer a strategy whereby he can degrade the performance significantly (i.e., $P_e$ can be forced to be inversely proportional to $E_b/\eta_0$ rather than exponential).

For tone jamming, the situation becomes somewhat more complicated than it is for noise jamming, especially for DS systems. This is because the system performance depends upon the location of the tone (or tones), and upon whether the period of the spreading sequence is equal to or greater than the duration of a data symbol. Oftentimes the effect

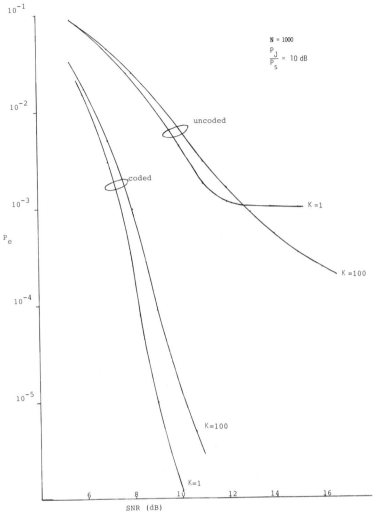

Fig. 12. Probability of error versus SNR.

of a despread tone is approximated as having arisen from an equivalent amount of Gaussian noise. In this case, the results presented above would be appropriate. However, the Gaussian approximation is not always justified, and some conditions for its usage are given in [20] and [27].

The situation is simpler in FH systems operating in the presence of partial-band tone jamming, and as shown, for example, in [24], the performance of a noncoherent FH–FSK system in partial-band tone jamming is often virtually the same as the performance in partial-band noise jamming. One important consideration in FH systems with either noise or tone jamming is the need for error-correction coding. This can be seen very simply by assuming that the jammer is much stronger than the desired signal, and that it chooses to put all of its power in a single slot (i.e., the jammer jams one out of $N$ slots). The $K = 1$ uncoded curve of Fig. 12 corresponds to this situation. Then with no error-correction coding, the system will make an error (with high probability)

every time it hops to a MARK frequency when the corresponding SPACE frequency is being jammed or vice versa. This will happen on the average one out of every $N$ hops, so that the probability of error of the system will be approximately $1/N$, independent of signal-to-noise ratio. This is readily seen to be the case in Fig. 12. The use of coding prevents a simple error as caused by a spot jammer from degrading the system performance. To illustrate this point, an error-correcting code (specifically a Golay code [2]) was used in conjunction with the system whose uncoded performance is shown in Fig. 12, and the performance of the coded system is also shown in Fig. 12. The advantage of using error-correction coding is obvious from comparing the corresponding curves.

Finally, there are, of course, many other types of common jamming signals besides broad-band noise or single frequency tones. These include swept-frequency jammers, pulse-burst jammers, and repeat jammers. No further discussion of these

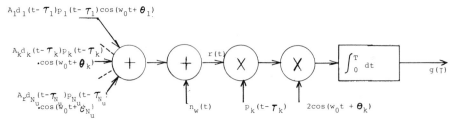

Fig. 13.   DS CDMA system.

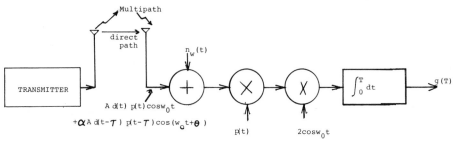

Fig. 14.   DS used to combat multipath.

jammers will be presented in this paper, but references such as [28]-[30] provide a reasonable description of how these jammers affect system performance.

## V. CODE DIVISION MULTIPLE ACCESS (CDMA)

As is well known, the two most common multiple access techniques are frequency division multiple access (FDMA) and time division multiple access (TDMA). In FDMA, all users transmit simultaneously, but use disjoint frequency bands. In TDMA, all users occupy the same RF bandwidth, but transmit sequentially in time. When users are allowed to transmit simultaneously in time and occupy the same RF bandwidth as well, some other means of separating the signals at the receiver must be available, and CDMA [also termed spread-spectrum multiple access (SSMA)] provides this necessary capability.

In DS CDMA [31]-[33], each user is given its own code, which is approximately orthogonal (i.e., has low cross correlation) with the codes of the other users. However, because CDMA systems typically are asynchronous (i.e., the transition times of the data symbols of the different users do not have to coincide), the design problem is much more complicated than that of having, say, $N_u$ spreading sequences with uniformly low cross correlations such as the Gold codes discussed in Section III and in the Appendix. As will be seen below, the key parameters in a DS CDMA system are both the cross-correlation and the partial-correlation functions, and the design and optimization of code sets with good partial-correlation properties can be found in many references such as [16], [34], and [35].

The system is shown in Fig. 13. The received signal is given by

$$r(t) = \sum_{i=1}^{N_u} A_i d_i(t - \tau_i) p_i(t - \tau_i) \cos(\omega_0 t + \theta_i) + n_w(t) \quad (38)$$

where

$d_i(t)$ = message of $i$th user and equals $\pm 1$
$p_i(t)$ = spreading sequence waveform of $i$th user
$A_i$ = amplitude of $i$th carrier
$\theta_i$ = random phase of $i$th carrier uniformly distributed in $[0, 2\pi]$
$\tau_i$ = random time delay of $i$th user uniformly distributed in $[0, T]$
$T$ = symbol duration
$n_w(t)$ = additive white Gaussian noise.

Assuming that the receiver is correctly synchronized to the $k$th signal, we can set both $\tau_k$ and $\theta_k$ to zero without losing any generality. The final test statistic out of the integrate-and-dump receiver of Fig. 14 is given by

$$g(T) = A_k + \frac{1}{T} \sum_{\substack{i=1 \\ i \neq k}}^{N_u} A_i \int_0^T d_i(t - \tau_i)$$

$$\cdot p_i(t - \tau_i) p_k(t) \cos(\theta_i) dt$$

$$+ \frac{2}{T} \int_0^T n_w(t) p_k(t) \cos(\omega_0 t) dt \quad (39)$$

where double frequency terms have been ignored.

Consider the second term on the RHS of (39). It is a sum of $N_u - 1$ terms of the form

$$A_i \cos(\theta_i) \int_0^T d_i(t - \tau_i) p_i(t - \tau_i) p_k(t) dt.$$

Notice that, because the $i$th signal is not, in general, in sync with the $k$th signal, $d_i(t - \tau_i)$ will change signs somewhere in the interval $[0, T]$ 50 percent of the time. Hence, the above integral will be the sum of two partial correlations of $p_i(t)$ and $p_k(t)$, rather than one total cross correlation. Therefore, (39) can be rewritten

$$g(T) = A_k + \sum_{\substack{i=1 \\ i \neq k}}^{N_u} A_i [\pm \rho_{ik}(\tau_i) \pm \hat{\rho}_{ik}(\tau_i)] \cos(\theta_i) + n(T) \quad (40)$$

where

$$\rho_{ik}(\tau_i) \triangleq \frac{1}{T} \int_0^{\tau_i} p_i(t - \tau_i) p_k(t) \, dt$$

$$\hat{\rho}_{ik}(\tau_i) \triangleq \frac{1}{T} \int_{\tau_i}^{T} p_i(t - \tau_i) p_k(t) \, dt$$

and

$$n(T) \triangleq \frac{2}{T} \int_0^T n_w(t) p_k(t) \cos \omega_0 t \, dt.$$

Notice that the coefficients in front of $\rho_{ik}(\tau_i)$ and $\hat{\rho}_{ik}(\tau_i)$ can independently have a plus or minus sign due to the data sequence of the $i$th signal. Also notice that $\rho_{ik}(\tau_i) + \hat{\rho}_{ik}(\tau_i)$ is the total cross correlation between the $i$th and $k$th spreading sequences. Finally, the continuous correlation functions $\rho_{ik}(\tau) \pm \hat{\rho}_{ik}(\tau)$ can be expressed in terms of the discrete even and odd cross-correlation functions, respectively, that were defined in Section III.

While the code design problem in CDMA is very crucial in determining system performance, of potentially greater importance in DS CDMA is the so-called "near-far problem." Since the $N_u$ users are typically geographically separated, a receiver trying to detect the $k$th signal might be much closer physically to, say, the $i$th transmitter rather than the $k$th transmitter. Therefore, if each user transmits with equal power, the signal from the $i$th transmitter will arrive at the receiver in question with a larger power than that of the $k$th signal. This particular problem is often so severe that DS CDMA cannot be used.

An alternative to DS CDMA, of course, is FH CDMA [36]-[40]. If each user is given a different hopping pattern, and if all hopping patterns are orthogonal, the near-far problem will be solved (except for possible spectral spillover from one slot into adjacent slots). However, the hopping patterns are never truly orthogonal. In particular, any time more than one signal uses the same frequency at a given instant of time, interference will result. Events of this type are sometimes referred to as "hits," and these hits become more and more of a problem as the number of users hopping over a fixed bandwidth increases. As is the case when FH is employed as an antijam technique, error-correction coding can be used to significant advantage when combined with FH CDMA.

FH CDMA systems have been considered using one hop per bit, multiple hops per bit (referred to as fast frequency hopping or FFH), and multiple bits per hop (referred to as slow frequency hopping or SFH). Oftentimes the characteristics of the channel over which the multiple users transmit play a significant role in influencing which type of hopping one employs. An example of this is the multipath channel, which is discussed in the next section.

It is clearly of interest to consider the relative capacity of a CDMA system compared to FDMA or TDMA. In a perfectly linear, perfectly synchronous system, the number of orthogonal users for all three systems is the same, since this number only depends upon the dimensionality of the overall signal space. In particular, if a given time-bandwidth product $G_P$ is divided up into, say, $G_P$ disjoint time intervals for TDMA, it can also be "divided" into $N$ binary orthogonal codes (assume that $G_P = 2^m$ for some positive integer $m$).

The differences between the three multiple-accessing techniques become apparent when various real-world constraints are imposed upon the ideal situation described above. For example, one attractive feature of CDMA is that it does not *require* the network synchronization that TDMA requires (i.e., if one is willing to give up something in performance, CDMA can be (and usually is) operated in an asynchronous manner). Another advantage of CDMA is that it is relatively easy to add additional users to the system. However, probably the dominant reason for considering CDMA is the need, in addition, for some type of external interference rejection capability such as multipath rejection or resistance to intentional jamming.

For an asynchronous system, even ignoring any near-far problem effects, the number of users the system can accommodate is markedly less than $G_P$. From [31] and [35], a rough rule-of-thumb appears to be that a system with processing gain $G_P$ can support approximately $G_P/10$ users. Indeed, from [31, eq. (17)], the peak signal voltage to rms noise voltage ratio, averaged over all phase shifts, time delays, and data symbols of the multiple users, is approximately given by

$$\overline{\text{SNR}} \simeq \left[ \frac{N_u - 1}{3G_P} + \frac{\eta_0}{2E_b} \right]^{-1/2}$$

where the overbar indicates an ensemble average. From this equation, it can be seen that, given a value of $E_b/\eta_0$, $(N_u - 1)/G_P$ should be in the vicinity of 0.1 in order not to have a noticeable effect on system performance.

Finally, other factors such as nonlinear receivers influence the performance of a multiple access system, and, for example, the effect of a hard limiter on a CDMA system is treated in [45].

## VI. MULTIPATH CHANNELS

Consider a DS binary PSK communication system operating over a channel which has more than one path linking the transmitter to the receiver. These different paths might consist of several discrete paths, each one with a different attenuation and time delay relative to the others, or it might consist of a continuum of paths. The RAKE system described in [1] is an example of a DS system designed to operate effectively in a multipath environment.

For simplicity, assume initially there are just two paths, a direct path and a single multipath. If we assume the time delay the signal incurs in propagating over the direct path is smaller than that incurred in propagating over the single multipath, and if it is assumed that the receiver is synchronized to the time delay and RF phase associated with the direct path, then the system is as shown in Fig. 14. The received signal is given by

$$r(t) = Ad(t)p(t) \cos \omega_0 t + \alpha Ad(t - \tau)p(t - \tau)$$

$$\cdot \cos (\omega_0 t + \theta) + n_w(t) \tag{41}$$

where $\tau$ is the differential time delay associated with the two paths and is assumed to be in the interval $0 \leqslant \tau \leqslant T$, $\theta$ is a random phase uniformly distributed in $[0, 2\pi]$, and $\alpha$ is the relative attenuation of the multipath relative to the direct path. The output of the integrate-and-dump detection filter is given by

$$g(T) = A + [\pm \alpha A \rho(\tau) + \alpha A \hat{\rho}(\tau)] \cos \theta \tag{42}$$

where $\rho(\tau)$ and $\hat{\rho}(\tau)$ are partial correlation functions of the spreading sequence $p(t)$ and are given by

$$\rho(\tau) \triangleq \frac{1}{T} \int_0^\tau p(t)p(t - \tau) \, dt \tag{43}$$

and

$$\hat{\rho}(\tau) \triangleq \frac{1}{T} \int_\tau^T p(t)p(t - \tau) \, dt. \tag{44}$$

Notice that the sign in front of the second term on the RHS of (42) can be plus or minus with equal probability because this term arises from the pulse preceding the pulse of interest (i.e., if the $i$th pulse is being detected, this term arises from the $i - 1$th pulse), and this latter pulse will be of the same polarity as the current pulse only 50 percent of the time. If the signs of these two pulses happen to be the same, and if $\tau > T_c$ where $T_c$ is the chip duration, then $\rho(\tau) + \hat{\rho}(\tau)$ equals the autocorrelation function of $p(t)$ (assuming that a full period of $p(t)$ is contained in each $T$ second symbol), and this latter quantity equals $-(1/L)$, where $L$ is the period of $p(t)$. In other words, the power in the undesired component of the received signal has been attenuated by a factor of $L^2$.

If the sign of the preceding pulse is opposite to that of the current pulse, the attenuation of the undesired signal will be less than $L^2$, and typically can be much less than $L^2$. This is analogous, of course, to the partial correlation problem in CDMA discussed in the previous section.

The case of more than two discrete paths (or a continuum of paths) results in qualitatively the same effects in that signals delayed by amounts outside of $\pm T_c$ seconds about a correlation peak in the autocorrelation function of $p(t)$ are attenuated by an amount determined by the processing gain of the system.

If FH is employed instead of DS spreading, improvement in system performance is again possible, but through a different mechanism. As was seen in the two previous sections, FH systems achieve their processing gain through interference avoidance, not interference attenuation (as in DS systems).

This same qualitative difference is true again if the interference is multipath. As long as the signal is hopping fast enough relative to the differential time delay between the desired signal and the multipath signal (or signals), all (or most) of the multipath energy will fall in slots that are orthogonal to the slot that the desired signal currently occupies.

Finally, the problems treated in this and the previous two sections are often all present in a given system, and so the use of an appropriate spectrum-spreading technique can alleviate all three problems at once. In [41] and [42], the joint problem of multipath and CDMA is treated, and in [43] and [44], the joint problem of multipath and intentional interference is analyzed. As indicated in Section V, if only multiple accessing capability is needed, there are systems other than CDMA that can be used (e.g., TDMA). However, when multipath is also a problem, the choice of CDMA as the multiple accessing technique is especially appropriate since the same signal design allows both many simultaneous users and improved performance of each user individually relative to the multipath channel.

In the case of signals transmitted over channels degraded by both multipath and intentional interference, either factor by itself suggests the consideration of a spectrum-spreading technique (in particular, of course, the intentional interference), and when all three sources of degradation are present simultaneously, spread spectrum is a virtual necessity.

## VII. ACQUISITION

As we have seen in the previous sections, pseudonoise modulation employing direct sequence, frequency hopping, and/or time hopping is used in spread-spectrum systems to achieve bandwidth spreading which is large compared to the bandwidth required by the information signal. These PN modulation techniques are typically characterized by their very low repetition-rate-to-bandwidth ratio and, as a result, synchronization of a receiver to a specified modulation constitutes a major problem in the design and operation of spread-spectrum communications systems [46]-[50].

It is possible, in principle, for spread-spectrum receivers to use matched filter or correlator structures to synchronize to the incoming waveform. Consider, for example, a direct-sequence amplitude modulation synchronization system as shown in Fig. 15(a). In this figure, the locally generated code $p(t)$ is available with delays spaced one-half of a chip $(T_c/2)$ apart to ensure correlation. If the region of uncertainty of the code phase is $N_c$ chips, $2N_c$ correlators are employed. If no information is available regarding the chip uncertainty and the PN sequence repeats every, say, 2047 chips, then 4094 correlators are employed. Each correlator is seen to examine $\lambda$ chips, after which the correlator outputs $V_0$, $V_1$, $\cdots$, $V_{2N_c-1}$ are compared and the largest output is chosen. As $\lambda$ increases, the probability of making an error in synchronization decreases; however, the acquisition time increases. Thus, $\lambda$ is usually chosen as a compromise between the probability of a synchronization error and the time to acquire PN phase.

A second example, in which FH synchronization is employed, is shown in Fig. 15(b). Here the spread-spectrum signal

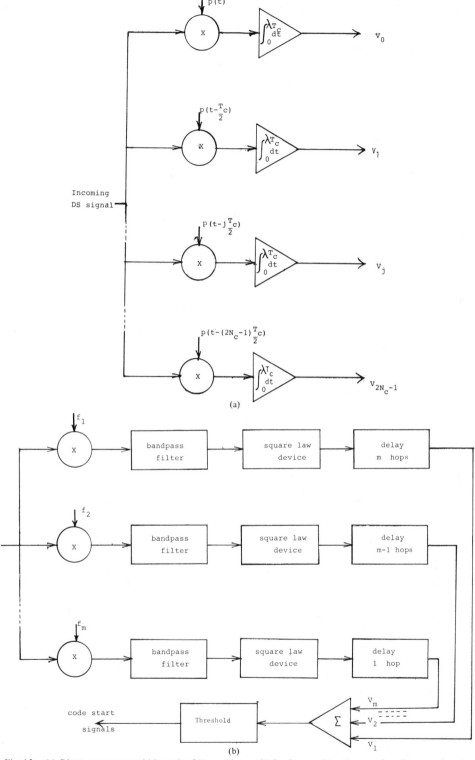

Fig. 15. (a) Direct sequence acquisition using $2N_c$ correlators. (b) Passive correlator structure for a frequency-hopping coarse acquisition scheme.

hops over, for example, $m = 500$ distinct frequencies. Assume that the frequency-hopping sequence is $f_1, f_2, \cdots, f_m$ and then repeats. The correlator then consists of $m = 500$ mixers, each followed by a bandpass filter and square law detector. The delays are inserted so that when the correct sequence appears, the voltages $V_1, V_2, \cdots, V_m$ will occur at the same instant of time at the adder and will, therefore, with high probability, exceed the threshold level indicating synchronization of the receiver to the signal.

While the above technique of using a bank of correlators or matched filters provides a means for rapid acquisition, a considerable reduction in complexity, size, and receiver cost can be achieved by using a single correlator or a single matched filter and repeating the procedure for each possible sequence shift. However, these reductions are paid for by the increased acquisition time needed when performing a serial rather than a parallel operation. One obvious question of interest is therefore the determination of the tradeoff between the number of parallel correlators (or matched filters) used and the cost and time to acquire. It is interesting to note that this tradeoff may become a moot point in several years as a result of the rapidly advancing VLSI technology.

No matter what synchronization technique is employed, the time to acquire depends on the "length" of the correlator. For example, in the system depicted in Fig. 15(a), the integration is performed over $\lambda$ chips where $\lambda$ depends on the desired probability of making a synchronization error (i.e., of deciding that a given sequence phase is correct when indeed it is not), the signal-to-thermal noise power ratio, and the signal-to-jammer power ratio. In addition, in the presence of fading, the fading characteristics affect the number of chips and hence the acquisition time.

The importance that one should attribute to acquisition time, complexity, and size depends upon the intended application. In tactical military communications systems, where users are mobile and push-to-talk radios are employed, rapid acquisition is needed. However, in applications where synchronization occurs once, say, each day, the time to synchronize is not a critical parameter. In either case, once acquisition has been achieved and the communication has begun, it is extremely important not to lose synchronization. Thus, while the acquisition process involves a search through the region of time–frequency uncertainty and a determination that the locally generated code and the incoming code are sufficiently aligned, the next step, called *tracking*, is needed to ensure that the close alignment is maintained. Fig. 16 shows the basic synchronization system. In this system, the incoming signal is first locked into the local PN signal generator using the acquisition circuit, and then kept in synchronism using the tracking circuit. Finally, the data are demodulated.

One popular method of acquisition is called the *sliding correlator* and is shown in Fig. 17. In this system, a single correlator is used rather than $L$ correlators. Initially, the output phase $k$ of the local PN generator is set to $k = 0$ and a partial correlation is performed by examining $\lambda$ chips. If the integrator output falls below the threshold and therefore is deemed too small, $k$ is set to $k = 1$ and the procedure is repeated. The determination that acquisition has taken place

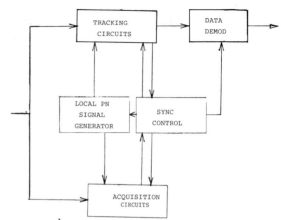

Fig. 16. Functional diagram of synchronization subsystem.

is made when the integrator output $V_I$ exceeds the threshold voltage $V_T(\lambda)$.

It should be clear that in the worst case, we may have to set $k = 0, 1, 2, \cdots,$ and $2N_c - 1$ before finding the correct value of $k$. If, during each correlation, $\lambda$ chips are examined, the worst case acquisition time (neglecting false-alarm and detection probabilities) is

$$T_{acq,max} = 2\lambda N_c T_c. \tag{45}$$

In the $2N_c$-correlator system, $T_{acq,max} = T_c\lambda$, and so we see that there is a time–complexity tradeoff.

Another technique, proposed by Ward [46], called rapid acquisition by sequential estimation, is illustrated in Fig. 18. When switch $S$ is in position 2, the shift register forms a PN generator and generates the same sequence as the input signal. Initially, in order to synchronize the PN generator to the incoming signal, switch $S$ is thrown to position 1. The first $r$ chips received at the input are loaded into the register. When the register is fully loaded, switch $S$ is thrown to position 2. Since the PN sequence generator generates the same sequence as the incoming waveform, the sequences at positions 1 and 2 must be identical. That such is the case is readily seen from Fig. 19 which shows how the code $p(t - jT_C)$ is initially generated. Comparing this code generator to the local generator shown in Fig. 18, we see that with the switch in position 1, once the register is filled, the outputs of both mod 2 adders are *identical*. Hence, the bit stream at positions 1 and 2 are the same and switch $S$ can be thrown to position 2. Once switch $S$ is thrown to position 2, correlation is begun between the incoming code $p(t - jT_c)$ in white noise and the locally generated PN sequence. This correlation is performed by first multiplying the two waveforms and then examining $\lambda$ chips in the integrator.

When no noise is present, the $N$ chips are correctly loaded into the shift register, and therefore the acquisition time is $T_{acq} = rT_c$. However, when *noise is present,* one or more chips may be incorrectly loaded into the register. The resulting waveform at 2 will then not be of the same phase as the sequence generated at 1. If the correlator output after $\lambda T_c$ ex-

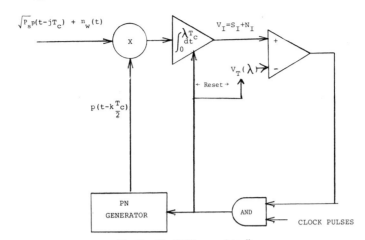

Fig. 17. The "sliding correlator."

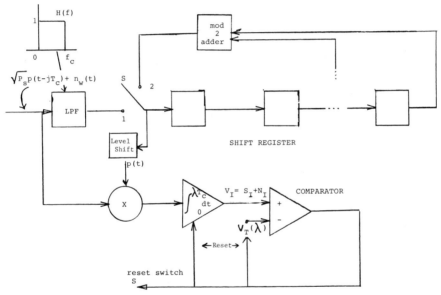

Fig. 18. Shift register acquisition circuit.

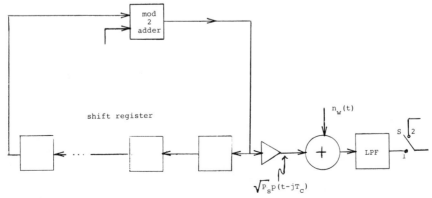

Fig. 19. The equivalent transmitter SRSG.

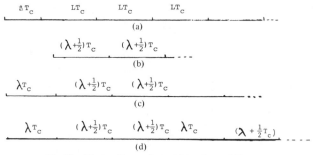

Fig. 20. Timing diagram for serial search acquisition.

ceeds the threshold voltage, we assume that synchronization has occurred. If, however, the output is less than the threshold voltage, switch $S$ is thrown to position 1, the register is re-loaded, and the procedure is repeated.

Note that in both Figs. 17 and 18, correlation occurs for a time $\lambda T_c$ before predicting whether or not synchronism has occurred. If, however, the correlator output is examined after a time $nT_c$ and a decision made at each $n \leqslant \lambda$ as to whether 1) synchronism has occurred, 2) synchronism has not occurred, or 3) a decision cannot be made with sufficient confidence and therefore an additional chip should be examined, then the average acquisition time can be reduced substantially.

One can approximately calculate the mean acquisition time of a parallel search acquisition system, such as the system shown in Fig. 15, by noting that after integrating over $\lambda$ chips, a correct decision will be made with probability $P_D$ where $P_D$ is called the probability of detection. If, however, an incorrect output is chosen, we will, after examining an additional $\lambda$ chips, again make a determination of the correct output. Thus, on the average, the acquisition time is

$$\bar{T}_{acq} = \lambda T_c P_D + 2\lambda T_c P_D(1-P_D) + 3\lambda T_c P_D(1-P_D)^2 + \cdots$$

$$= \frac{\lambda T_c}{P_D} \qquad (46)$$

where it is assumed that we continue searching every $\lambda$ chips even after a threshold has been exceeded. This is not, in general, the way an actual system would operate, but does allow a simple approximation to the true acquisition time.

Calculation of the mean acquisition time when using the "sliding correlator" shown in Fig. 17 can be accomplished in a similar manner (again making the approximation that we never stop searching) by noting that we are initially offset by a random number of chips $\Delta$ as shown in Fig. 20(a). After the correlator of Fig. 17 finally "slides" by these $\Delta$ chips, acquisition can be achieved with probability $P_D$. (Note that this $P_D$ differs from the $P_D$ of (46), since the latter $P_D$ accounts for false synchronizations due to a correlator matched to an incorrect phase having a larger output voltage than does the correlator matched to the correct phase.) If, due to an incorrect decision, synchronization is not achieved at that time, $L$ additional chips must then be examined before acquistion can be achieved (again with probability $P_D$).

We first calculate the average time needed to slide by the $\Delta$ chips. To see how this time can change, refer to Fig. 20(b) which indicates the time required if we are not synchronized. $\lambda$ chips are integrated, and if the integrator output $V_I < V_T$ (the threshold voltage), a $\frac{1}{2}$ chip delay is generated, and we then process an additional $\lambda$ chips, etc. We note that in order to slide $\Delta$ chips in $\frac{1}{2}$ chip intervals, this process must occur $2\Delta$ times. Since each repetition takes a time $(\lambda + \frac{1}{2})T_c$, the total elapsed time is $2\Delta(\lambda + \frac{1}{2})T_c$.

Fig. 20(b) assumes that at the end of each examination interval, $V_I < V_T$. However, if a false alarm occurs and $V_I > V_T$, no slide of $T_c/2$ will occur until after an additional $\lambda$ chips are searched. This is shown in Fig. 20(c). In this case, the total elapsed time is $2\Delta(\lambda + \frac{1}{2})T_c + \lambda T_c$. Fig. 20(d) shows the case where false alarms occurred twice. Clearly, neither the separation between these false alarms nor where they occur is relevant. The total elapsed time is now $2\Delta(\lambda + \frac{1}{2})T_c + 2\lambda T_c$.

In general, the average elapsed time to reach the correct synchronization phase is

$$\bar{T}_{s/\Delta} = \left[ 2\Delta(\lambda + \tfrac{1}{2})T_c + \lambda T_c P_F + 2\lambda T_c P_F^2 + \cdots \right]$$

$$= \left[ 2\Delta(\lambda + \tfrac{1}{2})T_c + \lambda T_c P_F \sum_{n=1}^{\infty} n P_F^{n-1} \right]$$

$$= \left[ 2\Delta(\lambda + \tfrac{1}{2})T_c + \frac{\lambda T_c P_F}{(1-P_F)^2} \right] \qquad (47)$$

where $P_F$ is the false alarm probability. Equation (47) is for a given value of $\Delta$. Since $\Delta$ is a random variable which is equally likely to take on any integer value from 0 to $L-1$, $\bar{T}_{s/\Delta}$ must be averaged over all $\Delta$. Therefore,

$$\bar{T}_s \triangleq \frac{1}{L} \sum_{\Delta=0}^{L-1} \bar{T}_{s/\Delta} = (L-1)\left[ (\lambda + \tfrac{1}{2})T_c + \frac{\lambda T_c P_F}{(1-P_F)^2} \right]. \qquad (48)$$

Equation (48) is the average time needed to slide through $\Delta$ chips. If, after sliding through $\Delta$ chips, we do not detect the correct phase, we must now slide through an additional $L$ chips. The mean time to do this is given by (47), with $\Delta$ replaced by $L$. We shall call this time $\bar{T}_{s/L}$:

$$\bar{T}_{s/L} = 2L\left[ (\lambda + \tfrac{1}{2})T_c + \frac{\lambda T_c P_F}{(1-P_F)^2} \right]. \qquad (49)$$

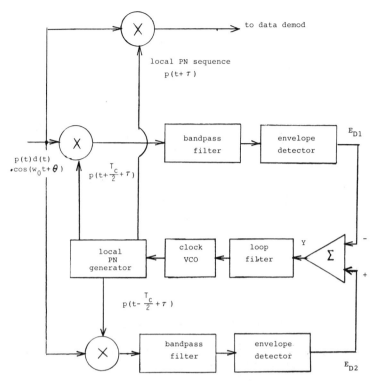

Fig. 21.  Delay-locked loop for tracking direct-sequence PN signals.

The mean time to acquire a signal can now be written as

$$\bar{T}_{\text{acq}} = \bar{T}_s + \bar{T}_{s/L}[P_D(1-P_D) + 2P_D(1-P_D)^2 + \cdots]$$

$$\bar{T}_s + \frac{1-P_D}{P_D}\bar{T}_{s/L}$$

or

$$\bar{T}_{\text{acq}} = (L-1)\left[(\lambda + \tfrac{1}{2})T_c + \frac{\lambda T_c P_F}{(1-P_F)^2}\right]$$

$$+ \frac{1-P_D}{P_D}\, 2L\left[(\lambda + \tfrac{1}{2})T_c + \frac{\lambda T_c P_F}{(1-P_F)^2}\right]. \quad (50)$$

## VIII. TRACKING

Once acquisition, or coarse synchronization, has been accomplished, tracking, or fine synchronization, takes place. Specifically, this must include chip synchronization and, for coherent systems, carrier phase locking. In many practical systems, no data are transmitted for a specified time, sufficiently long to ensure that acquisition has occurred. During tracking, data are transmitted and detected. Typical references for tracking loops are [51]–[54].

The basic tracking loop for a direct-sequence spread-spectrum system using PSK data transmission is shown in Fig. 21. The incoming carrier at frequency $f_0$ is amplitude modu-

lated by the product of the data $d(t)$ and the PN sequence $p(t)$. The tracking loop contains a local PN generator which is offset in phase from the incoming sequence $p(t)$ by a time $\tau$ which is less than one-half the chip time. To provide "fine" synchronization, the local PN generator generates two sequences, delayed from each other by one chip. The two bandpass filters are designed to have a two-sided bandwidth $B$ equal to twice the data bit rate, i.e.,

$$B = 2R = 2/T. \quad (51)$$

In this way the data are passed, but the product of the two PN sequences $p(t)$ and $p(t \mp T_c/2 + \tau)$ is *averaged*. The envelope detector eliminates the data since $|d(t)| = 1$. As a result, the output of each envelope detector is approximately given by

$$E_{D1,2} \cong \overline{\left| p(t)p\left(t \pm \frac{T_c}{2} + \tau\right)\right|} = \left| R_p\left(\tau \pm \frac{T_c}{2}\right)\right| \quad (52)$$

where $R_p(x)$ is the autocorrelation function of the PN waveform as shown in Fig. 7(a). [See Section III for a discussion of the characteristics of $R_p(x)$.]

The output of the adder $Y(t)$ is shown in Fig. 22. We see from this figure that, when $\tau$ is positive, a positive voltage, proportional to $Y$, instructs the VCO to increase its frequency, thereby forcing $\tau$ to decrease, while when $\tau$ is negative, a

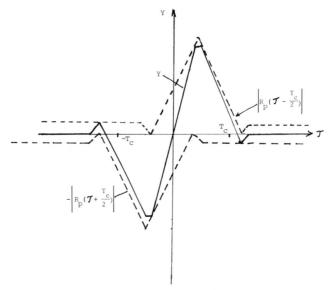

Fig. 22. Variation of $Y$ with $\tau$.

negative voltage instructs the VCO to reduce its frequency, thereby forcing $\tau$ to increase toward 0.

When the tracking error $\tau$ is made equal to zero, an output of the local PN generator $p(t + \tau) = p(t)$ is correlated with the input signal $p(t) \cdot d(t) \cos(\omega_0 t + \theta)$ to form

$$p^2(t)d(t) \cos(\omega_0 t + \theta) = d(t) \cos(\omega_0 t + \theta).$$

This despread PSK signal is inputted to the data demodulator where the data are detected.

An alternate technique for synchronization of a DS system is to use a tau-dither (TD) loop. This tracking loop is a delay-locked loop with only a single "arm," as shown in Fig. 23(a). The control (or gating) waveforms $g(t)$, $\bar{g}(t)$, and $g'(t)$ are shown in Fig. 23(b), and are used to generate both "arms" of the DLL even though only one arm is present. The TD loop is often used in lieu of the DLL because of its simplicity.

The operation of the loop is explained by observing that the control waveforms generate the signal

$$V_p(t) = g(t)p(t + \tau - T_c/2) + \bar{g}(t)p(t + \tau + T_c/2). \quad (53)$$

Note that either one or the other, but not both, of these waveforms occurs at each instant of time. The voltage $V_p(t)$ then multiplies the incoming signal

$$d(t)p(t) \cos(\omega_0 t + \theta).$$

The output of the bandpass filter is therefore

$$E_f(t) = \{d(t)\bar{g}(t)\overline{p(t)p(t + \tau + T_c/2)}$$
$$+ d(t)g(t)\overline{p(t)p(t + \tau - T_c/2)}\} \cos(\omega_0 t + \theta) \quad (54)$$

where, as before, the average occurs because the bandpass

filter is designed to pass the data and control signals, but cuts off well below the chip rate. The data are eliminated by the envelope detector, and (54) then yields

$$E_d(t) = \bar{g}(t) |R_p(\tau + T_c/2)| + g(t) |R_p(\tau - T_c/2)|. \quad (55)$$

The input $Y(t)$ to the loop filter is

$$Y(t) = E_d(t)g'(t)$$
$$= \bar{g}(t) |R_p(\tau - T_c/2)| - g(t) |R_p(\tau - T_c/2)| \quad (56)$$

where the "−" sign was introduced by the inversion caused by $g'(t)$.

The narrow-band loop filter now "averages" $Y(t)$. Since each term is zero half of the time, the voltage into the VCO clock is, as before,

$$V_c(t) = -|R_p(\tau - T_c/2)| + |R_p(\tau + T_c/2)|. \quad (57)$$

A typical tracking system for an FSK/FH spread-spectrum system is shown in Fig. 24. Waveforms are shown in Fig. 25. Once again, we have assumed that, although acquisition has occurred, there is still an error of $\tau$ seconds between transitions of the incoming signal's frequencies and the locally generated frequencies. The bandpass filter BPF is made sufficiently wide to pass the product signal $V_p(t)$ when $V_1(t)$ and $V_2(t)$ are at the same frequency $f_i$, but sufficiently narrow to reject $V_p(t)$ when $V_1(t)$ and $V_2(t)$ are at different frequencies $f_i$ and $f_{i+1}$. Thus, the output of the envelope detector $V_d(t)$, shown in Fig. 24, is unity when $V_1(t)$ and $V_2(t)$ are at the same frequency and is zero when $V_1(t)$ and $V_2(t)$ are at different frequencies. From Fig. 25, we see that $V_g(t) = V_d(t) V_c(t)$ and is a three-level signal. This three-level signal is filtered to form a dc voltage which, in this case, presents a negative voltage to the VCO.

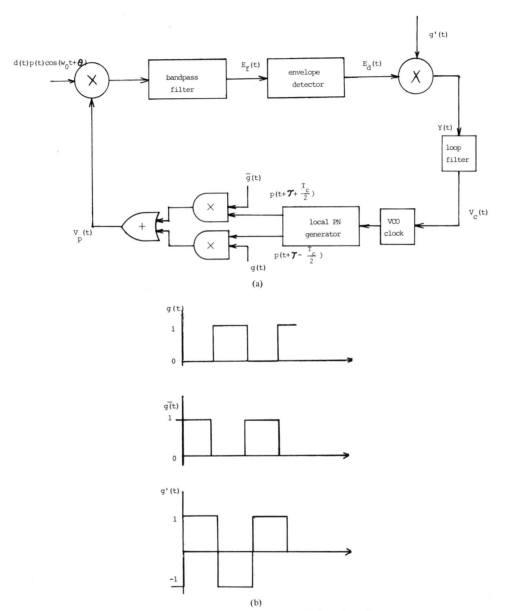

(a)

(b)

Fig. 23. The tau-dither loop. (a) Block diagram. (b) Control waveforms.

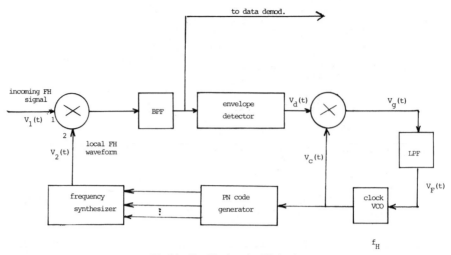

Fig. 24. Tracking loop for FH signals.

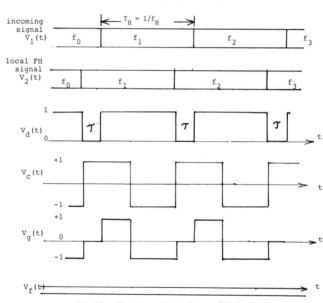

Fig. 25. Waveforms for tracking an FH signal.

It is readily seen that when $V_2(t)$ has frequency transitions which precede those of the incoming waveform $V_1(t)$, the voltage into the VCO will be negative, thereby delaying the transition, while if the local waveform frequency transitions occur after the incoming signal frequency transitions, the voltage into the VCO will be positive, thereby speeding up the transition.

The role of the tracking circuit is to keep the offset time $\tau$ small. However, even a relatively small $\tau$ can have a major impact on the probability of error of the received data. Referring to the DS system of Fig. 21, we see that if $\tau$ is not zero, the input to the data demodulator is $p(t)p(t + \tau)d(t) \cos(\omega_0 t + \theta)$ rather than $p^2(t)d(t) \cos(\omega_0 t + \theta) = d(t) \cos(\omega_0 t + \theta)$. The data demodulator removes the carrier and then averages the remaining signal, which in this case is

$p(t)p(t + \tau)d(t)$. The result is $\overline{p(t)p(t + \tau)d(t)}$. Thus, the amplitude of the data has been reduced by $\overline{p(t)p(t + \tau)} = R_p(\tau) \leqslant 1$. For example, if $\tau = T_c/10$, that data amplitude is reduced to 90 percent of its value, and the power is reduced to 0.81. Thus, the probability of error in correctly detecting the data is reduced from

$$P_e = Q\left(\sqrt{\frac{2E_b}{n_0}}\right)$$

to

$$P_e(\tau = T_c/10) = Q\left(\sqrt{\frac{1.62E_b}{n_0}}\right),$$

and at an $E_b/\eta_0$ of 9.6 dB, $P_e$ is increased from $10^{-5}$ to $10^{-4}$.

## IX. CONCLUSIONS

This tutorial paper looked at some of the theoretical issues involved in the design of a spread-spectrum communication system. The topics discussed included the characteristics of PN sequences, the resulting processing gain when using either direct-sequence or frequency-hopping antijam considerations, multiple access when using spread spectrum, multipath effects, and acquisition and tracking systems.

No attempt was made to present other than fundamental concepts; indeed, to adequately cover the spread-spectrum system completely is the task for an entire text [55], [56]. Furthermore, to keep this paper reasonably concise, the authors chose to ignore both practical system considerations such as those encountered when operating at, say, HF, VHF, or UHF, and technology considerations, such as the role of surface acoustic wave devices and charge-coupled devices in the design of spread-spectrum systems.

Spread spectrum has for far too long been considered a technique with very limited applicability. Such is not the case. In addition to military applications, spread spectrum is being considered for commercial applications such as mobile telephone and microwave communications in congested areas.

The authors hope that this tutorial will result in more engineers and educators becoming aware of the potential of spread spectrum, the dissemination of this information in the classroom, and the use of spread spectrum (where appropriate) in the design of communication systems.

## APPENDIX

### ALGEBRAIC PROPERTIES OF LINEAR FEEDBACK SHIFT REGISTER SEQUENCES

In order to fully appreciate the study of shift register sequences, it is desirable to introduce the polynomial representation (or generating function) of a sequence

$$C(x) = \sum_{i=0}^{\infty} C_i x^i \leftrightarrow (C_0, C_1, C_2, \cdots). \tag{A1}$$

If the sequence is periodic with period $L$, i.e.,

$$C_0, C_1, C_2, \cdots, C_{L-1} C_0 C_1, \cdots, C_{L-1}, C_0, \cdots,$$

then since $x^L C(x) \leftrightarrow (0, 0, \cdots, 0, C_0, C_1, C_2, \cdots)$,

$$C(x)(1 - x^L) = \sum_{i=0}^{L-1} C_i x^i \triangleq R(x) \tag{A2}$$

with $R(x)$ the (finite) polynomial representation of one period.

Thus, for any periodic sequence of period $L$,

$$C(x) = \frac{R(x)}{1 - x^L}; \qquad \deg R(x) < L. \tag{A3}$$

Next consider the periodic sequence generated by the LFSR recursion. Multiplying each side by $x^n$ and summing gives

$$\sum_{n=0}^{\infty} C_n x^n = \sum_{k=1}^{r} a_k \sum_{n=0}^{\infty} C_{n-k} x^n$$

$$= \sum_{k=1}^{r} a_k \sum_{l=0}^{k-1} C_{l-k} x^l + \sum_{k=1}^{r} a_k x^k \left( \sum_{n=0}^{\infty} C_n x^n \right).$$

The left-hand side is the generating function $C(x)$ of the sequence. The first term on the right is a polynomial of degree $< r$, call it $g(x)$, which depends only on the initial state of the register $C_{-1}, C_{-2}, C_{-3}, \cdots, C_{-r}$. Thus, the basic equation of the register sequence may be written as

$$C(x) = \frac{g(x)}{f(x)}; \qquad \deg g(x) < r \tag{A4}$$

where $f(x) \triangleq 1 - \Sigma_{k=1}^{r} a_k x^k$ is the characteristic polynomial[2] (or connection polynomial) of the register. Since $C(x)$ is the generating polynomial of a sequence of period $L = 2^r - 1$, it can be shown from (A3) and (A4) that $f(x)$ must divide $1 - x^L$. This is illustrated in the following example.

*Example*

The three-stage binary maximal length register with $f(x) = 1 + x + x^3$ has period 7. If the initial contents of the register are $C_{-3} = 1$, $C_{-2} = 0$, $C_{-1} = 0$, then $g(x) = a_3 = 1$ and $C(x) = 1/(1 + x + x^3)$. Long division (modulo 2) yields

$$C(x) = 1 + x + x^2 + x^4 + x^7 + x^9 + \cdots$$

which is the generating function of the periodic sequence

$$1\,1\,1\,0\,1\,0\,0\,\vdots\,1\,1\,1\,0\,1\cdots,$$

and which is precisely the sequence generated by the corresponding recursion

$$C_n = C_{n-1} + C_{n-3} \text{ (mod 2)}.$$

Observe that

$$(1 + x + x^3)(1 + x + x^2 + x^4) = 1 + x^7$$

so that $f(x)$ divides $1 + x^7$. Also, we may write

$$C(x) = \frac{1}{1 + x + x^3} \cdot \frac{1 + x + x^2 + x^4}{1 + x + x^2 + x^4} = \frac{1 + x + x^2 + x^4}{1 + x^7}$$

which is in the form of (A3).

---

[2] For binary sequences, all sums are modulo two and minus is the same as plus. The polynomials defining them have 0, 1 coefficients and are said to be polynomials over a finite field with two elements. A field is a set of elements, and two operations, say, + and $\cdot$, which obey the usual rules of arithmetic. A finite field with $q$ elements is called a Galois field and is designated as $GF(q)$.

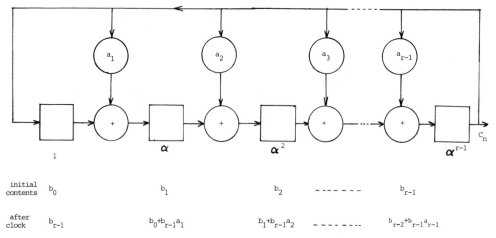

Fig. 26.   Binary modular shift register generator with polynomial
$$f_M(x) = 1 + a_1 x + a_2 x^2 + \cdots + a_{r-1} x^{r-1} + x^r.$$

It is easy to see (by multiplying and equating coefficients of like powers) that if

$$\frac{1}{f(x)} = \frac{1}{1 + a_1 x + a_2 x^2 + \cdots + a_r x^r}$$

$$= C_0 + C_1 x + C_2 x^2 + \cdots = C(x)$$

then

$$C_n = \sum_{k=1}^{r} a_k C_{n-k},$$

so that (except for initial conditions) $f(x)$ completely describes the maximal length sequence. Now what properties must $f(x)$ possess to ensure that the sequence is maximal length? Aside from the fact that $f(x)$ must divide $1 + x^L$, it is necessary (but not sufficient) that $f(x)$ be irreducible, i.e., $f(x) \neq f_1(x)$ $\cdot f_2(x)$. Suppose that $f(x) = f_1(x) f_2(x)$ with $f_1(x)$ of degree $r_1$, $f_2(x)$ of degree $r_2$, and $r_1 + r_2 = r$. Then we can write, by partial fractions,

$$\frac{1}{f(x)} = \frac{\alpha(x)}{f_1(x)} + \frac{\beta(x)}{f_2(x)} ; \qquad \begin{array}{l} \deg \alpha(x) < r_1 \\ \deg \beta(x) < r_2. \end{array}$$

The maximum period of the expansion of the first term is $2^{r_1}-1$ and that of the second term is $2^{r_2}-1$. Hence, the period of $1/f(x) \leqslant$ least common multiple of $(2^{r_1}-1, 2^{r_2}-1)$ $< 2^r - 3$. This is a contradiction, since if $f(x)$ were maximal length, the period of $1/f(x)$ would be $2^r - 1$. Thus, a *necessary* condition that the LFSR is maximal length is that $f(x)$ is irreducible.

A *sufficient* condition is that $f(x)$ is *primitive*. A primitive polynomial of degree $r$ over $GF(2)$ is simply one for which the period of the coefficients of $1/f(x)$ is $2^r - 1$. However, additional insight can be had by examining the roots of $f(x)$. Since $f(x)$ is irreducible over $GF(2)$, we must imagine that the

roots are elements of some larger (extension) field. Suppose that $\alpha$ is such an element and that $f(\alpha) = 0 = \alpha^r + a_{r-1} \alpha^{r-1} + \cdots + a_1 \alpha + 1$ or

$$\alpha^r = a_{r-1} \alpha^{r-1} + \cdots + a_1 \alpha + 1. \tag{A5}$$

We see that *all* powers of $\alpha$ can be expressed in terms of a linear combination of $\alpha^{r-1}$, $\alpha^{r-2}$, $\cdots$, $\alpha$, 1 since any powers larger than $r-1$ may be reduced using (A5). Specifically, suppose we have some power of $\alpha$ that we represent as

$$\beta \triangleq b_0 + b_1 \alpha + b_2 \alpha^2 + \cdots + b_{r-1} \alpha^{r-1}. \tag{A6}$$

Then if we multiply this $\beta$ by $\alpha$ and use (A5), we obtain

$$\beta\alpha = b_{r-1} + (b_0 + b_{r-1} a_1)\alpha + (b_1 + b_{r-1} a_2)\alpha^2 + \cdots$$
$$+ (b_{r-2} + b_{r-1} a_{r-1})\alpha^{r-1}. \tag{A7}$$

The observations above may be expressed in another, more physical way with the introduction of an LFSR in modular form [called a modular shift register generator (MSRG)] shown in Fig. 26. The feedback, modulo 2, is between the delay elements. The binary contents of the register at any time are shown as $b_0, b, \cdots, b_{r-1}$. This vector can be identified with $\beta$ as

$$\beta = b_0 + b_1 \alpha + \cdots + b_{r-1} \alpha^{r-1} \leftrightarrow [b_0, b_1, \cdots b_{r-1}],$$

the contents of the first stage being identified with the coefficient of $\alpha^0$, those of the second stage with the coefficient of $\alpha^1$, etc. After one clock pulse, it is seen that the register contents correspond to

$$\beta\alpha = b_{r-1} + (b_0 + b_{r-1})\alpha + \cdots$$
$$+ (b_{r-2} + b_{r-1} a_{r-1})\alpha^{r-1}$$
$$\leftrightarrow [b_{r-1}, \cdots, b_{r-2} + b_{r-1} a_{r-1}].$$

Thus, the MSRG is an $\alpha$-multiplier. Now if $\alpha^0, \alpha, \alpha^2, \alpha^3, \cdots,$ $\alpha^{L-1}$, $L = 2^r - 1$ are all *distinct*, we call $\alpha$ a *primitive* element of $GF(2^r)$. The register in Fig. 26 cycles through all states (starting in any nonzero state), and hence generates a maximal length sequence. Thus, another way of describing that the polynomial $f_M(x)$ is primitive (or maximal length) is that it has a primitive element in $GF(2^r)$ as a root.

There is an intimate relationship between the MSRG shown in Fig. 26 and the SSRG shown in Fig. 4. From Fig. 26 it is easily seen that the output sequence $C_n$ satisfies the recursion

$$C_n = \sum_{k=0}^{r-1} a_k C_{n-r+k}. \tag{A8}$$

Multiplying both sides by $x^n$ and summing yields

$$C(x) \triangleq \sum_{n=-\infty}^{\infty} C_n x^n = \sum_{k=0}^{r-1} a_k \sum_{n=0}^{\infty} C_{n-r+k} x^n$$

$$= \sum_{k=0}^{r-1} a_k x^{r-k} \sum_{l=-r+k}^{-1} C_l x^l$$

$$+ x^r \sum_{k=0}^{r-1} a_k x^{-k} \left( \sum_{n=0}^{\infty} C_n x^n \right) \tag{A9}$$

or

$$C(x) = g_M(x) + x^r \sum_{k=0}^{r-1} a_k x^{-k} C(x). \tag{A10}$$

$g_M(x)$ is the first term on the right-hand side of (A9) and is a polynomial of degree $< r$ which depends on the initial state. Then we have

$$C(x) = \frac{g_M(x)}{f_M(x)} \tag{A11}$$

where

$$f_M(x) = 1 - \{a_0 x^r + a_1 x^{r-1} + a_2 x^{r-2} + \cdots + a_{r-1}x\}$$

(recall that in $GF(2)$, minus is the same as plus) is the characteristic (or connection) polynomial of the MSRG. Since the sequence $C_n$ [of coefficients of $C(x)$] when $f_M(x)$ is primitive depends *only* on $f_M(x)$ (discounting phase), the relationship between the SSRG and the MSRG which generates the *same* sequence is

$$f(x) = x^r f_M\left(\frac{1}{x}\right). \tag{A12}$$

$f_M(x)$ is called the *reciprocal* polynomial of $f(x)$ and is obtained from $f(x)$ by reversing the order of the coefficients.

There are several good tables of irreducible and primitive polynomials available [2], [5], [6], and although the tables

TABLE I
THE NUMBER OF MAXIMAL LENGTH LINEAR SRG SEQUENCES
OF DEGREE $r = \lambda(r) = \phi(2^r - 1)/r$

| $r$ | $2^r - 1$ | $\lambda(r)$ |
|---|---|---|
| 1 | 1 | 1 |
| 2 | 3 | 1 |
| 3 | 7 | 2 |
| 4 | 15 | 2 |
| 5 | 31 | 6 |
| 6 | 63 | 6 |
| 7 | 127 | 18 |
| 8 | 255 | 16 |
| 9 | 511 | 48 |
| 10 | 1,023 | 60 |
| 11 | 2,047 | 176 |
| 12 | 4,095 | 144 |
| 13 | 8,191 | 630 |
| 14 | 16,383 | 756 |
| 15 | 32,767 | 1,800 |
| 16 | 65,535 | 2,048 |
| 17 | 131,071 | 7,710 |
| 18 | 262,143 | 8,064 |
| 19 | 524,287 | 27,594 |
| 20 | 1,048,575 | 24,000 |
| 21 | 2,097,151 | 87,672 |
| 22 | 4,194,303 | 120,032 |

do not list all the primitive polynomials, algorithms exist [7] which allow one to generate all primitive polynomials of a given degree if one of them is known. The number $\lambda(r)$ of primitive polynomials of degree $r$ is [4]

$$\lambda(r) = \frac{\phi(2^r - 1)}{r} \tag{A13}$$

where $\phi(m)$ is the number of integers less than $m$ which are relatively prime to $m$ (Euler totient function). The growth of this number with $r$ is shown in Table I.

The algebra of LFSR's is useful in constructing codes with uniformly low cross correlation known as Gold codes. The underlying principle of these codes is based on the following *theorem* [15].

If $f_1(x)$ is the minimal polynomial of the primitive element $\alpha \in GF(2^r)$ and $f_t(x)$ is the minimal polynomial of $\alpha^t$, where both $f_1(x)$ and $f_t(x)$ are of degree $r$ and

$$t = \begin{cases} 2^{\frac{r+1}{2}} + 1, & r \text{ odd} \\ 2^{\frac{r+2}{2}} + 1, & r \text{ even}, \end{cases}$$

then the product $f(x) \triangleq f_1(x)f_t(x)$ determines an LFSR which generates $2^r + 1$ different sequences (corresponding to the $2^r + 1$ states in distinct cycles) of period $2^r - 1$, and such that for any pair $C'$ and $C''$,

$$L |R_{C'C''}(\tau)| < t.$$

83

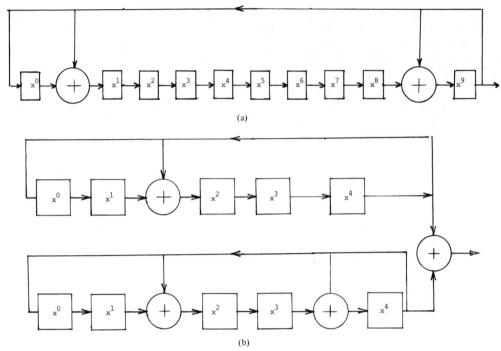

(a)

(b)

Fig. 27. Two implementations of LFSR which generate Gold codes of length $2^5 - 1 = 31$ with maximum cross correlation $t = 9$. (a) LFSR with $f(x) = 1 + x + x^3 + x^9 + x^{10}$. (b) LFSR which generates sequences corresponding to $f(x) = (1 + x^2 + x^5) \cdot (1 + x^2 + x^4 + x^5) = 1 + x + x^3 + x^9 + x^{10}$.

Futhermore, $R_{C'C''}(\tau)$ is only a three-valued function for any integer $\tau$.

A minimal polynomial of $\alpha$ is simply the smallest degree monic[3] polynomial for which $\alpha$ is a root. With the help of a table of primitive polynomials, we can identify minimal polynomials of powers of $\alpha$ and easily construct Gold codes. For example, if $r = 5$ and $t = 2^3 + 1 = 9$, using [2] we find that $f_1(x) = 1 + x^2 + x^5$ and $f_9(x) = 1 + x^2 + x^4 + x^5$. Then $f(x) = 1 + x + x^3 + x^9 + x^{10}$. The two ways to represent this LFSR (in MSRG form) are shown in Fig. 27. Fig. 27(a) shows one long nonmaximal length register of degree 10 which generates sequences of period $2^5 - 1 = 31$. Since there are $2^{10} - 1$ possible nonzero initial states, the number of initial states that result in distinct cycles is $(2^{10} - 1)/(2^5 - 1) = 2^5 + 1 = 33$. Each of these initial states specifies a different Gold code of length 31. Fig. 27(b) shows how the same result can be obtained by adding the outputs of the two MLFSR's of degree 5 together modulo two. This follows simply from the observation that the sequence(s) generated by $f(x)$ are just the coefficients in the expansion of $1/f(x) = 1/f_1(x) \cdot f_9(x)$. By using partial fractions, one can see that the resulting coefficients are the (modulo two) sum of the coefficients of like powers of $x$ in the expansion of $1/f_1(x)$ and $1/f_9(x)$. Naturally, the sequence resulting will depend on the relative *phases* of the two degree-5 registers. As before, there are $(2^{10} - 1)/(2^5 - 1) =$

$2^5 + 1 = 33$ relative phases which result in 33 different sequences satisfying the cross-correlation bound given by the theorem.

## GLOSSARY OF SYMBOLS

| | |
|---|---|
| $a_n$ | $\{0, 1\}$ feedback taps for LFSR. |
| $B_D$ | One-sided bandwidth (Hz) for data signal(s). |
| $B_{ss}$ | One-sided bandwidth (Hz) of baseband spread-spectrum signal. |
| $C(x)$ | Generating function of $C_n$; $C(x) = \sum_{n=0}^{\infty} C_n x^n$. |
| $C_n$ | $\{0, 1\}$ LFSR sequence. |
| $C_n'$ or $C_n''$ | $\{1, -1\}$ LFSR sequence. |
| $D$ | Dimensionality of underlying signal space. |
| $d(t)$ | Data sequence waveform. |
| $\Delta$ | Initial offset, in chips, of incoming signal and locally generated code. |
| DS | Direct sequence. |
| $E_b$ | Energy/information bit. |
| $E_J$ | Jammer energy over the correlation interval. |
| $E_s$ | Energy/symbol. |
| $f(x)$ | Characteristic (connection) polynomial of an LFSR, $f(x) = 1 + a_1 x + \cdots + a_{r-1} x^{r-1} + x^r$. |
| $f_c$ | Chip rate; $T_c = 1/f_c$. |
| FH | Frequency hopping. |
| $G_P$ | Processing gain. |
| $J(t)$ | Jammer signal waveform. |

[3] A monic polynomial is one whose coefficient of its highest power is unity.

| | |
|---|---|
| $K$ | Number of frequencies jammed by partial-band jammer. |
| $L = 2^r - 1$ | Period of PN sequence. |
| $L$ | Implementation losses. |
| $\lambda$ | Number of chips examined during each search in the process of acquisition. |
| $\lambda(r)$ | Number of binary maximal length PN codes of degree $r$ (length $L = 2^r - 1$). |
| $M$ | Signal alphabet size. |
| $M_J$ | Jamming margin. |
| $n$ | Number of chips/bit or number of dimensions of spread signal space. |
| $N$ | Number of frequencies in FH. |
| $N_c$ | Number of chips in uncertainty region at start of acquisition. |
| $N_u$ | Number of users in CDMA system. |
| $\eta_0$ | One-sided white noise power spectral density (W/Hz). |
| $\eta_{0J}$ | $P_J/2f_c$ = power density of jammer. |
| $n_w(t)$ | Additive white Gaussian noise (AWGN). |
| $p(t)$ | Spreading sequence waveform. |
| $P_D$ | Probability of detection. |
| $P_e$ | Probability of error. |
| $P_F$ | Probability of false alarm. |
| $P_J$ | Jammer power. |
| $P_N$ | Noise power. |
| $P_s$ | Signal power. |
| PN | Pseudonoise sequence. |
| $r$ | Number of stages of shift register. |
| $r(t)$ | Received waveform. |
| $R$ | Data rate (bits/s). |
| $R_p(\tau)$ | Autocorrelation function. |
| $R_{C'C''}(\tau)$ | Cross-correlation function of two (periodic) $\pm 1$ sequences $C_n', C_n''.$ |
| $\rho(\tau)$ | $(1/T) \int_0^\tau p(t) p(t - \tau)\, dt$ (partial correlation function). |
| $\hat{\rho}(\tau)$ | $(1/T) \int_\tau^T p(t) p(t - \tau)\, dt.$ |
| $S(t)$ | Transmitted signal waveform. |
| $S_p(f)$ | Power spectral density of spreading sequence waveform [also denoted $S_{ss}(f)$]. |
| SNR | Signal-to-noise power ratio. |
| SJR | Signal-to-jammer power ratio. |
| $T$ | Signal or symbol duration. |
| $T_c$ | Chip duration. |
| $T_H$ | Time to hop one frequency; $1/T_H$ = hopping rate. |
| $V$ | Correlator output voltage. |

## ACKNOWLEDGMENT

The authors wish to thank the anonymous reviewers for their constructive suggestions in the final preparation of this paper.

## REFERENCES

[1] R. A. Scholtz, "The origins of spread-spectrum communications," this issue, pp. 822–854.

[2] W. W. Peterson and E. J. Weldon, Jr., *Error Correcting Codes*, 2nd ed. Cambridge, MA: M.I.T. Press, 1972.

[3] J. M. Wozencraft and I. M. Jacobs, *Principles of Communication Engineering*. New York: Wiley, 1965.

[4] S. W. Golomb, *Shift Register Sequences*. San Francisco, CA: Holden Day, 1967.

[5] R. W. Marsh, *Table of Irreducible Polynomials over GF(2) Through Degree 19*. Washington, DC: NSA, 1957.

[6] W. Stahnke, "Primitive binary polynomials," *Math. Comput.*, vol. 27, pp. 977–980, Oct. 1973.

[7] E. R. Berlekamp, *Algebraic Coding Theory*. New York: McGraw-Hill, 1968.

[8] J. L. Massey, "Shift-register synthesis and BCH decoding," *IEEE Trans. Inform. Theory*, vol. IT-15, pp. 122–127, Jan. 1969.

[9] N. G. deBruijn, "A combinatorial problem," in *Koninklijke Nederlands Akademie Van Wetenschappen Proc.*, 1946, pp. 758–764.

[10] E. J. Groth, "Generation of binary sequences with controllable complexity," *IEEE Trans. Inform. Theory*, vol. IT-17, pp. 288–296, May 1971.

[11] E. L. Key, "An analysis of the structure and complexity of nonlinear binary sequence generators," *IEEE Trans. Inform. Theory*, vol. IT-22, pp. 732–736, Nov. 1976.

[12] H. Beker, "Multiplexed shift register sequences," presented at CRYPTO '81 Workshop, Santa Barbara, CA, 1981.

[13] J. L. Massey and J. J. Uhran, "Sub-baud coding," in *Proc. 13th Annu. Allerton Conf. Circuit and Syst. Theory*, Monticello, IL, Oct. 1975, pp. 539–547.

[14] J. H. Lindholm, "An analysis of the pseudo randomness properties of the subsequences of long *m*-sequences," *IEEE Trans. Inform. Theory*, vol. IT-14, 1968.

[15] R. Gold, "Optimal binary sequences for spread spectrum multiplexing," *IEEE Trans. Inform. Theory*, vol. IT-13, pp. 619–621, 1967.

[16] D. V. Sarwate and M. B. Pursley, "Cross correlation properties of pseudo-random and related sequences," *Proc. IEEE*, vol. 68, pp. 598–619, May 1980.

[17] A. Lempel, M. Cohn, and W. L. Eastman, "A new class of balanced binary sequences with optimal autocorrelation properties," IBM Res. Rep. RC 5632, Sept. 1975.

[18] A. G. Konheim, *Cryptography, A Primer*. New York: Wiley, 1981.

[19] D. L. Schilling, L. B. Milstein, R. L. Pickholtz, and R. Brown, "Optimization of the processing gain of an *M*-ary direct sequence spread spectrum communication system," *IEEE Trans. Commun.*, vol. COM-28, pp. 1389–1398, Aug. 1980.

[20] L. B. Milstein, S. Davidovici, and D. L. Schilling, "The effect of multiple-tone interfering signals on a direct sequence spread spectrum communication system," *IEEE Trans. Commun.*, vol. COM-30, pp. 436–446, Mar. 1982.

[21] S. W. Houston, "Tone and noise jamming performance of a spread spectrum *M*-ary FSK and 2, 4-ary DPSK waveforms," in *Proc. Nat. Aerosp. Electron. Conf.*, June 1975, pp. 51–58.

[22] G. K. Huth, "Optimization of coded spread spectrum systems performance," *IEEE Trans. Commun.*, vol. COM-25, pp. 763–770, Aug. 1977.

[23] R. H. Pettit, "A susceptibility analysis of frequency hopped *M*-ary NCPSK—Partial-band noise on CW tone jamming," presented at the Symp. Syst. Theory, May 1979.

[24] L. B. Milstein, R. L. Pickholtz, D. L. Schilling, "Optimization of the processing gain of an FSK-FH system," *IEEE Trans. Commun.*, vol. COM-28, pp. 1062–1079, July 1980.

[25] M. K. Simon and A. Polydoros, "Coherent detection of frequency-hopped quadrature modulations in the presence of jamming—Part I: QPSK and QASK; Part II: QPR class I modulation," *IEEE Trans. Commun.*, vol. COM-29, pp. 1644–1668, Nov. 1981.

[26] A. J. Viterbi and I. M. Jacobs, "Advances in coding and modulation for noncoherent channels affected by fading, partial band, and multiple access interference," in *Advances in Communication Systems*, vol. 4. New York: Academic, 1975.

[27] J. M. Aein and R. D. Turner, "Effect of co-channel interference on CPSK carriers," *IEEE Trans. Commun.*, vol. COM-21, pp. 783–790, July 1973.

[28] R. H. Pettit, "Error probability for NCFSK with linear FM jamming," *IEEE Trans. Aerosp. Electron. Syst.*, vol. AES-8, pp. 609–614, Sept. 1972.

[29] A. J. Viterbi, "Spread spectrum communications—Myths and realities," *IEEE Commun. Mag.*, pp. 11–18, May 1979.

[30] D. J. Torrieri, *Principles of Military Communication Systems*. Dedham, MA: Artech House, 1981.

[31] M. B. Pursley, "Performance evaluation for phase-coded spread spectrum multiple-access communication—Part I: System analysis," *IEEE Trans. Commun.*, vol. COM-25, pp. 795–799, Aug. 1977.

[32] K. Yao, "Error probability of asynchronous spread-spectrum multiple access communication systems," *IEEE Trans. Commun.*, vol. COM-25, pp. 803–809, Aug. 1977.

[33] C. L. Weber, G. K. Huth, and B. H. Batson, "Performance considerations of code division multiple access systems," *IEEE Trans. Veh. Technol.*, vol. VT-30, pp. 3–10, Feb. 1981.

[34] M. B. Pursley and D. V. Sarwate, "Performance evaluation for phase-coded spread spectrum multiple-access communication—Part II. Code sequence analysis," *IEEE Trans. Commun.*, vol. COM-25, pp. 800–803, Aug. 1977.

[35] M. B. Pursley and H. F. A. Roefs, "Numerical evaluation of correlation parameters for optimal phases of binary shift-register sequences," *IEEE Trans. Commun.*, vol. COM-25, pp. 1597–1604, Aug. 1977.

[36] G. Solomon, "Optimal frequency hopping sequences for multiple access," in *Proc. 1973 Symp. Spread Spectrum Commun.*, vol. 1, AD915852, pp. 33–35.

[37] D. V. Sarwate and M. B. Pursley, "Hopping patterns for frequency-hopped multiple-access communication," in *Proc. 1978 IEEE Int. Conf. Commun.*, vol. 1, pp. 7.4.1–7.4.3.

[38] P. S. Henry, "Spectrum efficiency of a frequency-hopped-DPSK spread spectrum mobile radio system," *IEEE Trans. Veh. Technol.*, vol. VT-28, pp. 327–329, Nov. 1979.

[39] O. C. Yue, "Hard-limited versus linear combining for frequency-hopping multiple-access systems in a Rayleigh fading environment," *IEEE Trans. Veh. Technol.*, vol. VT-30, pp. 10–14, Feb. 1981.

[40] R. W. Nettleton and G. R. Cooper, "Performance of a frequency-hopped differentially modulated spread-spectrum receiver in a Rayleigh fading channel," *IEEE Trans. Veh. Technol.*, vol. VT-30, pp. 14–29, Feb. 1981.

[41] D. E. Borth and M. B. Pursley, "Analysis of direct-sequence spread-spectrum multiple-access communication over Rician fading channels," *IEEE Trans. Commun.*, vol. COM-27, pp. 1566–1577, Oct. 1979.

[42] E. A. Geraniotis and M. B. Pursley, "Error probability bounds for slow frequency-hopped spread-spectrum multiple access communications over fading channels," in *Proc. 1981 Int. Conf. Commun.*

[43] L. B. Milstein and D. L. Schilling, "Performance of a spread spectrum communication system operating over a frequency-selective fading channel in the presence of tone interference," *IEEE Trans. Commun.*, vol. COM-30, pp. 240–247, Jan. 1982.

[44] ——, "The effect of frequency selective fading on a noncoherent FH-FSK system operating with partial-band interference," this issue, pp. 904–912.

[45] J. M. Aein and R. L. Pickholtz, "A simple unified phasor analysis for PN multiple access to limiting repeaters," this issue, pp. 1018–1026.

[46] R. B. Ward, "Acquisition of pseudonoise signals by sequential estimation," *IEEE Trans. Commun. Technol.*, vol. COM-13, pp. 474–483, Dec. 1965.

[47] R. B. Ward and K. P. Yiu, "Acquisition of pseudonoise signals by recursion-aided sequential estimation," *IEEE Trans. Commun.*, vol. COM-25, pp. 784–794, Aug. 1977.

[48] P. M. Hopkins, "A unified analysis of pseudonoise synchronization by envelope correlation," *IEEE Trans. Commun.*, vol. COM-25, pp. 770–778, Aug. 1977.

[49] J. K. Holmes and C. C. Chen, "Acquisition time performance of PN spread-spectrum systems," *IEEE Trans. Commun.*, vol. COM-25, pp. 778–783, Aug. 1977.

[50] S. S. Rappaport, "On practical setting of detection thresholds," *Proc. IEEE*, vol. 57, pp. 1420–1421, Aug. 1969.

[51] J. J. Spilker, Jr., "Delay-lock tracking of binary signals," *IEEE Trans. Space Electron. Telem.*, vol. SET-9, pp. 1–8, Mar. 1963.

[52] P. T. Nielson, "On the acquisition behavior of delay lock loops," *IEEE Trans. Aerosp. Electron. Syst.*, vol. AES-12, pp. 415–523, July 1976.

[53] ——, "On the acquisition behavior of delay lock loops," *IEEE Trans. Aerosp. Electron. Syst.*, vol. AES-11, pp. 415–417, May 1975.

[54] H. P. Hartman, "Analysis of the dithering loop for PN code tracking," *IEEE Trans. Aerosp. Electron. Syst.*, vol. AES-10, pp. 2–9, Jan. 1974.

[55] R. C. Dixon, *Spread Spectrum Systems*. New York: Wiley, 1976.

[56] J. K. Holmes, *Coherent Spread Spectrum Systems*. New York: Wiley, 1982.

# Optimal Detection of Digitally Modulated Signals

NORMAN F. KRASNER, MEMBER, IEEE

*Abstract*—In this paper optimal detectors are derived and analyzed for the general class of digitally modulated signals in which the sequence of symbols is unknown *a priori* and information data are not of interest. The detectors test the signal present condition in background white Gaussian noise versus the null condition of noise alone. Particular attention is focused upon cases in which the SNR per symbol is low compared to unity. The models employed herein are sufficiently general to include most forms of spread-spectrum signals as well as other digital type communication signals.

## I. INTRODUCTION

**T**HE class of digitally modulated signals in which the symbol sequence is unknown *a priori* is addressed in this paper. Such signals consist of a train of symbols, normally placed on a carrier, where each symbol is randomly chosen from a fixed symbol set. Optimal detectors are derived herein to test the hypothesis of the presence of such a signal plus noise against the null hypothesis noise alone. The presence of information data on the signal is of no interest to the receiver. Derivation and study of such detectors not only permits evaluation of the detectability of the presence of such signals, but permits easy determination of maximum-likelihood estimators for the various signals' parameters.

The signals of interest to this study arise in many digital communications systems and in recent years have been used extensively in spread-spectrum communications. Most of the work on optimal detection of this class of signals has been quite signal specific and has often been done in conjunction with derivation of maximum-likelihood estimation of signal parameters (see [1] and [2], for example). On the other hand, more general treatments of detectability have mainly been limited to evaluation of energy detection methods [3]-[5].

This paper shows that the class of digitally modulated signals can be treated in a very unified manner. Particular attention is placed on situations in which there is low received energy to noise power density per symbol and many symbols are observed. Under such circumstances it is shown that in some cases the best energy detection methods are quite poor compared to the optimal detectors, but in other cases such energy detectors do quite well. These properties are completely characterized in terms of properties of the symbol sets. The latter part of the paper derives optimal detectors for digitally modulated signals that have undergone filtering operations prior to transmittal.

Manuscript received July 1, 1981; revised February 4, 1982.
The author is with Probe Systems, Inc., Sunnyvale, CA 94086.

## II. DIGITAL TYPE SIGNALS, UNFILTERED

### A. Signal Format

Throughout the discussion we shall assume that the receiver observes a signal buried in additive, white Gaussian noise with spectral density (single-sided) $N_0$. In particular, when the signal is present the observation process is represented by

$$y(t) = s(t; p) + n(t) \qquad \text{for } t \in [0, T] \tag{1}$$

where $s(t; p)$ is the signal and $n(t)$ is the additive, white Gaussian noise. The term $p$ represents a vector of unknown parameters, which may be modeled as either deterministic or statistical. This vector may consist of, for example, carrier frequency, unknown additive phase angle, symbol sequence, etc. In this paper parameters are either known or modeled statistically. For the case of interest in this section, we shall be more specific and assume that $s(t; p)$ can be modeled as

$$s(t; p) = a(t; q) \cos (2\pi f_0 t + \theta)$$
$$- b(t; q) \sin (2\pi f_0 t + \theta) \tag{2}$$

where

$$a(t; q) = \sum_{k=0}^{[M]} a_{q_k}[t/(T/M) - k + t_0] \, \text{rect} \, (0, T) \tag{3}$$

and

$$b(t; q) = \sum_{k=0}^{[M]} b_{q_k}[t/(T/M) - k + t_0] \, \text{rect} \, (0, T) \tag{4}$$

where $q_k$ is a random process which has integral values in some set, say $q_k \in \{0, 1, \cdots, Q - 1\}$, with probability 1, $\text{rect} \, (0, T)$ is the function which is unity on the interval $[0, T]$ and zero elsewhere, and the notation $[M]$ means the smallest integer greater than or equal to number $M$. The functions $a_{q_k}(\cdot)$ and $b_{q_k}(\cdot)$ are for each $q_k$ nonzero only on the interval $[0, 1]$. The parameters of (2) and (3) are as follows: $f_0$ is unknown carrier frequency, $\theta$ is unknown but constant phase angle assumed uniformly distributed over $[0, 2\pi]$,[1] $M$ is an unknown number of symbols during a time interval of length $T$ ($M$ is generally not an integer) and $M$ is assumed uniformly distributed over an interval $[\breve{M}, \widehat{M}]$, $t_0$ is the unknown symbol transition time modulo one symbol period and is assumed

---

[1] Later the case is considered in which $\theta$ varies from one symbol to the next.

Reprinted from *IEEE Trans. Commun.*, vol. COM-30, pp. 885–895, May 1982.

87

uniformly distributed over $[0, 1]$ (note that the normalization of (3) makes this unknown transition time modulo $T/M$, the correct symbol period for $M$ symbols over $T$ seconds), and finally we shall assume that $q_k$ are independent and identically distributed (i.i.d.) and take on any of $Q$ integral values with equal probability $(Q^{-1})$ for every $k$. Basically, (3) merely represents a string of $M$ symbols of a digital type signal over a time duration $T$. Each symbol may take on any of $Q$ distinct waveforms with equal probability, and different symbols are statistically independent.

As a simple example of the above representation, consider the biphase signal in which we can set $b(t; q) = 0$ and allow $q_k$ to take only two values, 0 and 1. Furthermore, for this case we can set $a_0 = A$ rect $(0, 1)$ and $a_1 = -A$ rect $(0, 1)$ with $A$ the signal amplitude. Note in the expression (3) that the vector $q$ is identical with a vector whose components are $q_k$, the random symbol process, plus the random variables $M$ and $t_0$.

For the most part, we shall assume that the input signal-to-noise ratio, defined by

$$(S/N)_i = \frac{\max\limits_{\text{all } j} \left\{ \int_0^{T/\breve{M}} [a_j{}^2(t/(T/\breve{M})) + b_j{}^2(t/(T/\breve{M}))] \, dt \right\}}{N_0}$$

(5)

is small compared to unity. This is twice the maximum symbol energy to noise density, and its importance will be seen in the next section.

It will prove useful in the following discussions to consider various classes of digitally modulated signals, as specified by the nature of the symbol sets. Throughout we mean by a symbol any of the quantities $a_i(\cdot) + j\, b_i(\cdot)$, $i \in [0, Q-1]$, and inner product is defined in the usual way over $[0, 1]$. Symbol sets that have different symbols orthogonal are termed *orthogonal symbol sets*. Symbol sets such that for each symbol there is an identical symbol with opposite polarity are termed *balanced symbol sets*. Balanced symbol sets in which half the symbols form an orthogonal set are *biorthogonal sets* (including, as a degenerate case, antipodal signal sets), and biorthogonal sets with equal energy symbols are *biorthonormal sets*.

### B. Optimal Detectors–Coherent Carrier Case

The likelihood ratio for discriminating between (1) and the case of noise only is derived from the well-known equation [6]-[8]

$$L(y) = E_{q, f_0, t_0, \theta, M} \left[ \exp\left( \frac{2}{N_0} \int_0^T y(t) s(t; p) \, dt \right. \right.$$
$$\left. \left. - \frac{1}{N_0} \int_0^T s^2(t; p) \, dt \right) \right]$$

(6)

where $E$ denotes expectation with respect to the subscripted variables (assuming all such variables are statistically described).

We shall assume throughout that the second integral in the brackets of (6) is independent of the value of any parameters.[2] This is true if all signals have equal energy and otherwise may be a good approximation for large $M$. We can then lump this latter integral into a multiplicative constant $k_1$ and further reduce (6) to the form

$$L(y) = k_1 E_{q, f_0, t_0, \theta, M} \left\{ \exp\left[ \frac{2}{N_0} \int_0^T y(t) \right. \right.$$
$$\cdot \left[ \sum_{k=0}^{[M]} a_{q_k}(t/(T/M) - k + t_0) \cos(2\pi f_0 t + \theta) \right.$$
$$\left. \left. \left. - b_{q_k}(t/(T/M) - k + t_0) \sin(2\pi f_0 t + \theta) \right] dt \right] \right\}$$

(7)

$$= k_1 E_{q, f_0, t_0, \theta, M} \left\{ \prod_{k=0}^{[M]} \exp\left[ \frac{2}{N_0} \int_{(k-t_0)T/M}^{(k-t_0+1)T/M} y(t) \right. \right.$$
$$\cdot \text{rect}(0, T) [a_{q_k}(t/(T/M) - k + t_0) \cos(2\pi f_0 t + \theta)$$
$$\left. \left. - b_{q_k}(t/(T/M) - k + t_0) \sin(2\pi f_0 t + \theta)] \, dt \right] \right\}.$$

(8)

We can now average over $q$, noting that the assumptions of independence of different symbols allow expectations with respect to $q$ to factor in a multiplicative fashion.

$$L(y) = k_1 E_{f_0, t_0, \theta, M} \left\{ \prod_{k=0}^{[M]} \left\{ \sum_{j=0}^{Q-1} \frac{1}{Q} \right. \right.$$
$$\cdot \exp\left[ \frac{2}{N_0} \int_{(k-t_0)T/M}^{(k-t_0+1)T/M} y(t) \, \text{rect}(0, T) \right.$$
$$\cdot [a_j(t/(T/M) - k + t_0) \cos(2\pi f_0 t + \theta)$$
$$\left. \left. \left. - b_j(t/(T/M) - k + t_0) \sin(2\pi f_0 t + \theta)] \, dt \right] \right\} \right\}$$

(9)

$$= k_1 E_{f_0, t_0, \theta, M}$$
$$\cdot \left\{ \exp\left\{ \sum_{k=0}^{[M]} \ln \left\{ \sum_{j=0}^{Q-1} \frac{1}{Q} \right. \right. \right.$$
$$\cdot \exp\left[ \frac{2}{N_0} \int_{(k-t_0)T/M}^{(k-t_0+1)T/M} y(t) \, \text{rect}(0, T) \right.$$
$$\cdot [a_j(t/(T/M) - k + t_0) \cos(2\pi f_0 t + \theta)$$
$$\left. \left. \left. \left. - b_j(t/(T/M) - k + t_0) \sin(2\pi f_0 t + \theta)] \, dt \right] \right\} \right\} \right\}.$$

(10)

[2] We shall assume that signal energy is known (although this is seldom the case) since most of our results will not depend on this knowledge.

The reason for using the form given in (10) will be made clear in a moment. In order to perform the averaging operation over $t_0$, it is convenient to utilize filtering type nomenclature rather than the correlation operation of (9) or (10). Let us define the functions

$$\tilde{a}_j(t) = a_j(1-t) \tag{11a}$$

$$\tilde{b}_j(t) = b_j(1-t). \tag{11b}$$

In the following equations the notation * will mean the convolution operation. For simplicity, when no evaluation time is provided, it is assumed that the convolution is evaluated at time $t$ (the running variable). Otherwise, we shall use the notation $[d(t) * g(t)]_\tau$ to indicate that the convolution is evaluated at the time $\tau$. Using this nomenclature, we can recast (10) into the form

$$L(y) = k_1 E_{f_0, t_0, \theta, M} \left\{ \exp \left\{ \sum_{k=0}^{[M]} \ln \left\{ \sum_{j=0}^{Q-1} Q^{-1} \right. \right. \right.$$

$$\cdot \exp \left[ \frac{2}{N_0} y(t) \, \text{rect} \, (0, T) \right.$$

$$\cdot \cos(2\pi f_0 t + \theta) * \tilde{a}_j(t/(T/M))$$

$$- \frac{2}{N_0} y(t) \, \text{rect} \, (0, T) \sin(2\pi f_0 t + \theta)$$

$$\left. \left. \left. * \tilde{b}_j(t/(T/M)) \right]_{\frac{(k+1-t_0)T}{M}} \right\} \right\} \right\} \tag{12}$$

$$= k_1 E_{f_0, \theta, M} \left\{ (T/M)^{-1} \, \text{rect} \, (0, T/M) \right.$$

$$* \exp \left\{ \left( \sum_{k=0}^{[M]} \delta \left( t - \frac{kT}{M} \right) \right) \right.$$

$$* \ln \left\{ \sum_{j=0}^{Q-1} Q^{-1} \exp \left[ \frac{2}{N_0} y(t) \, \text{rect} \, (0, T) \right. \right.$$

$$\cdot \cos(2\pi f_0 t + \theta) * \tilde{a}_j(t/(T/M))$$

$$- \frac{2}{N_0} y(t) \, \text{rect} \, (0, T) \sin(2\pi f_0 t + \theta)$$

$$\left. \left. \left. * \tilde{b}_j(t/(T/M)) \right] \right\} \right\} \right\}_{T+\frac{T}{M}}. \tag{13}$$

Equation (13) is the final answer that is possible for the general case. Although it appears extremely complicated, it actually has a fairly easy interpretation if one refers to Fig. 1, which illustrates an implementation of this equation for known $f_0$, $\theta$, and $M$.

When $f_0$, $\theta$, and $M$ are not known, and these parameters are modeled as random variables, the optimum receiver consists of a bank of receivers like those of Fig. 1, which are matched to all possible choices of these three parameters and whose outputs are averaged and compared to a threshold at time $T + T/M$. Obviously, this is impossible to implement. It is expected that rather coarse quantization can be done for the parameters $\theta$ and $M$ without too serious a degradation in performance. However, accurate quantization of $f_0$ is required for good performance (relative to the optimal detector).

It appears from Fig. 1 that in the low input SNR case, simplification is possible by elimination of the first set of $\exp(\cdot)$ operations and the following $\ln(\cdot)$ operation. This can be seen as follows. Denote the square-bracketed portion of (13) by $d_j$, that is,

$$d_j = \frac{2}{N_0} y(t) \, \text{rect} \, (0, T) \cos(2\pi f_0 t + \theta) * \tilde{a}_j(t/(T/M))$$

$$- \frac{2}{N_0} y(t) \, \text{rect} \, (0, T) \sin(2\pi f_0 t + \theta) * \tilde{b}_j(t/(T/M)). \tag{14}$$

The mean value of $d_j$ is at most equal to the right-hand side of (5), which is small compared to unity for the low input SNR case. Also, the variance of $d_j$ is at most equal to (5), as some simple computation shows. Hence, one sees that with high degree of probability this expression is small compared to unity for low input SNR (e.g., $-15$ dB or less). Then one can obtain the following approximation to (13):

$$\ln \sum_{j=0}^{Q-1} Q^{-1} \exp d_j$$

$$\cong \ln \sum_{j=0}^{Q-1} Q^{-1} \left( 1 + d_j + \frac{d_j^2}{2} \right)$$

$$= \ln \left[ 1 + Q^{-1} \sum_{j=0}^{Q-1} \left( d_j + \frac{d_j^2}{2} \right) \right] \tag{15}$$

$$\cong Q^{-1} \sum_{j=0}^{Q-1} d_j + Q^{-1} \sum_{j=0}^{Q-1} \frac{d_j^2}{2}. \tag{16}$$

Recalling that for low input SNR, $d_j \ll 1$, we can then see that (16) reduces to the two results for balanced and orthogonal signals:

$$\ln \sum_{j=0}^{Q-1} Q^{-1} \exp d_j \cong Q^{-1} \sum_{j=0}^{Q-1} \frac{d_j^2}{2} \tag{17a}$$

for balanced symbols, and

$$\cong Q^{-1} \sum_{j=0}^{Q-1} d_j \tag{17b}$$

for orthogonal symbols. Block diagrams for these two cases are shown in Figs. 2 and 3.

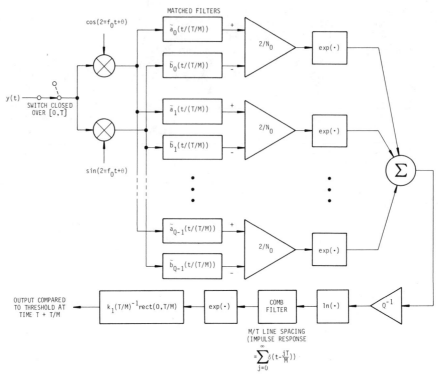

Fig. 1.   Optimum receiver for digitally modulated signal with known carrier phase and frequency and known symbol rate.

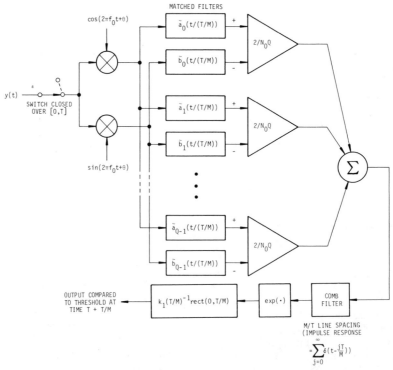

Fig. 2.   Approximation to optimum receiver of Fig. 1 for low input SNR when symbols are orthogonal to one another.

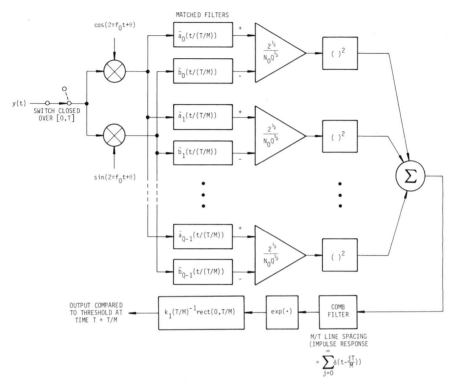

Fig. 3.   Approximation to optimum receiver of Fig. 1 for balanced symbol set and low input SNR.

In Figs. 1–3 the comb filter is designated with an infinite impulse response. This allows these structures to hold when the observation time $T$ is varied (the symbol duration ($T/M$) is held fixed). That is, these structures are recursive. This allows their use in the development of recursive maximum-likelihood estimates of carrier and symbol phases. The exp/summation/ln nonlinearity of Fig. 1 and the comb filter are responsible for the high performance of frequency doublers and other nonlinearities that are often used in conjunction with carrier extraction of suppressed carrier phase shift keyed signals. This nonlinearity also produces a comb of narrow tones at dc plus the symbol rate and its harmonics in a manner closely related to many proposed symbol synchronizers. The dc term is associated with the concept of energy detection or "radiometry."

### C. Optimal Detectors–Incoherent Carrier Case

Suppose now that instead of the carrier phase $\theta$ being constant, it is independent from one symbol to the next. Assume, however, that it is uniformly distributed over $[0, 2\pi]$ for each symbol period. In this case it is not hard to modify the previous analysis to arrive at the resulting likelihood ratio:

$$L(y) = k_1 E_{f_0, M} \left\{ (T/M)^{-1} \operatorname{rect}(0, T/M) \right.$$

$$\left. * \exp \left\{ \left( \sum_{k=0}^{\lfloor M \rfloor} \delta \left( t - \frac{kT}{M} \right) \right) * \ln \left\{ \sum_{j=0}^{Q-1} Q^{-1} \right. \right. \right.$$

$$\cdot I_0 \left[ \frac{2}{N_0} \left[ (L_{acj} - L_{bsj})^2 \right. \right.$$

$$\left. \left. \left. \left. + (L_{asj} + L_{bcj})^2 \right]^{1/2} \right] \right\} \right\}_{T + T/M} \tag{18a}$$

where

$$L_{acj} = y(t) \cos (2\pi f_0 t) * \tilde{a}_j(t/(T/M))$$

$$L_{bsj} = y(t) \sin (2\pi f_0 t) * \tilde{b}_j(t/(T/M))$$

$$L_{asj} = y(t) \sin (2\pi f_0 t) * \tilde{a}_j(t/(T/M))$$

$$L_{bcj} = y(t) \cos (2\pi f_0 t) * \tilde{b}_j(t/(T/M)). \tag{18b}$$

For small $x$ we can approximate $I_0(x)$ by $1 + x^2/4$ and then use this approximation in (18) plus the approximation for $\ln (1 + x) \cong x$ in order to remove two of the nonlinearities of (18) and obtain

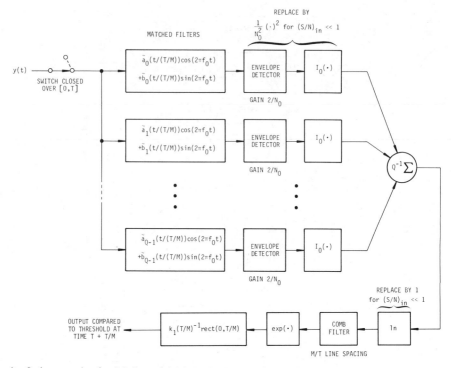

Fig. 4. Optimum receiver for digitally modulated signals with random phase from symbol to symbol and known carrier frequency and symbol rate (signal assumed band limited).

$$L(y) = k_1 E_{f_0,M} \left\{ (T/M)^{-1} \text{ rect } (0, T/M) \right.$$

$$* \exp \left\{ \left( \sum_{k=0}^{[M]} \delta \left( t - \frac{kT}{M} \right) \right) * \sum_{j=0}^{Q-1} \frac{1}{N_0^2 Q} \right.$$

$$\left. \left. \cdot \left[ (L_{acj} - L_{bsj})^2 + (L_{asj} + L_{bcj})^2 \right] \right\} \right\}_{T+T/M}.$$

$$(19)$$

A block diagram corresponding to (18) and (19) is shown in Fig. 4. In this diagram we have replaced the bracketed expressions of (18a) and (19) by filters and envelope detectors, under a band-limited assumption on $s(t; p)$.

### D. Reduced Realization of Detectors

As is evident in Figs. 3 and 4, for low input SNR when symbol sets are either balanced and/or carrier phase is incoherent, the primary portion of the detectors is a quadratic form. Accordingly, by making an appropriate unitary transformation we can replace the symbol set used in the detector by an orthogonal symbol set with a number of nonzero symbols equal to the dimensionality of the original symbol set. This will greatly reduce in many cases the complexity of the optimal detectors, and in addition will aid in the determination of error rates. For example, the number of symbols for a multiple phase shift keyed signal may be arbitrarily large,

but the dimensionality of the symbol set is two, i.e., two sinusoids in quadrature. Throughout this section we assume that $f_0$ is much greater than $M/T$, a fact which will allow us to neglect terms in the detector at $2f_0$.

Turning to the balanced symbol set case, let us assume that of the $Q$ symbols, the last $Q/2$ symbols are antipodal to the first $Q/2$. Let $s_a(t)$ be the column vector function of dimension $Q/2$ with components $\tilde{a}_0(t/(T/M))$, $\tilde{a}_1(t/(T/M))$, $\cdots$, $\tilde{a}_{Q/2-1}(t/(T/M))$ and let $s_b(t)$ be the column vector function of dimension $Q/2$ with components $\tilde{b}_0(t/(T/M))$, $\tilde{b}_1(t/(T/M))$, $\cdots$, $\tilde{b}_{Q/2-1}(t/(T/M))$. The output of the summer of Fig. 3 at time $t$, neglecting multiplicative constants, is then

$$\| y_c(t) * s_a(t) + y_s(t) * s_b(t) \|_t^2 \qquad (20)$$

where

$$y(t) = y_c(t) \cos (2\pi f_0 t + \theta) - y_s(t) \sin (2\pi f_0 t + \theta),$$

$\| \cdot \|_t^2$ is the usual norm defined by

$$\| x \|^2 = \sum_{i=0}^{Q/2-1} x_i^2$$

where $x_i$ is the $i$th component of $x$, and in the norm the subscript $t$ represents evaluation of the convolution operation at time $t$.

We now show that we can replace the symbol set *used in the detector* by an orthogonal symbol set. We form a new symbol set by the transformation $\hat{s}_a(t) = Ps_a(t)$ and $\hat{s}_b(t) = Ps_b(t)$, where $P$ is a $Q/2 \times Q/2$ dimensional matrix that satisfies the following equations:

$$P'P = I_{Q/2} \qquad \text{(i.e., } P \text{ is unitary)} \tag{21a}$$

$$PSP' \text{ is a diagonal matrix} \tag{21b}$$

where $S$ is the $Q/2 \times Q/2$ dimensional matrix whose $(i, j)$th element is given by

$$[S]_{ij} = \int_0^{T/M} [a_i(t)a_j(t) + b_i(t)b_j(t)] \, dt. \tag{21c}$$

A matrix $P$ that satisfies (21) is formed by setting its rows equal to the eigenvectors (orthonormalized) of the matrix $S$. Equation (21b) will then be the matrix with eigenvalues along the diagonal (diagonalization is always possible since $S$ is symmetric).

If we use $\hat{s}_a(t)$ and $\hat{s}_b(t)$ in (20) instead of $s_a(t)$ and $s_b(t)$, we see that the expression is still mathematically identical owing to the unitary property (21a). The transformed symbol set

$$\hat{a}_i(t/(T/M)) \cos (2\pi f_0 t) - \hat{b}_i(t/(T/M)) \sin (2\pi f_0 t)$$

is now orthogonal.

It is easy to see that the same argument holds for the incoherent symbol case. Instead of the quadratic form (20) one uses

$$\|y_c(t) * s_a(t) - y_s(t) * s_b(t)\|^2$$

$$+ \|y_c(t) * s_b(t) + y_s(t) * s_a(t)\|^2 \tag{22}$$

which follows from (19).

A second example of a highly reducible symbol set is that of a frequency-hopped biphase type signal in which there are $L$ phase transitions per hop and $Q_1$ hop frequencies. In this case there are a total of $2^L Q_1$ possible symbols (per hop duration), but a transformed symbol set exists with only $L Q_1$ symbols. This is the case since any sequence of $L$ phase transitions (biphase type) can be formed from a linear combination of $L$ separate pulses delayed relative to one another by multiples of the duration of a transition, the so-called "chip" length.

Actually, an even simpler detector implementation is possible, since this symbol set belongs to a class termed "concatenated" symbol sets. Such sets are characterized by symbols each of which is composed of a sequence of subsymbols. The optimum detector has a treelike structure with branches similar to Fig. 1. Details of this structure can be found in [9].

### E. Comparison of Optimal Detectors with Radiometers

In the constant carrier phase case with low input SNR the detector structures of Figs. 2 and 3 apply when $\theta$, $f_0$, and $M$

are known. If symbol phase is also assumed known, then the last two blocks of these figures may be deleted and one focuses attention on the outputs of the comb filters. For long integration times, i.e., large $M$, the statistics of these outputs are nearly Gaussian. Accordingly, useful measures of performance may be gotten for these situations in terms of the mean value and variance at the comb filter output for the signal present and signal absent conditions. Even when the abovementioned parameters are random variables, these same measures of performance are still important, as discussed later.

Assume then that all signal parameters are known except for the symbol sequence. In the following discussion we define the mean shift $\mu_s$ to be the difference in the mean value at the comb filter output between the signal present condition and the signal absent condition. The variance $\sigma_s^2$ is that at the comb filter output for the signal absent condition. For low input SNR situations the variance is only marginally altered if the signal is included. In all cases the comb filter output is sampled at time $T + T/M$. We define output signal-to-noise ratio $(S/N)_0$ to be the ratio $\mu_s^2/\sigma_s^2$.

For the orthonormal case it is easy to show from Fig. 2 that

$$\mu_s = \sigma_s^2 = \frac{E\left\{ \int_0^T [a^2(t; q) + b^2(t; q)] \, dt \right\}}{QN_0} \triangleq \frac{2E_s(T)}{QN_0} \tag{23}$$

and hence

$$(S/N)_0 = \mu_s^2/\sigma_s^2 = 2E_s(T)/QN_0. \tag{24}$$

Note that the output SNR is linearly related to the input SNR as defined by (5), that is,

$$(S/N)_i = 2E_s(T/M)/N_0 = 2E_s(T)/MN_0 \tag{25}$$

where $2E_s(T/M)$ is the numerator of (5). This linear relationship makes this class of signals highly detectable. It should be emphasized that this linear relationship only holds for the carrier phase continuous case.

For the case of biorthonormal symbol sets one can find from Fig. 3 that

$$\mu_s = \sigma_s^2 = 4E_s^2(T)/QN_0^2 M \tag{26}$$

and hence

$$(S/N)_0 = 4E_s^2(T)/QN_0^2 M. \tag{27}$$

In deriving (26) one must note that for any symbol of Fig. 3 there is an equal and opposite polarity symbol. Note that here the output SNR is proportional to the square of the input SNR, a fact which makes these signals much less detectable than the orthonormal symbol set case at low input SNR. In the balanced symbol set case with equal energy symbols the results (26) and (27) still apply if one uses for $Q$ twice the number of *linearly independent symbols*. This follows directly

93

from the reduced realization of detectors presented in Section II-D.

It is important to compare the above results to performance possible with an energy detector, or "radiometer." This device consists simply of a bandpass filter followed by a squaring device and integrator. It is not difficult to show that the optimum radiometer in the low input SNR case is one that has a transfer function whose magnitude squared is equal to the power spectrum of the signal in question. To simplify the discussion assume that the power spectrum of the signal is relatively flat over a bandwidth $W$. Then it is a simple matter to compute the SNR out of the radiometer assuming a signal is present at its input with low input SNR. The result is

$$(S/N)_0 = E_s^2(T)/N_0^2 WT. \tag{28}$$

In all cases this will be poor compared to the optimal detector for the orthonormal case if the input SNR is low. For the biorthonormal case one can find the ratio of the output SNR's of the radiometer and optimal detector:

$$\frac{(S/N)_0 \text{ radiometer}}{(S/N)_0 \text{ optimal}} \approx \frac{Q}{4(T/M)W}. \tag{29}$$

The term $(T/M)W$ in the denominator of (29) is the so-called "time-bandwidth product" of the signal,[3] which is greater than $Q/4$ for a biorthonormal symbol set with $Q$ symbols ($Q/2$ are orthogonal). Hence, the radiometer can have output SNR close to that of the optimal detector.

In order to examine the accuracy of (29), let us examine two cases in which optimal radiometer performance is readily evaluated. Equation (29) is based upon a flat spectrum approximation for the radiometer performance computation. Consider the case of quadriphase PN (biorthonormal) which has symbol set $A \cos(2\pi f_0 t + k\pi/4)$, $k = 0, 1, 2, 3$, and $t \in [0, T/M]$. The best radiometer utilizes a bandpass filter with impulse response $\cos(2\pi f_0 t) \text{ rect } (0, T/M)$. It is straightforward to compute the mean-shift, variance, and hence output SNR for this radiometer with result $(S/N)_0 = 2/3 \, E_s^2(T)/N_0^2 M$. Comparing this to (27) with $Q = 4$ shows that the ratio of output SNR for this radiometer to that for the optimal detector is 2/3. Hence, (29) is an upper bound for this case. Similarly, for the biphase case with the same radiometer just discussed, the ratio is 1/3 although (29) gives result 1/2.

Another example of biorthonormal signal sets is the frequency-hopping case in which allowable carrier frequencies are spaced relative to one another by multiples of the symbol rate $(M/T)$ and where for each frequency and symbol period there are four possible symbols. For large bandwidth $W$ there are then a total of $4W/(M/T) = 4TW/M$ symbols and, hence, the approximation in (29) becomes unity. For very large $TW/M$ one would expect that (29) would be quite accurate for the best radiometer since the power spectrum of the signal is quite close to rectangular in shape, as some simple calculation shows. It can be seen from the preceding that for the same bandwidth, quadriphase PN and frequency hopping have

approximately the same detectability, although a precise comparison depends upon the exact definition of bandwidth and other factors.

The ratio (29) shows that in some cases a radiometer is very poor compared to the optimal. For example, $Q$ might be 2 and $(T/M)W$ can be very large, say 1000. An example is the two symbol antipodal set in which each symbol is a linearly swept FM signal which sweeps over a bandwidth of $(1000/(T/M))$ during the symbol period $T/M$.

It is not hard to show that the approximate result (29) holds for the more general case of *equal energy* balanced symbol sets if $Q$ is interpreted as twice the dimension of the symbol space. This follows directly from the norm preserving transformation (21) discussed in Section II-D.

For the orthonormal symbol set case in which carrier phase is random from one symbol to the next it follows from Fig. 4 with low input SNR that $\mu_s = \sigma_s^2 = \mu_s^2/\sigma_s^2 = E_s^2/QN_0^2 M$. These quantities are defined as in the coherent case and we assume all quantities are known except symbol sequence. For this case we then have

$$\frac{(S/N)_0 \text{ radiometer}}{(S/N)_0 \text{ optimal}} \approx \frac{Q}{(T/M)W}. \tag{30}$$

Note that the $Q = 1$ incoherent case yields the same performance as the $Q = 4$ biorthonormal coherent case.

The output SNR's computed in this section allow development of good approximate bounds to probability of detection for fixed false alarm rate (Neyman–Pearson criterion), for low input SNR situations and long integration times ($M$ large). Such bounds are derived under the assumption that the comb filter outputs of Figs. 2–4 have Gaussian statistics, an assumption which is asymptotically correct as $M$ approaches infinity. Here, no assumption is made that parameters are known, but instead they are statistically described as in Section II-A. Upper bounds are easily found by evaluating the detection probability $P_d$ for the case of all parameters known except symbol sequence. Clearly, this $P_d$ must dominate that for the random parameter case. Lower bounds are found by evaluating the detection probability for a well-chosen suboptimal detector. This suboptimal detector is constructed from a set of "parallel channels," each one of which is "matched" to a set of hypothesized parameters. If the output of any channel exceeds a threshold, then a detection is made. By properly choosing such channels one can establish good approximate lower bounds. Details of these procedures may be found in [9]. For the coherent biphase case these bounds have shown that even with great uncertainty in all parameters, the optimal detector requires only an increase in signal energy of 2-3 dB to achieve the same $P_d$ as in the case with no uncertainty.

## III. FILTERED AND CORRELATED SIGNALS

In this section we consider the situation in which the signal of interest is formed by passing a signal of the form (2) through a bandpass filter, which may severely distort the waveform. Such filtering is often performed on spread-spectrum type signals to make them appear more "noise-like." In addition, there are many important signal types which consist of

---

[3] Measured over a symbol period.

a sequence of symbols which are correlated with one another to produce a desired effect, such as good spectral properties. Examples of these are staggered quadriphase shift keyed signals (SQPSK) and minimum shift keying (MSK). These may also be modeled as filtered signals of the type (2). We shall show in this section that the optimal detectors for these classes of signals are found by simply cascading the matched filters used in optimal detectors for the unfiltered case with a filter matched to that used in the signal production.

In order to make the following presentation more palatable, we shall first consider the case of biphase signals and furthermore assume that all signal parameters are known with the exception of the symbol sequence. In addition, we shall assume initially that the impulse response of the filter under consideration is finite in duration. All these assumptions will later be eliminated to yield the general situation.

The situation of interest is then one in which the signal has the form

$$s(t) = \sum_{k=0}^{M-1} (-1)^{\phi_k} A \cos(2\pi f_0 t) \, \text{rect} \, (kT/M, (k+1)T/M)$$

$$(31)$$

where $\phi_k$ are i.i.d. random variables assuming values 0 and 1 with equal probability, and that it is passed through a filter with impulse response $h(t)$ which is zero outside the interval $[0, DT/M]$, where $D$ is an integer. Hence, the signal that we deal with has the form

$$\tilde{s}(t) = [s(t) * h(t)]_t; \qquad t \in [0, T + DT/M]. \tag{32}$$

It will be convenient in the following to make the definitions

$$m(t) = h(t) * A \cos(2\pi f_0 t) \, \text{rect} \, (0, T/M) \tag{33a}$$

$$= \sum_{j=0}^{D} [m_j(t - jT/M) \cos(2\pi f_0 t)$$

$$+ \hat{m}_j(t - jT/M) \sin(2\pi f_0 t)] \tag{33b}$$

$$\triangleq m_c(t) \cos(2\pi f_0 t) + m_s(t) \sin(2\pi f_0 t) \tag{33c}$$

where (33b) follows from an assumption that $h(t)$ is a "narrow-band" function, and where the functions $m_j(t)$ and $\hat{m}_j(t)$ are zero outside $[0, T/M]$. Then (32) can be rewritten

$$\tilde{s}(t) = \left[ \sum_{k=0}^{M-1} \sum_{j=0}^{D} (-1)^{\phi_k} m_j(t - (j+k)T/M) \right] \cos(2\pi f_0 t)$$

$$+ \left[ \sum_{k=0}^{M-1} \sum_{j=0}^{D} (-1)^{\phi_k} \hat{m}_j(t - (j+k)T/M) \right] \sin(2\pi f_0 t).$$

$$(34)$$

Now, the detection problem is to discriminate between the situations of signal plus noise versus noise alone, with the sig-

nal format that of (34). If we knew the sequence $\phi_k$, then the likelihood ratio for the problem is completely straightforward:

$$L(y) = K_1 \exp \left\{ \left[ \frac{2}{N_0} \int_0^{T+DT/M} y(t) \cos(2\pi f_0 t) \right. \right.$$

$$\left. \cdot \sum_{k=0}^{M-1} \sum_{j=0}^{D} (-1)^{\phi_k} m_j(t - (j+k)T/M) \, dt \right]$$

$$+ \left[ \frac{2}{N_0} \int_0^{T+DT/M} y(t) \sin(2\pi f_0 t) \right.$$

$$\left. \left. \cdot \sum_{k=0}^{M-1} \sum_{j=0}^{D} (-1)^{\phi_k} \hat{m}_j(t - (j+k)T/M) \, dt \right] \right\} \tag{35a}$$

where

$$K_1 = \exp \left[ -\frac{1}{N_0} \int_0^{T+DT/M} \tilde{s}^2(t) \, dt \right]. \tag{35b}$$

We need only average over all possible sequences $\phi_k$ to arrive at the detector for the case of interest. This is straightforward and yields

$$L(y) = K_1 \prod_{k=0}^{M-1} \cosh \left\{ \left[ \frac{2}{N_0} \int_{kT/M}^{(k+D+1)T/M} y(t) \right. \right.$$

$$\left. \cdot \cos(2\pi f_0 t) \sum_{j=0}^{D} m_j(t - (j+k)T/M) \, dt \right]$$

$$+ \left[ \frac{2}{N_0} \int_{kT/M}^{(k+D+1)T/M} y(t) \sin(2\pi f_0 t) \right.$$

$$\left. \left. \cdot \sum_{j=0}^{D} \hat{m}_j(t - (j+k)T/M) \, dt \right] \right\} \tag{36a}$$

$$= K_1 \exp \left\{ \left\{ \ln \cosh \left[ \frac{2}{N_0} y(t) \cos(2\pi f_0 t) \right. \right. \right.$$

$$\left. * \tilde{m}_c(t) + \frac{2}{N_0} y(t) \sin(2\pi f_0 t) * \tilde{m}_s(t) \right] \right\}$$

$$\left. * \sum_{k=0}^{M-1} \delta(t - kT/M) \right\}_{T+(D+1)T/M} \tag{36b}$$

where

$$\tilde{m}_c(t) = m_c \left( \frac{(D+1)T}{M} - t \right), \quad \tilde{m}_s(t) = m_s \left( \frac{(D+1)T}{M} - t \right).$$

$$(36c)$$

As usual, we can incorporate the averaging over the unknown

parameters and obtain

$$L(y) = K_1 E_{f_0 \theta, M} \left\{ (M/T) \text{ rect } (0, T/M) \right.$$

$$* \exp \left\{ \sum_{k=0}^{[M]} \delta(t - kT/M) * \left\{ \ln \cosh \left[ \frac{2}{N_0} y(t) \right. \right. \right.$$

$$\left. \cdot \cos (2\pi f_0 t + \theta) * \tilde{m}_c(t) + \frac{2}{N_0} y(t) \right.$$

$$\left. \left. \left. \cdot \sin (2\pi f_0 t + \theta) * \tilde{m}_s(t) \right] \right\} \right\} \right\}_{T + (D+1)T/M} . \qquad (37)$$

Note that $m_c$ and $m_s$ are functions of $M$, although not explicitly shown.

This result is similar to that for the biphase case without the filtering except that, in the latter case, $\tilde{m}_c(t)$ would be replaced with $A$ rect $(0, T/M)$ and $\tilde{m}_s(t)$ would be set to zero. In the above, we assume that $y(t)$ is time limited to the interval $[0, T]$. Note the time at which the detection threshold is compared: $T + (D+1)T/M$.

The analysis for the general case in which $s(t)$ is given by (2) and (3) follows the same path as that of the preceding result and produces the likelihood ratio

$$L(y) = K_1 E_{f_0, \theta, M} \left\{ (T/M)^{-1} \text{ rect } (0, T/M) \right.$$

$$* \exp \left\{ \left( \sum_{k=0}^{[M]} \delta(t - kT/M) \right) \right.$$

$$* \ln \left\{ \sum_{q=0}^{Q-1} Q^{-1} \exp \left[ \frac{2}{N_0} y(t) \text{ rect } (0, T) \right. \right.$$

$$\left. \cdot \cos (2\pi f_0 t) * \tilde{m}_c{}^q(t) + \frac{2}{N_0} y(t) \text{ rect } (0, T) \right.$$

$$\left. \left. \left. \cdot \sin (2\pi f_0 t + \theta) * \tilde{m}_s{}^q(t) \right] \right\} \right\}_{T + (D+1)T/M}$$

$$(38a)$$

$$m^q(t) = h(t) * [a_q(t/(T/M)) \cos (2\pi f_0 t + \theta)$$

$$- b_q(t/(T/M)) \sin (2\pi f_0 t + \theta)] \text{ rect } (0, T/M) \quad (38b)$$

$$= m_c{}^q(t) \cos (2\pi f_0 t) + m_s{}^q(t) \sin (2\pi f_0 t) \qquad (38c)$$

$$\tilde{m}_c{}^q(t) = m_c{}^q((D+1)T/M - t)$$

$$\tilde{m}_s{}^q(T) = m_s{}^q((D+1)T/M - t). \qquad (38d)$$

Again, $m_c{}^q(t)$ and $m_s{}^q(t)$ are actually functions of $M$, although not explicitly denoted, and the functions $a_q(t/(T/M))$ and $b_q(t/(T/M))$ are nonzero only over the interval $[0, T/M]$.

Similarly, it is not difficult to find the likelihood ratio for filtered digital signals in which the phase angle is random from symbol to symbol. The result is (see Fig. 4)

$$L(y) = K_1 E_{f_0, M} \left\{ (T/M)^{-1} \text{ rect } (0, T/M) \right.$$

$$* \exp \left\{ \left( \sum_{k=0}^{[M]} \delta(t - kT/M) \right) * \ln \left\{ \sum_{q=0}^{Q-1} Q^{-1} \right. \right.$$

$$\left. \cdot I_0 \left[ \frac{2}{N_0} \text{ env } [y(t) \text{ rect } (0, T)] \right. \right.$$

$$\left. \left. \left. * \tilde{m}^q (t)] \right] \right\} \right\}_{T + (D+1)T/M} \qquad (39)$$

where $m^q(t)$ is given by (38) except that $\theta$ is set to zero and env is the envelope operation. In both of the above results $K_1$ is defined by (35b) and (32).

In principle, we can allow filter functions $h(t)$ with impulse responses infinitely long, i.e., $D$ approaching infinity. The results (38) and (39) still hold. However, it is seen that the sampling time of the detector is $T + (D+1)T/M$ and, hence, if we let $D$ approach infinity, we must wait infinitely long in order to perform optimally. In practice, however, the filter impulse response is usually negligibly small in amplitude after a period of time equal to several times the reciprocal of the filter bandwidth.

The results of this section also apply to a slightly different formulation of the problem, where instead of considering the signal to be formed in the manner shown in (32), we consider that the baseband symbols are first filtered *prior* to being placed on a carrier. That is, the formulation is

$$s(t) = [h_c(t) * a(t; q)] \cos (2\pi f_0 t + \theta)$$

$$- [h_s(t) * b(t; q)] \sin (2\pi f_0 t + \theta) \qquad (40a)$$

$$= m_c(t) \cos (2\pi f_0 t + \theta) - m_s(t) \sin (2\pi f_0 t + \theta) \quad (40b)$$

where the terminology follows (2) and (3). The individual symbols that make up $a(t; q)$ and $b(t; q)$ are then replaced by

$$m_c{}^q(t) = h_c(t) * a_q [t/(T/M)]$$

$$m_s{}^q(t) = h_s(t) * b_q [t/(T/M)]. \qquad (41)$$

We then assume that the impulse responses of $h_c(t)$ and $h_s(t)$ are nonzero only on $[0, DT/M]$. It is easy to see from the previous derivation that the results of (38) and (39) still hold, providing that we use the definitions of $m_c{}^q$ and $m_s{}^q$ in (41) rather than those used previously in (38).

We illustrate the use of the above results with examples of the staggered quadriphase and MSK signal types. Staggered quadriphase is formed by adding in quadrature two biphase signals with a relative delay of one-half symbol duration between the two. That is, a signal of the form $A_{2k} \cos (2\pi f_0 t)$ is added to one of the form $A_{2k+1} \sin (2\pi f_0 t)$, where $A_j$ are i.i.d. random variables assuming values $\pm A$ with equal probability, the transition times of $A_{2k}$ occur at multiples of $T_s$ and those of $A_{2k+1}$ at $T_s/2$ plus multiples of $T_s$. We can consider that this signal is generated from a symbol set of four symbols, $\pm A \cos{(2\pi f_0 t)} \text{ rect } (0, T_s) \pm A \sin{(2\pi f_0 t)} \text{ rect } (0, T_s)$, and that the baseband symbols corresponding to the $\sin{(2\pi f_0 t)}$ term are passed through a filter with impulse response $\delta(t - T_s/2)$, i.e., a delay of $T_s/2$. Accordingly, the optimum detector looks like that for a standard QPSK signal except the matched filters following the cos multiplier (cf. Fig. 1) are preceded by a delay of $T_s/2$.

In the MSK case the symbols $A_{2k}$ and $A_{2k+1}$ have the forms $\pm A \cos{(\pi t/T_s)}$ and $\pm A \sin{(\pi t/T_s)}$, respectively. Hence, we can follow the identical procedure as was used for staggered QPSK to arrive at the optimum detector for this class of signals.

## IV. CONCLUDING REMARKS

Extensions of the results of this paper can easily be made in several directions. First, the symbol set size can sometimes be infinite in those cases where reduced realizations are possible, since the optimal detector (low SNR per symbol case) need only contain a number of matched filters equal to the dimensionality of the set. Thus, the optimal detector for the MPSK case with an infinite number of phases (uniformly distributed over $[0, 2\pi]$) still only contains two matched filters.

The probability distribution of symbols over the symbol set need not be uniform. The modification required is simply weighting in a nonuniform manner the terms into the summer of Figs. 1 or 4.

One open question remaining concerns the performance of the optimal detectors in the filtered cases. No general results of this type have so far been obtained, although analysis is often possible in specific cases. One normally would expect the relative performance of a radiometer to improve if a heavy *noninvertible* filtering operation is employed on the symbol set.

## ACKNOWLEDGMENT

The author would like to thank the referees for their careful reading of the manuscript and for many suggestions that have greatly improved the clarity of this paper.

## REFERENCES

[1] W. C. Lindsey and M. K. Simon, *Telecommunications Systems Engineering.* Englewood Cliffs, NJ: Prentice-Hall, 1973, sect. 9.2.

[2] M. H. Meyers and L. E. Franks, "Joint carrier phase and symbol timing recovery for PAM systems," *IEEE Trans. Commun.*, vol. COM-28, pp. 1121–1129, Aug. 1980.

[3] H. Urkowitz, "Energy detection of unknown deterministic signals," *Proc. IEEE*, vol. 55, pp. 523–531, April 1967.

[4] K. Y. Park, "Performance evaluation of energy detectors," *IEEE Trans. Aerosp. Electron. Syst.*, vol. AES-14, pp. 237–241, Mar. 1978.

[5] R. A. Dillard, "Detectability of spread-spectrum signals," *IEEE Trans. Aerosp. Electron. Syst.*, vol. AES-15, pp. 526–537, July 1979.

[6] H. L. Van Trees, *Detection, Estimation, and Modulation Theory, Part 1.* New York: Wiley, 1968, sect. 4.2.1, 4.4.

[7] C. W. Helstrom, *Statistical Theory of Signal Detection*, 2nd ed. Long Island City, NY: Pergamon, 1968, ch. V.

[8] E. J. Kelley, I. S. Reed, and W. L. Root, "The detection of radar echoes in noise," *J. Soc. Ind. Appl. Math.*, vol. 8, pp. 309–341, 481–507, Sept. 1960.

[9] N. F. Krasner, "Optimal processing of digitally modulated signals, Part 1: Detection," Probe Systems, Inc., Sunnyvale, CA, Rep. PSI-ER-9348-01, Dec. 1981.

# Part II
# Interference Rejection

B Y far and away, the most important use of spread-spectrum communications is in the area of interference rejection, and more often than not, the interference is intentional, commonly referred to as jamming. While there are a variety of spreading techniques, such as direct sequence (DS), frequency hopping (FH), time hopping and chirp, as well as hybrids using two or more of the above techniques, most of the interest, at least in the open literature, is in the areas of DS or FH techniques. Each of these two has its own set of advantages and disadvantages with respect to the other, and hence neither is clearly superior to the other.

For example, it is known that DS systems provide more accurate range measurements, and DS techniques are more amenable to coherent detection receivers than are FH schemes, which are ordinarily used with noncoherent detectors. Also, FH systems are more susceptible to relatively simple partial-band jamming than are DS systems. The analog to partial-band jamming in an FH system is pulse burst jamming in a DS system. But pulse burst jamming requires a higher peak power to achieve the same average power of a continuous time jamming signal, and this places more difficult requirements on the technology used to generate the interference.

Alternately, FH signals can typically achieve wider spread bandwidths, are easier to acquire, and do not require a contiguous spectrum for spreading (specific frequencies or bands can be avoided). Other comparisons between the two spreading techniques are possible, but it should be clear that the specific conditions under which a given system has to operate must be considered in making the final choice as to which technique to use.

While intentional jamming has been emphasized above, jamming signals are, of course, not the only type of interfering signals. Another very common type of interference is that due to multipath. Multipath channels are often referred to as frequency selective fading channels, which means that different parts of the frequency spectrum of the signal fade differently from one another, and spread-spectrum signals are known to be effective in combating such multipath induced interference.

There are nine papers concerned with interference rejection. Both DS and FH systems are considered, and the material spanned by these nine papers is reasonably indicative of the many areas of research that fall into the category of "interference rejection."

The first part of this book, "Perspectives," referred to the importance of coding in achieving acceptable interference-rejection performance. Coded error probabilities of antijam systems are often difficult to evaluate. Omura and Levitt ("Coded Error Probability Evaluation for Antijam Communications Systems") decouple the coding aspects of antijam design from the rest of the system and derive coded error probability bounds in terms of the channel cutoff data rate parameter $R_0$. They compare numerically computable values of $R_0$ for several detection and decoder metrics to indicate a systematic general approach to the performance evaluation of antijam communications which includes error-correction coding. They also indicate how side information (i.e., varying channel parameters, knowledge of jammer characteristics, etc.) can be included in the evaluation.

Milstein and Schilling ("The Effect of Frequency-Selective Fading on a Noncoherent FH-FSK System Operating with Partial-Band Tone Interference") compare the probability of error with and without error-correction coding of an FH-FSK signal format in the presence of partial-band tone interference under frequency-selective Rician channel conditions. Their analysis examines the sensitivity of error rate to the number of interfering tones, the interference-to-signal ratio in each frequency slot, and the relative amount of power in the scatter component of the received signal compared to the specular component. They show that performance can degrade rapidly with increased power in the scatter component and that, similar to the nonfading channel, for a given set of system parameters there is a value for the number of interfering tones that causes the most degradation in the probability of error. This value occurs when the interference-to-signal ratio in each frequency slot interfered with is approximately 0 dB.

In many interference environments, the combination of coding and spread-spectrum processing may not be sufficient to reduce the interference effects to an acceptable level. Thus, ancillary techniques may be required to augment the system's interference-rejection performance. Ketchum and Proakis ("Adaptive Algorithms for Estimating and Suppressing Narrow-Band Interference in PN Spread-Spectrum Systems") describe the signal-to-interference ratio improvements that can be obtained by adaptive prefiltering of direct-sequence spread-spectrum signals corrupted by both single and multiple narrow-band interference. They compare the effectiveness of prefilter designs based on 1) spectral estimation of the receiver input signal plus interference, and 2) linear prediction that assumes that the interference can be modeled by white noise passed through an all-pole filter. The results achievable by both approaches are shown to be generally similar. However, the authors indicate that superior performance (up to 5 dB additional S/I improvement) is gained for the linear-prediction approach if the derived prefilter is followed by its own matched filter. Li and Milstein ("Rejection of Narrow-Band Interference in PN Spread-Spectrum Systems Using Transversal Filters") also consider the problem of using two-sided transversal filters and prediction-error filters to reject narrow-band interference. Their results indicate that when the interference tone is near the carrier of the spread-spectrum signal, the performance improvement of the two-sided transversal filter is better than that of the prediction-error filter having the same number of taps. In addition, results are presented suggesting that the two-sided transversal filter is also better for multiple-tone interference conditions.

The spatial filtering capability of antenna arrays provides an additional approach for augmenting interference-rejection performance. The combination of adaptive arrays with spread-spectrum communications has been a topic of research for many years now, and the paper by Winters ("Spread Spectrum in a Four-Phase Communication System Employing Adaptive Antennas") presents some new results in this area. In addition to providing the expected antijam capability, the technique is further enhanced by providing for rapid acquisition of the spread signal.

Matsumoto and Cooper ("Performance of a Nonlinear FH-DPSK Spread-Spectrum Receiver with Multiple Narrow-Band Interfering Signals") compare the performance in both fading and nonfading channels of an FH-DPSK signal structure for linear and nonlinear receivers employing differentially coherent matched filters. In the nonlinear receiver, a bandpass limiter follows the matched filter for each hopped frequency. The FH signal is designed for multipath applications that assume independent relatively slow fading in each FH channel. In this paper, the interference is similar to the desired signal, and is modeled as a summation of sinusoids with random phasing. The authors show that the nonlinear receiver is superior to the linear receiver for any signal-to-interference ratio and any number of interfering signals. In contrast to Matsumoto and Cooper, Lee and Miller ("Error Performance Analyses of Differential Phase-Shift-Keyed/Frequency-Hopping Spread-Spectrum Communication System in the Partial Band Jamming Environment") address the derivation of the probability of error for partial-band interference in the context of a pure FH spread-spectrum system in which multipath is not expected to be a significant factor. Here, earlier probability-of-error analyses are extended to cover all the different jamming effects in the decision process. Their results potentially have broader applicability than the specific implementation that they describe.

A third paper on FH-DPSK is by Simon ("The Performance of $M$-ary FH-DPSK in the Presence of Partial-Band Multitone Jamming"). This paper presents general results for the average probability of error and includes in the analysis the optimal jammer strategy (i.e., the one that maximizes the probability of error).

In the final paper in this section, Smith ("Tradeoff Between Processing Gain and Interference Immunity in Co-Site Multichannel Spread-Spectrum Communications") describes the susceptibility of spread-spectrum radios to mutual and cosite interference, identifies potential interference-rejection solutions (nonintersecting frequency sets, code division, spectrum shaping, etc.), and assesses their relative effectiveness. The primary emphasis is on FH-FSK systems, although PN-PSK systems are also examined. The analytical results are applied to tactical line-of-sight voice radio communications; the effects of multiple independent voice nets, multiple cosite radios, and antenna isolation on overall system performance are illustrated. The capabilities of alternative interference-control methods to provide efficient multichannel spectrum utilization (in terms of the number of independent channels allowable per megahertz of band allocation) are compared. Results are parameterized as a function of the differential loss in processing gain necessary to achieve a desired bit error rate.

# Coded Error Probability Evaluation for Antijam Communication Systems

JIM K. OMURA, FELLOW, IEEE, AND BARRY K. LEVITT, MEMBER, IEEEE

*Abstract*—We present a general union-Chernoff bound on the bit error probability for coded communication systems and apply it to examples of antijam systems. The key feature of this bound is the decoupling of the coding aspects of the system from the remaining part of the communication system which includes jamming, suboptimum detectors, and arbitrary decoding metrics which may or may not use jammer state knowledge.

## I. INTRODUCTION

**T**HE use of coding techniques is extremely important in antijam communication systems where coding gains are usually much greater than in conventional communication systems [1]. Evaluations of coded error probabilities for antijam communication systems are also more difficult since here the decoding decision metrics are generally no longer maximum likelihood metrics and there is a variety of detector forms that may be considered. Against jamming there are no optimum detectors. We are usually interested in robust detectors that are effective with all types of jammers. In addition, the receiver may have side information available such as knowledge of when a jammer signal is on or not during the transmission of a coded symbol. Side information may also include time-varying channel parameter values such as propagation conditions at HF frequencies measured with channel sounders or the number of other friendly active transmitters in a code division multiple access environment.

In this paper we shall consider the general antijam communication system illustrated in Fig. 1. We will derive a general coded bit error bound which will serve as the basis for evaluating the performance of all such complex communication systems. These bounds are easy to evaluate, whereas exact bit error probability evaluations are usually impractical to do. The key feature of this approach is the decoupling of the coding aspects of the system from the remaining part of the communication system. Specifically, we compute the cutoff rate parameter [1]-[4]

$$R_0 \text{ bits/channel use} \tag{1}$$

which represents the practically achievable reliable data rate per coded symbol where the coding channel is the part en-

Manuscript received August 4, 1981; revised January 25, 1982. This work was sponsored by the U.S. Army Communications and Electronics Command under Contract NAS 7-100.

J. K. Omura is with the Department of System Science, School of Engineering and Applied Science, University of California, Los Angeles, CA 90024.

B. K. Levitt is with the Communication Systems Research Section, Jet Propulsion Laboratory, Pasadena, CA 91103.

closed by dotted lines in Fig. 1. This cutoff rate will be a function of the equivalent channel symbol energy to noise ratio

$$\frac{E_c}{N_0} = r \left( \frac{E_b}{N_0} \right) \tag{2}$$

which is shown here to be directly related to the energy-per-bit to noise ratio usually given by (assuming jamming limits performance)[1]

$$\frac{E_b}{N_0} = \frac{PG}{J/S}. \tag{3}$$

Here $r$ is the data rate in bits per channel signal and we have the usual jamming parameters

$$
\begin{aligned}
PG \quad &= \frac{W}{R}, \text{ processing gain} \\
S \quad &= \text{signal power} \\
J \quad &= \text{jammer power} \\
W \quad &= \text{spread bandwidth}[2] \\
R \quad &= \text{data rate in bits per second.}
\end{aligned}
\tag{4}
$$

For conventional coded direct sequence (DS) coherent BPSK signals, $E_c$ is the energy per coded bit while for frequency-hopped (FH) noncoherent MFSK, $E_c$ is the energy of each hopped signal.

One of the key points of this paper is that for any specific code we can derive a bound on the coded bit error probability of the form

$$P_b \leqslant B(R_0) \tag{5}$$

which is only a function of the cutoff rate $R_0$. Since the function $B(R_0)$ is unique for each code and the cutoff rate parameter $R_0$ is independent of the code used, we are able to decouple the coding from the rest of the communication system. Thus, to evaluate various antijam communication systems we can first compare them using the cutoff rate parameter. Codes can then be evaluated separately. This decoupling of the code and the coding channel (shown enclosed by dotted lines

---

[1] We use this same definition for all spread-spectrum systems.
[2] Measured in hertz and centered about the carrier frequency.

Reprinted from *IEEE Trans. Commun.*, vol. COM-30, pp. 896–903, May 1982.

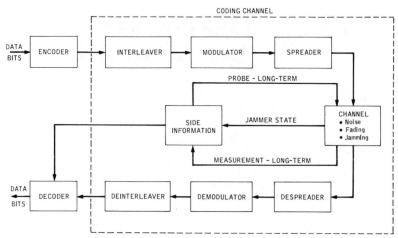

Fig. 1.   AJ system overview.

in Fig. 1) is possible for antipodal and orthogonal signals such as those commonly used in antijam communication systems.

## II. PAIRWISE ERROR CHERNOFF BOUND

Evaluation of coded bit error bounds are based on pairwise error probabilities [5]. Consider two coded sequences of $N$ symbols denoted

$$x = (x_1, x_2, \cdots, x_N) \tag{6}$$

and

$$\hat{x} = (\hat{x}_1, \hat{x}_2, \cdots, \hat{x}_N).$$

Assume the coded sequence $x$ is transmitted over the coding channel shown as everything inside the dotted line in Fig. 1. Let

$$y = (y_1, y_2, \cdots, y_N) \tag{7}$$

denote the channel output sequence. In addition, we can have side information in the form

$$z = (z_1, z_2, \cdots, z_N). \tag{8}$$

Throughout this paper we assume the coding channel is memoryless. That is,

$$P_N(y \mid x, z) = \prod_{n=1}^{N} P(y_n \mid x_n, z_n) \tag{9}$$

and

$$P_N(z) = \prod_{n=1}^{N} P(z_n). \tag{10}$$

It follows that we also have

$$P_N(y \mid x) = \sum_z P_N(y \mid x, z) P_N(z)$$

$$= \prod_{n=1}^{N} P(y_n \mid x_n). \tag{11}$$

Many coding channels cannot be modeled as memoryless channels unless the coded symbols are first interleaved and the detector output symbols are deinterleaved. We assume ideal interleaving in this paper.

Assuming that $x$ and $\hat{x}$ are the only two possible coded sequences, when the channel output sequence is $y$ we have the general decision rule. Choose $x$ if and only if

$$\sum_{n=1}^{N} m(y_n, x_n; z_n) > \sum_{n=1}^{N} m(y_n, \hat{x}_n; z_n) \tag{12}$$

where $m(y, x; z)$ is any function of $y$, $x$, and $z$ referred to as a metric. If $x$ is transmitted then the pairwise error probability is given by

$$\text{pr}\{x \to \hat{x}\}$$

$$= \text{pr}\left\{ \sum_{n=1}^{N} m(y_n, x_n; z_n) \le \sum_{n=1}^{N} n(y_n, \hat{x}_n; z_n) \Big| x \right\}. \tag{13}$$

Applying the Chernoff bound (see Jacobs [6]) we have

$$\text{pr}\{x \to \hat{x}\}$$

$$\le \min_{\lambda \ge 0} \prod_{n=1}^{N} E\{e^{\lambda[m(y_n, \hat{x}_n; z_n) - m(y_n, x_n; z_n)]} \mid x_n\}. \tag{14}$$

For the antipodal and orthogonal signals and all metrics used in antijam communication systems we have the property that

$$E\{e^{\lambda[m(y_n, \hat{x}_n; z_n) - m(y_n, x_n; z_n)]} | x_n\}.$$

$$= \begin{cases} D(\lambda), & \hat{x}_n \neq x_n \\ 1, & \hat{x}_n = x_n. \end{cases} \tag{15}$$

Thus

$$\text{pr}\{x \to \hat{x}\} \leq \min_{\lambda \geq 0} [D(\lambda)]^{w[x, \hat{x}]} \tag{16}$$

where $w(x, \hat{x})$ is the number of places where $\hat{x}_n \neq x_n$, $n = 1, 2, \cdots, N$.

Finally we define

$$D = \min_{\lambda \geq 0} D(\lambda)$$

$$= \min_{\substack{\lambda \geq 0 \\ \hat{x} \neq x}} E\{e^{\lambda[m(y, \hat{x}; z) - m(y, x; z)]} | x\} \tag{17}$$

resulting in the pairwise error bound

$$\text{pr}\{x \to \hat{x}\} \leq D^{w(x, \hat{x})}. \tag{18}$$

The pairwise error bound is the basis for error bound for the general case where there are many coded sequences. Let $C$ denote the set of all coded sequences. The bit error bound has the general form (see [5, sect. 4.4])

$$P_b \leq \sum_x \sum_{\substack{x \neq x \\ x, \hat{x} \in C}} a(x, \hat{x}) P_r\{x \to \hat{x}\} q(x)$$

$$\leq \sum_x \sum_{\substack{\hat{x} \neq x \\ x, \hat{x} \in C}} a(x, \hat{x}) D^{w(x, \hat{x})} \cdot q(x) \tag{19}$$

where $a(x, \hat{x})$ counts the number of bit errors occurring if $\hat{x}$ is chosen when $x$ is transmitted and $q(x)$ is the probability of sequence $x$ being transmitted. Thus, we have the form

$$P_b \leq G(D) \tag{20}$$

where $G(\cdot)$ is a function determined solely by the specific code.

In (20) we see the decoupling of the coding impact given by the function $G(\cdot)$, and the coding channel influence through the parameter $D$.

## III. CUTOFF RATE

Let the code symbol alphabet be $\{0, 1, \cdots, M-1\}$ where $M$ is the number of distinct code symbols. Suppose we randomly select the coded sequences $x$ and $\hat{x}$ where each component of $x$ and $\hat{x}$ is independently selected and each of the $M$ symbols is equally probable. The pairwise error probability averaged

over this random selection of $x$ and $\hat{x}$ is[3]

$$\overline{\text{pr}\{x \to \hat{x}\}} \leq \overline{D^{w(x, \hat{x})}}$$

$$= \prod_{n=1}^{N} \overline{D^{w(x_n, \hat{x}_n)}}$$

$$= \left\{ \frac{1 + (M-1)D}{M} \right\}^N \tag{21}$$

where the overbar indicates the average over the coded sequences.

We define the cutoff rate for this case as

$$R_0 = \log_2 M - \log_2 \{1 + (M-1)D\} \tag{22}$$

so that

$$\overline{\text{pr}\{x \to x\}} \leq 2^{-NR_0}. \tag{23}$$

This definition of the cutoff rate is the natural generalization of the usual definition where the maximum likelihood metric is assumed [4]. We generalize here to arbitrary metrics.

Note that there is a one-to-one relationship between $D$ and $R_0$. Thus, for a specific code we have the bit error bound of the form

$$P_b \leq G(D)$$

$$= G\left( \frac{M 2^{-R_0} - 1}{M - 1} \right)$$

$$= B(R_0) \tag{24}$$

where $B(\cdot)$ is a function determined solely by the specific code. In this final coded bit error bound we have the decoupling of the coding given by $B(\cdot)$ and the coding channel characterized by the cutoff rate parameter $R_0$. Here $R_0$ has the usual interpretation of representing the "practically achievable" data rate with coding.

## IV. CONVENTIONAL COHERENT BPSK

For the usual coherent BPSK modulation with the additive white Gaussian noise channel we have the coding channel model shown in Fig. 2(a) where the additive noise component $n$ is a zero-mean Gaussian random variable with variance

$$E\{n^2\} = \frac{N_0}{2}. \tag{25}$$

Here $N_0$ is the single-sided density of the additive white Gaussian noise process. The quantizer is necessary when decoding with a digital processor. Here a "0" coded symbol results in

[3] Note that

$$w(x, \hat{x}) = \sum_{n=1}^{N} w(x_n, \hat{x}_n) \quad \text{where } w(x, \hat{x}) = \begin{cases} 1, x \neq \hat{x} \\ 0, x = \hat{x}. \end{cases}$$

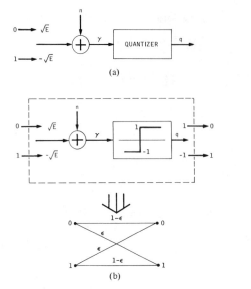

Fig. 2. (a) General BPSK coding channel. (b) Hard decision BPSK coding channel.

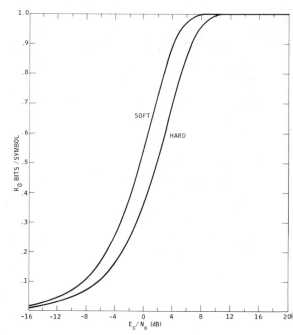

Fig. 3. $R_0$ for hard and soft decision detectors.

a cosine waveform of energy $E_c$ while a "1" coded symbol is a negative cosine waveform of energy $E_c$.

If there is no quantizer in Fig. 2(a) then the conventional maximum likelihood metric is

$$m(y, x) = yx. \tag{26}$$

This is referred to as a soft decision channel. If we assume $x$ is the transmitted symbol then

$$y = x + n \tag{27}$$

and

$$\begin{aligned} D(\lambda) &= E\{e^{\lambda[y\hat{x} - yx]}|x\}|_{\hat{x} \neq x} \\ &= E\{e^{\lambda(x+n)(\hat{x}-x)}\}|_{\hat{x} \neq x} \\ &= e^{-2\lambda E_c + \lambda^2 E_c N_0} \end{aligned} \tag{28}$$

or

$$\begin{aligned} D &= \min_{\lambda \geq 0} \{e^{-2\lambda E_c + \lambda^2 E_c N_0}\} \\ &= e^{-E_c/N_0}. \end{aligned} \tag{29}$$

Suppose next we assume a hard decision channel sketched in Fig. 2(b) where the quantizer forces a decision on each transmitted coded symbol. This results in the binary symmetric channel (BSC) where the coded symbol error probability is

$$\begin{aligned} \epsilon &= \text{pr}\{n \geq \sqrt{E_c}\} \\ &= Q\left(\sqrt{\frac{2E_c}{N_0}}\right) \end{aligned} \tag{30}$$

where

$$Q(x) = \int_x^\infty \frac{1}{\sqrt{2\pi}} e^{-t^2/2} \, dt. \tag{31}$$

For the hard decision channel the maximum likelihood metric is

$$m(y, x) = \begin{cases} 1, & y = x \\ 0, & y \neq x. \end{cases} \tag{32}$$

Hence

$$\begin{aligned} D(\lambda) &= E\{e^{\lambda[m(y,\hat{x}) - m(y,x)]}|x\}|_{\hat{x} \neq x} \\ &= \epsilon e^{\lambda} + (1 - \epsilon)e^{-\lambda} \end{aligned} \tag{33}$$

and

$$\begin{aligned} D &= \max_{\lambda \geq 0} \{\epsilon e^{\lambda} + (1 - \epsilon)e^{-\lambda}\} \\ &= \sqrt{4\epsilon(1 - \epsilon)}. \end{aligned} \tag{34}$$

In general, for binary symbols ($M = 2$) we have

$$R_0 = 1 - \log_2 (1 + D) \qquad \text{bits/symbol}. \tag{35}$$

Fig. 3 shows $R_0$ versus $E_c/N_0$ for the soft and hard decision detectors. Note that there is roughly a 2 dB difference for most values of $E_c/N_0$.

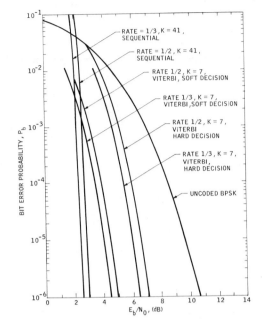

Fig. 4. Comparison of the bit error probability performances of rate 1/2 and 1/3, hard and soft Viterbi and sequential decoders.

The most commonly used code for coherent BPSK and QPSK modulations is the constraint length $K = 7$ rate $r = \frac{1}{2}$ convolutional code found by Odenwalder [7]. This code has the bit error bound[4]

$$P_b \leqslant \frac{1}{2} [36D^{10} + 211D^{12} + 1404D^{14} + 11633D^{16} + \cdots].$$
(36)

Another common code is the $K = 7, r = \frac{1}{3}$ convolutional code also found by Odenwalder where the bit error bound is given by

$$P_b \leqslant \frac{1}{2} [D^{14} + 20D^{16} + 53D^{18} + 184D^{20} + \cdots].$$
(37)

Fig. 4 shows these two bit error bounds for the hard and soft decision detectors along with $K = 41$ convolutional codes. We assume that the maximum likelihood Viterbi decoders [5] were used for the $K = 7$ convolutional codes. For the larger constraint lengths sequential decoding is required (see [5, ch. 6]).

Note that here, regardless of the code used, we see the same difference in $E_b/N_0$ between hard and soft decision detectors as seen in Fig. 3 for $R_0$ versus $E_c/N_0$. The difference in $E_c/N_0$ for fixed value of $R_0$ directly translates to the difference in $E_b/N_0$ for the corresponding bound on bit error probability. Here $E_c/N_0$ and $E_b/N_0$ are related by (2).

One can view the uncoded case as a special case of coding where the code rate is $r = 1$ and all coded sequences are of

4 The factor of one half applies to all maximum likelihood metrics considered here. See Jacobs [6].

length $N = 1$. For the maximum likelihood soft decision detector the general bound becomes

$$P_b \leqslant \frac{1}{2} D$$
$$= \frac{1}{2} e^{-E_b/N_0}$$
(38)

whereas the exact bit error probability for coherent BPSK is

$$P_b = Q\left(\sqrt{\frac{2E_b}{N_0}}\right).$$
(39)

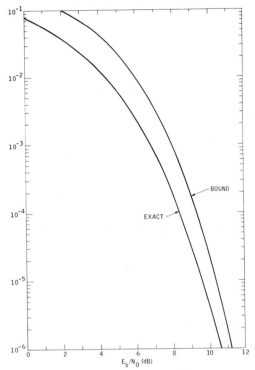

Fig. 5. BPSK bit error probability.

Fig. 5 shows both the exact bit error probability and this special case of the general coded bit error bound. For most values of $E_b/N_0$ there is less than a 1.5 dB difference between the exact uncoded bit error probability and the bound.

## V. DS/BPSK AND PULSE JAMMING

The previous results for the white Gaussian noise channel also apply to the case where there is continuous jamming of a direct sequence (DS) spread BPSK signal. The only difference is that $E_b/N_0$ is given by (3). Jamming, however, can take many forms. To illustrate the application of the general bounds we present two examples of pulse jamming of DS/BPSK signals.

Suppose there is a pulse jammer with average power $J/\rho$ for $\rho$ fraction of the time and zero power for $1 - \rho$ fraction of time. We assume that when the jammer is active the channel is like an additive white Gaussian noise channel with power spectral density (one-sided)

$$N_0 = \frac{J}{\rho W} \qquad (40)$$

where $W$ is the spread-spectrum signal bandwidth. During the transmission of a coded symbol we define the jammer state random variable $z$ where

$$P(z = 1) = \rho$$

and

$$P(z = 0) = 1 - \rho. \qquad (41)$$

This is the jammer state side information that may be available at the receiver where $z = 1$ indicates a jammer is transmitting during a coded symbol time while $z = 0$ indicates there is no jammer signal. With $z$ available at the receiver the metric we consider is

$$m(y, x; z) = C(z)yx \qquad (42)$$

which is a weighted correlation metric.

Next we compute the parameter

$$D(\lambda) = E\{e^{\lambda[C(z)y(\hat{x}-x)]}|x\}|_{\hat{x} \neq x}$$
$$= \rho \exp\{-2\lambda C(1)E_c + \lambda^2 C(1)E_c N_0/\rho\}$$
$$+ (1 - \rho) \exp\{-2\lambda C(0)E_c\}. \qquad (43)$$

If the receiver has jammer state side information (knowledge of $z$) then the metric can have $C(0)$ as large as possible to make the second term in (43) negligibly small. We can also normalize $C(1) = 1$ to obtain

$$D(\lambda) = \rho \exp\{-2\lambda E_c + \lambda^2 E_c N_0/\rho\} \qquad (44)$$

and

$$D = \min_{\lambda \geqslant 0}\ [\rho \exp\{-2\lambda E_c + \lambda^2 E_c N_0/\rho\}]$$
$$= \rho \exp\left\{-\rho\left(\frac{E_c}{N_0}\right)\right\}. \qquad (45)$$

Suppose the receiver has no side information. Then the metric has weighting $C(1) = C(0)$ which we can normalize to unity. Then

$$D(\lambda) = \rho \exp\{-2\lambda E_c + \lambda^2 E_c N_0/\rho\}$$
$$+ (1 - \rho) \exp\{-2\lambda E_c\}$$
$$= e^{-2\lambda E_c}[\rho e^{\lambda^2 E_c N_0/\rho} + 1 - \rho] \qquad (46)$$

and $D$ is obtained by minimizing $D(\lambda)$ over $\lambda \geqslant 0$.

Both (44) and (46) correspond to soft decision detectors. The only difference between these two is the availability of side information concerning the jammer state that is available to the first decoder. The difference is shown in Figs. 6 and 7 where $R_0 = 1 - \log_2(1 + D)$ is sketched versus $E_c/N_0$ for various values of $\rho$. Clearly, having jammer state knowledge

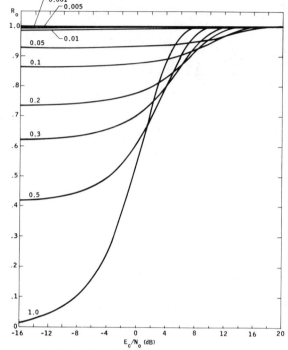

Fig. 6.   Soft decision with jammer state knowledge.

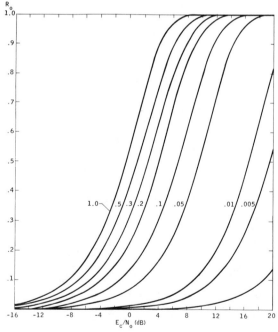

Fig. 7.   Soft decision with no jammer state knowledge.

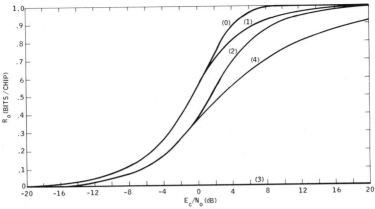

Fig. 8. $R_0$ for DS/BPSK examples.

helps improve the overall bit error probability. The special case where $\rho = 1$ coincides with the conventional soft decision curve shown in Fig. 3.

For hard decision channels we have corresponding results

$$D = \rho\sqrt{4\epsilon(1 - \epsilon)} \qquad (47)$$

when the receiver has jammer state information and

$$D = \sqrt{4\rho\epsilon(1 - \rho\epsilon)} \qquad (48)$$

when the receiver has no jammer state information. Here

$$\epsilon = Q\left(\sqrt{\frac{2\rho E_c}{N_0}}\right). \qquad (49)$$

In Fig. 8 we plot $R_0 = 1 - \log_2(1 + D)$ for $D$ in (29) denoted "0," $D$ for (45) denoted "1," $D$ for (46) denoted "3," $D$ for (47) denoted, "2," and $D$ for (48) denoted "4." In each case we selected the worst case value of $\rho$ which maximized $D$ for each value of $E_c/N_0$.

## VI. TRANSLATION OF CODED ERROR BOUNDS

The numerical evaluation of $R_0$ versus $E_c/N_0$ is straightforward for most of the coding channels encountered in antijam communication systems. This cutoff rate parameter is independent of the specific code used but can now be used to evaluate the coded bit error probability for any code whose standard bit error probability curve is available. For example, in Fig. 4 we have the coded bit error bounds for several codes used over the additive white Gaussian noise channel. The basic modulation here is coherent BPSK or QPSK. Such curves are typically available in textbooks and published papers. For the standard additive white Gaussian noise channel the standard cutoff rates are shown in Fig. 3 for coherent BPSK modulation.

Suppose we now want to evaluate the coded bit error bound for the constraint length $K = 7$ rate $r = \frac{1}{2}$ convolutional code used in a DS/BPSK antijam system where there is

a pulse jammer with $\rho = 0.05$. Also suppose the detector used was a soft decision detector and no jammer state information is available. Here we assume the metric used is simply

$$m(y, x) = yx. \qquad (50)$$

Fig. 7 shows the $R_0$ versus $E_c/N_0$ for this case. We can now translate the standard curve for this code shown in Fig. 4 to determine the bit error bound for this same code for the new coding channel described above.

Note that for $10^{-6}$ coded bit error probability the standard curve in Fig. 4 shows a required

$$(E_b/N_0)_0 = 5 \text{ dB} \qquad (51)$$

or since $E_c = E_b/2$ for $r = \frac{1}{2}$

$$(E_c/N_0)_0 = 2 \text{ dB}. \qquad (52)$$

Next for this choice of $E_c/N_0$ from Fig. 3 we have the cutoff rate required

$$R_0 = 0.74 \quad \text{bits/symbol.} \qquad (53)$$

Recall that the coded bit error bound can always be expressed as a function of $R_0$ as shown in (5) or, more specifically for this code in (36) where $D$ is related to $R_0$, by (35). Thus, if we have another coding channel with the same value of the cutoff rate parameter $R_0$ then the bit error bound is also the same, namely in this case $10^{-6}$. For our new coded channel we determine from Fig. 7 for $R_0$ given by (53) and $\rho = 0.05$, the new required

$$E_c/N_0 = 12 \text{ dB} \qquad (54)$$

or

$$E_b/N_0 = 15 \text{ dB}. \qquad (55)$$

Thus in this pulse jamming example we require 15 dB of $E_b/N_0$ defined by (3) to achieve $10^{-6}$ coded bit error probability with the given convolutional code. By continuing this translation for several bit error probabilities we obtain the complete coded bit error bound for the case of interest by translating the standard coded bit error bound curve.

The translation of standard coded bit error bounds to obtain corresponding bit error bounds for different coding channels applies to all binary input coding channels. It also generalizes to $M$-ary input coding channels of the kind that arise in most spread-spectrum systems. We show this with frequency-hopped noncoherent MFSK singals with partial band jamming in [8]. For antijam communication systems where jamming and receiver structures can take many possible forms, the general approach of first obtaining numerically computable values of the cutoff rate parameters and then translating standard error rate curves serves as a systematic general approach to the analysis of antijam communication systems.

## VII. DISCUSSION

For the coding channel (shown as that part of the communication system of Fig. 1 that is inside the dotted line) we can easily evaluate the cutoff rate $R_0$ versus $E_c/N_0$, the effective coded symbol-to-noise ratio. For most antijam systems of interest the coded bit error bounds can be expressed directly in terms of this cutoff rate parameter. Thus, we can compare coded bit error bounds for a variety of detector and decoder metrics by comparing the corresponding cutoff rate values.

We illustrated this with DS/BPSK examples. Similarly, examples for a frequency-hopped (FH) noncoherent MFSK system with various types of jammers are presented in [8]. Results for the fading HF channel are given by Avidor [9].

## REFERENCES

[1] A. J. Viterbi, "Spread spectrum communications—Myths and realities," *IEEE Commun. Soc. Mag.*, vol. 17, pp. 11–18, May 1979.

[2] J. M. Wozencraft and R. S. Kennedy, "Modulation and demodulation for probabilistic coding," *IEEE Trans. Inform. Theory*, vol. IT-12, pp. 291–297, July 1966.

[3] J. L. Massey, "Coding and modulation in digital communications," in *Proc. Int. Zurich Sem. Digital Commun.*, Zurich, Switzerland, Mar. 12–15, 1974.

[4] L. Biederman, J. K. Omura, and P. C. Jain, "Decoding with approximate channel statistics for bandlimited nonlinear satellite channels," *IEEE Trans. Inform. Theory*, vol. IT-27, Nov. 1981, pp. 697–708.

[5] A. J. Viterbi and J. K. Omura, *Principles of Digital Communication and Coding.* New York: McGraw-Hill, 1979.

[6] I. M. Jacobs, "Probability of error bounds for binary transmission on the slowly fading Rician channel," *IEEE Trans. Inform. Theory*, vol. IT-12, pp. 431–441, Oct. 1966.

[7] J. P. Odenwalder, "Optimum decoding of convolutional codes," Doctoral dissertation, School Eng. Appl. Sci., Univ. California, Los Angeles, 1970, p. 64.

[8] J. K. Omura, B. Levitt, and R. Stokey, "FH/MFSK performance in a partial band jamming environment," *IEEE Trans. Commun.*, to be published.

[9] D. Avidor, "Anti-jam analysis of frequency hopping $M$-ary FSK communication systems in HF Rayleigh fading channels," Doctoral dissertation, School Eng. Appl. Sci., Univ. California, Los Angeles, 1981.

# The Effect of Frequency-Selective Fading on a Noncoherent FH–FSK System Operating with Partial-Band Tone Interference

LAURENCE B. MILSTEIN, SENIOR MEMBER, IEEE, AND DONALD L. SCHILLING, FELLOW, IEEE

*Abstract*—The performance of a binary noncoherent frequency-hopped–frequency-shift-keyed system operating in the presence of partial-band tone interference and transmitted over a frequency-selective Rician fading channel is derived. The probability of error for the system is derived both with and without the use of error-correction coding, and the two cases are compared on the basis of equal information rates and equal spread bandwidths.

## I. INTRODUCTION

IN [1], the performance of a binary noncoherent frequency-hopped–frequency-shift-keyed (FH-FSK) system operating in the presence of partial band interference was analyzed. The use of error-correction coding was employed, and for a given spread bandwidth, the number of slots over which the signal could hop was traded off versus the information rate of the code.

This paper will extend the results of [1] by assuming that both the desired signal and the interfering signals are transmitted through a frequency-selective Rician fading channel. The channel model will be the one described in [2]-[6], namely, the wide-sense-stationary-uncorrelated-scattering (WSSUS) model. References [2] and [3] consider the performance of a direct-sequence spread-spectrum communication system operating over a WSSUS channel, and so the results presented in this paper will complement those results in the sense of providing similar information for a frequency-hopped system.

The paper is divided into four sections, with Section II containing the analysis, Section III containing numerical results, and Section IV containing the final conclusions.

## II. ANALYSIS

The system to be analyzed is shown in Fig. 1. The desired signal and the interfering signals are assumed to fade independently of one another since they originate from geographically isolated points. It is further assumed that the frequency separation of the MARK and SPACE filters, taken to be $2/T$ where $T$ is the symbol duration, is large enough to

Manuscript received June 9, 1981; revised October 14, 1981. This paper was presented at the National Telecommunications Conference, New Orleans, LA, December 1981.

L. B. Milstein is with the Department of Electrical Engineering and Computer Sciences, University of California at San Diego, La Jolla, CA 92093.

D. L. Schilling is with the Department of Electrical Engineering, City College of New York, New York, NY 10031.

result in independent fading of signals located at the respective center frequencies. This latter assumption is made only to simplify some of the succeeding analysis; results similar to those to be presented can be obtained without making this assumption.

The transmitted signal is given by

$$S(t) = \frac{A}{2}(1 + d(t)) \sum_j \cos((\omega_M + \omega_j)t + \theta_{Mj})P(t - jT)$$
$$+ \frac{A}{2}(1 - d(t)) \sum_j \cos((\omega_S + \omega_j)t + \theta_{Sj})P(t - jT) \tag{1}$$

where

$$P(t) \triangleq \begin{cases} 1 & 0 \leqslant t \leqslant T \\ 0 & \text{elsewhere} \end{cases} \tag{2}$$

and where $d(t)$ is a random binary data sequence taking on values $\pm 1$ with equal probability, $A$ is a constant amplitude, $\omega_M$ and $\omega_S$ are the MARK and SPACE frequencies, respectively, $\omega_j$ is the frequency out of the synthesizer at the transmitter on the $j$th hop, and $\theta_{Mj}$ and $\theta_{Sj}$ are random phases in the MARK and SPACE channels, respectively. It is assumed that the hopping rate is one hop per bit, so that the frequency $\omega_j$ can change every $T$ seconds.

From [2]-[5], the output of the channel due to the signal (i.e., the signal component of the receiver input) is given by

$$S_r(t)$$
$$= \text{Re}\left\{\frac{A}{2} \sum_j \left[\gamma(1 + d_j) \int_{-\infty}^{\infty} \beta_M(\tau)P(t - jT - \tau)\,d\tau \right.\right.$$
$$\left.\left. + (1 + d_j)P(t - jT)\right] \exp j[(\omega_M + \omega_j)t + \theta_{Mj}]\right\}$$
$$+ \text{Re}\left\{\frac{A}{2} \sum_j \left[\gamma(1 - d_j) \int_{-\infty}^{\infty} \beta_S(\tau)P(t - jT - \tau)\,d\tau \right.\right.$$
$$\left.\left. + (1 - d_j)P(t - jT)\right] \cdot \exp j[(\omega_S + \omega_j)t + \theta_{Sj}]\right\} \tag{3}$$

Reprinted from *IEEE Trans. Commun.*, vol. COM-30, pp. 904–912, May 1982.

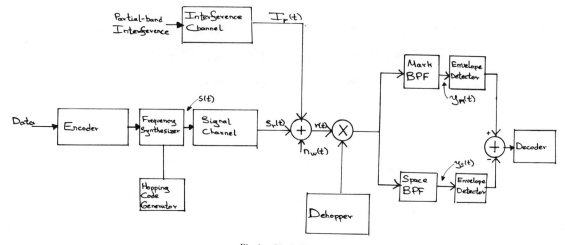

Fig. 1.   Block diagram.

where $\gamma^2$ is a measure of the ratio of the power in the scatter component to the power in the specular component, $d_j$ is the data symbol during the interval $jT \leqslant t \leqslant (j + 1)T$, and $\beta_M(\tau)$ and $\beta_S(\tau)$ are independent, zero-mean, complex Gaussian random processes representing the low-pass equivalent impulse responses of the frequency selective channels that the MARK and SPACE signals see, respectively. Notice from (3) that there is both a specular and a scatter component to the received signal (i.e., the channel is assumed to be Rician, not Rayleigh).

$\beta_M(\tau)$ and $\beta_S(\tau)$ are described by their correlation functions. Denoting either by $\beta(\tau)$, from [2] or [3],

$$\frac{1}{2} E\{\beta(\tau_1)\beta^*(\tau_2)\} = \rho(\tau_1)\delta(\tau_1 - \tau_2) \tag{4}$$

where $\rho(\tau)$ is a real function of $\tau$ and $\delta(\tau)$ is the Dirac delta function. As in [2] and [3], the following two assumptions are made:

$$E\{\beta(\tau_1)\beta(\tau_2)\} = 0 \tag{5}$$

and

$$\rho(\tau) \equiv 0 \, |\tau| \geqslant T. \tag{6}$$

The former assumption implies that the in-phase and quadrature components of $\beta(\tau)$ have the same statistical properties and are uncorrelated with one another, and the latter assumption limits the intersymbol interference that the channel generates to just adjacent symbols.

In addition to $S_r(t)$, the received signal $r(t)$ will consist of additive white Gaussian noise of two-sided spectral density $\eta_0/2$ plus partial-band tone interference. That is, if there are a total of $N$ slots available over which to hop the signal, it will be assumed that $K$ slots have an interfering tone present in them, where $1 \leqslant K \leqslant N$. The outputs of the channel due to

tones at $\omega_M + \omega_j$ and $\omega_S + \omega_j$ are given by

$$I_{r_M}(t) = \lambda_M \, \text{Re} \left\{ \left[ A_I + \gamma_I A_I \int_{-\infty}^{\infty} \beta_{I_M}(\tau) \, d\tau \right] \right.$$
$$\left. \cdot \exp j[(\omega_M + \omega_j)t + \varphi_{Mj}] \right\} \tag{7}$$

and

$$I_{r_S}(t) = \lambda_S \, \text{Re} \left\{ \left[ A_1 + \gamma_I A_I \int_{-\infty}^{\infty} \beta_{I_S}(\tau) \, d\tau \right] \right.$$
$$\left. \cdot \exp j[(\omega_S + \omega_j)t + \varphi_{Sj}] \right\}, \tag{8}$$

respectively. In (7) and (8), $A_I$ and $\gamma_I$ are real constants, $\beta_{I_M}(\tau)$ and $\beta_{I_S}(\tau)$ are the low-pass equivalent impulse responses that the interfering tones see and which have the same statistical properties as $\beta_M(\tau)$ and $\beta_S(\tau)$, and $\varphi_{Mj}$ and $\varphi_{Sj}$ are random phases uniformly distributed in $[0, 2\pi]$. They are independent of one another and are independent of the signal phases. Finally, $\lambda_M$ and $\lambda_S$ are either zero or one, depending upon whether or not the respective channel has one of the $K$ tones present in it.

The receiver consists of a dehopper which is assumed to be operating in synchronism with the hopping pattern of the received signal (but not coherent in phase on any individual hop), followed by a standard noncoherent FSK detector. The MARK and SPACE bandpass filters are taken to be ideal rectangular filters centered at $\omega_M$ and $\omega_S = \omega_M + 2(2\pi/T)$, respectively, and each has bandwidth $2(2\pi/T)$. It will be assumed that when either a MARK or a SPACE is transmitted, it produces a nonzero output only from its own BPF.

Consider the output of the MARK BPF. The low-pass

110

equivalent impulse response of this filter is given by

$$h(t) = \frac{\sin \omega_c t}{\pi t} \qquad (9)$$

where $\omega_c \triangleq (2\pi/T)$. The filter output is given by

$$y_M(t) = \text{Re} \left\{ [z_M(t) * h(t)] \exp [j\omega_M t] \right\}$$
$$+ n_{c_M}(t) \cos \omega_M t - n_{s_M}(t) \sin \omega_M t \qquad (10)$$

where, over the interval, say, $0 \leqslant t \leqslant T$,

$$z_M(t) \triangleq \frac{A}{2}(1 + d_0) + \frac{\gamma A}{2} \sum_j (1 + d_j)$$

$$\cdot \int_{-\infty}^{\infty} \beta_M(\tau) P(t - jT - \tau) \, d\tau \delta \omega_0 \omega_j$$

$$+ \left[ \lambda_M A_I + \lambda_M \gamma_I A_I \int_{-\infty}^{\infty} \beta_{I_M}(\tau) \, d\tau \right]$$

$$\cdot e^{j\varphi_M} \sum_j P(t - jT)\delta \omega_0 \omega_j. \qquad (11)$$

In (10), $n_{c_M}(t)$ and $n_{s_M}(t)$ are the in-phase and quadrature components, respectively, of the MARK BPF output due to the thermal noise, and $\delta \omega_0 \omega_j$ represents the effect of the dehopper on the signal and the interference. It equals one if the frequency at the $j$th hop equals what it was on the zeroth hop, and it equals zero otherwise. Notice that if there is a large number of frequencies over which the signal can hop, the event $\{\delta \omega_0 \omega_j = 1\}$ (for $j \neq 0$) is a very unlikely event. Also notice that the phase of the desired signal has been referenced to zero, and the phase of the tone interference has been made independent of $j$. This again represents the action of the dehopper, since the only nonzero terms in the above sums correspond to $\omega_j = \omega_0$, and hence constant phases for both the signal and the interference.

Using (9) and (11) in (10) yields

$$y_M(t) = \frac{A}{2\pi}(1 + d_0)[Si(\omega_c t) - Si(\omega_c(t - T))]$$

$$\cdot \cos \omega_M t + \lambda_M \frac{A_I}{\pi} \cos (\omega_M t + \varphi_M)$$

$$\cdot [Si(\omega_c t) - Si(\omega_c(t - T))] + S_{F_M}(t) + I_{F_M}(t)$$

$$+ n_{c_M}(t) \cos \omega_M t - n_{s_M}(t) \sin \omega_M t \qquad (12)$$

where

$$S_{F_M}(t) \triangleq \text{Re} \left[ \left\{ \left[ \frac{\gamma A}{2} \int_{-\infty}^{\infty} \beta_M(\tau) \sum_j (1 + d_j)\delta \omega_0 \omega_j \right. \right. \right.$$

$$\left. \left. \left. \cdot P(t - jT - \tau) \, d\tau \right] * \frac{\sin \omega_c T}{\pi t} \right\} e^{j\omega_M t} \right] \qquad (13)$$

and

$$I_{F_M}(t) \triangleq \lambda_M \, \text{Re} \left[ \left\{ \left[ A_I \gamma_I \int_{-\infty}^{\infty} \beta_{I_M}(\tau) \, d\tau \right. \right. \right.$$

$$\left. \left. \left. \cdot \sum_j P(t - jT)\delta \omega_0 \omega_j \right] * \frac{\sin \omega_c t}{\pi t} \right\} e^{j(\omega_M t + \varphi_M)} \right]. \qquad (14)$$

In (12), the function $Si(\cdot)$ is the sine integral defined as

$$Si(x) \triangleq \int_0^x \frac{\sin y}{y} \, dy.$$

Notice that at a given instant of time $t$, both $S_F(t)$ and $I_F(t)$ are conditional zero-mean Gaussian random variables, since they both result from a linear operation on a Gaussian random process. Therefore, they are completely characterized once their variances are known.

The calculation of the variances is presented in the Appendix. In order to arrive at a final answer, a specific $\rho(\tau)$ defined in (4) has to be chosen, and as was the case in [2] and [3], the following $\rho(\tau)$ will be used:

$$\rho(\tau) = \begin{cases} \dfrac{1}{T} \left( 1 - \dfrac{|\tau|}{T} \right) & |\tau| \leqslant T \\ 0 & \text{elsewhere.} \end{cases} \qquad (15)$$

Physically, $\rho(\tau)$ is the so-called delay-multipath spread of the channel, and the $\rho(\tau)$ of (15) is representative of channels whose time dispersion is no greater than plus-or-minus one symbol duration.

Using the results from the Appendix, it is now possible to evaluate the average probability of error of the system. From either [1] or [6], conditioned on $\varphi_M$, $d_1$, $d_{-1}$, $\lambda_M$, $\lambda_S$, $\omega_{-1}$, and $\omega_1$, and assuming that a MARK is transmitted (i.e., $d_0 = 1$), then

$$P(e \mid \lambda_M, \lambda_S, \varphi_M, d_1, d_{-1}, \omega_{-1}, \omega_1)$$

$$= \frac{1}{2} [1 - Q(\sqrt{b_i}, \sqrt{a_i}) + Q(\sqrt{a_i}, \sqrt{b_i})]$$

$$- \frac{A_i}{2} \exp \left[ -\frac{a_i + b_i}{2} \right] I_0(\sqrt{a_i b_i}) \qquad (16)$$

where

$$Q(a, b) \triangleq \int_b^{\infty} x \exp \left( -\frac{a^2 + x^2}{2} \right) I_0(ax) \, dx \qquad (17)$$

$$I_0(x) \triangleq \frac{1}{2\pi} \int_0^{2\pi} e^{x \cos \theta} \, d\theta \qquad (18)$$

$$A_i \triangleq \frac{\dfrac{I_{M_i} - I_{S_i}}{4\pi^2 T} + \gamma_I^2 \dfrac{4}{\pi^2} \dfrac{J}{S} (\lambda_M - \lambda_S) I_{J_i}}{N/S + \dfrac{I_{M_i} + I_{S_i}}{4\pi^2 T} + \gamma_I^2 \dfrac{4}{\pi^2} \dfrac{J}{S} (\lambda_M + \lambda_S) I_{J_i}} \qquad (19)$$

$$a_i \triangleq \frac{4}{\pi^2} Si^2 \left( \frac{\omega_c T}{2} \right)$$

$$\cdot \frac{\lambda_S \dfrac{J}{S}}{\dfrac{N}{S} + \dfrac{I_{M_i} + I_{S_i}}{4\pi^2 T} + \gamma_I^2 \dfrac{4}{\pi^2} \dfrac{J}{S} (\lambda_M + \lambda_S) I_{J_i}} \quad (20)$$

$$b_i \triangleq \frac{4}{\pi^2} Si^2 \left( \frac{\omega_c T}{2} \right)$$

$$\cdot \frac{\left( 1 + \lambda_M \sqrt{\dfrac{J}{S}} \cos \varphi_M \right)^2 + \left( \lambda_M \sqrt{\dfrac{J}{S}} \sin \varphi_M \right)^2}{\dfrac{N}{S} + \dfrac{I_{M_i} + I_{S_i}}{4\pi^2 T} + \gamma_I^2 \dfrac{4}{\pi^2} \dfrac{J}{S} (\lambda_M + \lambda_S) I_{J_i}}$$

$$(21)$$

$S/N \triangleq$ signal-to-noise ratio in BPF bandwidth

$$= \frac{A^2}{4\eta_0 f_c} \quad (22)$$

and

$$\frac{J}{S} \triangleq \frac{A_I^2}{A^2}. \quad (23)$$

Averaging (16) over the data yields

$$P(e \mid \lambda_M, \lambda_S, \varphi_M, \omega_{-1}, \omega_1)$$

$$= \frac{1}{4} \sum_{d_{-1}} \sum_{d_1} P(e \mid \lambda_M, \lambda_S, \varphi_M, d_1, d_{-1}, \omega_{-1}, \omega_1)$$

$$(24)$$

and averaging (24) over $\lambda_M$, $\lambda_S$, and $\varphi_M$ yields (see [1, Appendix A])

$$P(e \mid \omega_{-1}, \omega_1)$$

$$= \frac{(N-K)(N-K-1)}{N(N-1)} P(e \mid \lambda_M = 0, \lambda_S = 0, \varphi_M, \omega_{-1}, \omega_1)$$

$$+ \frac{K(N-K)}{N(N-1)} \left[ \frac{1}{2\pi} \int_0^{2\pi} P(e \mid \lambda_M = 1, \lambda_S = 0, \right.$$

$$\varphi_M, \omega_{-1}, \omega_1) \, d\varphi_M$$

$$\left. + P(e \mid \lambda_M = 0, \lambda_S = 1, \varphi_M, \omega_{-1}, \omega_1) \right]$$

$$+ \frac{K(K-1)}{N(N-1)} \frac{1}{2\pi} \int_0^{2\pi} P(e \mid \lambda_M = 1, \lambda_S = 1,$$

$$\varphi_M, \omega_{-1}, \omega_1) \, d\varphi_M. \quad (25)$$

Notice that when $\lambda_M = 0$, the resulting probability of error expression is independent of $\varphi_M$, and so no averaging over this latter quantity has to be done.

The average probability of error is now given by

$$P_e = \left( \frac{N_1 - 1}{N_1} \right)^2 P(e \mid \omega_{-1} \neq \omega_0, \omega_1 \neq \omega_0)$$

$$+ \frac{N_1 - 1}{N_1^2} [P(e \mid \omega_{-1} = \omega_0, \omega_1 \neq \omega_0)$$

$$+ P(e \mid \omega_{-1} \neq \omega_0, \omega_1 = \omega_0)]$$

$$+ \frac{1}{N_1^2} P(e \mid \omega_{-1} = \omega_0, \omega_1 = \omega_0) \quad (26)$$

where

$$N_1 \triangleq \frac{N}{2}. \quad (27)$$

This last relation follows because if there is a total of $N$ slots to hop over, there are a total of $N/2$ possible frequency outputs from the hopping code generator (assuming that MARK and SPACE slots are contiguous). Notice that in (26) the first term on the RHS is the dominant term if $N_1 \gg 1$.

Finally, none of the above results included the effect of error-correction coding. To allow for the use of such coding while keeping the information rate, spread bandwidth, and alphabet size constant requires that the number of slots available for hopping decrease by the information rate of the code [1]. In particular, for an $(n, k)$ block code, if there are $N$ uncoded slots, there are $(k/n)N$ coded slots. Also, from [7], the average probability of bit error, given an error-correcting code capable of correcting all combinations of $e$ and fewer errors, is approximately given by

$$P_b \simeq \frac{1}{n} \sum_{i=e+1}^{n} i \binom{n}{i} P_e^i (1 - P_e)^{n-i} \quad (28)$$

where $P_e$ is given by (26) except that the $S/N$ term is adjusted for the new noise bandwidth (i.e., $S/N$ is decreased by the rate of the code).

## III. NUMERICAL RESULTS

In this section, the results of evaluating the expressions for the probability of error derived above are presented. Specifically, results are calculated for both an uncoded system and one employing a (23, 12) Golay code when the desired signal is received in the presence of partial band interference which falls in $K$ of the available $N$ slots. The Golay code was chosen because it is representative of codes that have a fair amount of error-correction capability and for which implementable decoding algorithms are available. Results are shown for $K$ taking many different values from 1 to 200, and $N$ is taken either as 511 or 1000 (uncoded)

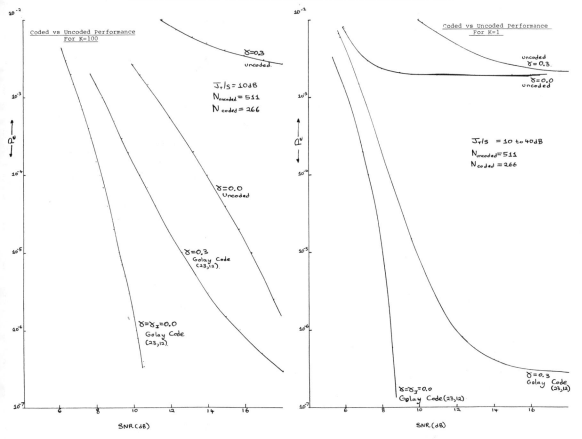

Fig. 2.   Coded versus uncoded performance for $K = 1$.

Fig. 3.   Coded versus uncoded performance for $K = 100$.

so that the number of slots the coded system has available for hopping is 266 or 521, respectively.

In all the figures showing performance results, the abscissa is the signal-to-noise ratio defined as

$$\text{SNR} \triangleq \frac{A^2}{2(2\eta_0 f_c)}. \tag{29}$$

It is seen that SNR is the ratio of the transmitted signal power to the noise power in a bandwidth equal to that of either the MARK or SPACE BPF. Therefore, to meaningfully compare the coded and the uncoded curves, the coded curves have to be shifted by $10 \log (23/12) = 2.83$ dB to the right, since for equal average powers and equal information rates in the two systems, the energy-per-symbol of the coded system is 12/23 of the energy per symbol of the uncoded system. Finally, if there is more than one interfering tone present, the total interference power is kept constant, and is equally divided among the $K$ tones. Therefore, if $J_T$ is the total interference power, $J_T = KJ$ where $J = A_I^2/2$. In Fig. 2, results for $\gamma = \gamma_I = 0.0$ and 0.3 are presented. The initial number of slots available for hopping is $N = 511$, and a single interfering tone is present. As expected, the probability of error of the uncoded system "bottoms out" at an error rate of $1/N$, and the curves are essentially identical for any $J_T/S$ in the range 10–40 dB. The encoded system displays markedly superior performance, but the sensitivity of even the coded system to increased power in the scatter component of the received signal is clearly seen from Fig. 2.

If the number of interfering tones is increased to $K = 100$, the results for $J_T/S = 10$ dB are shown in Fig. 3. With the exception of the uncoded system when $\gamma = \gamma_I = 0.0$, all other systems perform worse for $K = 100$ than for $K = 1$. Also, all systems are much more sensitive to $J_T/S$ when $K$ is increased to 100. For example, at $J_T/S = 30$ dB, the error rates are all greater than 0.1.

When the number of uncoded slots is increased to 1000, results for $K = 1$ and $K = 100$ are shown in Figs. 4 and 5, respectively. In these two figures, the $\gamma$'s range from $\gamma = \gamma_I = 0.0$ to 0.5, and the same general observations as noted with respect to Figs. 2 and 3 apply to those latter two figures.

The effect of varying $K$, the number of interfering tones, is explicitly illustrated in Figs. 6–9. Fig. 6 corresponds to $\gamma = \gamma_I = 0.0$, and Fig. 7 corresponds to $\gamma = \gamma_I = 0.3$. In both figures, $J_T/S = 10$ dB. It can be seen that $K$ somewhere

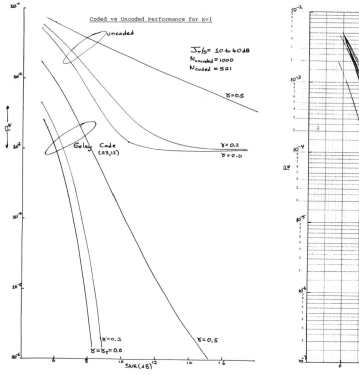

Fig. 4.   Coded versus uncoded performance for $K = 1$.

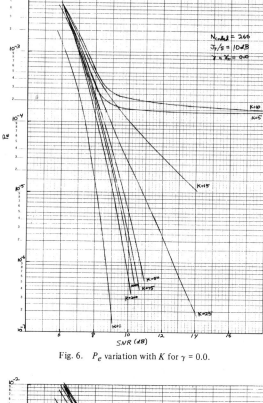

Fig. 6.   $P_e$ variation with $K$ for $\gamma = 0.0$.

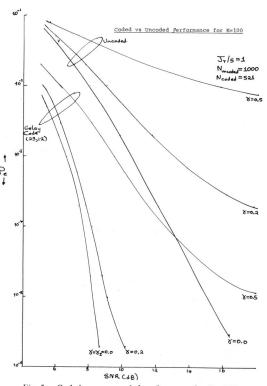

Fig. 5.   Coded versus uncoded performance for $K = 100$.

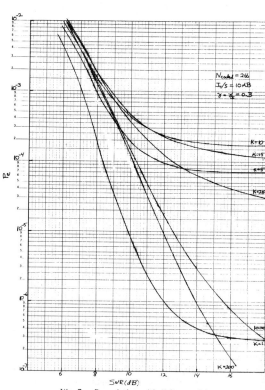

Fig. 7.   $P_e$ variation with $K$ for $\gamma = 0.3$.

114

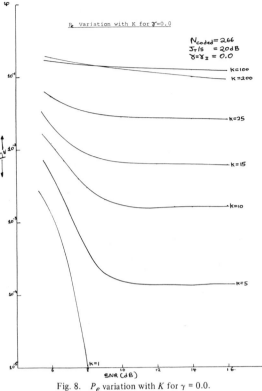

Fig. 8.   $P_e$ variation with $K$ for $\gamma = 0.0$.

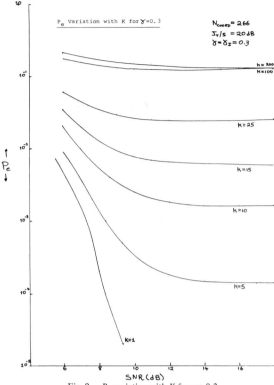

Fig. 9.   $P_e$ variation with $K$ for $\gamma = 0.3$.

in the range of 10 degrades the system performance the most. If $J_T/S$ is increased to 20 dB, Figs. 8 and 9 show the corresponding results. Now $K$ in the vicinity of 100 corresponds to worst case performance. This, of course, is reasonable, since with large $E/\eta_0$, the interference power per slot does not have to be much greater than 0 dB (relative to the signal power) to cause an error with high probability. In other words, from the point of view of worst case performance, the number of interfering tones should be roughly equal to that number which results in a $J/S$ per slot of 0 dB. Note, however, that if the scatter component becomes too large, this result will no longer be valid since then there is a nonnegligible random component to the received waveform even when $E/\eta_0$ is very large.

With respect to the effect of the scatter components, the curves of Fig. 10 illustrate the sensitivity of the system to $\gamma_I$. There are two sets of curves in this figure, corresponding to $\gamma = 0.5$ and $\gamma = 1$, respectively. In each set of curves, there is one curve for $\gamma_I = \gamma$, and another curve for $\gamma_I = 0$. These curves all correspond to $N = 266$, $J_T/S = 10$ dB and $K = 100$. The effect of varying $\gamma$ can be seen by comparing the two curves in Fig. 10 for which $\gamma_I = 0$. As a final observation, note that while the scatter component of the desired signal tends to degrade the system performance for the range of values of $\gamma$ illustrated here, at some point as $\gamma$ increases, the performance of the system will improve. This can be seen by noting that as $\gamma$ becomes very large, the channel approaches a Rayleigh channel. Hence, the only "signal power" is that which arrives as scatter terms. In this

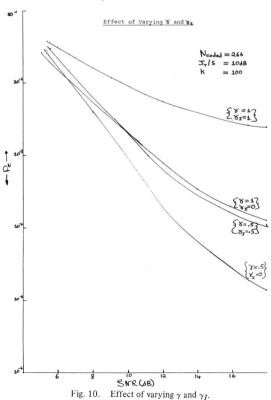

Fig. 10.   Effect of varying $\gamma$ and $\gamma_I$.

case, the definition of SNR used in this paper is no longer meaningful, since from (29), SNR is a measure of specular signal power to thermal noise power, and for large enough $\gamma$, the relative amount of specular signal power is negligible.

## IV. CONCLUSION

An analysis of the performance of a noncoherent FH-FSK system operating over a frequency-selective Rician fading channel has been presented. The FH signal was received in the presence of partial-band tone interference, and the receiver employed a (23, 12) Golay code for error correction.

The average probability of error of the system was determined, and the sensitivity of the error rate to such factors as number of interfering tones, ratio of interference power to signal power, and relative amount of received power in the scatter component of the frequency-selective channel to that in the specular component was investigated. It was seen that the performance of the system can degrade rapidly with increasing power in the scatter component and that, as is the case for a nonfading channel, for a given set of system parameters, there is a value for $K$, the number of interfering tones, which causes the most degradation in probability of error. That value of $K$ results when $J/S$ [not $J_T/S$] approximately equals 0 dB.

Also, it can be seen upon comparing Figs. 6 and 7 that the worst case performance of the system is not as sensitive to the presence of the scatter component as is the performance of the system for suboptimal values of $K$. For example, the $K = 10$ curves of Figs. 6 and 7 are quite similar, but the $K = 25$ curves are markedly different. Likewise, the $K = 100$ curves of Figs. 8 and 9 are not much different from one another (other than the location of the crossover points of the $K = 100$ and $K = 200$ curves). However, for these latter figures, most of the corresponding curves are quite similar. This is just indicative of the fact that the interference power is now large enough to almost completely dominate the behavior of the system, and hence adding or not adding a nominal amount of scatter power to the system does not cause a significant change in performance.

## APPENDIX

It is desired to derive the conditional variances of $S_F(T)$ and $I_F(T)$. Consider $S_F(T)$ first. Its conditional variance is given by

$$
\mathrm{var}\,(S_{F_N}(t))
$$
$$
\triangleq \sigma_M^{\,2}
$$
$$
= E\left\{\mathrm{Re}\left[\left\{\left[\frac{\gamma A}{2}\int_{-\infty}^{\infty}\beta_M(\tau_1)\sum_j (1+d_j)\delta\omega_0\omega_j\right.\right.\right.\right.
$$
$$
\left.\cdot\, P(t-jT-\tau_1)\,d\tau_1\right]*\frac{\sin\omega_c t}{\pi t}\bigg\}e^{j\omega_M t}\right]
$$
$$
\cdot\,\mathrm{Re}\left[\left\{\left[\frac{\gamma A}{2}\int_{-\infty}^{\infty}\beta_M(\tau_2)\sum_k (1+d_k)\delta\omega_0\omega_k\right.\right.\right.
$$
$$
\left.\left.\left.\cdot\, P(t-kT-\tau_2)\,d\tau_2\right]*\frac{\sin\omega_c t}{\pi t}\bigg\}e^{j\omega_M t}\right]\right\}
$$

$$
=\frac{\gamma^2 A^2}{4}\int_{-\infty}^{\infty}\int_{-\infty}^{\infty}\int_{-\infty}^{\infty}\int_{-\infty}^{\infty}\rho(\tau)\delta(\tau_1-\tau_2)
$$
$$
\cdot\,\sum_j (1+d_j)\delta\omega_0\omega_j P(\lambda_1-jT-\tau_1)
$$
$$
\cdot\,\sum_k (1+d_k)\delta\omega_0\omega_k P(\lambda_2-kT-\tau_2)
$$
$$
\cdot\,\frac{\sin\omega_c(t-\lambda_1)}{\pi(t-\lambda_1)}\frac{\sin\omega_c(t-\lambda_2)}{\pi(t-\lambda_2)}\,d\tau_1\,d\tau_2\,d\lambda_1\,d\lambda_2
$$
$$
=\frac{\gamma^2 A^2}{4}\int_T^T \rho(\tau)\left[\sum_j (1+d_j)\delta\omega_0\omega_j\right.
$$
$$
\left.\cdot\int_\infty^\infty P(\lambda-jT-\tau)\frac{\sin\omega_c(t-\lambda)}{\pi(t-\lambda)}\,d\lambda\right]^2 d\tau \qquad \text{(A1)}
$$

where use has been made of (4), (5), (6), and the following relation used in [3]:

$$
\mathrm{Re}\,(a)\,\mathrm{Re}\,(b)=\tfrac{1}{2}\,\mathrm{Re}\,(ab^*)+\tfrac{1}{2}\,\mathrm{Re}\,(ab). \qquad \text{(A2)}
$$

Using (15), and assuming that intersymbol interference is limited to just adjacent pulses, the sum of the RHS of (A1) will be limited to the terms $j = -1, 0$, and 1. With this assumption, there are four situations of interest. These correspond to $(\omega_{-1}=\omega_0,\ \omega_1=\omega_0)$, $(\omega_{-1}\neq\omega_0,\ \omega_1\neq\omega_0)$, $(\omega_{-1}\neq \omega_0,\ \omega_1=\omega_0)$, and $(\omega_{-1}=\omega_0,\ \omega_1\neq\omega_0)$. If these four cases are referred to as cases 1, 2, 3, and 4, respectively, and if $\sigma_M^{\,2}$ is now denoted

$$
\sigma_{M_i}^{\,2}=\frac{A^2}{4\pi^2 T}\,I_{M_i} \qquad \text{(A3)}
$$

where $i = 1, 2, 3, 4$, then from (A1) we have the following four values for the $I_{M_i}$, assuming that the sampling time of the system is $t = T/2$:

$$
I_{M_1}
$$
$$
=\gamma^2\int_0^T \left(1-\frac{\tau}{T}\right)\left\{(1+d_{-1})\right.
$$
$$
\cdot\left[Si\left(\omega_c\left(\tau-\frac{T}{2}\right)\right)-Si\left(\omega_c\left(\tau-\frac{3T}{2}\right)\right)\right]
$$
$$
+(1+d_0)\left[Si\left(\omega_c\left(\tau+\frac{T}{2}\right)\right)-Si\left(\omega_c\left(\tau-\frac{T}{2}\right)\right)\right]
$$
$$
+(1+d_1)\left[Si(\tfrac{3}{2}\omega_c T)-Si\left(\omega_c\left(\tau+\frac{T}{2}\right)\right)\right]\bigg\}^2 d\tau
$$
$$
+\gamma^2\int_{-T}^0 \left(1+\frac{\tau}{T}\right)\left\{(1+d_{-1})\left[Si\left(\omega_c\left(\tau-\frac{T}{2}\right)\right)\right.\right.
$$
$$
\left.+Si(\tfrac{3}{2}\omega_c T)\right]+(1+d_0)\left[Si\left(\omega_c\left(\tau+\frac{T}{2}\right)\right)\right.
$$
$$
\left.-Si\left(\omega_c\left(\tau-\frac{T}{2}\right)\right)\right]+(1+d_1)\left[Si\left(\omega_c\left(\tau+\frac{3T}{2}\right)\right)\right.
$$
$$
\left.\left.-Si\left(\omega_c\left(\tau+\frac{T}{2}\right)\right)\right]\right\}^2 d\tau \qquad \text{(A4)}
$$

$I_{M_2}$

$$= \gamma^2 \int_0^T \left(1 - \frac{\tau}{T}\right)(1 + d_0)^2 \left[ Si\left(\frac{\omega_c T}{2}\right) \right.$$

$$\left. - Si\left(\omega_c\left(\tau - \frac{T}{2}\right)\right)\right]^2 d\tau + \gamma^2(1 + d_0)^2 \int_{-T}^0 \left(1 + \frac{\tau}{T}\right)$$

$$\cdot \left[ Si\left(\frac{\omega_c T}{2}\right) + Si\left(\omega_c\left(\tau + \frac{T}{2}\right)\right)\right]^2 d\tau \tag{A5}$$

$I_{M_3}$

$$= \gamma^2 \int_0^T \left(1 - \frac{\tau}{T}\right)\left[(1 + d_0)\left[ Si\left(\omega_c\left(\frac{T}{2} + \tau\right)\right)\right.\right.$$

$$\left. - Si\left(\omega_c\left(\tau - \frac{T}{2}\right)\right)\right] + (1 + d_1)\left[ Si\left(\frac{3\omega_c T}{2}\right)\right.$$

$$\left.\left. - Si\left(\omega_c\left(\tau + \frac{T}{2}\right)\right)\right]\right]^2 d\tau + \gamma^2 \int_{-T}^0 \left(1 + \frac{\tau}{T}\right)$$

$$\cdot \left[ (1 + d_0)\left[ Si\left(\omega_c\left(\tau + \frac{T}{2}\right)\right) + Si\left(\frac{\omega_c T}{2}\right)\right]\right.$$

$$+ (1 + d_1)\left[ Si\left(\omega_c\left(\tau + \frac{3T}{2}\right)\right)\right.$$

$$\left.\left. - Si\left(\omega_c\left(\tau + \frac{T}{2}\right)\right)\right]\right]^2 d\tau \tag{A6}$$

$I_{M_4}$

$$= \gamma^2 \int_0^T \left(1 - \frac{\tau}{T}\right)\left[(1 + d_{-1})\left[ Si\left(\omega_c\left(\tau - \frac{T}{2}\right)\right)\right.\right.$$

$$\left. - Si\left(\omega_c\left(\tau - \frac{3T}{2}\right)\right)\right] + (1 + d_0)\left[ Si\left(\frac{\omega_c T}{2}\right)\right.$$

$$\left.\left. - Si\left(\omega_c\left(\tau - \frac{T}{2}\right)\right)\right]\right]^2 d\tau$$

$$+ \gamma^2 \int_{-T}^0 \left(1 + \frac{\tau}{T}\right)\left[(1 + d_{-1})\left[ Si\left(\omega_c\left(\tau - \frac{T}{2}\right)\right)\right.\right.$$

$$+ Si\left(3\frac{\omega_c T}{2}\right)\right] + (1 + d_0)\left[ Si\left(\omega_c\left(\tau + \frac{T}{2}\right)\right)\right.$$

$$\left.\left. - Si\left(\omega_c\left(\tau - \frac{T}{2}\right)\right)\right]\right]^2 d\tau. \tag{A7}$$

In a similar manner, it can be shown that

$$\text{var}\left(I_{F_M}\left(\frac{T}{2}\right)\right) \triangleq \sigma_{I_M}^2 = 4\frac{\lambda_M \gamma_I^2 A_I^2}{\pi^2} I_{J_i} \tag{A8}$$

where

$$I_{J_1} \triangleq \left[ Si\left(\frac{3\omega_c T}{2}\right)\right]^2 \tag{A9}$$

$$I_{J_2} \triangleq Si^2\left(\frac{\omega_c T}{2}\right) \tag{A10}$$

$$I_{J_3} \triangleq \frac{1}{4}\left[ Si\left(\frac{\omega_c T}{2}\right) + Si\left(\frac{3\omega_c T}{2}\right)\right]^2 \tag{A11}$$

and

$$I_{J_4} = I_{J_3}. \tag{A12}$$

If the SPACE channel is now considered, then $\sigma_{S_i}^2$ will equal $\sigma_{M_i}^2$ if all terms of the form $(1 + d_j)$ are replaced with $(1 - d_j)$. Also, $\sigma_{I_S}^2$ will equal $\sigma_{I_M}^2$ if $\lambda_M$ is replaced by $\lambda_S$.

## ACKNOWLEDGMENT

The authors would like to gratefully acknowledge the help of Dr. M. Braff for generating all the numerical results.

## REFERENCES

[1] L. B. Milstein, R. L. Pickholtz, and D. L. Schilling, "Optimization of the processing gain of an FSK-FH system," *IEEE Trans. Commun.*, vol. COM-28, pp. 1062–1079, July 1980.

[2] L. B. Milstein and D. L. Schilling, "Performance of a spread spectrum communication system operating over a frequency-selective fading channel in the presence of tone interference," *IEEE Trans. Commun.*, vol. COM-30, pp. 240–247, Jan. 1982.

[3] D. E. Borth and M. B. Pursley, "Analysis of direct-sequence spread spectrum multiple-access communications over a Rician fading channel," *IEEE Trans. Commun.*, vol. COM-27, pp. 1566–1577, Oct. 1979.

[4] R. S. Kennedy, *Fading Dispersive Communication Channels.* New York: Wiley, 1969.

[5] P. A. Bello and B. P. Nelin, "The effect of frequency selective fading on the binary error probabilities of incoherent and differentially coherent matched filter receivers," *IEEE Trans. Commun. Syst.*, vol. CS-11, pp. 170–180, June 1963.

[6] M. Schwartz, W. R. Bennett, and S. Stein, *Communications Systems and Techniques.* New York: McGraw-Hill, 1966, ch. 9, 10.

[7] *Error Correction Handbook*, Final Rep. prepared by Linkabit Corp. for U. S. Air Force, Contract FYY 620-76-C-0056, July 15, 1976.

# Adaptive Algorithms for Estimating and Suppressing Narrow-Band Interference in PN Spread-Spectrum Systems

JOHN W. KETCHUM AND JOHN G. PROAKIS, MEMBER, IEEE

*Abstract*—Results of an investigation of algorithms for estimating and suppressing narrow-band interference in pseudonoise (PN) spread-spectrum digital communication systems are presented. Techniques for determining the coefficients of a linear, interference suppression filter are described, which are based on linear prediction and conventional spectral analysis methods. Numerical results are presented on the characteristics of the linear filter and on its effectiveness in suppressing the interference. The error rate performance of a receiver that employs the interference suppression filter followed by a PN correlator is determined based on the assumption that the combined noise and residual interference at the output of the PN correlator is Gaussian distributed. Performance results obtained by Monte Carlo simulation are also presented and compared with theoretical results.

## I. INTRODUCTION

SPREAD-SPECTRUM, direct sequence, or pseudonoise (PN) modulation is employed in digital communication systems to reduce the effects of interference due to other users and intentional jamming. When the interference is narrow-band the cross correlation of the received signal with the replica of the PN code sequence reduces the level of the interference by spreading it across the frequency band occupied by the PN signal. Thus, the interference is rendered equivalent to a lower level noise with a relatively flat spectrum. Simultaneously, the cross-correlation operation collapses the desired signal to the bandwidth occupied by the information signal prior to spreading.

The interference immunity of a PN spread-spectrum communication system corrupted by narrow-band interference can be further improved by filtering the signal prior to cross correlation, where the objective is to reduce the level of the interference at the expense of introducing some distortion on the desired signal. This filtering can be accomplished by exploiting the wide-band spectral characteristics of the desired PN signal and the narrow-band characteristic of the interference. Since the spectrum of the PN signal is relatively flat across the signal frequency band, a strong narrow-band interference is easily recognized and estimated.

Manuscript received June 24, 1981; revised January 14, 1982. This paper was presented at the IEEE International Symposium on Information Theory, Santa Monica, CA, February 1981. This work was supported by the Rome Air Development Center under Contract F30602-78-C-0102.

J. W. Ketchum was with the Department of Electrical Engineering, Northeastern University, Boston, MA 02115. He is now with the Communications Products Technology Laboratory, GTE Laboratories, Waltham, MA.

J. G. Proakis is with the Department of Electrical Engineering, Northeastern University, Boston, MA 02115.

In this paper, algorithms are considered for estimating and suppressing the narrow-band interference. The algorithms are classified into two general categories. Algorithms in the first category employ the fast Fourier transform (FFT) algorithm for performing a spectral analysis on the received signal. On the basis of the spectral estimate a transversal filter is designed for suppressing the interference. These algorithms are termed nonparametric, since no prior knowledge of the characteristics of the interference is assumed in forming the estimate. The algorithms in the second category are based on linear prediction and are termed parametric. That is, the interference is modeled as having been generated by passing white noise through an all-pole filter. Linear prediction is used to estimate the coefficients of the all-pole model. The estimated coefficients specify an appropriate noise whitening, all-zero (transversal) filter through which the received signal is passed for the purpose of suppressing the narrow-band interference. The algorithms are described in Section II.

In Section III the effectiveness of the interference suppression algorithms is assessed by evaluating the performance of the receiver with and without the interference suppression filter. The signal-to-noise ratio (SNR) at the output of the PN correlator is one performance index used in the evaluation of the improvement provided by the interference suppression algorithms. Prior work on this problem has been done by Hsu and Giordano [1]. Our work extends their results by considering both single-band and multiple-band interference in the evaluation of performance. An important aspect in this investigation is the size of the interference suppression filter required to handle multiple-band interference. In addition to its performance, the frequency response of the interference suppression filter is determined and illustrated. Finally, some results are presented on the performance of the receiver as measured in terms of the probability of error, which were obtained both by applying a Gaussian assumption on the total residual noise and interference and by Monte Carlo simulation.

*Mathematical Model of PN Spread-Spectrum Communication System*

The transmitted signal is generated as shown in the block diagram of Fig. 1. The number of PN chips per information bit is $L$. Thus, the (equivalent low-pass) signal for the $k$th information bit can be expressed as

$$b_k(t) = \sum_{j=1}^{L} p_{kj} q(t - j\tau_c), \tag{1}$$

Reprinted from *IEEE Trans. Commun.*, vol. COM-30, pp. 913–924, May 1982.

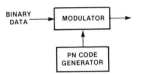

Fig. 1.   Direct sequence spread-spectrum transmitter.

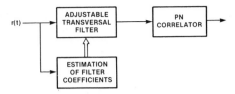

Fig. 2.   Direct sequence spread-spectrum receiver with adaptive interference suppression.

where $\{p_{kj}\}$ represent the output sequence from the PN code generator for the $k$th information bit and $q(t)$ is a rectangular pulse of duration $\tau_c$ and unit energy. The total transmitted singal may be expressed in the form

$$s(t) = \sum_k I_k b_k(t - kT_b), \tag{2}$$

where $\{I_k\}$ represents the binary information sequence and $T_b = L\tau_c$ is the bit interval (reciprocal of the bit rate).

The signal transmitted over the channel is corrputed by additive white Gaussian noise $n(t)$ and by narrow-band interference denoted by $i(t)$. The narrow-band interference is modeled as consisting of either a number of CW tones, i.e.,

$$i(t) = \sum_{m=1}^{Q} A_m \cos (2\pi f_m t + \phi_m), \tag{3}$$

or a filtered narrow-band Gaussian noise process.

From the description given above, the received signal has the form

$$r(t) = s(t) + i(t) + n(t). \tag{4}$$

The receiver attempts to suppress the interference $i(t)$ and then to recover the information sequence by cross correlation with a replica of the PN sequence. The estimation and suppression of the narrow-band interference is accomplished prior to signal demodulation, as illustrated in general terms in the block diagram shown in Fig. 2. The estimation of the filter coefficients may be accomplished by means of a linear prediction algorithm or by means of a spectral analysis algorithm based on the FFT algorithm. In any case, the objective is to design an adaptive transversal filter that highly attenuates the received signal in those frequency bands which contain strong interference.

## II. ALGORITHMS FOR ESTIMATION AND SUPPRESSION OF NARROW-BAND INTERFERENCE

In this section we present a number of algorithms for estimating and suppressing a narrow-band interference embedded in a wide-band PN spread-spectrum signal. The algorithms may be classified into two general categories. The algorithms in the first category employ the FFT algorithm for performing a spectral analysis from which an appropriate transversal filter is specified. These algorithms may be termed nonparametric, since no prior knowledge of the characteristics of the interference is assumed in forming the estimate. The algorithms in the second category are based on linear prediction and may be termed parametric. That is, the interference is modeled as having been generated by passing white noise through an all-pole filter [2].

*Interference Suppression Based on Nonparametric Spectral Estimates*

The basis for this method is that the power density spectrum of the PN sequence is relatively flat while the spectrum of the narrow-band interference is highly peaked. The first step in this method is to estimate the power spectral density of the received signal. The spectral estimate can be obtained by any one of the well-known spectral analysis techniques described in [3]. For illustrative purposes, we have selected the Welch method and we have made use of the computer program listed in [4] to generate the numerical results presented in Section III.

Once the power spectral density of the received signal is estimated, the interference suppression filter can be designed. A transversal filter is an appropriate filter structure for this application, since we desire to use a filter that contains notches (zeros) in the frequency range occupied by the interference. A relatively simple method for designing the transversal filter in the discrete-time (sampled-data) domain is to select its discrete Fourier transform (DFT) to be the reciprocal of the square root of the power spectral density at equally spaced frequencies. To elaborate, suppose that the transversal filter has $K$ taps. The problem is to specify the $K$ tap coefficients $\{h(n)\}$ or, equivalently, the DFT $H(k)$, defined as

$$H(k) = \sum_{n=0}^{K-1} h(n)e^{-j\frac{2\pi}{K}nk}, \quad k = 0, 1, \cdots, K-1. \tag{5}$$

The DFT $H(k), k = 0, 1, \cdots, K-1$ is selected as

$$H(k) = \frac{1}{\sqrt{P\left(\frac{k}{K}R_s\right)}} e^{-j\frac{2\pi}{K}(\frac{K-1}{2})k} \tag{6}$$

where $P(f)$, $0 \leqslant f \leqslant R_s$, denotes the estimate of the power spectral density and $R_s$ denotes the sampling rate. We note that $P(f)$ is periodic with period $R_s$ and that $R_s/2$ is the folding frequency [3].

It is desirable to have a transversal filter that has linear phase. This is achieved if the impulse response $h(n)$ is real and satisfies the symmetry condition[1]

$$h(n) = h(K - 1 - n). \tag{7}$$

---

[1] See, for example, Oppenheim and Schafer [3, pp. 157–158].

The symmetry condition in (7) is satisfied if $H(k) = H^*(K - k)$. But $H(k)$ as defined in (6) does satisfy this condition since $P(f) = P(R_s - f)$. Hence, $h(n)$ is symmetric.

In effect, the filter characteristic obtained from (6) attempts to approximate an inverse filter to the power spectral density. That is, the interference suppression filter attempts to whiten the spectrum of the incoming signal. Thus, the filter will have a large attenuation in the frequency range occupied by the interference and a relatively small attenuation elsewhere.

### Interference Suppression Based on Linear Prediction

In contrast to the nonparametric spectral analysis method described above, the following method for estimating the narrow-band interference is based on modeling the interference as white noise passed through an all-pole filter. That is, instead of using the received signal to estimate the spectrum directly, the signal is used to estimate the pole positions. This estimation is accomplished by means of linear prediction. An estimate of the power spectral density is easily obtained from the all-pole model. However, this step can be omitted. That is, the power density spectrum need not be computed explicitly for the purpose of designing the suppression filter. The interference suppression filter is simply a transversal (all-zero) filter having zero positions that coincide with the estimated pole positions. Thus, the spectrum of the signal at the output of the transversal filter is rendered white.

In order to develop the mathematical formulation for the all-pole model, we begin with the received signal

$$r(t) = s(t) + i(t) + n(t) \tag{8}$$

where $s(t)$ is the information-bearing signal, $i(t)$ denotes the narrow-band interference, and $n(t)$ is assumed to be a sample function of a white Gaussian noise process. For convenience, we assume that $r(t)$ is sampled at the chip rate of the PN sequence. Thus, (8) can be expressed as

$$r(k) = s(k) + i(k) + n(k), \qquad k = 1, 2, \cdots. \tag{9}$$

We assume that $s(k)$, $i(k)$, and $n(k)$ are mutually uncorrelated.

An estimate of the interference $i(t)$ is formed from $r(k)$. Assume for the moment that the statistics of $i(t)$ are known and are stationary. Then, we can predict $i(k)$ from $r(k - 1), r(k - 2), \cdots, r(k - m)$. That is,

$$\hat{i}(k) = \sum_{l=1}^{m} a_l r(k - l) \tag{10}$$

where $\{a_l\}$ are the coefficients of the linear predictor. It should be emphasized that (10) predicts the interference but not the signal $s(k)$, because $s(k)$ is uncorrelated with $r(k - l)$ for $l = 1, 2, \cdots, m$,[2] as a consequence of the sampling being done at the chip rate.

---

[2] This is true provided that $m$ is less than the PN sequence length.

The coefficients in (10) are determined by minimizing the mean square error between $r(k)$ and $i(k)$, which is defined as

$$\mathscr{E}(m) = E[r(k) - \hat{i}(k)]^2$$
$$= E\left[ r(k) - \sum_{l=1}^{m} a_l r(k - l) \right]^2. \tag{11}$$

Minimization of $\mathscr{E}(m)$ with respect to the predictor coefficients $\{a_l\}$ can be easily accomplished by invoking the orthogonality principle in mean square estimation [5]. This leads to the set of linear equations

$$\sum_{l=1}^{m} a_l \rho(k - l) = \rho(k), \qquad k = 1, 2, \cdots, m \tag{12}$$

where

$$\rho(k) = E[r(m)r(k + m)] \tag{13}$$

is the autocorrelation function of the received signal $r(k)$. The equations in (12) are usually called the Yule–Walker equations [2]. They can be written in matrix form as

$$R_m a_m = b_m \tag{14}$$

where $R_m$ is the $m \times m$ autocorrelation matrix, $a_m$ is the vector of filter coefficients, and $b_m$ is a vector of autocorrelation coefficients $\rho(k)$, $1 \leq k \leq m$. The matrix $R_m$ is a Toeplitz matrix, which is efficiently inverted by use of the Levinson–Durbin algorithm [6], [7]. This algorithm is an order recursive method for solving a set of linear equations when the coefficient matrix is Toeplitz. The recursive relations are

$$a_{11} = \frac{\rho(1)}{\rho(0)}$$

$$a_{mm} = \frac{\rho(m) - \beta_{m-1}{}^T a_{m-1}}{\mathscr{E}_{min}(m - 1)}, \qquad m > 1$$

$$a_{mk} = a_{m-1,k} - a_{mm} a_{m-1,m-k}, \qquad 1 \leq k \leq m - 1$$

$$\mathscr{E}_{min}(m) = \rho(0) - b_m{}^T a_m$$
$$= (1 - a_{mm}{}^2)\mathscr{E}_{min}(m - 1), \qquad m \geq 1$$

$$\mathscr{E}_{min}(0) = \rho(0) \tag{15}$$

where $\beta_m$ is the vector $b_m$ in reversed order. Thus, (14) is initially solved for $m = 1$, then the solution is used to obtain the solution for $m = 2$, and the process is repeated until the solution for the $m$th-order predictor is obtained.

Once the prediction coefficients are determined, the estimate $\hat{i}(k)$ of the interference, given by (10), is subtracted from $r(k)$ and the difference signal is processed further in order to extract the digital information. Thus, the equivalent transversal filter for suppressing the interference is described by the transfer function

$$A_m(z) = 1 - \sum_{k=1}^{m} a_{mk} z^{-k} \qquad (16)$$

where $z^{-1}$ denotes a unit of delay. The corresponding all-pole model for the interference signal is $1/A_m(z)$. An alternative structure which follows naturally from the Levinson recursion is the lattice structure [8].

The solution of (12) for the coefficients $\{a_{mk}\}$ of the prediction filter requires knowledge of the autocorrelation function $\rho(k)$. In practice, the autocorrelation function of $i(k)$ and, hence, $r(k)$ is unknown and it may also be slowly varying in time. Consequently, one must consider methods for obtaining the predictor coefficients directly from the received signal $\{r(k)\}$. This may be accomplished in a number of ways. In this investigation, three different methods were considered. In all cases, we obtained the predictor coefficients by using a block of $N$ samples of $\{r(k)\}$. The three methods are described below.

### Direct Application of the Levinson Algorithm

The first method is simply based on the direct estimation of $\rho(k)$ from the block of $N$ samples. The estimate of $\rho(k)$ is

$$\hat{\rho}(k) = \sum_{n=0}^{N-k} r(n)r(n+k), \qquad k = 0, 1, \cdots, m. \qquad (17)$$

The estimate $\hat{\rho}(k)$ may then be substituted in (12) in place of $\rho(k)$ and the Levinson–Durbin algorithm can be used to solve the equations efficiently. Thus, the recursive relations given in the previous section apply with $\rho(k)$ replaced by $\hat{\rho}(k)$.

### Burg Algorithm

The second method considered for obtaining the prediction coefficients is the Burg algorithm [2], [9]. Basically, the Burg algorithm may be viewed as an order-recursive least squares algorithm in which the Levinson recursion is used in each iteration. The performance index used in the Burg algorithm is

$$\mathcal{E}_B(m) = \sum_{i=m+1}^{N} [f_m{}^2(i) + b_m{}^2(i)] \qquad (18)$$

where $f_m(i)$ and $b_m(i)$ are the forward and backward errors in an $m$th-order predictor, which are defined as

$$f_m(i) = r(i) - \sum_{k=1}^{m} a_{mk} r(i-k)$$

$$b_m(i) = r(i-m) - \sum_{k=1}^{m} a_{mk} r(i-m+k). \qquad (19)$$

The predictor coefficients $a_{mk}$ for $1 \leqslant k \leqslant m-1$ are forced to satisfy the Levinson–Durbin recursion given by (15). As a con-

sequence of this constraint, the forward and backward errors satisfy the recursive relations

$$f_m(i) = f_{m-1}(i) - a_{mm} b_{m-1}(i-1)$$

$$b_m(i) = b_{m-1}(i-1) - a_{mm} f_{m-1}(i)$$

$$f_0(i) = b_0(i) = r(i). \qquad (20)$$

The relations in (20) are substituted into (18) and $\mathcal{E}_B(m)$ is minimized with respect to $a_{mm}$. The result of this minimization is

$$a_{mm} = \frac{2 \displaystyle\sum_{i=m+1}^{N} f_{m-1}(i) b_{m-1}(i-1)}{\displaystyle\sum_{i=m+1}^{N} [f_{m-1}{}^2(i) + b_{m-1}{}^2(i-1)]},$$

$$m \geqslant 1. \qquad (21)$$

The relations given in (20) and (21) along with the Levinson–Durbin recursion for $\{a_{mk}\}$, $1 \leqslant k \leqslant m-1$ given in (15) constitute the Burg algorithm for obtaining the prediction coefficients directly from the data.

### Least Squares Algorithm

As we indicated above, the Burg algorithm is basically a least squares algorithm with the added constraint that the predictor coefficients satisfy the Levinson recursion. As a result of this constraint, an increase in the order of the predictor requires only a single parameter optimization at each stage. In contrast to this approach, we considered an unconstrained least squares algorithm in which the optimum predictor coefficients, in the sense of least squares, are obtained by minimizing (18) over the entire set of predictor coefficients. This minimization process yields the following set of equations:

$$\sum_{k=1}^{m} a_{mk} \phi(l, k) = \phi(l, 0), \qquad l = 1, 2, \cdots, m \qquad (22)$$

where

$$\phi(l, k) = \sum_{i=m+1}^{N} [r(i-k)r(i-l) + r(i-m+k)r(i-m+l)]. \qquad (23)$$

The linear equations in (22) can be expressed in matrix form as

$$\Phi_m a_m = \phi_m, \qquad (24)$$

where the matrix $\Phi_m$ is an $(m \times m)$ autocorrelation matrix with elements $\{\phi(k, l)\}$ and $\phi_m$ is an $m$-dimensional vector with elements $\phi(l, 0)$, $l = 1, 2, \cdots, m$. The matrix $\Phi_m$ is symmetric. However, in contrast to the autocorrelation matrix

121

$R_m$ in (14), which is Toeplitz, the matrix $\Phi_m$ is not Toeplitz. Consequently, the Levinson algorithm cannot be used to solve (24) recursively. In spite of the fact that $\Phi_m$ is not Toeplitz, it is still possible to derive a recursive algorithm for the predictor coefficients based on the least squares performance index. Morf *et al.* [10]-[13] have developed such a recursive least squares algorithm that not only allows one to recursively increase the order of the predictor, but also allows one to update the predictor coefficients recursively in time for a predictor of any given order. A detailed development of this recursive least squares algorithm has been given by Pack and Satorius [14] and, for the sake of brevity, will not be repeated here. An order-recursive version of this algorithm has also been described recently by Marple [15]. In short, the prediction coefficients based on the least squares criterion can be solved efficiently by means of an algorithm that is both recursive in order and in time.

We have used these algorithms in Monte Carlo simulations to estimate the coefficients of the interference suppression filter. When the narrow-band interference is stationary, a large number of data samples may be used to obtain a good estimate of the interference. However, in a jamming environment, where the jammer has the ability to change his frequencies dynamically, the adaptive algorithms will be effective only if they use a small sample size. From simulation results which are described in the following section, we illustrate that good estimates are obtained from the linear prediction algorithms with as few as 50 data samples (50 PN chips) and a filter length of fifteen taps (15 coefficients).

In discussions with other colleagues who have used the Levinson and Burg algorithms we were told that good estimates of the predictor coefficients can be obtained with as few samples as twice the number of predictor coefficients. The Levinson algorithm appears to require a few more samples. We expect the least squares algorithm to be comparable to the Burg algorithm in its performance and its sample size requirements. We did not perform a thorough study of the sample size requirements for obtaining good estimates, however. Additional quantitative results are needed, so this topic merits further investigation.

## III. PERFORMANCE RESULTS ON INTERFERENCE SUPPRESSION

This section deals with the performance of the interference suppression filter. First, we determine the improvement provided by interference suppression as measured in terms of the SNR at the output of the PN correlator. The output SNR is by far the most convenient performance index for obtaining numerical results. This performance index is used to assess the improvement in performance obtained by an interference suppression filter. In addition, we describe the frequency response characteristics of the interference suppression filter. Finally, we present some Monte Carlo simulation results on the performance of the receiver as measured in terms of the probability of error, as well as some error rate results based on the assumption that the total residual noise and interference is Gaussian distributed.

### SNR Improvement Factor Resulting from Interference Suppression

The received signal, sampled at the chip rate, can be represented as

$$r(k) = p(k) + i(k) + n(k), \qquad k = 1, 2, \cdots, \tag{25}$$

where the binary sequence $\{p(k)\}$ represents the PN chips with values $\pm 1$, $\{i(k)\}$ represents the sequence of samples of the narrow-band interference and $\{n(k)\}$ represents the sequence of wide-band noise samples.

In order to demonstrate the effectiveness of the interference suppression algorithms, we shall compare the performance of the receiver with and without the suppression filter. Three types of suppression filters are considered. One is specified on the basis of the nonparametric spectral algorithm and its coefficients are denoted as $h(k)$, $0 \leqslant k \leqslant K$. This filter has $K + 1$ taps. Also, it has linear phase[3] and, hence if we select $K$ to be even, the impulse response is symmetric with respect to the center tap $h(K/2)$. For mathematical convenience we normalize $h(K/2)$ to unity and use it as the reference tap in the PN cross correlation that follows the suppression filter. Thus, the filter introduces a delay of $K/2$ chips.

The second type of interference suppression filter is based on the parametric spectral estimation algorithms and its impulse response is $h(0) = 1$, $h(k) = -a_{mk}$, $k = 1, 2, \cdots, m$ where the $\{a_{mk}\}$ are the prediction coefficients. The reference tap is now $h(0)$, so there is no delay through this filter.

The third type of interference suppression filter considered may be viewed as two filters in cascade. One filter is obtained from linear prediction (second type of filter) as described in the previous section. Following this filter is its matched filter. The cascade of the two filters results in a linear phase filter with an odd number of taps, say $K + 1$, where $K$ is even, and symmetric about its center tap, which we assume to be normalized to unity. This tap is used as the reference tap for the cross correlation that follows. Hence, there is a delay of $K/2$ chips introduced by the two filters in cascade. This configuration of a noise-whitening filter followed by its matched filter and then the PN correlator maximizes the output SNR.

In the following derivation we focus on the second type of filter for which $h(0) = 1$, $h(k) = -a_{mk}$ and $K = m$. The results of the derivation are, then, modified to apply to the two linear phase filters. The input to the filter is $r(k)$ and its output is

$$y(k) = \sum_{l=0}^{K} h(l)r(k-l), \qquad k = 1, 2, \cdots$$

$$= \sum_{l=0}^{K} h(l)[p(k-l) + i(k-l) + n(k-l)] . \tag{26}$$

The signal $\{y(k)\}$ is then fed to the PN correlator. The output of the PN correlator, which is the decision variable for recover-

---

[3] Since the filter has $K + 1$ taps the condition for linear phase is $h(n) = h(K - n)$.

ing the binary information, is expressed as

$$U = \sum_{k=1}^{L} y(k)p(k) \qquad (27)$$

where $L$ represents the number of chips per information bit or the processing gain.

By substituting (26) into (27), the decision variable can be expressed in the form

$$U = \sum_{k=1}^{L} p(k) \left\{ \sum_{l=0}^{K} h(l)[p(k-l) + i(k-l) + n(k-l)] \right\}$$

$$= \sum_{k=1}^{L} p^2(k) + \sum_{l=1}^{K} h(l) \sum_{k=1}^{L} p(k)p(k-l)$$

$$+ \sum_{k=1}^{L} \sum_{l=0}^{K} h(l)p(k)[i(k-l) + n(k-l)]$$

$$= L + \sum_{l=1}^{K} h(l) \sum_{k=1}^{L} p(k)p(k-l)$$

$$+ \sum_{k=1}^{L} \sum_{l=0}^{K} h(l)p(k)i(k-l)$$

$$+ \sum_{k=1}^{L} \sum_{l=0}^{K} h(l)p(k)n(k-l). \qquad (28)$$

The first term on the right-hand side of (28) represents the desired signal component, the second term represents the self-noise caused by the dispersive characteristic of the filter, the third term represents the residual narrow-band interference at the output of the PN correlator, and the last term represents the additive wide-band noise.

For the comparison that we wish to make, the SNR at the output of the PN correlator is a mathematically tractable performance index. To determine the expression for the SNR we must compute the mean and variance of $U$. We assume that the binary PN sequence is white, the interference $i(k)$ has zero mean and autocorrelation function $\rho_i(k)$, and the additive noise $n(k)$ is white with variance $\sigma_n^2$. Then, the mean of $U$ is

$$E(U) = L \qquad (29)$$

and the variance is

$$\mathrm{var}\,(U) = L \sum_{l=1}^{K} h^2(l) + L \sum_{l=0}^{K} \sum_{m=0}^{K} h(l)h(m)\rho_i(l-m)$$

$$+ L\sigma_n^2 \sum_{l=0}^{K} h^2(l). \qquad (30)$$

The first term on the right-hand side of the expression for the variance represents the mean square value of the self-noise due to the time dispersion introduced by the interference suppression filter. The second term is the mean square

value of the residual narrow-band interference. The last term is the mean square value of the wide-band noise.

The SNR at the output of the correlator is defined as the ratio of the square of the mean to the variance. Thus,

$SNR_0$

$$= \frac{L}{\displaystyle\sum_{l=1}^{K} h^2(l) + \sum_{l=0}^{K} \sum_{m=0}^{K} h(l)h(m)\rho_i(l-m) + \sigma_n^2 \sum_{l=0}^{K} h^2(l)}. \qquad (31)$$

If there is no suppression filter, $h(l) = 1$ for $l = 0$ and zero otherwise. Therefore, the corresponding output SNR is

$$SNR_{no} = \frac{L}{\rho_i(0) + \sigma_n^2} \qquad (32)$$

where $\rho_i(0)$ represents the total power of the narrow-band interference.

The ratio of the SNR in (31) to the SNR in (32) represents the improvement in performance due to the use of the interference suppression filter. This ratio, denoted by $\eta$, is

$$\eta = \frac{\rho_i(0) + \sigma_n^2}{\displaystyle\sum_{l=1}^{K} h^2(l) + \sum_{l=0}^{K} \sum_{m=0}^{K} h(l)h(m)\rho_i(l-m) + \sigma_n^2 \sum_{l=0}^{K} h^2(l)}. \qquad (33)$$

We observe that $\eta$ is independent of the processing gain $L$.

The SNR expressions in (31) and (33) must be modified when the interference suppression filter is designed to have linear phase. For such a filter, having $K + 1$ taps, where $K$ is even, and a center tap $h(K/2) = 1$, the self-noise at its output is proportional to

$$2 \sum_{l=0}^{\frac{K}{2}-1} \left(2 - \frac{l}{L}\right) h^2(l).$$

Therefore, the denominators of (31) and (33) are modified by the substitution of this factor in place of the factor

$$\sum_{l=1}^{K} h^2(l).$$

In plotting the improvement factor, it is convenient to use a logarithmic scale. Thus, we define

$$\eta_{dB} = 10 \log \eta. \qquad (34)$$

This factor will be plotted against the normalized SNR at the output of the PN correlator when there is no suppression filter. In other words, the abscissa is SNR per chip without filtering, defined as

$$\frac{SNR_{no}}{L} = \frac{1}{\rho_i(0) + \sigma_n^2}. \qquad (35)$$

As a consequence, the graphs of $\eta_{dB}$ versus $SNR_{no}/L$ are universal plots in the sense that they apply to any PN spread-spectrum system with arbitrary processing gain for a given $\rho_i(0)$ and $\sigma_n^2$.

### Characteristics and Performance of the Interference Suppression Filter Based on Linear Prediction

In this section we discuss the characteristics of the interference suppression filter and we illustrate its performance as measured in terms of the improvement factor $\eta_{dB}$. The filter considered first is the one having the impulse response $h(0) = 1$, $h(k) = -a_{mk}$ where $\{a_{mk}\}$ are the coefficients of an $m$th-order linear predictor. These coefficients may be determined from (12) using the exact autocorrelation function or by one of the adaptive algorithms described in Section II. Initially, we use the exact autocorrelation function to assess the performance of the suppression filter. Then, we compare the performance results with those obtained from use of the adaptive algorithms.

For illustrative purposes, the model that we use for narrow-band interference consists of a sum of equally spaced sinusoids covering 20 percent of the signal band. A second model was also investigated, which consisted of filtered white noise. However, we obtained results which were nearly identical to those obtained with the first model, as long as the white noise was filtered sharply enough, for example, with a six-pole Butterworth filter.

In particular, for the sum of sinusoids model, $i(k)$ is expressed as

$$i(k) = \sum_{m=1}^{Q} A_m \cos\left(2\pi f_m k + \phi_m\right) \tag{36}$$

where the amplitudes $\{A_m\}$ were selected to be identical and the phases are uniformly distributed on the interval $(0, 2\pi)$. The autocorrelation function of $i(k)$ is

$$\rho(k) = \frac{1}{2} \sum_{m=1}^{Q} A_m^2 \cos 2\pi f_m k. \tag{37}$$

The number of tones used in $i(k)$ was either $Q = 100$ or $Q = 10$.

Fig. 3 illustrates the improvement factor plotted versus SNR per chip for 4-tap and 15-tap filters for sinusoidal interference with $Q = 10$ and $Q = 100$ tones. In all cases $\sigma_n^2 = 0.01$ (signal-to-Gaussian noise ratio is 20 dB) and the interference power $\rho_i(0)$ is varied. Since the abscissa is $10 \log 1/(\sigma_n^2 + \rho_i(0))$ and $\rho_i(0) \gg \sigma_n^2$ for the range of values used in Fig. 3, it follows that the abscissa is approximately $10 \log 1/\rho_i(0)$. It is evident from this figure that there is little difference in performance between $Q = 10$ and $Q = 100$. Furthermore, there is very little gain in performance when the number of taps is increased from 4 to 15. The major conclusion that we have reached from the above results is that the model for the interference is not critical. Consequently, in most of our numerical results we used sinusoidal interference with $Q = 100$ tones. We also observe that the gain achieved by the sup-

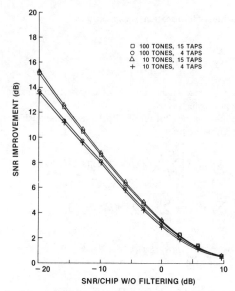

Fig. 3. Improvement for single-band interference. Filter coefficients computed using exact autocorrelation coefficients, with $\sigma_n^2 = 0.01$.

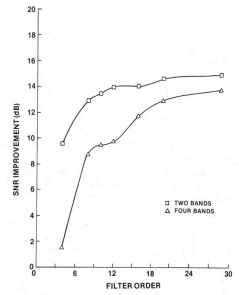

Fig. 4. Improvement factor for two-band and four-band interference as a function of filter order. Filter coefficients computed using exact autocorrelation coefficients with $\sigma_n^2 = 0.01$, and SNR = −20 dB/chip without filtering.

pression filter when the interference-to-Gaussian noise ratio is above 5–10 dB is relatively small.

Second, we have investigated the length of the interference suppression filter required to achieve good performance. In this computation we maintained the 20 percent bandwidth occupancy for the interference, but we distributed it equally in either two or four nonoverlapping frequency bands. Fig. 4 illustrates the improvement factor as a function of the number of filter taps when the SNR per chip without filtering is −20 dB. From the graphs we observe that a filter having eight or more taps performs well when the interference is split into four frequency bands.

The frequency response characteristics of the filters with 16-tap and 29-tap predictors with four bands of interference and an SIR/chip of −20 dB are shown in Fig. 5. It appears that the 16-tap filter introduces some distortion in the frequency range between notches. If the filter order is increased, however, the frequency response improves, as is demonstrated by the 29-tap filter.

Similar results were obtained when the interference was split into a larger number of bands. The conclusion we have reached from observation of these results is that the filter will suppress the multiband interference provided that it has enough degrees of freedom to assign at least one complex-conjugate pair of zeros to each band.

More specifically, these results indicate that a prediction filter having a number of coefficients that is fewer than twice the number of interference bands is useless, in the sense that it does nothing. Apparently, this is a limitation of the mean square error criterion used to design the prediction filter. If one knows that the number of degrees of freedom is fewer than twice the number of interference bands, an ad hoc scheme such as arbitrarily assigning a complex-conjugate pair of zeros to each band, up to the maximum number of bands that can be suppressed with the given number of degrees of freedom, may be better.

Next, we consider the linear phase counterpart of the suppression filter obtained through linear prediction. Its consideration may be justified if we view the combined narrowband interference plus wide-band noise as an equivalent colored noise process added to the desired PN signal. Now in detection of a signal $S(f)$ in colored noise with power spectral density $P(f)$, the output SNR at the receiver is maximized when the receiver consists of a noise-whitening filter, say $H(f)$, followed by a filter matched to $H(f)S(f)$. Thus, a matched filter with frequency response characteristic $H^*(f)S^*(f)$ will maximize the output SNR. If $H(f)$ represents the interference suppression filter with impulse response $h(t)$, then $H^*(f)$ represents a filter with impulse response $h(-t)$. Thus, the cascade of these two filters is a filter having an even impulse response. Since we have determined the coefficients of $H(f)$ by means of linear prediction, the coefficients of $H^*(f)$ are simply the time reverse of those obtained for $H(f)$, and the cascade of $H(f)$ and $H^*(f)$ is a linear phase filter. Use of such a filter prior to the PN correlator improves performance. This is illustrated in Fig. 6 for a 4-tap and a 15-tap predictor. We observe that at −20 dB per chip SNR the 4-tap predictor in cascade with its matched filter provides about 18.5 dB of improvement. The 15-tap predictor with its matched filter provides about 21.5 dB of improvement. In comparison, the 4-tap predictor without its matched filter provides about 13.5 dB of improvement. Therefore, the inclusion of the matched filter has resulted in about 5 dB gain at a −20 dB SNR per chip. Such a large gain is highly significant and suggests that the use of the matched filter is very desirable.

Finally, we turn our attention to the performance characteristics of the filters designed directly from the data by means

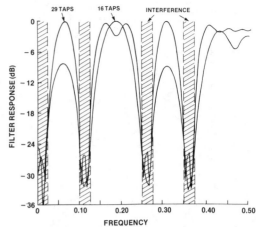

Fig. 5. Frequency response of 16-tap and 29-tap filters with four bands of interference. Filter coefficients computed using exact autocorrelation coefficients, with $\sigma_n^2 = 0.1$ and SNR = −20 dB/chip without filtering.

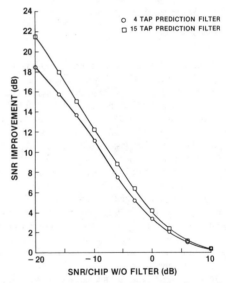

Fig. 6. Improvement factor for suppression filter in cascade with its matched filter. Filter coefficients computed using exact autocorrelation coefficients, with $\sigma_n^2 = 0.01$.

of the adaptive prediction algorithms described in Section II. For this discussion it suffices to consider only single-band interference and only the whitening filter, with no matched filter. The main point that we wish to make with regard to the performance of the algorithms is best illustrated with the results shown in Fig. 7. We used a block of 50 data samples to compute the coefficients of a fourth-order predictor by means of the least squares algorithm, the Burg algorithm and the Levinson algorithm. For the latter, the data were used to generate the estimate of the autocorrelation function. The predictor coefficients obtained from the data were used in the computation of the improvement factor given by (33). The graphs indicate that all three algorithms perform equally well.

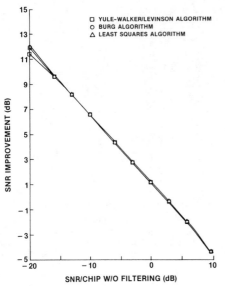

Fig. 7. Improvement factor obtained with linear prediction algorithms. Filter coefficients for 4-tap predictor computed using 50 signal samples, with $\sigma_n^2 = 0.01$.

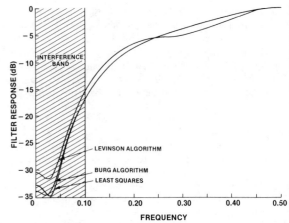

Fig. 8. Frequency response of filters designed from least squares, Burg, and Levinson–Durbin algorithms. Filter coefficients for 4-tap predictor computed using 50 signal samples, with $\sigma_n^2 = 0.01$ and SNR = $-20$ dB/chip without filtering.

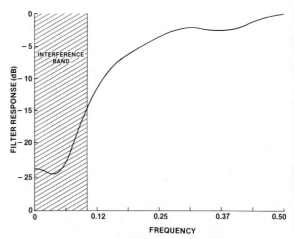

Fig. 9. Frequency response of 15-tap filter for SNR/chip = $-20$ dB. Filter coefficients computed using Welch method, with $\sigma_n^2 = 0.01$.

In other words, the difference in performance among the three algorithms is insignificant. This behavior is further substantiated by observing the corresponding frequency response characteristics of the suppression filter. For example, Fig. 8 illustrates the frequency response characteristics of the suppression filter designed from 50 samples of data on the basis of the three algorithms. Here, we also observe very minor differences in the frequency response characteristics.

We have not investigated the quality of the estimates of the prediction coefficients for sample sizes smaller than 50. However, when the order of the predictor is a large fraction ($\frac{1}{2}$ or greater) of the sample size $N$, the quality of the autocorrelation estimate for large lags, which is used in the Levinson algorithm, is poor. In such a case, the Burg algorithm and the least squares algorithm are expected to yield better performance than the Levinson algorithm.

The results of these simulations agree very well with the analytical results shown in Fig. 3 except at low values of interference where the improvement factor approaches 0 dB theoretically, but simulation data indicate a small loss in performance. This suggests that for small interference, it is best to arbitrarily set the predictor coefficients to 0.

*Performance of Interference Suppression Filter Based on Nonparametric Spectral Estimates*

In Section II we described a method for designing an interference suppression filter based on conventional, nonparametric methods for spectral estimation. As an illustration of the effectiveness of this approach, we computed estimates of the power spectral density from simulated received data and used the resultant estimates to specify a filter characteristic in accordance with (6). The Welch method [3] was used to generate the estimates of the power spectral density. For this computation, the FFT size selected was 64 points. The resultant interference suppression filter consists of 15 taps. The number of data points used to generate the spectral estimate is 992.

Fig. 9 illustrates the frequency response of the 15-tap interference suppression filter which is obtained when the SNR per chip is $-20$ dB and the interference occupies 20 percent of the signal band. The resulting filter contains a notch in the desired frequency band. The behavior of this type of filter in varying amounts of interference is similar to that obtained by means of linear prediction, with the exception that a larger filter is required to obtain a comparable notch in the interference band.

As a final computation, the coefficients $h(n)$ of the interference suppression filter were substituted into (33) and the improvement factor was evaluated. Fig. 10 illustrates the improvement factor as a function of the SNR per chip without

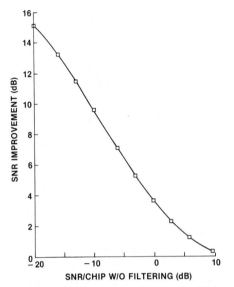

Fig. 10. Improvement factor for 15-tap filter based on Welch method. Filter coefficients computed using Welch method, with $\sigma_n^2 = 0.01$.

filtering. When compared with our previous results for single-band interference using linear prediction, we find that the performance improvement has similar characteristics, although the linear prediction filter $H(f)$ followed by its matched filter $H^*(f)$ result in better performance, especially at low SNR per chip. In general, however, it appears that the nonparametric method for spectral estimation coupled with the filter design formula in (6) provides a viable means for suppressing narrow-band interference in a wide-band signal. The one disadvantage of this method is the relatively large sample size required to generate a good spectral estimate.

*Bit Error Rate Results*

The results given above on SNR improvement indicate that significant performance gains can be achieved by using the adaptive interference cancellation techniques outlined in Section II. The bit error rate results which are given below confirm this conclusion. The error rate results were obtained by applying a Gaussian assumption to SNR expressions given previously, and from a Monte Carlo simulation on a digital computer.

An approximation to the bit error rate can be found by assuming that the decision variable $U$ given in (28) has a Gaussian distribution with mean and variance given by (29) and (30). This is equivalent to assuming that the performance of this system is the same as that of binary PSK signaling corrupted only by white Gaussian noise with variance equal to the total noise due to white noise, interference, and self-noise, at the output of the PN correlator. We observe that the computations which resulted in (29) and (30) do not include the effect of intersymbol interference caused by the time-dispersive interference suppression filter. This type of noise is negligible when the processing gain $L$ is much greater than the length of the suppression filter.

Under these assumptions, the bit error rate is given by

$$P_b = P(U < 0) = \int_{-\infty}^{0} \frac{1}{\sqrt{2\pi}\sigma} e^{-(U-\mu)^2/2\sigma^2} \, dU \qquad (38)$$

where $\sigma^2 = \text{var}(U)$ and $\mu = E(U)$. Thus,

$$P_b = \tfrac{1}{2} \, \text{erfc} \sqrt{\gamma_b}, \qquad (39)$$

where the SNR per bit $\gamma_b$ is given by

$$\gamma_b = \frac{\mu^2}{2\sigma^2} = \text{SNR}_0/2 \qquad (40)$$

and $\text{SNR}_0$ is given in (31).

In order to compare the performance of the receiver when the excision filter is used to suppress the narrow-band interference with the performance of the optimum receiver without any interference, the bit error rate is plotted against the SNR without interference, i.e., the signal-to-white Gaussian noise ratio $E_b/N_0$. When there is no interference all the filter coefficients are zero except that $h(0) = 1$. Hence, the SNR per bit without interference is

$$\frac{E_b}{N_0} = \frac{L}{2\sigma_n^2} \qquad (41)$$

where $\sigma_n^2 = N_0 W/2$, $N_0$ is the single-sided spectral density of the Gaussian noise and $W$ is the signal bandwidth, which has been normalized to unity. Since

$$\frac{E_b}{N_0} \geqslant \gamma_b = \tfrac{1}{2} \, \text{SNR}_0, \qquad (42)$$

the difference between $E_b/N_0$ and $\gamma_b$ represents the loss in SNR due to 1) the presence of residual interference, 2) the self-noise resulting from the time-dispersive interference suppression filter, and 3) the amplification of the additive Gaussian noise by the filter.

Fig. 11 illustrates the error rate performance based on the Gaussian approximation for a receiver which employs a 4-tap predictor to suppress the interference consisting of 100 tones occupying 20 percent of the frequency band of the signal. The signal-to-interference ratio per chip is −20 dB. We do not bother to plot the performance of a spread-spectrum receiver that does not suppress the interference by means of adaptive filtering, because its error rate is in the vicinity of $\frac{1}{2}$ for all processing gains shown in Fig. 11. In spite of the interference suppression filter, there is a significant degradation in performance due to the presence of the interference.

The addition of the matched filter following the interference suppression filter results in a significant improvement in the performance of the receiver as illustrated in Fig. 12. At large processing gains, the performance of the receiver is degraded by only a few decibels from the performance of ideal binary PSK without interference.

Finally, Fig. 13 illustrates the performance results obtained from a Monte Carlo simulation. The conditions which were

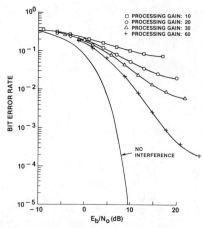

Fig. 11.   Bit error probability under the Gaussian assumption for 4-tap predictor with no matched filter. Filter coefficients computed using exact autocorrelation coefficients with −20 dB SIR, and nondispersive channel.

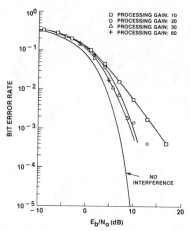

Fig. 13.   Bit error probability from simulation results for 4-tap predictor with matched filter.

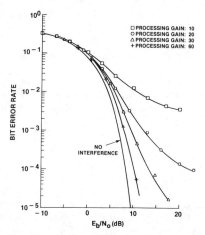

Fig. 12.   Bit error probability under the Gaussian assumption for 4-tap predictor with matched filter. Filter coefficients computed using exact autocorrelation coefficients with −20 dB SIR, and nondispersive channel.

simulated were identical to those for which the Gaussian approximation is given in Fig. 12. The simulation processed a simulated received sequence approximately 100 000 chips long, corrupted by white Gaussian noise and interference consisting of a sum of 10 sinusoids with random phase uniformly distributed over $(0, 2\pi)$ and occupying 20 percent of the band. The received sequence was processed in blocks of 1200 chips; excision filter coefficients were computed from each block of 1200 samples, using the least squares algorithm, and the 1200 samples were processed using the resulting filter coefficients. The processing of each block was overlapped in such a way that there was no loss of data at the ends of the blocks. Thus, the error rate was estimated from 10 000 bits at a processing gain of 10; 5000 bits at a processing gain of 20; etc. The results given in Fig. 13 confirm the validity of the Gaussian assumption.

In conclusion, we have demonstrated that a narrow-band interference embedded in a PN spread-spectrum signal can be estimated and suppressed by means of an adaptive filter that precedes the PN correlator. The estimation of the interference can be accomplished by linear prediction or by nonparametric methods based on the FFT algorithm. Linear prediction has the advantage of requiring fewer data points in arriving at the estimate. We have also investigated the characteristics of the interference suppression filter and derived the performance gain achieved by it. The error rate performance of the receiver was approximated by assuming that the output of the PN correlator is Gaussian. The validity of this approximation was demonstrated by Monte Carlo simulation of the PN spread-spectrum receiver.

## ACKNOWLEDGMENT

The authors with to express their appreciation to Dr. J. T. Gamble for the technical support and useful suggestions that he provided during the course of this work.

## REFERENCES

[1]   F. M. Hsu and A. A. Giordano, "Digital whitening techniques for improving spread spectrum communications performance in the presence of narrow-band jamming and interference," *IEEE Trans. Commun.*, vol. COM-26, pp. 209–216, Feb. 1978.

[2]   T. J. Ulrych and T. N. Bishop, "Maximum entropy spectral analysis and autoregressive decomposition," *Rev. Geophys. Space Phys.*, vol. 13, pp. 183–200, Feb. 1975.

[3]   A. V. Oppenheim and R. W. Schafer, *Digital Signal Processing.*   Englewood Cliffs, NJ: Prentice-Hall, 1975.

[4]   *Programs for Digital Signal Processing.*   New York: IEEE Press, 1979.

[5]   A. Papoulis, *Probability, Random Variables and Stochastic Processes.*   New York: McGraw-Hill, 1965.

[6]   H. Levinson, "The Wiener rms error criterion in filter design and prediction," *J. Math. Phys.*, vol. 25, pp. 261–278, 1947.

[7]   J. Durbin, "Efficient estimation of parameters in moving-average models," *Biometrika*, vol. 46, pp. 306–316, 1959.

[8]   J. Makhoul, "A class of all-zero lattice digital filters: Properties and applications," *IEEE Trans. Acoust., Speech, Signal Processing*, vol. ASSP-26, pp. 304–314, Aug. 1978.

[9]   J. P. Burg, "Maximum entropy spectral analysis," in *Proc. 37th Meet. Soc. Exploration Geophysicists*, 1967; also reprinted in

*Modern Spectrum Analysis,* D. G. Childers, Ed. New York: IEEE Press, 1978, pp. 34–41.

[10] M. Morf, A. Vieira, and D. T. Lee, "Ladder forms for identification and speech processing," in *Proc. IEEE Conf. Decision Contr.,* New Orleans, LA, Dec. 1977, pp. 1074–1078.

[11] M. Morf and D. T. Lee, "Recursive least squares ladder forms for fast parameter tracking," in *Proc. IEEE Conf. Decision Contr.,* San Diego, CA, Jan. 10–12, 1978, pp. 1362–1367.

[12] M. Morf, D. T. Lee, J. R. Nicholls, and A. Vieira, "A classification of algorithms for ARMA models and ladder realizations," in *Proc. IEEE Int. Conf. Acoust., Speech, Signal Processing,* Hartford, CT, May 1977, pp. 13–19.

[13] M. Morf, B. Dickinson, T. Kailath, and A. Vieira, "Efficient solution of covariance equations for linear prediction," *IEEE Trans. Acoust., Speech, Signal Processing,* vol. ASSP-25, pp. 429–433, Oct. 1977.

[14] J. D. Pack and E. H. Satorius, "Least squares, adaptive lattice algorithms," Naval Ocean Syst. Center, San Diego, CA, Tech. Rep. 423, Apr. 1979.

[15] L. Marple, "A new autoregressive spectrum analysis algorithm," *IEEE Trans. Acoust., Speech, Signal Processing,* vol. ASSP-28, pp. 441–454, Aug. 1980.

# Rejection of Narrow-Band Interference in PN Spread-Spectrum Systems Using Transversal Filters

LOH-MING LI AND LAURENCE B. MILSTEIN, SENIOR MEMBER, IEEE

*Abstract*—In the presence of narrow-band interference, the performance of PN spread-spectrum communication systems can be improved by using various interference rejection filters. In this paper, both prediction error filters and transversal filters with two-sided taps are used and the performance of each is analyzed. Analytic expressions for the signal-to-noise ratio improvement factors are obtained. If the frequency of the jamming tone is near the carrier frequency of the spread-spectrum signal, the performance of the transversal filter with two-sided taps is better than that of the prediction error filter with the same number of taps.

## I. INTRODUCTION

IN a spread-spectrum communication system employing a direct-sequence (DS) pseudonoise spreading signal, the effect of narrow-band interference is reduced due to the inherent processing gain of the system. When the processing gain does not provide sufficient improvement due to bandwidth restriction, the performance of the system can be further improved by using some method of interference rejection [1] – [3]. Fig. 1 shows the baseband part of a BPSK-DS receiver using an adaptive transversal filter. Reference [3] uses the prediction error filter shown in Fig. 2 for filtering.

If the spread-spectrum signal samples taken at different taps are not correlated, and if there is only white noise interference, the tap weights will be zero to maintain minimum output error power. If there is additional narrow-band interference, the tap weights will be adjusted to predict the input signal so that the resulting mean squared error is minimized. The level of the interference is reduced at the expense of introducing some distortion on the desired signal.

The prediction error filter uses previous signal samples only. In this paper, we will analyze the performance of the transversal filter with two-sided taps shown in Fig. 3, which uses future samples as well as previous samples to reduce the interference. We will compare its performance with that of the prediction error filter. When the interference is stationary, due to symmetry of the filter with two-sided taps, the number of multiplications can be reduced so the circuit can be simplified (see Section V).

Manuscript received May 14, 1981; revised November 17, 1981. This paper was presented at the National Telecommunications Conference, New Orleans, LA, December 1981.

L.-M. Li is with the Department of Electrical Communication, Chengdu Institute of Radio Engineering, Chengdu, China, on leave at the Department of Electrical Engineering and Computer Sciences, University of California at San Diego, La Jolla, CA 92093.

L. B. Milstein is with the Department of Electrical Engineering and Computer Sciences, University of California at San Diego, La Jolla, CA 92093.

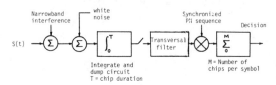

Fig. 1. Baseband part of the receiver.

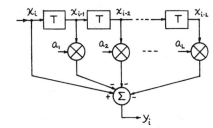

Fig. 2. Prediction error filter.

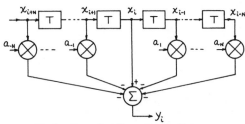

Fig. 3. Transversal filter with two-sided taps.

## II. PERFORMANCE OF THE TWO-SIDED ADAPTIVE TRANSVERSAL FILTER UNDER SINGLE TONE JAMMING

In this situation, the sample on the central tap of the filter shown in Fig. 3 at time $iT$ can be written as

$$x_i = d_i + V \cos (\Omega iT + \theta) + n_i \qquad (1)$$

where $T$ is the sampling interval and is equal to the chip duration, $d_i$ is the signal amplitude with random value $+\sqrt{S}$ or $-\sqrt{S}$ where $S$ is the signal power, $V$ and $\Omega$ are the amplitude and the angular frequency, respectively, of the jamming tone, $\theta$ is a random phase angle with uniform distribution over the interval 0 to $2\pi$, and $n_i$ is the noise sample due to thermal noise.

The tap weights $a_{-N}, \cdots, a_{-1}, a_1, \cdots, a_N$ are adjusted to obtain minimum $E[y_i{}^2]$, where $y_i$ is the output sample at

Reprinted from *IEEE Trans. Commun.*, vol. COM-30, pp. 925–928, May 1982.

time $iT$. The Wiener solution of the optimum tap weights can be obtained by solving the following $2N$ equations:

$$\sum_{\substack{k=-N \\ k \neq 0}}^{N} r(l-k)a_{k\,opt} = r(l), \qquad l = -N, \cdots, -1, 1, \cdots, N$$

(2)

where

$$r(m) = E[x_i x_{i-m}].$$

(3)

The minimum $E[y_i^2]$ is

$$E[y_i^2]_{\min} = r(0) - \sum_{\substack{k=-N \\ k \neq 0}}^{N} r(k)a_{k\,opt}.$$

(4)

Assume that the period of the PN sequence is sufficiently long so that the PN signal samples at different taps are approximately uncorrelated. Then from (1) and (3)

$$r(m) = (S + \sigma_n^2)\delta(m) + J \cos m\Omega T$$

(5)

where $\delta(m)$ is the Kronecker delta function, $J$ is the jamming tone power, and $\sigma_n^2$ is the power due to the thermal noise.

Because of the special form for (2) and (5), we can use a simple method described in [4] to solve (2). Assume

$$a_{k\,opt} = 2A \cos k\Omega T = A(e^{jk\Omega T} + e^{-jk\Omega T})$$

(6)

where $A$ is a coefficient to be determined. The validity of this assumption will be clear later. Substituting (6) into the $l$th equation of (2) and setting the coefficient of $e^{jl\Omega T}$ on both sides to be equal, we obtain

$$2NAJ + AJ\left[\sum_{k=1}^{N} e^{j2k\Omega T} + \sum_{k=-1}^{-N} e^{j2k\Omega T}\right] + 2A(s + \sigma_n^2)$$

$$= J.$$

(7)

From (7), we obtain

$$A = \frac{J}{2(S + \sigma_n^2) + J\left[2N - 1 + \dfrac{\sin(2N+1)\Omega T}{\sin \Omega T}\right]}.$$

(8)

Setting the coefficient of $e^{-jl\Omega T}$ on both sides of the $l$th equation of (2) to be equal yields the same $A$. Since (8) is independent of $l$, the value of $A$ given by (8) is the final result.

The minimum power of the filter output can be written as

$$E[y_i^2]_{\min} = S + E[e^2]_{\min}$$

(9)

where $S$ is the desired signal power and $E[e^2]_{\min}$ is the power of the error or noise component. From (4), (5), (6), and (9),

we obtain after some algebraic manipulation

$$E[e^2]_{\min} = \frac{J}{1 + \dfrac{J}{2(S + \sigma_n^2)}\left[2N - 1 + \dfrac{\sin(2N+1)\Omega T}{\sin \Omega T}\right]}$$

$$+ \sigma_n^2.$$

(10)

The signal-to-noise ratio improvement $G_2$ is

$$G_2 = \frac{(S/N)_{out}}{(S/N)_{in}}$$

$$= \frac{J + \sigma_n^2}{\dfrac{J}{1 + \dfrac{J}{2(S + \sigma_n^2)}\left[2N - 1 + \dfrac{\sin(2N+1)\Omega T}{\sin \Omega T}\right]} + \sigma_n^2}.$$

(11)

In the extreme case, if $\sigma_n^2 = 0$

$$G_2 = 1 + \frac{J}{2S}\left[2N - 1 + \frac{\sin(2N+1)\Omega T}{\sin \Omega T}\right].$$

(12)

From (12), we can see that $G_2$ increases either as the number of taps increases or as the input interference-to-signal power ratio $J/S$ increases. $G_2$ also depends on $\Omega T$. Fig. 4 shows an example where $2N = 10$, $\sigma_n^2 = 0$, and $J/S = 100$ (20 dB).

It is interesting to analyze the frequency response of the transversal filter with the optimum tap weights in order to get an insight into the interference rejection process. The transfer function of the transversal filter is

$$H(\omega) = e^{-j\omega NT}\left[1 - \sum_{\substack{k=-N \\ k \neq 0}}^{N} a_{k\,opt}e^{j\omega kT}\right].$$

(13)

$e^{-j\omega NT}$ implies a constant delay and can be neglected when the signal-to-noise ratio is considered. Using (6) and (8) in (13), it can be shown that

$$H(\omega) =$$

$$1 - \frac{J\left[\dfrac{\sin\dfrac{(2N+1)(\omega+\Omega)T}{2}}{\sin\dfrac{(\omega+\Omega)T}{2}} + \dfrac{\sin\dfrac{(2N+1)(\omega-\Omega)T}{2}}{\sin\dfrac{(\omega-\Omega)T}{2}} - 2\right]}{2(S + \sigma_n^2) + J\left[2N - 1 + \dfrac{\sin(2N+1)\Omega T}{\sin \Omega T}\right]}.$$

(14)

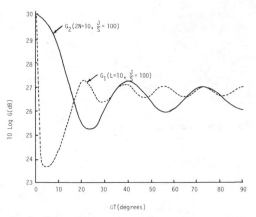

Fig. 4.   Output signal-to-noise ratio improvement versus $\Omega T$.

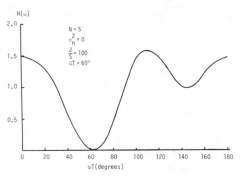

Fig. 5.   Frequency response of the transversal filter.

Fig. 5 shows an example where $N = 5$, $\sigma_n^2 = 0$, $J/S = 100$, and $\Omega T = 60°$. It can be seen that $H(\omega)$ behaves as a notch filter.

### III. PERFORMANCE OF THE PREDICTION ERROR FILTER UNDER SINGLE TONE JAMMING

The filter with $L$ taps is shown in Fig. 2. We use the same method of analysis as in Section II. The optimum tap weights $a_{k\,\text{opt}}$ ($k = 1, 2, \cdots, L$) satisfy the following equations:

$$\sum_{k=1}^{L} r(l-k)a_{k\,\text{opt}} = r(l), \qquad l = 1, 2, \cdots, L \tag{15}$$

where $r(m)$ is defined by (5). After calculating the $a_{k\,\text{opt}}$ [4], we obtain the signal-to-noise ratio improvement $G_1$ as

If $\sigma_n^2 = 0$,

$$G_1 = \left[ 1 + \frac{J}{2S} \left( L + \frac{\sin L\Omega T}{\sin \Omega T} \right) \right]$$

$$\cdot \frac{L + \dfrac{2S}{J} - \dfrac{\sin L\Omega T}{\sin \Omega T}}{L + \dfrac{2S}{J} - \dfrac{\sin L\Omega T}{\sin \Omega T} \cos(L+1)\Omega T}. \tag{17}$$

Fig. 4 shows an example where $L = 10$, $J/S = 100$ (20 dB), and $\sigma_n^2 = 0$. From Fig. 4, we can see that when $\Omega T$ is small, i.e., the frequency of the jamming tone is near the carrier frequency of the PSK signal, the performance of the transversal filter with two-sided taps is better than that of the prediction error filter with the same number of taps.

The reason for this can be roughly explained as follows. If $L\Omega T$ is not large, the correlation between a side tap and the main tap decreases as their time separation increases. The taps of the two-sided filter are grouped around the main tap more closely than those of the prediction error filter with the same number of taps, so it is easier to adjust the tap weights to reject the jamming tone without introducing too much excess distortion.

If $\sigma_n^2 = 0$, $J/S \gg 1$, and $2N = L \gg 1$, then from (12) and (17) we obtain

$$\frac{G_1}{G_2} \cong \frac{2\left( L - \dfrac{\sin L\Omega T}{\sin \Omega T} \right)}{(2L+1) - \dfrac{\sin(2L+1)\Omega T}{\sin \Omega T}}. \tag{18}$$

If $(2L+1)\Omega T \ll 1$, using $\sin\theta \cong \theta - (\theta^3/6)$ we obtain

$$\frac{G_1}{G_2} \cong \frac{L-1}{2(2L+1)} \cong \frac{1}{4} \qquad (-6 \text{ dB}). \tag{19}$$

When $S/J$ cannot be neglected, the difference will be less than 6 dB. Alternately, it can be seen from Fig. 4 that for at least some frequencies, the performance of the prediction error filter is slightly better than that of the two-sided filter.

### IV. PERFORMANCE UNDER MULTIPLE TONE JAMMING

Under multiple tone jamming, we can use the same method of analysis as used in the previous sections, but now

$$r(m) = (S + \sigma_n^2)\delta(m) + \sum_{k=1}^{K} J_k \cos m\Omega_k T \tag{20}$$

$$G_1 = \frac{J + \sigma_n^2}{2(S + \sigma_n^2)\dfrac{\left[ L + \dfrac{2(S + \sigma_n^2)}{J} \right] - \dfrac{\sin L\Omega T}{\sin \Omega T}\cos(L+1)\Omega T}{\left[ L + \dfrac{2(S + \sigma_n^2)}{J} \right]^2 - \dfrac{\sin^2 L\Omega T}{\sin^2 \Omega T}} + \sigma_n^2}. \tag{16}$$

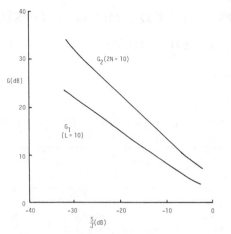

Fig. 6. Output signal-to-noise ratio improvement for multiple tone jamming example (see text).

where $K$ is the number of jamming tones. Except for some special cases (e.g., double tone jamming or $\Omega_k$ satisfying some special conditions [4]), it is difficult to obtain explicit expressions for the signal-to-noise ratio improvement in general. However, we can use numerical calculations, and Fig. 6 shows an example where $\sigma_n^2/S = 0.1$, $J_k = J/10$, $\Omega_k T = (k-1)3.6°$, $k = 1, 2, \cdots, 10$ (ten jamming tones with equal power), and the number of taps for both filters equals 10. From Fig. 6 we can see that in this case, the performance of the two-sided transversal filter ($G_2$) is better than that of the prediction error filter ($G_1$). This same result was obtained for other jamming tone locations as well.

## V. ALGORITHMS FOR TAP WEIGHT ADAPTATION

Initially, the frequency and the amplitude of the jamming tones are unknown to the receiver, i.e., the $r(m)$ in (5) or (20) are unknown, so that the tap weights cannot be adjusted according to (2) or (15). There are alternative methods to adjust the tap weights. One of them is to estimate the $r(m)$ from a finite number of received samples [3]. Alternatively, we may use some recursive algorithm to update the tap weights. The simplest one is the Widrow-Hoff LMS algorithm [5], [6].

When the interference is stationary, because of the sym-metry of the two-sided transversal filter, $a_{-k} = a_k$, $k = 1, 2,$ $\cdots, N$, and the samples on the $-k$th and $k$th taps can be added first and then multiplied by $a_k$. The adaptation algorithm can be expressed as

$$a_k^{(i+1)} = a_k^{(i)} + \alpha(x_{i-k} + x_{i+k})y_i \quad k = 1, 2, \cdots, N \quad (21)$$

where $a_k^{(i)}$ is the value of $a_k$ at time $iT$, and $\alpha$ is a sufficiently small real positive value to ensure convergence [5], [6]. Because the number of variable weights is halved, the number of multiplications is reduced. Algorithm (21) has been simulated by a digital computer, and the tap weights converged approximately to $a_{k\text{opt}}$.

## VI. CONCLUSIONS

We have analyzed the performance of both the transversal filter with two-sided taps and the prediction error filter under the condition of narrow-band interference. Analytical expressions of the signal-to-noise ratio improvement were obtained. If the frequency of the jamming tone is near the carrier frequency of the spread-spectrum signal, the performance of the transversal filter with two-sided taps is better than that of the prediction error filter with the same number of taps (in some cases by approximately 6 dB). For some other frequencies, the performance of the prediction error filter is slightly better, but the two-sided transversal filter has the advantage of simplification due to symmetry.

## REFERENCES

[1] M. J. Bouvier, Jr., "The rejection of large CW interferers in spread spectrum systems," *IEEE Trans. Commun.*, vol. COM-26, pp. 254–256, Feb. 1978.
[2] D. Shklarsky, P. Das, and L. B. Milstein, "Adaptive narrow-band interference suppression," in *Proc. Nat. Telecommun. Conf.*, 1979, pp. 15.2.1–15.2.3.
[3] F. M. Hsu and A. A. Giordano, "Digital whitening techniques for improving spread spectrum communications performance in the presence of narrowband jamming and interference," *IEEE Trans. Commun.*, vol. COM-26, pp. 209–216, Feb. 1978.
[4] J. R. Zeidler, E. H. Satorius, and D. M. Chabries, "Adaptive enhancement of multiple sinusoids in uncorrelated noise," *IEEE Trans. Acoust., Speech, Signal Processing*, vol. ASSP-26, pp. 240–254, June 1978.
[5] J. G. Proakis, "Adaptive filters," in *Proc. NATO Advanced Study Inst. Commun. Syst., Random Process Theory*, 1977, pp. 661–678.
[6] J. M. McCool and B. Widrow, "Principles and applications of adaptive filters: A tutorial review," in *Proc. IEEE Int. Symp. Circuits Syst.*, 1980, pp. 1143–1157.

# Spread Spectrum in a Four-Phase Communication System Employing Adaptive Antennas

JACK H. WINTERS, MEMBER, IEEE

*Abstract*—This paper discusses the use of spread spectrum in a four-phase communicaton system employing adaptive antennas. A system is described that provides protection against both conventional (i.e., noise and CW) and smart (in particular, repeat) jamming with rapid acquisition of the signal at the receiver. A method is shown for generating reference signals required by the adaptive array through the use of spread spectrum. With these reference signals, the received antenna pattern can be adapted to maximize desired signal to interference and noise power ratio at the receiver. The signal acquisition technique is also described and analyzed. Analytical and experimental results demonstrate both the rapid acquisition and protection against jamming with this system.

## I. INTRODUCTION

AN ADAPTIVE antenna is an array of antenna elements whose pattern is automatically controlled [1], [2]. The signal from each element of the array is multiplied by a controllable weight, which adjusts the amplitude and phase of that signal. The pattern of an adaptive antenna is automatically changed to null interfering signals and optimized desired signal reception.

Adaptive antennas can be combined with spread-spectrum communication techniques to yield even greater interference rejection capabilities than either one alone. A system combining the temporal processing of spread spectrum with the spatial processing of adaptive antennas can provide protection against a wide variety of jamming techniques.

The weights in an adaptive array may be controlled by several techniques [1], [2]. In particular, the technique used in the LMS array [1] is considered here. In the LMS array, the weights are adjusted to obtain the least mean-square error between the array output and a reference signal. This reference signal is a locally generated signal that allows the array feedback to differentiate between the desired signal and interference. It must be a signal correlated with the desired signal and uncorrelated with any interference.

The major problem in the development of a communication system using an adaptive array is the generation of the reference signal. A method must be developed for the acquisition of the signal by the receiver. This method includes the code timing acquisition, if the timing of the pseudonoise code used to spread the signal spectrum must be acquired by the receiver.

A reference signal generation technique has been previously

Manuscript received July 6, 1981; revised February 4, 1982. This work was supported in part by the Rome Air Development Center under Contract F30602-79-C-0068 and in part by The Ohio State University Research Foundation.

The author was with the Department of Electrical Engineering, Ohio State University, Columbus, OH 43210. He is now with Bell Laboratories, Holmdel, NJ 07733.

described for spread-spectrum signals using biphase modulation [3]. The timing of the pseudonoise code used in the system is acquired at the receiver by a slewing method. To determine the correct code timing, the code generated at the receiver is correlated with the received signal for all possible timing offsets [4]. The reference signal generation technique for the biphase system works well, but does have two shortcomings. First, it is vulnerable to repeat jammers with biphase remodulation. Second, short codes must be used to achieve reasonable acquisition times, and short codes may not provide adequate security for many applications.

This paper presents a four-phase communication system that may be used with an LMS adaptive antenna. The system is a spread-spectrum communication system. It is assumed that the location of the desired transmitter is not known at the receiver and that system code timing must be acquired by the receiver. A four-phase system has been developed that overcomes the shortcomings of the previous biphase system [3], without sacrificing the system's rapid acquisition and conventional (i.e., noise and CW) jamming protection capabilities. This four-phase system will now be described.

The four-phase signal consists of two orthogonal biphase signals. One signal contains a short code for rapid acquisition. The other contains a long code to be used for protection against smart jammers (jammers that the biphase system is vulnerable to). The reference signal generation technique and signal acquisition technique are described. Both rapid acquisition and protection against conventional and smart jamming are demonstrated by analytical and experimental results.

We first describe in Section II the four-phase signal and the LMS adaptive array used in the system. We then discuss in Section III reference signal generation for the array. Next, in Section IV the code timing acquisition process is discussed. Finally, in Section V we describe an experimental system, discuss experimental results, and compare them to theoretical results.

## II. SYSTEM DESCRIPTION

The transmitted carrier is modulated by two codes plus data. The one code is a pseudonoise code, i.e., a maximal length linear shift register sequence. This code has a short length (on the order of 1000 symbols) to permit its timing to be rapidly acquired by the receiver. The other code has a very long length (e.g., greater than $10^9$ symbols). The code may be either a pseudonoise (linear) or a nonlinear code [5], i.e., it is generated from a shift register with nonlinear feedback logic. This code is used to provide greater communication security in the system as shown later.

Reprinted from *IEEE Trans. Commun.*, vol. COM-30, pp. 929–936, May 1982.

The codes plus data modulate the carrier in the following way. Let the $m$th short code symbol and long code symbol be labeled $a_m$ and $b_m$ (equal to 0 or 1), respectively, and have a duration of $\Delta$ seconds. The $i$th data, or useful information, symbol is labeled $d_i$ (equal to 0 or 1) and has a duration of $T_b$ seconds. The data symbol duration is greater than the code symbol duration by an integer multiple $k$, the spreading ratio. With these symbols differentially encoded, the transmitted signal is given by

$$s(t) = \frac{A}{\sqrt{2}} \sin\left(\omega_1 t + \zeta(t)\right) + \frac{A}{\sqrt{2}} \cos\left(\omega_1 t + \phi(t)\right) \qquad (1)$$

where

$$\zeta(t) = \theta(t) + \gamma(t), \qquad (2)$$

$$\phi(t) = \phi_m = \phi_{m-1} + \pi b_m$$

$$\text{for } (m-1)\Delta \leqslant t < m\Delta \qquad (3)$$

and

$$\theta(t) = \theta_m = \theta_{m-1} + \pi a_m$$

$$\text{for } (m-1)\Delta \leqslant t < m\Delta, \qquad (4)$$

$$\gamma(t) = \gamma_i = \gamma_{i-1} + \pi d_i$$

$$(i-1)T_b \leqslant t < iT_b. \qquad (5)$$

In the above equations, $A$ is an amplitude constant, $\omega_1$ is the carrier frequency, and $\gamma_0$, $\theta_0$, and $\phi_0$ are equal to 0. The data symbol transitions coincide with the code symbol transitions. Thus, the signal is a four-phase differential phase shift keyed (DPSK) signal consisting of two orthogonal binary DPSK signals. One binary signal contains a short code plus data (as in the biphase system), and the other contains a long code.

A block diagram of an $N$ element LMS adaptive array [1] is shown in Fig. 1. The signal received by the $i$th element $y_i(t)$ is split with a quadrature hybrid into an in-phase signal $x_{I_i}(t)$ and a quadrature signal $x_{Q_i}(t)$. These signals are then multiplied by a controllable weight $w_{I_i}$ or $w_{Q_i}$. The weighted signals are then summed to form the array output $s_0(t)$. The array output is subtracted from a reference signal (described below) $r(t)$ to form the error signal $e(t)$. The element weights are generated from the error signal and the $x_{I_i}(t)$ and $x_{Q_i}(t)$ signals by using the correlation feedback loops as shown in Fig. 2.

The purpose of the reference signal is to make the array track the desired signal. The reference signal must be a signal correlated with the desired signal and uncorrelated with any interference. Generation of a reference signal from a four-phase signal is described below.

Two different reference signals can be generated from the four-phase signal. These two signals are the two binary DPSK signal components of the four-phase signal. Thus, one reference signal contains a short code, and the other a long code.

A reference signal can be generated from the four-phase signal using the loop shown in Fig. 3. This is the same loop

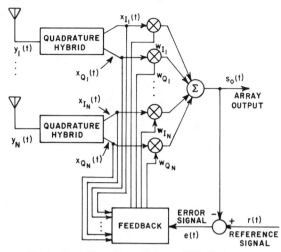

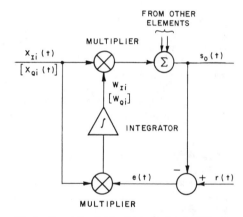

Fig. 1. Block diagram of an $N$ element LMS adaptive array.

Fig. 2. Correlation feedback loop for the adaptive array.

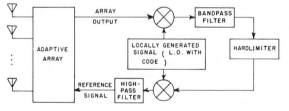

Fig. 3. Reference signal generation loop with the adaptive array.

that was used with the biphase system [3]. The array output is first mixed with a locally generated signal modulated by either the short or long code. When the codes of the locally generated signal and the array output signal are synchronized, the array output signal's spectrum is collapsed. It is collapsed to the data bandwidth when the locally generated signal is modulated by the short code and a single frequency component when the locally generated signal is modulated by the long code. The mixed output is then passed through a filter with either the data bandwidth or a very narrow bandwidth

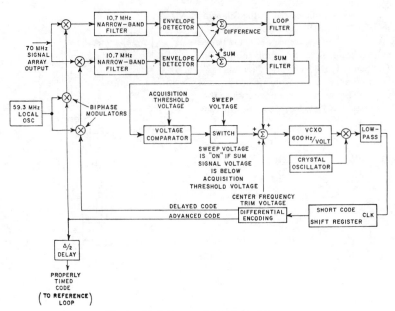

Fig. 4.   Block diagram of the delay lock loop.

depending on the code involved. The biphase desired signal component is, therefore, unchanged by the filter. The filter output is then hard limited so that the reference signal will have constant amplitude. The hard-limiter output is mixed with the locally generated signal to produce a biphase reference signal. The reference signal is, therefore, an amplitude scaled replica of one of the biphase components of the four-phase desired signal.

Any interference signal without the proper code has its waveform drastically altered by the reference loop. When the coded locally generated signal is mixed with the interference, the interference spectrum is spread by the code bandwidth. The bandpass filter further changes the interference component out of the mixer. As a result, the interference at the array output is uncorrelated with the reference signal.

## III. ACQUISITION OF THE SIGNAL

The acquisition of the signal by the receiver will now be described. In particular, we describe the method for obtaining the timing of the codes by the receiver. The method presented here combines a sequential search method and the rapid acquisition by sequential estimation (RASE) [6] method for code timing acquisition with the use of an adaptive array.

The acquisition method consists of several steps. The first step is the acquisition of the short code timing. The short code timing is acquired and then tracked by the delay lock loop shown in Fig. 4. In the delay lock loop the difference voltage from the two envelope detectors is used to track the code timing [3], and the sum voltage is used for timing acquisition. The sum voltage is used for timing acquisition because it indicates the alignment (maximum correlation) of the received and locally generated codes.

The acquisition method for the short code timing is similar

to that used in the biphase system [3], i.e., a sequential search (slewing) method is used. To acquire the short code timing, the code generator at the receiver is run faster than the received signal's code. When the two codes begin to align, the sum voltage increases. When the sum voltage exceeds the acquisition threshold, the sweep voltage is turned off, and code tracking begins. Also, during the short code acquisition process, the short code is used to generate the reference signal in the adaptive array. Thus, the reference code is also slewed. Before acquisition, the array nulls interference and the desired signal. When the codes begin to align, the array pulls the desired signal out of the noise. The desired signal power into the delay lock loop, therefore, also increases when the codes align. This process is described further in [3].

With this slewing method, the code generated at the receiver is correlated with the received signal for as many as all possible timing offsets. Thus, the acquisition time is proportional to the code length, and the code length must be short for reasonable acquisition times.

The second step in the acquisition method is the acquisition of the long code timing. During this step the short code is used to generate the reference signal in the array. Because the array has nulled all signals except the signal acquired (including CW and noise jammers), the array output consists mainly of the desired signal. Therefore, the RASE [6] method can be used to quickly acquire the long code timing. In the RASE method the timing for a code generated from an $n$ stage feedback shift register is obtained by detecting and loading $n$ consecutive code symbols into a similar shift register at the receiver.

Fig. 5 shows how the RASE method can be used to determine the long code timing in the four-phase system. The array output containing the desired four-phase signal is differentially

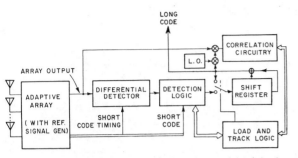

Fig. 5.   RASE with an adaptive array for four-phase modulated signal.

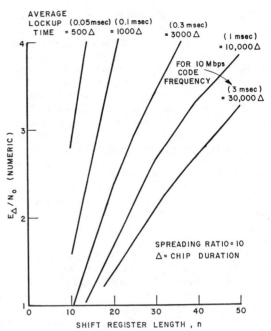

Fig. 6.   Required $E_\Delta/N_0$ versus shift register length for given average lockup times, with a long code correlation time of $20\,n\Delta$.

detected using the symbol transition timing from the tracking of the short code. The differential detector output (relating to the phase transitions in the signal) is then processed by the detection logic. These phase transitions depend on the long code, the short code, and the data symbols. Since the short code symbols are known, the detection logic can determine the long code symbols and, in some cases, also determine if errors have been made by the differential detector. If no errors are detected, the long code symbols are loaded into the shift register. The feedback loop in the shift register is then connected and the output of the shift register correlated for a short time with the array output to verify code timing. If the correlation of the two signals exceeds a threshold value, the output of the shift register is used to generate the reference signal for the array. Otherwise, the shift register is reloaded and the process repeated until code synchronization is obtained.

It should be noted that with the long code acquisition method the only requirement for the shift register network generating the code is that the only feedback connection must be to the first shift register stage. Thus, the timing for nonlinear codes (which are more secure than linear codes [7]) that are generated in this way [5] can also be obtained by the RASE method.

Fig. 6 shows the results of a simulation study of the long code acquisition process. This simulation study and additional analytical results are discussed in [8]. The Appendix presents a summary of some of these analytical results. In Fig. 6, the required energy per chip (code symbol interval) to noise density ratio $E_\Delta/N_0$ at the processor input (array output) is plotted versus the long code shift register length for various acquisition times. The results show that very long codes can be quickly acquired if $E_\Delta/N_0$ is greater than about 3. For example, with a 10 Mbit/s code modulation frequency, a $10^{12}$ length code ($n$ equal to 40) can be acquired in less than 0.001 s if $E_\Delta/N_0$ is greater than 3. In general, $E_\Delta/N_0$ at the array output will be at least 3 because of two factors. First, the energy per data bit to noise density ratio at the array output must be fairly high (usually greater than 15) for low bit error rates. Second, the spreading ratio is usually as small as possible to minimize the signal bandwidth and can be as small as 5 [3]. Thus, with these two conditions, the $E_\Delta/N_0$ at the array output will be greater than 3.

The four-phase system as described so far provides the same protection against conventional (i.e., noise and CW) jamming

as the biphase system [3]. With the biphase system, as much as a 35 dB improvement in signal-to-interference ratio can be achieved with a spreading ratio of only 5.

An additional step in the acquisition method provides protection against smart jammers (i.e., jammers using only the short code or repeat jammers with remodulation). If the long code timing is not acquired in a short period of time, the short code timing is changed and the acquisition procedure is repeated. The reason this provides smart jammer protection is described below.

## IV. SMART JAMMER PROTECTION

The long code is present to add greater security to the system. In the previous biphase system [3], the signal contained no long code, only a biphase signal with a short code plus data. Therefore, this system had the two shortcomings listed below.

First, in the biphase system data are modulated on the transmitted signal by the introduction of additional 180° phase shifts. Therefore, the signal from a jammer that repeats and adds 180° phase shifts (at a rate equal to or less than the data rate) to the desired signal cannot be distinguished from the desired signal by the array. The array may acquire the jammer's signal and null the desired signal. To overcome this problem, the data modulation must be different than biphase.

Second, short codes must be used for reasonable acquisition times. Short codes, however, may not provide adequate security for many applications. The short code repeats often during transmission of the signal, and, therefore, the code can easily be determined and used by a jammer. When the jamming signal contains the code, the receiver is unable to distinguish

137

the jamming signal from the desired signal. To overcome this problem, codes with very long periods are required.

Because the four-phase signal contains an orthogonal biphase signal with the long code, the biphase system shortcomings are overcome. First, the data modulation technique is no longer biphase. Examination of (1) shows that the additional phase transition due to the data bit "1" is either +90° or −90°. Also, the phase transition depends upon both the short and long code symbols. Thus, the data modulation method cannot be easily duplicated by a jammer. Furthermore, biphase remodulation by the jammer changes the long code symbols and, thus, can be detected. Second, a long code for communication security has been combined with the short code for rapid acquisition. Since both codes are acquired by the receiver, a jammer must use both codes to jam effectively. However, because the long code does not repeat for a long time and may even be nonlinear, it is very difficult for a jammer to determine and use this code [7].

Although there are other jamming strategies, the above two examples point out that for a smart jammer to effectively jam the four-phase system, it needs to determine the long code. In particular, the jammer must determine the length and feedback connections of the shift register generating the long code. With this system, protection against smart jamming depends on the security of the long code. Since very long (and nonlinear [5]) codes can be used, the codes can be very secure (i.e., it can be almost impossible for a jammer to determine the code feedback shift register from the signal). Thus, the system can provide significant protection against smart jamming.

We will now describe the process through which protection is provided against smart jammers. To begin, we will assume that the short code timing on a jammer signal is different from that of the desired signal.[1] Thus, during the first step in the acquisition method, the short code timing of either the desired signal or a jammer may be acquired by the receiver. If the desired signal's short code is acquired, then the jammer will be nulled and the acquisition process can be completed. If the jammer's short code is acquired, however, the long code timing will not be acquired in a short time, because the long code symbols are either not present or changed due to remodulation by the jammer. The short code timing at the receiver will then be changed, and the acquisition process repeated. Since the receiver's code timing is slewed sequentially during acquisition, all possible code timing offsets will be examined before the smart jammer's short code is acquired again. Thus, if the desired signal is present it will be acquired before acquisition of the smart jammer is attempted again. Thus, with smart jamming the acquisition time may be increased, but acquisition of the desired signal cannot be prevented.

## V. AN EXPERIMENTAL SYSTEM

An experimental system was developed to verify analytical results and demonstrate conventional and smart jamming pro-

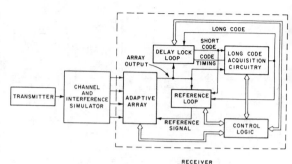

Fig. 7. Block diagram of the experimental four-phase communication system with an adaptive array at the receiver.

tection with rapid acquisition. A block diagram of the system is shown in Fig. 7. The transmitted four-phase signal is received by the adaptive array through a channel and interference simulator. This simulator was used to generate the received signals for each antenna element. The signals correspond to those received with a desired signal and interference arriving at the receiving array from different angles.[2] From the desired signal in the array output the short code timing is determined and tracked by the delay lock loop. Long code timing is acquired by the long code acquisition circuitry. The control logic manages the acquisition procedure steps, including which codes are used in the reference loop to generate the reference signal.

Some of the parameters for the system are listed in Table I. The code modulation frequency and the spreading ratio were chosen to be compatible with an existing four-element array [9]. The maximum allowed acquisition time is the time allowed for acquisition of the long code. If the long code is not acquired in this time, the short code timing is changed and the acquisition process repeated.

We will now consider the acquisition time as a function of the received $E_\Delta/N_0$ (array input), $E_\Delta/N_0 \mid_{\text{IN}}$. Fig. 8 shows the average total acquisition time versus $E_\Delta/N_0 \mid_{\text{IN}}$ for the system (see the Appendix for analytical results). The figure shows that the signal can be rapidly acquired even when $E_\Delta/N_0 \mid_{\text{IN}}$ is near 0 dB. It should be noted that with a four-element adaptive array the array output $E_\Delta/N_0$ will be 6 dB higher than $E_\Delta/N_0 \mid_{\text{IN}}$.

Figs. 9 and 10 show the CW interference suppression of the four-phase system. The CW signal is 20 dB stronger than the desired signal at the array input, yet very little interference power can be seen in the output. For the experimental system, this was the maximum jammer-to-signal ratio ($J/S$) for which acquisition could occur. However, the experimental system was designed to show concept feasibility, and not necessarily a large jammer rejection. Systems can be designed with a maximum $J/S$ of 40 dB or more. The reason for this limit in the experimental system is described below.

The maximum $J/S$ is dependent on the code modulation

---

[1] This is a reasonable assumption. For a repeat jammer with remodulation, the signal will be delayed by the jammer. If the jammer is using only the short code, it is unlikely that the jamming signal would arrive at the receiver with the same code timing as the desired signal.

[2] For the experimental results, the desired signal arrives from broadside and the interference has a 60° element-to-element phase shift, corresponding to an angle of arrival of 19.5° from broadside for half wavelength element spacing.

TABLE I
EXPERIMENTAL FOUR-PHASE SYSTEM PARAMETERS

| | |
|---|---|
| LMS Adaptive Array | 4 antenna elements |
| Code Modulation Frequency | 175.2 kbits/s |
| Spreading Ratio | 16 |
| Data Rate | 10.95 kbits/s |
| Minimum Received $E_\Delta/N_o$ | 0 dB |
| Average Acquisition Time | 0.23 s |
| Maximum Allowed Acquisition Time | 0.55 s |
| Short Code Length | 255 symbols |
| Long Code Length | 1.72 * 10 ** 10 symbols ($n = 34$) |

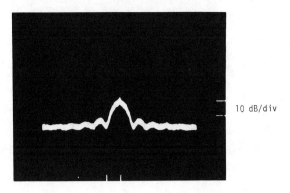

200 kHz/div

Fig. 10. Array output power density spectrum for input as shown in Fig. 9.

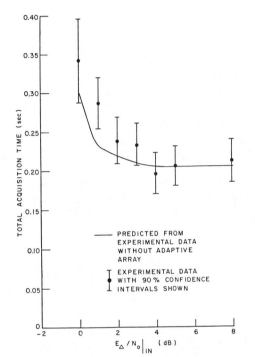

Fig. 8. Total acquisition time versus received energy per chip to noise density ratio $E_\Delta/N_0|_{\rm IN}$.

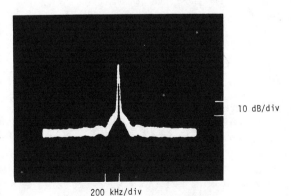

200 kHz/div

Fig. 9. Received power density spectrum at array input, $E_\Delta/N_0|_{\rm IN}$ equal to 8 dB and $J/S$ equal to 20 dB.

frequency and the acquisition time. The reason is that the rate of response of the array weights is proportional to the signal strength (in the LMS array). For the strongest interfering signal, the weights must respond slower than 0.2 times the code modulation frequency [10]. Otherwise, the weights will begin to modulate the interference to look like the reference signal. For the weakest desired signal, the weights must respond faster than the slewing speed during acquisition. Otherwise, during acquisition the desired signal will not be pulled out of the noise and the acquisition threshold in the delay lock loop exceeded. For the experimental system the ratio of 0.2 times the code modulation frequency to the sweep speed was approximately 20 dB (see the Appendix), as shown experimentally. It should be noted that a much higher maximum $J/S$ can be achieved with higher code modulation frequency. Furthermore, adaptive array algorithms are currently being studied [11] to eliminate the weight response problem.

Fig. 11 shows the average acquisition time with the four-phase system when a repeat jammer with biphase remodulation is present at the receiver. As seen in the figure, when $J/S$ is greater than −1 dB, the jammer signal's short code can be acquired by the receiver for a brief time, and, therefore, the average acquisition time for the desired signal is increased (see the Appendix). However, acquisition of the desired signal is not prevented until $J/S$ is greater than 20 dB. Thus, the smart jammer is no more effective in preventing acquisition than a CW jammer.

## VI. CONCLUSIONS

In this paper we have described a four-phase spread-spectrum communication system with an adaptive antenna. Reference signal generation for the adaptive array and a signal acquisition procedure were described. The system was shown to provide protection against both conventional and smart jammers (in particular, repeat jammers) with rapid acquisition of the signal at the receiver.

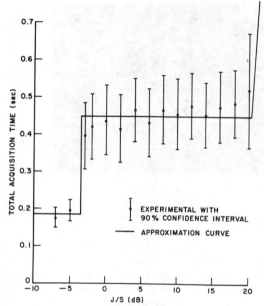

Fig. 11.  Total acquisition time versus jammer to signal power ratio for repeat jammer with biphase remodulation.

## APPENDIX
## SUMMARY OF SOME THEORETICAL
## RESULTS FROM [8]

*Average Long Code Acquisition Time*

The average long code acquisition time $T_{avg}$ can be determined theoretically if we make the following approximations for the acquisition process.

1) The long code can be differentially detected by itself (without the short code plus data).

2) The bit error probability for each code symbol is independent of other symbols.

3) The bit error probability is given by

$$P_E = \tfrac{1}{2} \exp\left[-E_\Delta/2N_0\right]$$

$$(\text{for } 0 \leqslant E_\Delta/N_0 \leqslant 4). \tag{A1}$$

These approximations are shown to be reasonably accurate in [8].

The long code acquisition scheme involves the loading of a shift register with the detected code symbols followed by the correlation of the output of the shift register with the received signal. With the approximations described previously, the probability of fully loading an $n$ stage shift register with error-free code symbols is given by

$$P_n = (1 - P_E)^n \tag{A2}$$

where $P_E$ is given in (A1). The trials of fully loading the shift register are independent Bernoulli trials. Thus, the number of trials required for an error-free loading (i.e., for acquisition) has a geometrical distribution. The probability of success on the $X$th trial is then given by

$$P_r(x = X) = P_n(1 - P_n)^{X-1} \tag{A3}$$

and the average number of trials required is the reciprocal of $P_n$.

For each loading of the shift register, the output of the shift register is correlated with the received signal over a multiple $M$ of the number of bits in the shift register $n$. The average long code acquisition time is, therefore, given by

$$T_{avg} = \frac{Mn\Delta}{(1 - P_E)^n}. \tag{A4}$$

For $M$ equal to 20, as in Fig. 6, theoretical results [from (A4)] show about 1 dB less $E_\Delta/N_0$ for the same $T_{avg}$ as compared to computer simulation results.

*Maximum Acquisition Time*

For the short code acquisition the acquisition time has a uniform probability density with the maximum acquisition time equal to twice the average. However, for the long code, the acquisition time has a geometrical distribution as shown in (A3). Thus, the probability of more than $X$ trials being required to achieve lockup can easily be shown to be given by [6]

$$P_r(x > X) = (1 - P_n)^X. \tag{A5}$$

Therefore, there is no maximum lockup time and a probability of acquisition $P_{acq}$ within a specified number of trials must be considered. From (A5) the probability of acquisition in $X$ trials is given by

$$P_{acq} = 1 - (1 - P_n)^X. \tag{A6}$$

Since $P_n$ is, in general, small (for large $P_n$ the long code acquisition time is negligible compared to the short code acquisition time), (A6) may be approximated by

$$P_{acq} \doteq 1 - e^{-XP_n} \tag{A7}$$

From the above equations we can determine the probability of long code acquisition in a given time (a multiple of the average acquisition time). For example, from (A7), there is a 99 percent probability of acquisition in 4.6 times the average long code acquisition time.

*Acquisition Time for the Experimental System*

For the experimental system, the maximum allowed acquisition time was chosen to be 0.55 s (see Table I) for a received $E_\Delta/N_0$ equal to 0 dB. We arbitrarily allow 0.4 s for short code acquisition and 0.15 s for long code acquisition.

140

Thus, for a 99 percent probability of acquisition, the average long code acquisition time is given by [from (A7)]

$$T_{avg} = \frac{0.15}{4.6} \doteq 0.033 \text{ s.} \qquad (A8)$$

The average short code acquisition time $t_{avg}$ is half the maximum or 0.2 s. Thus, the total average acquisition time is given by

$$T_{tot} = T_{avg} + t_{avg} \doteq 0.23 \text{ s} \qquad (A9)$$

as shown in Table I.

We can determine the average acquisition time for other received $E_\Delta/N_0$ values from (A4). Thus, for the experimental system,

$$T_{tot} \doteq \frac{20 \times 34/(175.2 \times 10^3)}{(1 - \exp(-E_\Delta/(2N_0)))^{34}} + 0.2 \text{ (s).} \qquad (A10)$$

Theoretical results [from (A10)] show about 1 dB greater $E_\Delta/N_0$ for the same $T_{avg}$ as compared to experimental results (Fig. 8; note that $E_\Delta/N_0$ in (A10) is four times $E_\Delta/N_0|_{IN}$).

*Maximum Jammer to Signal Power Ratio*

As discussed in the text, the maximum jammer to signal power ratio for acquisition is given by the ratio of 0.2 times the code modulation rate to the slewing speed. Since the code at the receiver must slew past an entire code cycle in the maximum short code acquisition time $t_{max}$ the slewing speed is simply the short code length $N$ divided by $t_{max}$. Thus,

$$J/S|_{max} = \frac{0.2/\Delta}{N/t_{max}} = \frac{0.2 t_{max}}{N\Delta}. \qquad (A11)$$

For the experimental system,

$$J/S|_{max} = \frac{0.2 \times 0.4}{255/(175.2 \times 10^3)} \qquad (A12)$$

$$\doteq 55 \text{ (17 dB).} \qquad (A13)$$

The weight response speed with the weakest desired signal can be somewhat less than the sweep speed, although the probability of not acquiring the short code in the maximum time $P_{miss}$ is increased. In the experimental system, the response speed for the weakest signal was about one-half the sweep speed. In this case $P_{miss}$ is still very small and $J/S|_{max}$ is increased to 20 dB (as shown in Figs. 9 and 10).

Note that $J/S|_{max}$ is increased further with higher code rates, longer acquisition times, and shorter length short codes.

*Acquisition Time with Smart Jamming (Experimental System)*

With a smart jammer present, the receiver will first acquire either the smart jammer's signal or the desired signal. If the jammer's signal is acquired first, 0.55 s will elapse before the receiver nulls the jammer and searches for the desired signal. The receiver then takes an average of about 0.2 s to acquire the desired signal. If the desired signal is acquired first, the average acquisition time is also about 0.2 s. Thus, with a smart jammer present, the average total acquisition time is given by

$$T_{tot} \doteq \frac{1}{2}(0.55 + 0.2) + \frac{1}{2}(0.2) = 0.475 \text{ s} \qquad (A14)$$

as shown in Fig. 11.

## ACKNOWLEDGMENT

I am grateful for the guidance of Dr. R. T. Compton, Jr. and A. A. Ksienski during the course of this work.

## REFERENCES

[1] B. Widrow, P. E. Mantey, L. J. Griffiths, and B. B. Goode, "Adaptive antenna systems," *Proc. IEEE*, vol. 55, p. 2143, Dec. 1967.

[2] S. R. Applebaum, "Adaptive arrays," *IEEE Trans. Antennas Propagat.*, vol. AP-24, p. 585, Sept. 1976.

[3] R. T. Compton, Jr., "An adaptive array in a spread-spectrum communication system," *Proc. IEEE*, vol. 66, p. 289, Mar. 1978.

[4] J. J. Spilker, Jr., "Delay-lock tracking of binary signals," *IEEE Trans. Space Electron. Telem.*, vol. SET-9, p. 1, Mar. 1963.

[5] S. W. Golomb, *Shift Register Sequences*. San Francisco, CA: Holden-Day, 1967.

[6] R. B. Ward, "Acquisition of pseudonoise signals by sequential estimation," *IEEE Trans. Commun.*, vol. COM-13, p. 475, Dec. 1965.

[7] W. Diffie and M. E. Hellman, "Privacy and authentication: An introduction to cryptography," *Proc. IEEE*, vol. 67, pp. 407–408, Mar. 1979.

[8] J. H. Winters, "A four-phase modulation system for use with an adaptive array," Ph.D. dissertation, Ohio State Univ., July 1981.

[9] T. W. Miller, R. Caldecott, and R. J. Huff, "A satellite simulator with a TDMA-system compatible adaptive array," ElectroScience Lab., Dep. Elec. Eng., Ohio State Univ., Rep. 3364-4, prepared under Contract F30602-72-C-0162 for Rome Air Develop. Center, Jan. 1976.

[10] T. W. Miller, "The transient response of adaptive arrays in TDMA systems," ElectroScience Lab., Dep. Elec. Eng., Ohio State Univ., Rep. 4116-1, prepared under Contract F30602-75-C-0061 for Rome Air Develop. Center, p. 287, June 1976.

[11] R. T. Compton, Jr., "An improved feedback loop for adaptive arrays," ElectroScience Lab., Dep. Elec. Eng., Ohio State Univ., Rep. 710929-3, prepared under Contract N00019-78-C-0131 for Naval Air Syst. Command, July 1978.

# Performance of a Nonlinear FH-DPSK Spread-Spectrum Receiver with Multiple Narrow-Band Interfering Signals

MASAO MATSUMOTO AND GEORGE R. COOPER, FELLOW, IEEE

*Abstract*—An analysis of the effect of multiple narrow-band interfering signals on the probability of error in a frequency-hopped differential phase shift keyed (FH-DPSK) spread-spectrum communication system with a nonlinear receiver is given. In this analysis it is assumed that the sum of the outputs from an $n$-subchannel receiver is a Gaussian random variable. Subject to this assumption, numerical evaluations of the probability of error for both nonfading and Rayleigh fading channels are presented and compared with those of a linear receiver. The results indicate that the performance of a nonlinear receiver is better than that of a linear receiver for any signal-to-interference ratio.

## INTRODUCTION

**A** FREQUENCY-HOPPED, differential phase shift keyed (FH-DPSK) spread-spectrum communication system has been proposed as a possible way to provide mobile radio service to a large number of customers [1]. It is said that such a spread-spectrum system may coexist in the same frequency band with conventional narrow-band systems without excessive mutual interference. In this paper we discuss the performance of the spread-spectrum system with a nonlinear receiver operating in the presence of multiple narrow-band interferers assuming a simplified, multiple-narrow-band (MNB) channel [1]. The results are also compared with those of a linear receiver [2].

## SYSTEM DESCRIPTION

*Transmitter*

A schematic diagram of a transmitter in an FH-DPSK is shown in Fig. 1 [4]. The messages are composed of sequences of symbols taken from a set of $n$ symbols and the $i$th symbol is, for example, represented as $s_i = [0 \cdots 010 \cdots 0]$ where 1 occurs only in the $i$th element of the vector. In order to increase resistance to interference, a transmitted signal corresponding to a message signal is selected from $n$ orthogonal signals by [3]

$$y_i = s_i \widetilde{H} \tag{1}$$

where $\widetilde{H}$ is an $n \times n$ orthogonal matrix in which all elements have the same magnitude. The form of $\widetilde{H}$ used in this paper is

Manuscript received July 1, 1981; revised February 3, 1982.
M. Matsumoto is with the Radio Regulatory Bureau, Ministry of Posts and Telecommunications, Tokyo, Japan.
G. R. Cooper is with the School of Electrical Engineering, Purdue University, West Lafayette, IN 47907.

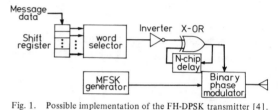

Fig. 1. Possible implementation of the FH-DPSK transmitter [4].

the nonnormalized Hadamard matrix, as shown for $n = 4$:

$$\widetilde{H} = \begin{bmatrix} 1 & 1 & 1 & 1 \\ 1 & -1 & 1 & -1 \\ 1 & 1 & -1 & -1 \\ 1 & -1 & -1 & 1 \end{bmatrix}.$$

It is clear that the vector $y_i$ is simply the $i$th row of the matrix $\widetilde{H}$. The elements of $y_i$ are transmitted (in effect) through separate subchannels by virtue of frequency hopping.

*Receiver*

In the receiver the array of $t_1$-second delay lines and bandpass filters, as shown in Fig. 2, selects the desired waveform from the total incoming signal. Each bandpass filter, which is followed by a bandpass limiter, is matched to a rectangular chip of duration of $t_1$. The linear combiner performs the orthogonal matrix operation expressed by

$$\hat{s} = \hat{y}\widetilde{H}^T \tag{2}$$

where $\hat{y}$ is the output of the low-pass filters and $\hat{s}$ is the output of the linear combiner. The $n$ elements of $\hat{s}$ are then compared to decide which symbol was transmitted.

## NONLINEAR RECEIVER INTERFERENCE ANALYSIS

*Assumptions and Notations*

The transmitted signal has the row vector form [3]

$$s(t) = [s_1(t), s_2(t), \cdots, s_n(t)], \qquad 0 \leqslant t \leqslant T \tag{3}$$

where

$$s_j(t) = c_j(t)\epsilon_j(t)\sqrt{2S} \cos(\omega_j t),$$

$$(j-1)t_1 \leqslant t \leqslant jt_1$$

Reprinted from *IEEE Trans. Commun.*, vol. COM-30, pp. 937–942, May 1982.

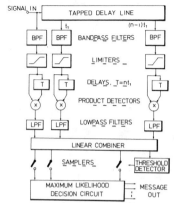

Fig. 2.   Block diagram of the nonlinear FH-DPSK receiver.

$n$ = number of time chips in one waveform

$t_1$ = duration of one time chip

$c_j(t)$ = differential signal ($\pm 1$) corresponding to the $j$th time chip

$\epsilon_j(t)$ = envelope of each pulse transmitted signal

$\omega_j$ = frequency assigned to the $j$th time chip

$T$ = duration of one waveform.

$c_j(t)$ is chosen according to the table below.

| $c_j(t-T)$ | Input Signal | $c_j(t)$ |
|---|---|---|
| $-1$ | $1$ | $-1$ |
| $-1$ | $-1$ | $1$ |
| $1$ | $1$ | $1$ |
| $1$ | $-1$ | $-1$ |

It is assumed that each time chip has a rectangular waveform so that

$$\epsilon_j(t) = 1, \qquad (j-1)t_1 \leqslant t < jt_1$$
$$= 0, \qquad \text{elsewhere.} \tag{4}$$

It is also assumed that the signal represented by (3) is transmitted over a simplified, multiple-narrow-band (MNB) channel so that the received signal is simply expressed by the product of the signal row vector (3) and the fading coefficient matrix $\tilde{\Phi}$ [3]:

$$\tilde{\Phi} = \begin{bmatrix} \phi_1 & & & & \\ & \phi_2 & & 0 & \\ & & \ddots & & \\ 0 & & & & \\ & & & & \phi_n \end{bmatrix}$$

where the $\phi_i$'s are scalar multipliers and are independently Rayleigh distributed.

On the other hand, the interfering signal that affects the $j$th subchannel is represented as

$$z_j = \sqrt{2 \overline{Z}_j} \cos (\omega_{zj} t + \theta_{zj}) \tag{6}$$

where $\omega_{zj}$ is the carrier frequency of the signal and $\theta_{zj}$ is the phase associated with $\omega_{zj}$. The quantity $\sqrt{\overline{Z}_j}$ is either a constant or a Rayleigh-distributed random variable. By way of contrast, the situation for impulsive interference is described in [7].

*Detector Output*

The output of the $j$th matched filter $r_j(t)$ is the sum of signal plus interference. Since this filter is matched to a rectangular envelope signal, its impulse response is

$$h(t) = \frac{2}{t_1} \cos \omega_j(t_1 - t), \qquad 0 \leqslant t \leqslant t_1$$
$$= 0, \qquad \text{elsewhere.} \tag{7}$$

Because of the tapped delay line, every time chip enters the filter to which it is matched at the same time. Thus, it is convenient to shift the time origin so that the $j$th signal is defined as

$$s_j(t) = \sqrt{2S} \, c_j(t) \cos \omega_j t, \qquad 0 \leqslant t \leqslant t_1$$
$$= 0, \qquad \text{elsewhere.} \tag{8}$$

If it were not for the differential modulation, $c_j(t)$ would be the $j$th element of the $i$th row of $\tilde{H}$ (i.e., $H_{ij}$) when $s_i$ is transmitted. However, because of the differential modulation, the phase of the same time chip in the previous waveform must be considered. From the table relating $c_j(t)$ and $c_j(t-T)$, it is clear that

$$c_j(t) = H_{ij} c_j(t-T).$$

Since both $s_j(t)$ and $h(t)$ have rectangular envelopes, when they are convolved to determine the output signal, the resulting envelope will be triangular with the form

$$\beta(t) = t/t_1, \qquad 0 \leqslant t \leqslant t_1$$
$$= 2 - t/t_1, \quad t_1 \leqslant t \leqslant 2t_1 \tag{9}$$
$$= 0, \qquad 2t_1 \leqslant t \leqslant T.$$

This waveform, of course, repeats periodically with period $T$.

The interfering signal may still be modeled as (6) and creates a steady-state sinusoid at the output of the matched filter. The amplitude of this sinusoid depends upon the frequency difference $\Delta\omega_j = \omega_{zj} - \omega_j$ and the frequency response of the matched filter, which has a form proportional to

$$S_a\left(\frac{\Delta\omega_j t_1}{2}\right) = \frac{\sin \dfrac{\Delta\omega_j t_1}{2}}{\dfrac{\Delta\omega_j t_1}{2}}. \tag{10}$$

It is now simply a matter of convolving $h(t)$ with the sum of $s_j(t)$ and $z_j(t)$ to obtain the filter output as

$$r_j(t) = H_{ij}c_j(t - T)\beta(t)\sqrt{2S}\phi_j \cos \omega_j(t - t_1)$$
$$+ \sqrt{2Z_j}S_a(\theta_j) \cos [\omega_{zj}(t - t_1) + \theta_{zj} + \theta_j]$$
$$\triangleq A_{rj}(t) \cos [\omega_j(t - t_1) + \phi_{rj}(t)] \qquad (11)$$

where

$$A_{rj}(t) \cos \phi_{rj}(t)$$
$$= H_{ij}c_j(t - T)\beta(t)\sqrt{2S}\phi_j$$
$$+ \sqrt{2Z_j}S_a(\theta_j) \cos [\Delta\omega_j(t - t_1) + \theta_{zj} + \theta_j] \qquad (12)$$

$$A_{rj}(t) \sin \phi_{rj}(t)$$
$$= \sqrt{2Z_j}S_a(\theta_j) \sin [\Delta\omega_j(t - t_1) + \theta_{zj} + \theta_j] \qquad (13)$$

$$\theta_j = \frac{\Delta\omega_j t_1}{2}. \qquad (14)$$

The output of a bandpass limiter (BPL) with an ideal hard limiter whose input-output function is shown in Fig. 3 is given by

$$v_j(t) = \frac{4A}{\pi} \cos [\omega_j(t - t_1) + \phi_{rj}(t)]. \qquad (15)$$

Then the output of the low-pass filter following the product detector is (assuming $C_j(t - T) = 1$ for all $j$)

$$y_j(t) = \frac{8A^2}{\pi^2} \cos [\phi_{rj}(t) - \phi_{rj}(t - T)] \qquad (16)$$

where

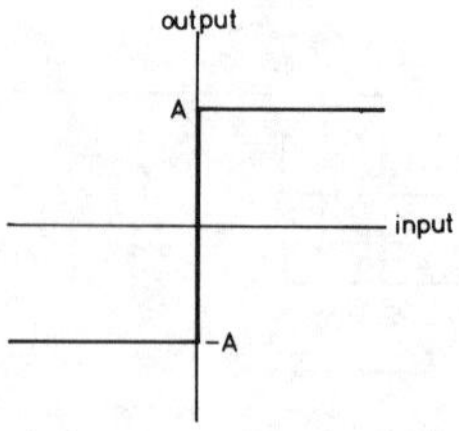

$$\phi_{rj}(t) = \tan^{-1} \frac{\sqrt{2Z_j}S_a(\theta_j) \sin [\Delta\omega_j(t - t_1) + \theta_{zj} + \theta_j]}{\sqrt{2S}\phi_j H_{ij}\beta(t) + \sqrt{2Z_j} S_a(\theta_j) \cos [\Delta\omega_j(t - t_1) + \theta_{zj} + \theta_j]} \qquad (17)$$

$$\phi_{rj}(t - T) = \tan^{-1} \frac{\sqrt{2Z_j}S_a(\theta_j) \sin [\Delta\omega_j(t - t_1) + \theta_{zj} - (2n - 1)\theta_j]}{\sqrt{2S}\phi_j H_{ij}\beta(t - T) + \sqrt{2Z_j} S_a(\theta_j) \cos [\Delta\omega_j(t - t_1) + \theta_{zj} - (2n - 1)\theta_j]}. \qquad (18)$$

Here it is assumed that $\theta_{zj}$ and $Z_j$ are constant over the period $T$ and that the instantaneous phase of the transmitted signal at the beginning of each time chip is the same for every repetition (i.e., $\omega_j T = 2\pi m$, $m$ an integer) in order to recover the maximum signal energy at the output of the product detector. $\theta_j$ determines the relative location of $\omega_{zj}$ with respect to $\omega_j$. Later in this paper, we assume that $\theta_j$'s are uniformly distributed over the interval $(-\pi, \pi)$, meaning that each interferer is uniformly distributed over the main lobe of each matched filter, i.e., $(\omega_j - (2\pi/t_1), \omega_j + (2\pi/t_1))$.

The conditional mean and variance of each $y_j$ are obtained from (16) assuming the $\theta_{zj}$ is a random variable uniformly

Fig. 3.   Input-output function of limiters.

distributed over $[-\pi, \pi]$. That is,

$$E[y_j | \Phi, Z, \Theta] = \frac{1}{2\pi} \int_{-\pi}^{\pi} y_j(t) \, d\theta_{zj} \qquad (19)$$

$$\text{var} [y_j | \Phi, Z, \Theta] = \frac{1}{2\pi} \int_{-\pi}^{\pi} y_j^2(t) \, d\theta_{zj} - E^2[y_j | \Phi, Z, \Theta] \qquad (20)$$

where

$$\Phi = [\phi_1, \phi_2, \cdots, \phi_n] \qquad (21)$$

$$Z = [Z_1, Z_2, \cdots, Z_n] \qquad (22)$$

$$\Theta = [\theta_1, \theta_2, \cdots, \theta_n]. \qquad (23)$$

The received vector $y = [y_1(t), y_2(t), \cdots, y_n(t)]$ is now subject to the linear transformation by the linear combiner which consists of multiplying $y$ by the Hadamard matrix. Thus, the $k$th output of the linear combiner is

$$x_k = \sum_{j=1}^{n} H_{kj}y_j. \qquad (24)$$

Since $y_j(t)$'s are independent, the conditional means and variances are

$$E[x_k | \Phi, Z, \Theta] = \sum_{j=1}^{n} H_{kj} E[y_j | \Phi, Z, \Theta] \qquad (25)$$

$$\text{var} [x | \Phi, Z, \Theta] = \sum_{j=1}^{n} \text{var} [y_j | \Phi, Z, \Theta]. \qquad (26)$$

The probability of an error in decoding is the probability that an incorrect linear combiner sample value exceeds the correct one. Thus

$$P_{e|s_i} = 1 - \text{Prob} \,[x_i > \text{all other } x_i's \,|\, s_i]$$

$$= 1 - \int_{-\infty}^{\infty} \int_{-\infty}^{x_i} \cdots \int_{-\infty}^{x_i} f(x_1, x_2, \cdots, x_n | s_i)$$

$$\cdot \, dx_1 \, dx_2 \cdots dx_n \tag{27}$$

where $P_{e|s_i}$ is the probability of decoding the incorrect word when $s_i$ is transmitted and $f(x_1, x_2, \cdots, x_n | s_i)$ is the conditional joint density of $x = [x_1, x_2, \cdots, x_n]$, given $s_i$ is received. The probability of making a word error is

$$\text{PWE} = \frac{1}{n} \sum_{i=1}^{n} P_{e|s_i}. \tag{28}$$

To evaluate $P_e | s_i$, the conditional joint density $f(x_1, x_2, \cdots, x_n | s_i)$ has to be obtained. Since $x_k$ is given by (24), the conditional density of $x_k$ [denoted by $f(x_k | \Phi, Z, H, s_i)$] is obtained by convolving the densities of $H_{kj} y_j$'s. However, even though $\theta_{zj}$'s are uniformly distributed in the interval $(-\pi, \pi)$ it is not easy to evaluate the densities of $y_i$'s. Therefore, we approximate the conditional density of $x_k$ by Gaussian with the mean and the variance given by (25) and (26). Since the accuracy of this approximation depends upon both the behaviors of the densities of $y_i$'s and the number of subchannels interfered $m$, it is difficult to estimate the accuracy of the approximation. However, it is said that if the densities of $m$ independent random variables are reasonably concentrated near their mean values then the Gaussian is a close approximation even for moderate values of $m$ [6]. In the evaluation of Figs. 4-6, the number of subchannels interfered is equal to that of subchannels (i.e., $m = n$), which seems to be reasonable for this approximation.

Therefore, given $s_1$ is transmitted, the probability that $x_k$ is greater than $x_1$ is

$$\text{Prob} \,[x_k > x_1 \,|\, \Phi, Z, \theta, \Theta, s_1]$$

$$= \int_{x_1}^{\infty} f(x_k | \Phi, Z, \Theta, s_1) \, dx_k$$

$$= Q[\mu + d(x_1, x_k)] \tag{29}$$

where

$$Q(x) = \frac{1}{\sqrt{2\pi}} \int_{x}^{\infty} e^{-t^2/2} \, dt \tag{30}$$

$$d(x_1, x_k) = \frac{E[x_1 | \Phi, Z, \Theta, s_1] - E[x_k | \Phi, Z, \Theta, s_1]}{\sqrt{\text{var} \,[x | \Phi, Z, \Theta, s_1]}} \tag{31}$$

$$\mu = \frac{x_1(t) - E[x_1 | \Phi, Z, \Theta, s_1]}{\sqrt{\text{var} \,[x | \Phi, Z, \Theta, s_1]}}. \tag{32}$$

$\mu$ is the normalized difference between the expected value and

its actual value. Although we assumed that $x_k$'s are Gaussian and they are uncorrelated, the probabilities given by (29) are not necessarily independent. Therefore, we make a further assumption that the $x_k$'s are jointly normal. Then the probability given by (27) is written as

$$P_{e|s_1} = 1 - \int_{-\infty}^{\infty} f(\mu) \prod_{k=2}^{n} \{ Q[\mu + d(x_1, x_k)] \} \, d\mu \tag{33}$$

where

$$f(\mu) = \frac{1}{\sqrt{2\pi}} e^{-\mu^2/2}. \tag{34}$$

The evaluation of $P_{e|s_1}$ is done at the sampling time $t = t_1$ when the bandpass filter's response to the signal is maximum (i.e., $\beta(t_1) = \beta(t_1 - T) = 1$). That is,

$$y_j(t_1) = \frac{8A^2}{\pi^2} \cos \,[\phi_{rj}(t_1) - \phi_{rj}(t_1 - T)] \tag{35}$$

where

$$\phi_{rj}(t_1) = \tan^{-1} \frac{\sqrt{2Z_j} S_a(\theta_j) \sin (\theta_{zj} + \theta_j)}{\sqrt{2S}\phi_j + \sqrt{2Z_j} S_a(\theta_j) \cos (\theta_{zj} + \theta_j)} \tag{36}$$

$$\phi_{rj}(t_1 - T)$$

$$= \tan^{-1} \frac{\sqrt{2Z_j} S_a(\theta_j) \sin (\theta_{zj} + \theta_j - 2n\theta_j)}{\sqrt{2S}\phi_j + \sqrt{2Z_j} S_a(\theta_j) \cos (\theta_{zj} + \theta_j - 2n\theta_j)}. \tag{37}$$

*Error in the Nonfading Channel*

The probability of error for the nonfading channel is obtained by substituting the identity matrix $\tilde{I}$ for the fading matrix $\tilde{\Phi}$. Fig. 4 shows $P_{e|s_1}$ for $n = 16, 32$ for the case in which $\theta_j$'s are independent random variables with identical uniform distribution over $(-\pi, \pi)$. That is,

$$P_e = 1 - \frac{1}{(2\pi)^n} \int_{-\pi}^{\pi} \cdots \int_{-\pi}^{\pi} [1 - P_{e|s_1}] \, d\theta_1 \cdots d\theta_n \tag{38}$$

$$\Phi = [1, 1, \cdots, 1]$$

$$Z = Z[1, 1, \cdots, 1] \qquad Z = \text{const.}$$

The above result is evaluated numerically by the Monte Carlo method [5]. For this evaluation, in addition to 300 Gaussian random variates, 300 random vectors of $\Theta$ were generated and each vector required the generation of $n$ random variates uniformly distributed over $(-\pi, \pi)$. The dashed lines in Fig. 4 show the $P_e$ for the linear receiver with same conditions. There seems to be no significant improvement in performance between the two systems.

*Error in the Simplified MNB Fading Channel*

In the simplified, multiple-narrow-band (MNB) fading channel, the $\phi_i^2$'s are independent and the joint density function

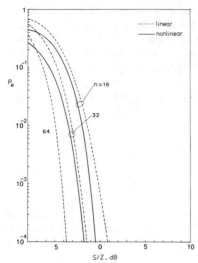

Fig. 4.   Probability of error versus signal-to-interference ratio; nonfading channel and constant interfering siglans.

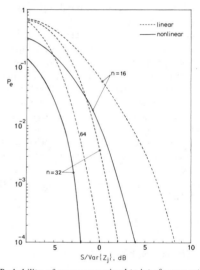

Fig. 5.   Probability of error versus signal-to-interference ratio; simplified MNB fading channel and constant interfering signals.

of $\phi_i^2$'s is written as

$$f(\phi_1^2, \phi_2^2, \cdots, \phi_n^2) = \prod_{k=1}^{n} f(\phi_k^2) \tag{39}$$

where

$$f(\phi_k^2) = e^{-\phi_k^2}, \qquad \phi_k^2 \geqslant 0$$

$$= 0, \qquad \phi_k^2 < 0. \tag{40}$$

Fig. 5 shows $P_e$ in a simplified MNB fading channel for the case in which $\theta_j$'s are independent random variables with identical uniform distributions over $(-\pi, \pi)$ and the $Z_j$'s are constant.

$$P_e = 1 - \frac{1}{(2n)^n} \int_{-\pi}^{\pi} \cdots \int_{-\pi}^{\pi} \int_{0}^{\infty} \cdots \int_{0}^{\infty} \prod_{k=1}^{n} f(\phi_k^2)$$

$$\cdot [1 - P_{e|s_1}] \, d\Theta \, d\Phi^2 \tag{41}$$

where

$$d\Phi^2 = d\phi_1^2 \cdots d\phi_n^2 \tag{42}$$

$$d\Theta = d\theta_1 \cdots d\theta_n \tag{43}$$

$$Z = Z[1, 1, \cdots, 1] \qquad Z = \text{const.} \tag{44}$$

Compared with the nonfading channel, it can be said that the degradation of performance is smaller for larger $n$. However, compared with the linear receiver, the performance is considerably improved for the region of interest, especially when $n = 16$.

Fig. 6 shows $P_e$ in a simplified MNB channel with $Z_j$'s that are exponentially distributed random variables with

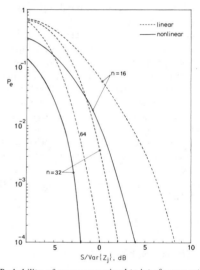

Fig. 6.   Probability of error versus signal-to-interference ratio; simplified MNB fading channels and Rayleigh distributed interfering signals.

density functions

$$f(Z_j) = \frac{1}{\sigma_z^2} e^{-Z_j/\sigma_z^2}, \qquad Z_j \geqslant 0$$

$$= 0, \qquad Z_j < 0 \tag{45}$$

and

$$P_e = 1 - \frac{1}{(2\pi)^n} \int_{-\pi}^{\pi} \cdots \int_{-\pi}^{\pi} \int_{0}^{\infty} \cdots \int_{0}^{\infty} \prod_{k=1}^{n} f(\phi_k^2)$$

$$\cdot f(Z_j)[1 - P_{e|s_1}] \, d\Theta \, d\Phi \, dZ \tag{46}$$

146

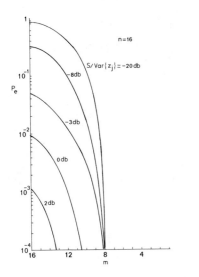

Fig. 7.   Probability of error versus number of subchannels interfered; nonlinear receiver.

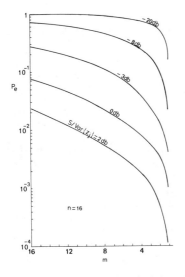

Fig. 8.   Probability of error versus number of subchannels interfered; linear receiver.

where

$$dZ = dZ_1 \cdots dZ_n.$$

Compared with the linear receiver, the performance is improved about 6 dB for both $n = 16$ and 32 for the region of interest.

Finally, we look at the case when the interfering signals exists in some subchannels and the rest are not interfered. Fig. 7 shows $P_e$ in a simplified MNB channel for the case in which the interfering signals appear in subchannels from the first to the $m$th, assuming all the interfering signals have density function defined by (45). Although in Fig. 7, the evaluations of $P_e$ are done even for small number of $m$, it should be noted that since the Gaussian approximation of the densities of $x_k$'s in (29) does not hold for small numbers of $m$, the evaluated $P_e$ for such numbers of $m$ are only for reference. Compared with the linear receiver shown in Fig. 8, we can expect vast improvement in the performance by using bandpass limiters as $m$ gets smaller. Similar improvement is described in [8].

## CONCLUSION

In this paper we have discussed the performance of an FH-DPSK spread-spectrum communication system affected by interfering signals. As the result, it can be said that the performance of a nonlinear receiver using bandpass limiters after matched filters is superior to the linear receiver for any signal-to-interference ratio and especially for a small number of subchannels interfered we can expect significant improvement.

## REFERENCES

[1]  G. R. Cooper and R. W. Nettleton, "A spread-spectrum technique for high-capacity mobile communications," *IEEE Trans. Veh. Technol.*, vol. VT-27, pp. 264–275, Nov. 1978.
[2]  M. Matsumoto and G. R. Cooper, "Multiple narrow-band interferers in an FH-DPSK spread-spectrum communication system," *IEEE Trans. Veh. Technol.*, vol. VT-30, pp. 37–42, Feb. 1981.
[3]  G. R. Cooper and R. W. Nettleton, "Spectral efficiency in cellular land-mobile communications: A spread-spectrum approach," TR-EE 78-44, Final Rep., NSF Grant ENG 76-80536, Oct. 31, 1978.
[4]  P. S. Henry, "Spectrum efficiency of a frequency-hopped-DPSK spread spectrum mobile radio system," *IEEE Trans. Veh. Technol.*, vol. VT-28, Nov. 1979.
[5]  Y. A. Schreider (G. T. Tee, Transl.), *The Monte Carlo Method.* New York: Pergamon, 1966.
[6]  A. Papoulis, *Probability, Random Variables, and Stochastic Processes.* New York: McGraw-Hill, 1965.
[7]  P. A. Bello and R. Esposito, "Error probabilities due to impulsive noise in linear and hard-limited DPSK systems," *IEEE Trans. Commun. Technol.*, vol. COM-19, Feb. 1971.
[8]  O. C. Yue, "Hard-limited versus linear combining for frequency-hopping multiple-access systems in a Rayleigh fading environment," *IEEE Trans. Veh. Technol.*, vol. VT-30, Feb. 1981.

# Error Performance Analyses of Differential Phase-Shift-Keyed/Frequency-Hopping Spread-Spectrum Communication System in the Partial-Band Jamming Environments

JHONG S. LEE, MEMBER, IEEE, AND LEONARD E. MILLER, MEMBER, IEEE

*Abstract*—This paper presents a complete analysis for the derivation of the probability of error for a hypothetical pure frequency-hopping (FH) spread-spectrum communication system, employing a binary differential phase-shift-keyed (DPSK) modulation and differentially coherent demodulation, in a partial-band jamming environment. The postulated DPSK/FH spread-spectrum system could be a practical system if the operational environment is multipath free. The optimum partial-band and wide-band jamming performance measures are obtained and compared under certain parametric conditions on the communicator's power levels. The present analysis is based on the application of the author's recent results on the derivation of the probability density functions (pdf) for the output of a cross correlator with arbitrary bandpass inputs of different power levels.

## I. INTRODUCTION

*System Description*

A FREQUENCY-hopping (FH) communication system is a class of spread-spectrum systems which can employ extremely wide bandwidth much greater than that actually required for communication. Two fundamental techniques for spread-spectrum systems are direct-sequence (DS) pseudo-noise modulation and FH. While the DS spread-spectrum system requires phase coherence over the wide bandwidth, the FH spread-spectrum system does not.

In FH systems the carrier is switched (or hopped) to a new frequency occupying a new frequency cell of bandwidth, say $B$ Hz, which is a small fraction of the total spread-spectrum bandwidth, say $W$ Hz, where $B \ll W$.

We consider a particular form of FH system where the carrier is phase modulated by a differentially encoded binary bit stream and each hopped carrier conveys 1 bit of information. This type of FH system is referred to as a pure FH or slow hopping system [1]. The rate of FH is identical to the information sequence bit rate. The modulation and frequency hopping scheme is depicted in Fig. 1(a), where it is shown that the differentially encoded sequence phase modulates the baseband carrier of frequency $f_o$ and then it is hopped to a new carrier whose hopping pattern is controlled by the frequency synthesizer commanded by a pseudorandom noise sequence generator (PNG). We assume a coherently frequency-hopped system in which the carrier phase remains continuous when the frequency is changed [1], [2].

The receiver structure is depicted in Fig. 1(b). When the re-

Manuscript received June 1, 1981; revised January 22, 1982. This work was supported by the Office of Naval Research, Statistics and Probability Program, under Contract N00014-80-C-0753.

The authors are with J. S. Lee Associates, Inc., Arlington, VA 22202.

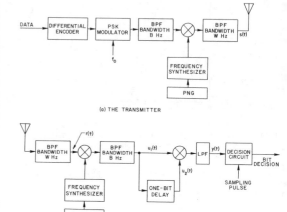

(a) THE TRANSMITTER

(b) THE RECEIVER

Fig. 1. A DPSK/FH spread-spectrum system's transmitter and receiver block diagrams.

ceived carrier is down-converted (dehopped) by the frequency synthesizer commanded by the PNG in synchronism with that in the transmitter side, the information-bearing noisy baseband signal is recovered. We consider differentially coherent demodulation for the extraction of the information sequences.

The postulated differential phase-shift-keyed/frequency-hopping (DPSK/FH) system has a limited application in a multipath environment. In fact, FH systems employing DPSK modulation and differentially coherent detection are not commonly employed in practice. In an earlier paper Simon and Polydoros [7] assume that future coherent frequency synthesizer design will permit correction of the inevitable phase errors arising from multipath reception. Nettleton and Cooper [8], on the other hand, assume that correction for the randomized phase structure created by the multipath is a difficult problem; they describe an FH system in which differential phase comparisons are made on the $j$th hops of succeeding waveforms so that there is no dependence on phase coherence between hops of different frequencies. The system described here can, of course, be employed in a multipath-free environment such as a ground-to-air link. To the extent that the projection of Simon and Polydoros may be realized, the postulated DPSK/FH system will have broader application. It is the purpose of this paper to obtain the performance

Reprinted from *IEEE Trans. Commun.*, vol. COM-30, pp. 943–952, May 1982.

measures of this system under partial-band jamming and wide-band jamming assumptions.

### The Statement of the Problem

In the DPSK/FH system just described, a binary DPSK system operates over a transmission channel of total bandwidth $W$ Hz with hopping carrier with information bandwidth $B$ Hz. In this system a processing gain of $W/B$ is attained when the jamming power is uniformly spread over $W$ Hz bandwidth. From the jammer's viewpoint, this is not an effective strategy since only $(B/W)$ of the total jamming power, say $J$ watts, affects the information bit. The effective jamming strategy would be to distribute the total jamming power of $J$ watts over a selected fraction of the frequency cells (channels). This type of jamming will produce errors on the signals which hop into the jammed cells to an intolerable level. To combat this type of jamming strategy, which is usually referred to as partial-band jamming, the communicator usually employs an error-control scheme.

Since differentially coherent demodulation of DPSK signals employs two symbol (bit) pulses in a bit decision interval, each of the pulses may or may not have been subjected to the jamming interference in a partial-band jamming strategy situation. Thus, the demodulation takes place in four possible circumstances. From an analytical standpoint, the case where only one of the two pulses is jammed presented a difficulty in the past. Houston [3] has considered the problem of assessing the performance of DPSK/FH systems in a partial-band jamming environment. His analysis, however, is applicable to the case where both pulses are always assumed to be jammed, which is true only when the entire system band is jammed. The error probability obtained under such simplifying assumptions would be a very loose upper bound.

The purpose of this paper is to show a complete analysis of the error mechanism for the DPSK/FH system, taking into due consideration all of the different jamming effects in the decision process. This analysis is based on the application of the authors' recent results [4], whereby we have developed a method that allows consideration of all cases where statistics change at the boundary between the signaling elements.

The error performance of a DPSK/FH system under near optimum partial-band jamming is computed and plotted in graphical form in comparison with less effective wide-band jamming performance for different cases of communicator's transmitter power level normalized by the assumed jamming power.

## II. THE ANALYSIS

It was stated earlier that we shall consider an effective jamming strategy in which the jammer uses his total available power of $J$ watts over a fraction $\gamma$ of the total spread-spectrum bandwidth $W$ Hz. We will assume that all the frequency cells (channels) are equally likely to be selected by the communicator since the FH is accomplished pseudorandomly.

Suppose that the jamming noise power is distributed uniformly over a fraction of the total frequency cells, each of which has bandwidth $B$ Hz, as depicted in Fig. 2. Then the

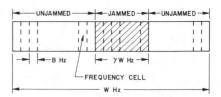

Fig. 2. Partial-band jamming model.

jamming power in the $i$th jammed cell is given by

$$\sigma_{Ji}^2 = \left( \frac{B}{\gamma W} \right) J \text{ watts,} \tag{1}$$

and this is the jamming power to which each information bit is subjected with probability $\gamma$. The probability that a given transmitted bit is not affected by the jamming power is $1 - \gamma$.

In Fig. 1(b), the "present" signal $u_1(t)$ and the "earlier" (delayed) signal $u_2(t)$ are given by

$$u_i(t) = s_i(t) + n_i(t) + j_i(t), \qquad i = 1, 2 \tag{2}$$

where $s_i(t)$, to be specified later, denotes the information carrying signal, whereas $n_i(t)$ and $j_i(t)$ represent system noise and jamming noise, respectively. In a partial-band jamming strategy, $j_i(t)$ in (2) may or may not be absent, depending upon whether $s_i(t)$ in (2) was unjammed or jammed. Let us define a joint event $(j_1, j_2)$, where $j_i = 0$ or $1$, $i = 1, 2$. The event $j_i = 0$ denotes the situation where $j_i(t)$ is absent in $u_i(t)$, whereas the event $j_i = 1$ denotes the presence of $j_i(t)$ in $u_i(t)$.

The analysis to follow will lead us to the point where we can specify the probability of error for the DPSK/FH system. As is common in the analysis of binary communication systems, of which the DPSK/FH system is a particular case, the error probability is computed from

$$P(e) = \tfrac{1}{2} P_s(e) + \tfrac{1}{2} P_m(e) \tag{3}$$

where $P_s(e)$ and $P_m(e)$ are the conditional probabilities of error, given, respectively, that "space" and "mark" are transmitted. For differentially coherent demodulation of the binary DPSK system under consideration, the conditional probabilities $P_s(e)$ and $P_m(e)$ can be obtained in a manner described below.

$$P_s(e) = P_m(e)$$
$$= \pi(0, 0)P_s[e \,|\, (0, 0)] + \pi(0, 1)P_s[e \,|\, (0, 1)]$$
$$+ \pi(1, 0)P_s[e \,|\, (1, 0)] + \pi(1, 1)P_s[e \,|\, (1, 1)] \tag{4}$$

where

$$\pi(j_1, j_2) = \text{probability of the joint event } (j_1, j_2).$$

It is plain that

$$\pi(0, 0) = (1 - \gamma)^2$$
$$\pi(0, 1) = \pi(1, 0) = \gamma(1 - \gamma)$$
$$\pi(1, 1) = \gamma^2. \tag{5}$$

It thus behooves us to compute the conditional probabilities appearing in (4) in order to get the final total error probability for the DPSK/FH system.

*The General Expressions for $P_s$ $[e \mid (j_1, j_2)]$*

From Fig. 1(b) and (2) we assume that the received information-carrying signals are given by

$$s_1(t) = \sqrt{2S_1} \cos(\omega_0 t - \theta_1) \tag{6}$$

and

$$s_2(t) = \sqrt{2S_2} \cos[\omega_0(t - T) - \theta_2]$$

$$= \sqrt{2S_2} \cos(\omega_0 t - \theta_2), \tag{7}$$

where the bit duration $T$ and the carrier frequency $f_o$ are such that $\omega_0 T = 2\pi k$, $k$ integer, and $S_i$ is the received signal power. The phases $\theta_1$ and $\theta_2$ contain the binary information through the values of $0°$ or $180°$. We assume that the jamming in any selected cell is a Gaussian noise with a given amount of power (mean square value) as shown in (1). We may thus express the noise $n_i(t)$ and the jamming noise $j_i(t)$ in a Rician notation as follows:

$$n_i(t) = X_{ci}(t) \cos \omega_0 t + X_{si}(t) \sin \omega_0 t \tag{8a}$$

$$j_i(t) = Y_{ci}(t) \cos \omega_0 t + Y_{si}(t) \sin \omega_0 t, \qquad i = 1, 2 \tag{8b}$$

where

$$E\{n_i^2(t)\} = E\{X_{ci}^2(t)\} = E\{X_{si}^2(t)\} = \sigma_{Ni}^2 \tag{8c}$$

$$E\{j_i^2(t)\} = E\{Y_{ci}^2(t)\} = E\{Y_{si}^2(t)\} = \sigma_{Ji}^2, \qquad i = 1, 2. \tag{8d}$$

Note that the index $i$ is associated with the two inputs to the multiplier at the DPSK demodulator. At this juncture we recognize that the DPSK demodulator, which consists of a multiplier (phase detector) followed by a low-pass filter (LPF), accomplishes the demodulation by a cross correlation with a delayed reference and sampling operation. The calculation of the conditional probability of error requires the expression of the probability density function of $y(t)$ at the output of the LPF [see Fig. 1(b)].

The pdf of $y(t)$ at the output of the LPF [Fig. 1(b)] can be shown to be [see the Appendix]

$$p(y) = \frac{1}{\sigma_1 \sigma_2} \exp\left\{-h_1^2 - h_2^2 - \frac{2|y|}{\sigma_1 \sigma_2}\right\}$$

$$\times \left\{ \begin{array}{l} \displaystyle\sum_{k=0}^{\infty} \left[\frac{1}{4}\sqrt{\frac{\sigma_1 \sigma_2 \lambda_1}{y}}\right]^k I_k\left(2\sqrt{\frac{y\lambda_1}{\sigma_1 \sigma_2}}\right) \\[6pt] \quad \cdot {}_1F_1\left(k + 1; 1; \dfrac{\lambda_2}{4}\right), \qquad y \geqslant 0 \\[12pt] \displaystyle\sum_{k=0}^{\infty} \left[\frac{1}{4}\sqrt{\frac{\sigma_1 \sigma_2 \lambda_2}{-y}}\right]^k I_k\left(2\sqrt{\frac{-y\lambda_2}{\sigma_1 \sigma_2}}\right) \\[6pt] \quad \cdot {}_1F_1\left(k + 1; 1; \dfrac{\lambda_1}{4}\right), \qquad y < 0 \end{array} \right. \tag{9a}$$

where ${}_1F_1(\cdot)$ is the confluent hypergeometric function and $I_k(\cdot)$ is the modified Bessel function of the first kind of order $k$, and where

$$h_i^2 \triangleq \frac{S_i}{\sigma_i^2}, \qquad \sigma_i^2 = \sigma_{Ni}^2 + \sigma_{Ji}^2, \qquad i = 1, 2 \tag{9b}$$

and

$$\lambda_{1,2} = h_1^2 + h_2^2 \pm 2h_1 h_2 \cos(\theta_1 - \theta_2). \tag{9c}$$

The conditional probabilities $P_s(e)$ and $P_m(e)$ are obtained from

$$P_s(e) \equiv P(e \mid \text{space}) = \Pr\{y < 0 \mid \theta_1 - \theta_2 = 0\}$$

$$= \int_{-\infty}^{0} p(y \mid \theta_1 - \theta_2 = 0)\, dy \tag{10a}$$

and

$$P_m(e) \equiv P(e \mid \text{mark}) = \Pr\{y > 0 \mid \theta_1 - \theta_2 = \pi\}$$

$$= \int_{0}^{\infty} p(y \mid \theta_1 - \theta_2 = \pi)\, dy. \tag{10b}$$

From (9c) we note that

$$\lambda_{1,2} = (h_1 \pm h_2)^2 \qquad \text{for space} \tag{11a}$$

and

$$\lambda_{1,2} = (h_1 \mp h_2)^2 \qquad \text{for mark.} \tag{11b}$$

Using (11), (9b) and (9a) and carrying out the integrations indicated in (10a) and (10b) we obtain the following results, using the notations

$$\alpha_i \triangleq \frac{S_i}{\sigma_{Ni}^2}, \qquad \beta_i \triangleq \frac{S_i}{\sigma_{Ji}^2}, \qquad i = 1, 2. \tag{12}$$

$$P_s[e \mid (0, 0)]$$

$$= \tfrac{1}{2} \exp\{-(\alpha_1 + \alpha_2)\} \times \sum_{k=0}^{\infty} \frac{4^{-k}}{k!} (\sqrt{\alpha_1} - \sqrt{\alpha_2})^{2k}$$

$$\times {}_1F_1[1; k + 1; \tfrac{1}{2}(\sqrt{\alpha_1} - \sqrt{\alpha_2})^2]$$

$$\times {}_1F_1[k + 1; 1; \tfrac{1}{4}(\sqrt{\alpha_1} + \sqrt{\alpha_2})^2]$$

$$\equiv P_m[e \mid (0, 0)] \tag{13a}$$

150

$P_s[e \mid (0, 1)]$

$$= \tfrac{1}{2} \exp \left\{ - \left( \alpha_1 + \frac{\alpha_2 \beta_2}{\alpha_2 + \beta_2} \right) \right\}$$

$$\times \sum_{k=0}^{\infty} \frac{4^{-k}}{k!} \left( \sqrt{\alpha_1} - \sqrt{\frac{\alpha_2 \beta_2}{\alpha_2 + \beta_2}} \right)^{2k}$$

$$\times {}_1F_1 \left[ 1; k+1; \frac{1}{2} \left( \sqrt{\alpha_1} - \sqrt{\frac{\alpha_2 \beta_2}{\alpha_2 + \beta_2}} \right)^2 \right]$$

$$\times {}_1F_1 \left[ k+1; 1; \frac{1}{4} \left( \sqrt{\alpha_1} + \sqrt{\frac{\alpha_2 \beta_2}{\alpha_2 + \beta_2}} \right)^2 \right]$$

$$= P_m[e \mid (0, 1)] \tag{13b}$$

and

$P_s[e \mid (1, 1)]$

$$= \tfrac{1}{2} \exp \left\{ - \left( \frac{\alpha_1 \beta_1}{\alpha_1 + \beta_1} + \frac{\alpha_2 \beta_2}{\alpha_2 + \beta_2} \right) \right\}$$

$$\times \sum_{k=0}^{\infty} \frac{4^{-k}}{k!} \left( \sqrt{\frac{\alpha_1 \beta_1}{\alpha_1 + \beta_1}} - \sqrt{\frac{\alpha_2 \beta_2}{\alpha_2 + \beta_2}} \right)^{2k}$$

$$\times {}_1F_1 \left[ 1; k+1; \frac{1}{2} \left( \sqrt{\frac{\alpha_1 \beta_1}{\alpha_1 + \beta_1}} - \sqrt{\frac{\alpha_2 \beta_2}{\alpha_2 + \beta_2}} \right)^2 \right]$$

$$\times {}_1F_1 \left[ k+1; 1; \frac{1}{4} \left( \sqrt{\frac{\alpha_1 \beta_1}{\alpha_1 + \beta_1}} + \sqrt{\frac{\alpha_2 \beta_2}{\alpha_2 + \beta_2}} \right)^2 \right]$$

$$\equiv P_m[e \mid (1, 1)]. \tag{13c}$$

It is apparent from symmetry arguments that

$$P_s[e \mid (0, 1)] = P_s[e \mid (1, 0)] \tag{14a}$$

$$P_m[e \mid (0, 1)] = P_m[e \mid (1, 0)]. \tag{14b}$$

The conditional probabilities of error given in (13) above are the general expressions in that the signal-to-noise power ratio and the signal-to-jamming power ratio in $u_1(t)$ and $u_2(t)$ are assumed to be different. The case of interest is one in which $\alpha_1 = \alpha_2$ and $\beta_1 = \beta_2$. This is equivalent to the assumption that, for $i = 1$ and 2,

$$\sigma_{Ni}{}^2 = \sigma_N{}^2, \quad \sigma_{Ji}{}^2 = \sigma_J{}^2, \quad S_i = S \tag{15a}$$

and

$$\alpha_i = \alpha, \quad \beta_i = \beta. \tag{15b}$$

From (1) and (15a) we have

$$\beta = \frac{S}{\sigma_J{}^2} = R\gamma \tag{16a}$$

where

$$R \triangleq \frac{S}{(B/W)J} = \frac{\text{signal power per cell}}{\text{average jamming power per cell}}. \tag{16b}$$

### The Effective Jamming Strategy

The most effective jamming strategy is one which concentrates the available jamming power of $J$ watts over a fractional bandwidth $\gamma W$ Hz within the total bandwidth $W$ Hz, as depicted[1] in Fig. 2. There will be an optimum value of $\gamma$ which will maximize the total error probability. The determination of the optimum fraction is based on the usual method of partial differentiation of the total error probability expression with respect to $\gamma$ and set the result equal to 0 to find the root. If we define $\gamma_0'$ to be the optimum fraction, the problem of finding $\gamma_0'$ reduces to solving the equation

$$\left\{ \frac{\partial}{\partial \gamma} \sum_{i=0}^{1} \sum_{j=0}^{1} \pi(i, j) P_s[e \mid (i, j)] \right\}_{\gamma = \gamma_0'} = 0 \tag{17}$$

where $\pi(i, j)$ and $P_s[e \mid (i, j)]$ are given in (5) and (13), respectively. The exact solution for the root of (17) is a formidable task, and it does not seem possible to find the root of (17) analytically. Clearly, therefore, the effort to find "the optimum" fraction $\gamma_0'$ of $\gamma$ must be abandoned, and one must seek to find a "near-optimum" fraction $\gamma_0$ of $\gamma$. Our approach in finding a near-optimum fraction $\gamma_0$ was to consider only the term which corresponds to the case where both symbols are jammed. Clearly, it is the term which is dominant among the four terms involved in (17). In this regard, our method of finding the fraction $\gamma_0$ is identical to that of Houston [3]. From (13c) with $\alpha_1 = \alpha_2 = \alpha$ and $\beta_1 = \beta_2 = \beta$, we thus obtain a near-optimum fraction $\gamma_0$ from the solution[2] of

$$\left\{ \frac{\partial}{\partial \gamma} \pi(1, 1) P_s[e \mid (1, 1)] \right\}_{\gamma = \gamma_0} = \left\{ \tfrac{1}{2} \gamma^2 \exp \left[ - \frac{\alpha R \gamma}{\alpha + R\gamma} \right] \right\}_{\gamma = \gamma_0}$$

$$= 0 \tag{18}$$

where we used the identity of $\beta = R\gamma$ from (16a). The solution of (18) is subject to the constraint, $0 < \gamma_0 \leqslant 1$. It can be easily shown that as long as $\alpha \equiv$ SNR is greater than 9 dB, the solution of (18) is given by

$$\gamma_0 = \frac{2}{R} \tag{19}$$

where $R$ was defined in (16b).

As will be shown in the graphical results of the performance comparison (Section IV), when the jammer concentrates his available power over the band $(\gamma_0 W)$ Hz $\equiv [(2/R)W]$ Hz, it produces much higher error probability than the wide-band jamming error probability.

---

1 Since every frequency cell is equally likely to be selected by the communicator, the probability of an individual hop being jammed depends only on the fraction $\gamma$ regardless of whether $\gamma$ is contiguous or not.

2 How near $\gamma_0$ is to the true optimum fraction $\gamma_0'$ cannot be known until the exact solution of (17) is obtained.

## III. THE BEHAVIOR OF THE DECISION VARIABLES

Before taking on the task of computing the error performance measures, which follow in the next section, it would be of interest to observe the behavior of decision variables under various jamming events of $(j_1, j_2)$; namely, $(0, 0), (0, 1) \equiv (1, 0)$, and $(1, 1)$. For this purpose we return to general expression of the pdf of the decision variable $y(t)$ given in (9) [see Fig. 1(b)]. As implied already by the pdf's appearing in the integrands of (10a) and (10b), the conditional pdf's are given by

$$p(y \mid \text{space}) = p(y \mid \theta_1 - \theta_2 = 0) \tag{20a}$$

$$p(y \mid \text{mark}) = p(y \mid \theta_1 - \theta_2 = \pi). \tag{20b}$$

From (9a)-(9c) we observe the "symmetry" properties to be

$$p(y \mid \text{space}) = p(-y \mid \text{mark}). \tag{21}$$

To show a graphical view of the conditional pdf of $y$, we define a convenient normalized variable by

$$z = y/\sigma_1{}^2. \tag{22}$$

In view of (21) it suffices to show the behavior of the conditional pdf of

$$p(z \mid \text{space}) = p(z \mid \theta_1 - \theta_1 = 0). \tag{23}$$

Assuming equal signal power $(S_1 = S_2 = S)$ of the successive symbols, (11a) with the substitution of (9b) results in

$$\lambda_1 = (h_1 + h_2)^2 = h_1{}^2 \left(1 + \frac{\sigma_1}{\sigma_2}\right)^2 \tag{24a}$$

$$\lambda_2 = (h_1 - h_2)^2 = h_1{}^2 \left(1 - \frac{\sigma_1}{\sigma_2}\right)^2. \tag{24b}$$

The pdf of the normalized decision variable $z$, given that "space" is transmitted, then becomes

$$p(z \mid \text{space}) = \sigma_1{}^2 p(y = \sigma_1{}^2 z \mid \text{space})$$

$$= \xi \exp\{-h_1{}^2(1 + \xi^2) - 2\xi |z|\}$$

$$\times \left\{ \begin{array}{l} \displaystyle\sum_{k=0}^{\infty} \left[\frac{h_1(1 + \xi)}{4\sqrt{\xi z}}\right]^k I_k[2h_1(1 + \xi)\sqrt{\xi z}] \\[4pt] \quad \cdot {}_1F_1\left[k + 1; 1; \frac{h_1{}^2}{4}(1 - \xi)^2\right], \quad z \geqslant 0 \\[10pt] \displaystyle\sum_{k=0}^{\infty} \left[\frac{h_1(1 - \xi)}{4\sqrt{-\xi z}}\right]^k I_k[2h_1(1 - \xi)\sqrt{-\xi z}] \\[4pt] \quad \cdot {}_1F_1\left[k + 1; 1; \frac{h_1{}^2}{4}(1 + \xi)^2\right], \quad z < 0 \end{array} \right. \tag{25a}$$

where

$$\xi \triangleq \sigma_1/\sigma_2. \tag{25b}$$

The pdf of (25) can now be observed under various jamming events $(j_1, j_2)$ as follows.

*Both Symbols Are Not Jammed; the Event of (0, 0)*—For

this case,

$$\sigma_1{}^2 = \sigma_2{}^2,$$

$$\xi = 1,$$

$$h_1{}^2 = h_2{}^2 = \alpha.$$

The conditional pdf then becomes

$$p(z \mid \text{space}, (0, 0)) = \exp\{-2\alpha - 2|z|\}$$

$$\times \left\{ \begin{array}{ll} \displaystyle\sum_{k=0}^{\infty} \left[\frac{1}{2}\sqrt{\frac{\alpha}{z}}\right]^k I_k[4\sqrt{\alpha z}], & z \geqslant 0 \\[10pt] e^\alpha, & z < 0. \end{array} \right. \tag{26}$$

*One of the Two Symbols Is Jammed; the Event of (0, 1) or (1, 0)*—For this case, let

$$\sigma_1{}^2 = \sigma_N{}^2, \quad \sigma_2{}^2 = \sigma_N{}^2 + \sigma_J{}^2$$

then

$$h_2{}^2 = \frac{\alpha \sigma_N{}^2}{\sigma_N{}^2 + \sigma_J{}^2} = \frac{\alpha\beta}{\alpha + \beta}$$

and

$$\xi = \sqrt{\frac{\alpha\beta}{\alpha + \beta}}.$$

The conditional pdf is then given by

$$p(z \mid \text{space}, (0, 1))$$

$$= p(z \mid \text{space}, (1, 0))$$

$$= \sqrt{\frac{\beta}{\alpha + \beta}} \exp\left\{-\alpha - \frac{\alpha\beta}{\alpha + \beta} - 2|z|\sqrt{\frac{\beta}{\alpha + \beta}}\right\}$$

$$\times \left\{ \begin{array}{l} \displaystyle\sum_{k=0}^{\infty} \left[\frac{1}{4}\left(\sqrt{\alpha} + \sqrt{\frac{\alpha\beta}{\alpha + \beta}}\right)\sqrt{\frac{\sqrt{1 + \alpha/\beta}}{z}}\right]^k \\[10pt] \quad \cdot I_k\left[2\left(\sqrt{\alpha} + \sqrt{\frac{\alpha\beta}{\alpha + \beta}}\right)\sqrt{\frac{z}{\sqrt{1 + \frac{\alpha}{\beta}}}}\right] \\[10pt] \quad \times {}_1F_1\left[k + 1; 1; \frac{1}{4}\left(\sqrt{\alpha} - \sqrt{\frac{\alpha\beta}{\alpha + \beta}}\right)^2\right], \\[6pt] \hspace{6cm} z \geqslant 0 \\[14pt] \displaystyle\sum_{k=0}^{\infty} \left[\frac{1}{4}\left(\sqrt{\alpha} - \sqrt{\frac{\alpha\beta}{\alpha + \beta}}\right)\sqrt{\frac{\sqrt{1 + \alpha/\beta}}{-z}}\right]^k \\[10pt] \quad \cdot I_k\left[2\left(\sqrt{\alpha} - \sqrt{\frac{\alpha\beta}{\alpha + \beta}}\right)\sqrt{\frac{-z}{\sqrt{1 + \frac{\alpha}{\beta}}}}\right] \\[10pt] \quad \times {}_1F_1\left[k + 1; 1; \frac{1}{4}\left(\sqrt{\alpha} + \sqrt{\frac{\alpha\beta}{\alpha + \beta}}\right)^2\right], \\[6pt] \hspace{6cm} z < 0. \end{array} \right. \tag{27}$$

*Both Symbols Are Jammed; the Event of (1, 1)*—For this case

$$h_1{}^2 = h_2{}^2 = \frac{\alpha\beta}{\alpha+\beta}$$

$$\xi = 1,$$

and thus the conditional pdf becomes

$$p(z \mid \text{space}, (1, 1))$$

$$= \exp\left\{-\frac{2\alpha\beta}{\alpha+\beta} - 2|z|\right\}$$

$$\times \begin{cases} \displaystyle\sum_{k=0}^{\infty} \left[\frac{1}{2}\sqrt{\frac{\alpha\beta}{(\alpha+\beta)z}}\right]^k I_k\left[4\sqrt{\frac{\alpha\beta z}{\alpha+\beta}}\right], & z \geq 0 \\ \exp\left(\dfrac{\alpha\beta}{\alpha+\beta}\right), & z < 0. \end{cases}$$

$$(28)$$

To observe the behavior of the decision variable under various jamming environments, we have chosen to fix the transmitter power to provide the probability of bit error of $10^{-4}$ under jam-free condition, and then varied the signal-to-jamming power ratios. The curves in Fig. 3(a)–(c) are the graphical presentations of (26)–(28) for SNR of $\alpha = 9.3$ dB (corresponding to BER of $10^{-4}$ under jam-free condition) and $\beta$, signal-to-jamming power ratio, of 10 dB, 0 dB, and $-3$ dB, under various jamming events $(j_1, j_2)$ as indicated. The difference in the behavior of the decision variable between the jamming event of $(0, 1)$ and $(1, 1)$ is quite evident. Each event defines a different random variable. The influence of signal-to-jamming power ratio is also clearly observed from these pdf's.

## IV. THE ERROR PROBABILITIES

The probability of error expressions for the DPSK/FH system can now be stated for both near-optimum jamming and wide-band jamming assumptions.

Using (19), (16a), (15b), (13), (5) and (4), the error probability expression for near-optimum jamming is given by

$$P(e; \gamma = \gamma_0)$$

$$= \frac{1}{2}\left(1 - \frac{2}{R}\right)^2 \exp(-\alpha)$$

$$+ \frac{2}{R}\left(1 - \frac{2}{R}\right)\exp\left\{-\left[\alpha + \frac{2\alpha}{\alpha+2}\right]\right\}$$

$$\times \sum_{k=0}^{\infty} \frac{\alpha^k}{4^k k!}\left(1 - \sqrt{\frac{2}{\alpha+2}}\right)^{2k}$$

$$\times {}_1F_1\left[1; k+1; \frac{\alpha}{2}\left(1 - \sqrt{\frac{2}{\alpha+2}}\right)^2\right]$$

$$\times {}_1F_1\left[k+1; 1; \frac{\alpha}{4}\left(1 + \sqrt{\frac{2}{\alpha+2}}\right)^2\right]$$

$$+ \frac{1}{2}\left(\frac{2}{R}\right)^2 \exp\left(-\frac{2\alpha}{\alpha+2}\right), \qquad R > 2 \qquad (29)$$

where $\alpha$ and $R$ were defined earlier in (12) and (16b), respectively.

If the total jamming power is assumed to be applied uniformly over the total system bandwidth $W$ Hz, every hop will be subject to a jamming with probability of 1. As stated earlier, this is an ineffective jamming strategy from the jammer's point of view. In this case the expression reduces to

$$P(e; \gamma = 1) = \frac{1}{2}\exp\left\{-\frac{\alpha R}{\alpha+R}\right\}, \qquad 0 < R < \infty. \qquad (30)$$

For comparison purposes near-optimum jamming performance (29) and wide-band jamming performance (30) are plotted in Figs. 4 and 5 for selected values of SNR 9.3 dB and 10.3 dB. These values of SNR correspond to DPSK/FH performance of error rates of $10^{-4}$ and $10^{-5}$, respectively, when jamming is assumed to be absent.

It should be noted that we have used SNR, rather than the conventional "energy-to-noise density," in the above expressions since our analysis is that of a single-sample detection model; the result can be easily extended to include the effect of postdetection integration over the bit (or symbol) duration. Had we assumed ideal postdetection integration the result would have been

$$\alpha = \frac{S}{N} = \frac{E_b}{N_0} \qquad (31)$$

where $E_b$ is the signal energy per bit and $N_0$ is the one-sided noise power spectral density.

## V. CONCLUSIONS

An analysis has been presented for the derivation of the probability of error for a hypothetical pure FH spread-spectrum system in a partial-band jamming environment, employing DSPK modulation with differentially coherent detection. Near-optimum partial-band jamming and wide-band jamming performance were derived and compared graphically. The main contribution of this paper was to show a complete derivation for the probability of error by including all possible error events in a DPSK/FH system under a partial-band jamming assumption. The calculation for the total unconditional error probability did include all possible conditional error probability terms of $P[e \mid (i, j)]$, $i, j = 0, 1$. The determination of the optimum fraction of the partial band was based on approximate solution, however, due to mathematical difficulties involved in finding the root of (17). Thus, the term "near-optimum" is used rather than "the optimum." Previous analyses have always ignored (due to analytical difficulty) the error phenomena of one of the two bits being jammed in a DPSK/FH system. This amounted to having excluded the conditional error probabilities $P[e \mid (0, 1)]$ and $P[e \mid (1, 0)]$ in error probability calculations. Our manner of graphical presentations for the jamming performance shows the quantitative extent of the advantages optimum or near-optimum jamming can have over the wide-band jamming strategy under some parametric assumptions on the communicator's power levels (or error rate requirements).

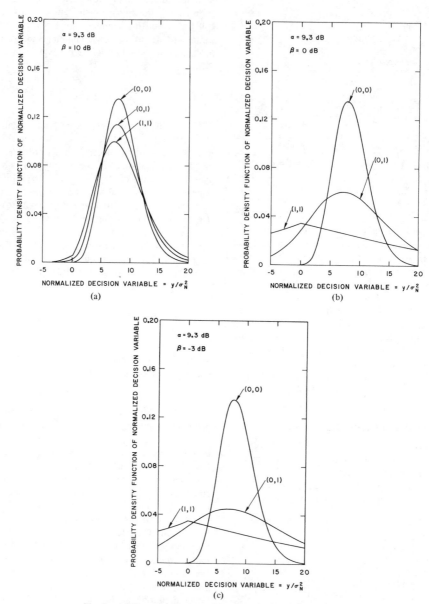

Fig. 3. (a) The normalized conditional pdf $p(z \mid \text{space}, (j_1, j_2))$ of the decision variable under various jamming events $(j_1, j_2)$ when SNR = 9.3 dB (for BER = $10^{-4}$ in jam-free condition) and $\beta$ = signal-to-jamming power ratio = 10 dB. (b) The normalized conditional pdf $p(z \mid \text{space}, (j_1, j_2))$ of the decision variable under various jamming events $(j_1, j_2)$ where SNR = 9.3 dB (for BER = $10^{-4}$ in jam-free condition) and $\beta$ = signal-to-jamming power ratio = 0 dB. (c) The normalized conditional pdf $p(z \mid \text{space}, (j_1, j_2))$ of the decision variable under various jamming events $(j_1, j_2)$ where SNR = 9.3 dB (for BER = $10^{-4}$ in jam-free condition) and $\beta$ = signal-to-jamming power ratio = −3 dB.

154

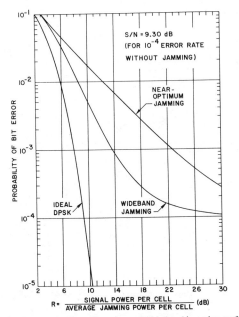

Fig. 4. The near-optimum jamming and wide-band jamming performances for a DPSK/FH system when the communicator's SNR = 9.3 dB (for BER = $10^{-4}$ in jam-free condition) in comparison with unjammed ideal system performance (for ideal DPSK curve, the abscissa reads SNR).

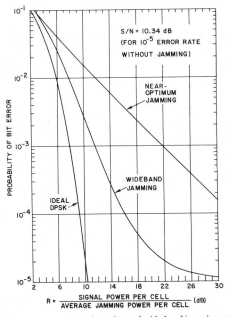

Fig. 5. The near-optimum jamming and wide-band jamming performances for a DPSK/FH system when the communicator's SNR = 10.34 dB (for BER = $10^{-5}$ in jam-free condition) in comparison with unjammed ideal system performance (for ideal DPSK curve, the abscissa reads SNR).

## APPENDIX

We will summarize briefly how the new pdf (9a) is derived.

Recognizing that the differentially coherent demodulator shown in Fig. 1(b) performs a cross correlation, let us consider the situation where the input waveforms $u_1(t)$ and $u_2(t)$ are given by

$$u_i(t) = A_i \cos(\omega_0 t - \theta_i) + n_i(t), \qquad i = 1, 2 \tag{A1}$$

where

$$n_i(t) = n_{ic}(t) \cos \omega_0 t + n_{is}(t) \sin \omega_0 t, \qquad i = 1, 2. \tag{A2}$$

If the bandpass waveforms $u_1(t)$ and $u_2(t)$ are written in Rician notation, that is,

$$u_1(t) = x_1(t) \cos \omega_0 t + x_2(t) \sin \omega_0 t$$

$$u_2(t) = x_3(t) \cos \omega_0 t + x_4(t) \sin \omega_0 t \tag{A2a}$$

then we may identify the components as

$$x = \begin{bmatrix} x_1(t) \\ x_2(t) \\ x_3(t) \\ x_4(t) \end{bmatrix} = \begin{bmatrix} s_{1c}(t) + n_{1c}(t) \\ s_{1s}(t) + n_{1s}(t) \\ s_{2c}(t) + n_{2c}(t) \\ s_{2s}(t) + n_{2s}(t) \end{bmatrix}$$

$$= \begin{bmatrix} A_1(t) \cos \theta_1(t) + n_{1c}(t) \\ A_1(t) \sin \theta_1(t) + n_{1s}(t) \\ A_2(t) \cos \theta_2(t) + n_{2c}(t) \\ A_2(t) \sin \theta_2(t) + n_{2s}(t) \end{bmatrix} \tag{A2b}$$

Now if the signal components are deterministic, and the noise is from a zero-mean, stationary Gaussian random process, then the components in (A2b) are jointly Gaussian random variables at a given instant, with mean and covariance matrix given by

$$m_x = E\{x\} = (A_1 \cos \theta_1, A_1 \sin \theta_1, A_2 \cos \theta_2,$$

$$A_2 \sin \theta_2)' \tag{A3}$$

and

$$K_x = E\{(x - m_x)(x - m_x)'\}$$

$$= \begin{bmatrix} \sigma_1^2 & 0 & \rho \sigma_1 \sigma_2 & r \sigma_1 \sigma_2 \\ 0 & \sigma_1^2 & -r \sigma_1 \sigma_2 & \rho \sigma_1 \sigma_2 \\ \rho \sigma_1 \sigma_2 & -r \sigma_1 \sigma_2 & \sigma_2^2 & 0 \\ r \sigma_1 \sigma_2 & \rho \sigma_1 \sigma_2 & 0 & \sigma_2^2 \end{bmatrix} \tag{A4}$$

The independence of $(x_1, x_2)$ and of $(x_3, x_4)$ shown in (A4) is a property of the Rician decomposition, while the $\rho$ is the correlation coefficient between $x$ components which are in phase, and the parameters $\pm r$ are the correlation coefficients between $x$ components which are $90°$ out of phase.

In terms of these $x$ components, the output $y(t)$ of the low-pass filter [see Fig. 1(b)] is given by

$$y(t) = \tfrac{1}{2}\{x_1(t)x_3(t) + x_2(t)x_4(t)\}. \tag{A5}$$

In terms of matrix operations, we observe that $y$ is the quadratic form

$$y = \tfrac{1}{2}x'Q_x x, \qquad Q_x = \tfrac{1}{2}\begin{bmatrix} 0 & 0 & 1 & 0 \\ 0 & 0 & 0 & 1 \\ 1 & 0 & 0 & 0 \\ 0 & 1 & 0 & 0 \end{bmatrix}. \tag{A6}$$

It is desired to know the pdf for $y(t)$ at a given time instant $t$. In previous work [5], [6] the pdf for $y(t)$ under general case described in (A3) and (A4) was obtained, but the resulting expression is so complicated that its application was almost exclusively limited to the special case of $r = 0$ and $\sigma_1 = \sigma_2$. These conditions[3] correspond to the requirements that the noise spectrums be even about the center frequency and of equal power.

Recently, it has been found that even under general conditions of $r \neq 0$ and $\sigma_1 \neq \sigma_2$, the output of the low-pass filter $y(t)$ can be expressed as the difference between two scaled noncentral chi-squared, independent random variables with two degrees of freedom. Discovery of this fact has facilitated the derivation of the new results for the pdf in several easily computable compact forms [4].

If the $x$ components are transformed linearly into new components $w$ by the relation

$$w = Ax, \tag{A7}$$

such that

$$E\{ww'\} = K_w = AK_x A' = \begin{bmatrix} v_1 & 0 & 0 & 0 \\ 0 & v_1 & 0 & 0 \\ 0 & 0 & v_2 & 0 \\ 0 & 0 & 0 & v_2 \end{bmatrix} \tag{A8}$$

and the output $y(t)$ is expressible

$$y = \tfrac{1}{2}w'Q_w w \tag{A9}$$

where

$$Q_w = (A^{-1})'Q_x A^{-1} = k\begin{bmatrix} 1 & 0 & 0 & 0 \\ 0 & 1 & 0 & 0 \\ 0 & 0 & -1 & 0 \\ 0 & 0 & 0 & -1 \end{bmatrix} \tag{A10}$$

then there exists a transformation $A$, as shown in (A7), which will give $y(t)$, (A9), as

$$y = \frac{k}{2}(w_1{}^2 + w_2{}^2) - \frac{k}{2}(w_3{}^2 + w_4{}^2)$$

$$= \frac{kv_1}{2}\chi_1{}^2(2;\lambda_1) - \frac{kv_2}{2}\chi_2{}^2(2;\lambda_2) \tag{A11}$$

where $\chi_1{}^2$ and $\chi_2{}^2$ are independent chi-square variables.

It is shown in [4] that the transformation $A$ is given by

$$A = \begin{bmatrix} \sqrt{\dfrac{\sigma_2}{\sigma_1}} & 0 & \sqrt{\dfrac{\sigma_1}{\sigma_2}}\sqrt{1-r^2} & -r\sqrt{\dfrac{\sigma_1}{\sigma_2}} \\[2ex] 0 & \sqrt{\dfrac{\sigma_2}{\sigma_1}} & r\sqrt{\dfrac{\sigma_1}{\sigma_2}} & \sqrt{\dfrac{\sigma_1}{\sigma_2}}\sqrt{1-r^2} \\[2ex] \sqrt{\dfrac{\sigma_2}{\sigma_1}}\sqrt{1-r^2} & -r\sqrt{\dfrac{\sigma_2}{\sigma_1}} & -\sqrt{\dfrac{\sigma_1}{\sigma_2}} & 0 \\[2ex] r\sqrt{\dfrac{\sigma_2}{\sigma_1}} & \sqrt{\dfrac{\sigma_2}{\sigma_1}}\sqrt{1-r^2} & 0 & -\sqrt{\dfrac{\sigma_1}{\sigma_2}} \end{bmatrix} \tag{A12}$$

and the parameters in (A11) are given by

$$k = \frac{1}{4\sqrt{1-r^2}} \tag{A13}$$

$$v_1 = 2\sigma_1\sigma_2\sqrt{1-r^2}\,(\sqrt{1-r^2} + \rho)$$

$$v_2 = 2\sigma_1\sigma_2\sqrt{1-r^2}\,(\sqrt{1-r^2} - \rho). \tag{A14}$$

[3] The reader would recognize that the pdf derived under the condition $\sigma_1 = \sigma_2$ will not be useful for its application to the situation where one bit is jammed while the other is not.

The noncentrality parameters $\lambda_1$ and $\lambda_2$ in (A11) are given by

$$\lambda_1 = \frac{h_1^2 + h_2^2 + 2\sqrt{1-r^2}\, h_1 h_2 \cos(\theta_1 - \theta_2) + 2rh_1 h_2 \sin(\theta_1 - \theta_2)}{\sqrt{1-r^2}\,(\sqrt{1-r^2} + \rho)} \tag{A15}$$

$$\lambda_2 = \frac{h_1^2 + h_2^2 - 2\sqrt{1-r^2}\, h_1 h_2 \cos(\theta_1 - \theta_2) + 2rh_1 h_2 \sin(\theta_1 - \theta_2)}{\sqrt{1-r^2}(\sqrt{1-r^2} - \rho)} \tag{A16}$$

and

$$h_1^2 \triangleq \frac{A_1^2}{2\sigma_1^2} \text{ and } h_2^2 \triangleq \frac{A_2^2}{2\sigma_2^2}. \tag{A17}$$

The characteristic function method is then used to derive the pdf for $y(t)$, given in (A11), and a resultant expression for pdf is given by

$p(y)$

$$= \begin{cases} \dfrac{1}{k(v_1 + v_2)} \exp\left\{ -\dfrac{\lambda_1}{2} - \dfrac{\lambda_2}{2} - \dfrac{y}{kv_1} \right\} \\[2mm] \quad \times \displaystyle\sum_{m=0}^{\infty} \left[ \dfrac{v_2}{v_1 + v_2} \sqrt{\dfrac{kv_1 \lambda_1}{2y}} \right]^m I_m\left( \sqrt{\dfrac{2\lambda_1 y}{kv_1}} \right) \\[2mm] \quad \cdot {}_1F_1\left( 1; m+1; \dfrac{\lambda_2}{2} \cdot \dfrac{v_1}{v_1 + v_2} \right) \quad y \geqslant 0 \\[4mm] \hfill (\text{A18a}) \\[4mm] \dfrac{1}{k(v_1 + v_2)} \exp\left\{ -\dfrac{\lambda_2}{2} - \dfrac{\lambda_1}{2} + \dfrac{y}{kv_2} \right\} \\[2mm] \quad \times \displaystyle\sum_{m=0}^{\infty} \left[ \dfrac{v_1}{v_1 + v_2} \sqrt{\dfrac{kv_2 \lambda_2}{-2y}} \right]^m I_m\left( \sqrt{\dfrac{-2\lambda_2 y}{kv_2}} \right) \\[2mm] \quad \cdot {}_1F_1\left( 1; m+1; \dfrac{\lambda_1}{2} \cdot \dfrac{v_2}{v_1 + v_2} \right) \quad y < 0 \quad (\text{A18b}) \end{cases}$$

in which the ${}_1F_1(\cdot)$ are confluent hypergeometric functions and the $I_m(\cdot)$ are modified Bessel functions of the first kind, each of which can be computed recursively. Equations (A18a) and (A18b) with $\rho = 0$ and $r = 0$ are the pdf's given in (9).

## REFERENCES

[1] J. D. Edell, "Wideband, noncoherent, frequency-hopped waveforms and their hybrids in low probability-of-intercept communications," Naval Res. Lab., Washington, DC, NRL Rep. 8025, Nov. 8, 1976.

[2] J. Gorski-Popiel, Ed., *Frequency Synthesis: Techniques and Applications.* New York: IEEE Press, 1975, p. 43.

[3] S. W. Houston, "Modulation techniques for communications—Part I: Tone and noise jamming performance of spread spectrum *M*-ary FSK and 2, 4-ary DPSK waveforms," in *Proc. IEEE 1975 Nat. Aerosp. Electron. Conf.*, June 10–12, 1975, pp. 51–58.

[4] L. E. Miller and J. S. Lee, "Bandpass correlator analysis for general input assumptions," *IEEE Trans. Inform. Theory,* to be published.

[5] ——, "The probability density function for the output of an analog cross-correlator with correlated bandpass inputs," *IEEE Trans. Inform. Theory,* vol. IT-20, pp. 433–440, July 1974.

[6] L. C. Andrews, "The probability density function for the output of a cross-correlator with bandpass inputs," *IEEE Trans. Inform. Theory,* vol. IT-19, pp. 13–19, Jan. 1973.

[7] M. K. Simon and A. Polydoros, "Coherent detection of frequency-hopped quadrature modulations in the presence of jamming—Part I: QPSK and QASK modulations," *IEEE Trans. Commun.,* vol. COM-29, pp. 1644–1660, Nov. 1981.

[8] R. W. Nettleton and G. R. Cooper, "Performance of a frequency-hopped differentially modulated spread-spectrum receiver in a Rayleigh fading channel," *IEEE Trans. Veh. Technol.,* vol. VT-30, pp. 14–29, Feb. 1981.

# The Performance of *M*-ary FH-DPSK in the Presence of Partial-Band Multitone Jamming

MARVIN K. SIMON, FELLOW, IEEE

*Abstract*—Using a geometric approach, the performance of *M*-ary FH-DPSK in the presence of partial-band multitone jamming is evaluated. The optimal jamming strategy is determined as a function of the number of signaling levels *M* and the ensuing results are used to determine worst case bit error probability performance as a function of this same parameter. It is demonstrated that, for $M = 2^m$, the best performance is obtained for $M = 4$.

## I. INTRODUCTION

IN a previous paper [1], Houston considered the performance of 2-ary and 4-ary FH-DPSK modulations in the presence of partial-band multitone jamming. Using a geometric approach, closed-form expressions were obtained for the system bit error probability and the optimal jamming strategy was determined in each case.

This paper presents a generalization of the geometric approach taken in [1] which allows determination of the performance of *M*-ary FH-DPSK in the presence of partial-band multitone jamming. Typically, $M = 2^m$, with *m* integer, and these are the only cases we shall consider. Also, as was done in [1], we shall evaluate the optimal jamming strategy which is equivalent to determining the tone interference-to-signal ratio which maximizes the system bit error probability. Wherever possible, we shall use the same notation as in [1] so as to allow comparison of our general results with the specific cases treated there.

## II. MULTITONE JAMMER MODEL

The total spread-spectrum bandwidth *W* is divided equally into frequency bins or hop bandwidths. The width of these bins is assumed to be equal to the symbol rate of the *M*-ary DPSK modulation. If $R_c = 1/T_c$ denotes the bit rate, the symbol rate is then given by $R_c/\log_2 M$ and, hence, the number of frequency bins in the total spread bandwidth is

$$n = \frac{W}{R_c/\log_2 M} = \frac{W \log_2 M}{R_c}. \tag{1}$$

The jammer is assumed to have perfect knowledge of system operation except for the hopping sequence. Thus, although he might know the location of the spread bandwidth in the frequency domain, he does not know *a priori* where, within this bandwidth, to concentrate his efforts. Thus, the best strategy for the multitone jammer is to dis-

Manuscript received June 30, 1981; revised December 21, 1981. This work was performed by the author as a consultant to Axiomatix, Inc., Los Angeles, CA.

The author is with the Jet Propulsion Laboratory, Pasadena, CA 91103.

tribute his power among *q* random phase tones which are contiguous and spaced in frequency by the symbol rate (i.e., one tone per hop bandwidth) and vary the number of jam tones *q* to maximize the system error probability. Equivalently, for a fixed number of frequency bins *n* as in (1), the jammer varies

$$P_H \triangleq \frac{q}{n} \tag{2}$$

to maximize error probability where $P_H$ of (2) is the probability that the jammer hits a frequency bin in any one hop.

It is convenient to renormalize the problem in terms of the ratio of jamming power per tone $J_0 = J/q$ to signal power *S*. Letting $\rho^2$ denote this ratio, i.e.,

$$\rho^2 = \frac{J/q}{S} \tag{3}$$

then, combining (1), (2), and (3), we have

$$P_H = \frac{J/\rho^2 S}{W \log_2 M/R_c} = \frac{1}{(\log_2 M) X \rho^2} \tag{4}$$

where

$$X \triangleq \left(\frac{S}{J}\right)\left(\frac{W}{R_c}\right) = \frac{ST_c}{N_J} = \frac{E_c}{N_J} \tag{5}$$

is the channel bit energy-to-jam noise spectral density ratio, the latter being defined by

$$N_J \triangleq \frac{J}{W}. \tag{6}$$

## III. ERROR PROBABILITY PERFORMANCE

In a differentially encoded *M*-ary DPSK modulation system, the information to be transmitted in the *i*th signaling interval is conveyed by appropriately selecting one of *M* phases

$$\theta_m = \frac{(2m-1)\pi}{M}; \qquad m = 1, 2, \cdots, M \tag{7}$$

and adding it to the total accumulated phase in the $(i-1)$st signaling interval of a constant-amplitude $(A)$, fixed-frequency (assumed known at the receiver) sinusoid. The transmitted signal $s^{(i)}(t)$ in the *i*th signaling interval is then conveniently

Reprinted from *IEEE Trans. Commun.*, vol. COM-30, pp. 953–958, May 1982.

represented in complex form by

$$\overline{S}^{(i)} = Ae^{j(\theta^{(i)} + \theta_T^{(i-1)})} \tag{8}$$

where $\theta_T^{(i-1)}$ is the total accumulated phase in the $(i-1)$ st signaling interval and $\theta^{(i)}$ ranges over the set $\{\theta_m\}$ of (7).

In the presence of multitone jamming interference as characterized in the previous section, a jamming tone $\tilde{j}(t)$, constant in both phase and magnitude (amplitude), is added to the transmitted signal. Since, when the jammer "hits" a frequency bin, he is assumed[1] to be of the same frequency as the signal, then we may also represent the jammer in complex form, namely

$$\tilde{J} = Ie^{j\phi} \tag{9}$$

where $\phi$ is a random phase uniformly distributed in the interval $(0, 2\pi)$. Thus, in any hop interval which is hit by the jammer [the probability of this occurring is $P_H$ of (4)], the total number of signals on which a decision for the $i$th signaling interval is to be based are given (in complex form) by

$$\overline{Y}^{(i-1)} = Ae^{j\theta_T^{(i-1)}} + Ie^{j\phi}$$

$$\overline{Y}^{(i)} = Ae^{j(\theta^{(i)} + \theta_T^{(i-1)})} + Ie^{j\phi}. \tag{10}$$

Assuming a receiver structure that is optimum in the absence of the jammer, i.e., it employs the optimum decision rule for $M$-ary DPSK against wide-band noise, then in the presence of the on-tune tone jammer this rule would result in the estimate

$$\hat{\theta}^{(i)} = \theta_k \tag{11}$$

where $k$ is such that

$$|\arg(\overline{Y}^{(i)} - \overline{Y}^{(i-1)}) - \theta_k| \leqslant \frac{\pi}{M}. \tag{12}$$

Then if $\theta_n$ is indeed the true value of $\theta^{(i)}$, a symbol (phase) error is made, i.e., $\hat{\theta}^{(i)} \neq \theta^{(i)}$ whenever

$$|\arg(\overline{Y}^{(i)} - \overline{Y}^{(i-1)}) - \theta_n| > \frac{\pi}{M}. \tag{13}$$

Letting[2] $Q_{2\pi n/M}$; $n = 0, \pm 1, \pm 2, \cdots, \pm((M-2)/2), M/2$ denote the probability of this error event, namely,

$$Q_{2\pi n/M} = \text{Prob}\left\{|\arg(Y^{(i)} - Y^{(i-1)}) - \theta_n| > \frac{\pi}{M}\right\} \tag{14}$$

---

[1] The assumption of on-tune tone jamming is made solely to simplify the analysis as others [1] have done in the past. Both the analytical technique and the sensitivity of the results that follow from its application depend heavily on this assumption. Specific evidence of this statement will be discussed at the conclusion of the paper.

[2] As was done in [1], we shall rotate the actual transmitted signal vectors by $\pi/M$ rads for convenience so that the possible transmitted signal phases of (7) become

$$\theta_m = \frac{2\pi m}{M}; \quad m = 0, \pm 1, \pm 2, \cdots, \pm\left(\frac{M-2}{2}\right), \frac{M}{2}.$$

---

then, analogous to [1, eq. (40)], the average symbol error probability for $M$-ary DPSK in the presence of multitone jamming is given by[3]

$$P_E(M, X) = \frac{P_H}{M} \sum_n Q_{2\pi n/M} \tag{15}$$

where the summation on $n$ ranges over the set $n = 0, \pm 1, \pm 2, \cdots, \pm((M-2)/2), M/2$. Since $\phi$ is uniformly distributed, we can recognize the symmetry

$$Q_{2\pi n/M} = Q_{-2\pi n/M}; \quad n = 1, 2, \cdots, \frac{M-2}{2}. \tag{16}$$

Furthermore, as demonstrated in [1],

$$Q_0 = 0. \tag{17}$$

Thus, using (16) and (17), $P_E(M)$ of (15) simplifies to

$$P_E(M, X) = \frac{P_H}{M}\left[Q_\pi + 2\sum_{n=1}^{\frac{M-2}{2}} Q_{2\pi n/M}\right]. \tag{18}$$

Finally, recalling from [1] the relation between average symbol and bit error probabilities, namely[4]

$$P_B(M, X) = \left[\frac{M}{2(M-1)}\right] P_E(M, X) \tag{19}$$

then the average bit error probability for $M$-ary DPSK in the presence of multitone jamming is given by

$$P_B(M, X) = \frac{P_H}{2(M-1)}\left[Q_\pi + 2\sum_{n=1}^{\frac{M-2}{2}} Q_{2\pi n/M}\right]$$

$$= \frac{1}{2(M-1)(\log_2 M)X\rho^2}$$

$$\times \left[Q_\pi + 2\sum_{n=1}^{\frac{M-2}{2}} Q_{2\pi n/M}\right]. \tag{20}$$

Clearly, then, evaluation of (20) rests on deriving expressions for $Q_{2\pi n/M}$; $n = 1, 2, \cdots, M/2$ as functions of $M$ and the jamming-to-signal power ratio $\rho^2$ of (3). Observe that, in terms of the vector definitions of the signal and tone jamming interference as in (8) and (9), the signal power and jammer power

---

[3] Note that, since we have the absence of additive background noise, the probability of error in hop intervals which are not hit by the jammer is zero.

[4] Actually, the relation in (19) holds as an equality only for orthogonal signal sets [2]. However, for low signal-to-jammer noise ratios, the right-hand side of (19) becomes a tight upper bound for the average bit error probability performance of $M$-ary FH-DPSK. For binary DPSK ($M = 2$), the equality in (19) is exact.

per tone are given by

$$S = \frac{A^2}{2} \; ; \quad \frac{J}{q} = \frac{I^2}{2}. \tag{21}$$

Thus, equivalently from (3), we have that

$$\rho^2 = \frac{I^2}{A^2}. \tag{22}$$

### A. Evaluation of $Q_{2\pi n/M}$

In view of (22), $Q_{2\pi n/M}$ of (14) may be restated in the normalized form

$$Q_{2\pi n/M} = \text{Prob} \{ | \arg (\bar{Z}^{(i)} - \bar{Z}^{(i-1)}) - 2n\theta | > \theta \} \tag{23}$$

where

$$\theta \triangleq \frac{\pi}{M} \tag{24}$$

and

$$\bar{Z}^{(i-1)} = e^{-jn\theta} + \rho e^{j\phi} \triangleq R_1 e^{j(-n\theta + \psi_1)}$$

$$\bar{Z}^{(i)} = e^{jn\theta} + \rho e^{j\phi} \triangleq R_2 e^{j(n\theta + \psi_2)}. \tag{25}$$

Note that, in obtaining (25) from (10), we have substituted for $\theta^{(i)}$ its assumed true value, namely, $\theta_n = 2\pi n/M = 2n\theta$ and, since $\phi$ is uniformly distributed, we have arbitrarily established the symmetry $\theta_T^{(i-1)} = -\pi n/M = -n\theta$. Fig. 1 is a graphic representation of (24) where we have further introduced the notation

$$\psi \triangleq \arg (\bar{Z}^{(i)} - \bar{Z}^{(i-1)}) \tag{26}$$

Thus, using (25) and (26)

$$\arg (\bar{Z}^{(i)} - \bar{Z}^{(i-1)}) - 2n\theta = \psi - 2n\theta$$

$$= n\theta + \psi_2 - (-n\theta + \psi_1) - 2n\theta$$

$$= \psi_2 - \psi_1 \tag{27}$$

and hence

$$Q_{2\pi n/M} = \text{Prob} \{ | \psi_2 - \psi_1 | > \theta \}$$

$$= 1 - \text{Prob} \{ | \psi_2 - \psi_1 | \leq \theta \}. \tag{28}$$

Consider the product (asterisk denotes complex conjugate)

$$(\bar{Z}^{(i-1)})* \bar{Z}^{(i)} e^{-j2n\theta}$$

$$= R_1 R_2 e^{j(n\theta - \psi_1)} e^{j(n\theta + \psi_2)} e^{-j2n\theta}$$

$$= R_1 R_2 e^{j(\psi_2 - \psi_1)}. \tag{29}$$

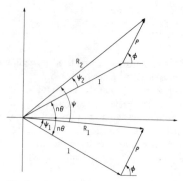

Fig. 1. A graphical representation of (24).

The above product can also be written in the form

$$(\bar{Z}^{(i-1)})* \bar{Z}^{(i)} e^{-j2n\theta}$$

$$= (e^{jn\theta} + \rho e^{-j\phi})(e^{jn\theta} + \rho e^{j\phi}) e^{-j2n\theta}$$

$$= [e^{j2n\theta} + \rho^2 + \rho e^{jn\theta} (e^{j\phi} + e^{-j\phi})] e^{-j2n\theta}$$

$$= 1 + \rho^2 e^{-j2n\theta} + 2\rho \cos \phi e^{-jn\theta}. \tag{30}$$

Thus, using (29) and (30) in (28) results in the equivalent relation

$$Q_{2\pi n/M} = 1 - \text{Prob} \{ -\theta \leq \arg [(\bar{Z}^{(i-1)})*$$

$$\times \bar{Z}^{(i)} e^{-j2n\theta}] \leq \theta \}$$

$$= 1 - \text{Prob} \{ -\theta \leq \arg [1 + \rho^2 e^{-j2n\theta}$$

$$+ 2\rho \cos \phi \, e^{-jn\theta}] \leq \theta \}. \tag{31}$$

Equation (31) can be given a geometric interpretation as in Fig. 2. Here the vector $\overline{OQ}$ (a line drawn from point $O$ to either point $Q$) represents the complex number whose argument is required in (31), i.e.,

$$\overline{OQ} = 1 + \rho^2 e^{-j2n\theta} + 2\rho \cos \, \phi \, e^{jn\theta}. \tag{32}$$

Thus, in terms of the geometry in Fig. 2, (31) may be written in the alternate form

$$Q_{2\pi n/M} = 1 - \text{Prob} \{ \text{point } Q \text{ is within the } 2\theta \text{ wedge} \}$$

$$= 1 - \text{Prob} \{ \text{point } Q \text{ lies along the line } AB \}. \tag{33}$$

Considering separately the cases where point $P$ falls outside and inside the $2\theta$ wedge, which, expressed mathematically, corresponds to the inequalities (see Fig. 3)

$$\rho^2 \gtrless \frac{\sin \theta}{\sin [(2n-1)\theta]} \tag{34}$$

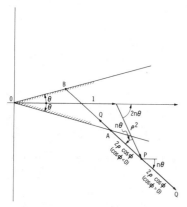

Fig. 2. A geometric interpretation of (31).

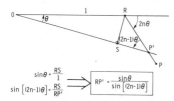

$$\sin\theta = \frac{RS}{1}$$

$$\sin\left[(2n-1)\theta\right] = \frac{RS}{RP'} \longrightarrow \boxed{RP' = \frac{\sin\theta}{\sin\left[(2n-1)\theta\right]}}$$

Fig. 3. The geometry needed to establish (34).

then after much routine trigonometry, it can be shown that

$$
Q_{2\pi n/M} = Q_{2n\theta}
$$

$$
= \begin{cases}
\frac{1}{\pi}\cos^{-1}\left[\dfrac{\rho^2\sin\left[(2n+1)\theta\right] + \sin\theta}{2\rho\sin\left[(n+1)\theta\right]}\right] \\[2mm]
\quad\times u(\rho - \rho_n) \\[2mm]
+\frac{1}{\pi}\cos^{-1}\left[\dfrac{\sin\theta - \rho^2\sin\left[(2n-1)\theta\right]}{2\rho\sin\left[(n-1)\theta\right]}\right] \\[2mm]
\quad\times u(\rho - \rho_{n-1}); \qquad 0 < \rho < 1 \\[2mm]
1; \qquad\qquad\qquad\qquad \rho \geqslant 1, \\[2mm]
\qquad\qquad n = 2, 3, \cdots, \dfrac{M}{2} - 1
\end{cases}
$$

(35)

where $u(\rho)$ is the unit step function and

$$
\rho_{n-1} \triangleq \frac{-\sin\left[(n-1)\theta\right] + \sin n\theta}{\sin\left[(2n-1)\theta\right]}.
$$

(36)

Note that

$$
\lim_{\rho\to 1} Q_{2n\theta} = \frac{1}{\pi}\left\{\cos^{-1}\left[\frac{\sin\left[(2n+1)\theta\right] + \sin\theta}{2\sin\left[(n+1)\theta\right]}\right]\right.
$$

$$
\left. + \cos^{-1}\left[\frac{\sin\theta - \sin\left[(2n-1)\theta\right]}{2\sin\left[(n-1)\theta\right]}\right]\right\}
$$

$$
= 1.
$$

(37)

For $n = 1$, the appropriate result analogous to (35) is

$$
Q_{2\theta} = \begin{cases}
0; & 0 < \rho < \rho_1 \\[2mm]
\dfrac{1}{\pi}\cos^{-1}\left[\dfrac{\rho^2\sin 3\theta + \sin\theta}{2\rho\sin 2\theta}\right]; & \rho_1 \leqslant \rho < 1 \\[2mm]
1; & \rho \geqslant 1
\end{cases}
$$

$$
\rho_1 = \frac{\sin 2\theta - \sin\theta}{\sin 3\theta} = \frac{\sin\dfrac{2\pi}{M} - \sin\dfrac{\pi}{M}}{\sin\dfrac{3\pi}{M}}.
$$

(38)

Here

$$
\lim_{\rho\to 1} Q_{2\theta} = \frac{1}{\pi}\cos^{-1}\left[\frac{\sin 3\theta + \sin\theta}{2\sin 2\theta}\right] < 1.
$$

(39)

Finally, for $n = M/2$, we have the result

$$
Q_{M\theta} = Q_{\pi}
$$

$$
= \begin{cases}
0; & 0 < \rho < \rho_{(M/2)-1} \\[2mm]
\dfrac{2}{\pi}\cos^{-1}\left[\dfrac{\left(\sin\dfrac{\pi}{M}\right)(1-\rho^2)}{2\rho\cos\dfrac{\pi}{M}}\right]; & \rho_{(M/2)-1} \leqslant \rho < 1 \\[2mm]
1; & \rho \geqslant 1
\end{cases}
$$

$$
\rho_{(M/2)-1} = \frac{1 - \cos\dfrac{\pi}{M}}{\sin\dfrac{\pi}{M}}.
$$

(40)

Also,

$$
\lim_{\rho\to 1} Q_{\pi} = 1.
$$

(41)

As an example, Fig. 4 is a plot of $Q_{2n\theta}$; $n = 1, 2, 3, 4, 8$ versus $\rho$ for $M = 16$. These probabilities are computed from (35), (38), and (40). Using these results in (20), Fig. 5 illustrates the product $XP_B(16, X)$ versus $\rho$. This curve has a maximum value of 1.457 at $\rho = 0.1614$, which, from (4) and (5), corresponds to the optimal (worst case) jamming strategy.

$$
P_H = \begin{cases}
\dfrac{9.597}{E_c/N_J}; & E_c/N_J \geqslant 9.597 \\[2mm]
1; & E_c/N_J < 9.597.
\end{cases}
$$

(42)

Thus, the average bit error probability performance of 16-ary FH-DPSK in the presence of the worst case tone jammer is

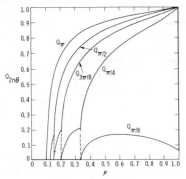

Fig. 4.   Individual signal point error probability components as a function of square root of jamming (per tone) to signal power ratio.

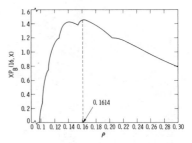

Fig. 5.   Bit error probability performance of 16-ary DPSK as a function of square root of jamming (per tone) to signal power ratio.

given by

$$P_{B_{max}}$$

$$= \begin{cases} \dfrac{1.457}{E_c/N_J}; & E_c/N_J \geq 9.597 \\[2ex] \dfrac{1}{30}\left[Q_\pi + 2\displaystyle\sum_{n=1}^{7} Q_{n\pi/8}\right]\Bigg|_{\rho=1/(2\sqrt{E_c/N_J})}; \\[1ex] & 0.25 < E_c/N_J \leq 9.597 \\[2ex] 0.5; & E_c/N_J < 0.25. \end{cases}$$

$$(43)$$

Similar results can be obtained[5] for *M*-ary DPSK with $M = 2, 4, 8$. The asymptotic behavior or these results (i.e., $P_H$ and $P_B$ inversely related to $E_c/N_J$) is given below.

Using the results in Table I and the fact that for any *M*

$$P_{B_{max}} = P_B(M, E_c/N_J)|_{\rho=1/\sqrt{(\log_2 M)E_c/N_J}};$$

$$E_c/N_J \leq K_\alpha \qquad (44)$$

with $P_B(M, E_c/N_J)$ given by (20). Fig. 6 is an illustration of

[5] It should be noted here that the results in [1] for the performance of FH-DPSK ($M = 4$) in the presence of the worst case tone jammer are partially incorrect. In particular, Houston finds $\rho = 0.52$ as the maximizing value. However, since the fraction of the band jammed, which is given by $P_H = 1/2X\rho^2$, cannot exceed 1, the value of $\rho = 0.52$ can only be achieved if $X > 1/2(0.52)^2 = 1.85$. For smaller values of $X$, $\rho = 1/\sqrt{2X}$ must be used. Thus, we arrive at the following corrected results for $M = 4$:

$$P_{B_{max}} = \begin{cases} \dfrac{0.2592}{E_c/N_J}; & E_c/N_J > 1.85 \\[2ex] \dfrac{1}{3\pi}\left[\cos^{-1}\left(\dfrac{2E_c/N_J - 1}{2\sqrt{2E_c/N_J}}\right)\right. \\[2ex] \left. + \cos^{-1}\left(\dfrac{2E_c/N_J + 1}{4\sqrt{E_c/N_J}}\right)\right]; & 0.5 < E_c/N_J \leq 1.85 \\[2ex] 0.5; & E_c/N_J \leq 0.5. \end{cases}$$

TABLE I
ASYMPTOTIC PERFORMANCE OF *M*-ARY FH-DPSK FOR
WORST CASE PARTIAL-BAND MULTITONE JAMMING

| WORST CASE PARTIAL–BAND FRACTION | | | $P_H = \dfrac{K_\alpha}{E_c/N_J}$ |
|---|---|---|---|
| MAXIMUM AVERAGE BIT ERROR PROBABILITY | | | $P_{B_{max}} = \dfrac{K_p}{E_c/N_J}$ |
| $M$ | $\rho$ | $K_\alpha$ | $K_p$ |
| 2 | 1 | 1 | 0.50 |
| 4 | 0.5220 | 1.835 | 0.2593 |
| 8 | 0.2760 | 4.376 | 0.5280 |
| 16 | 0.1614 | 9.597 | 1.457 |

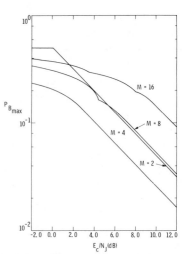

Fig. 6.   Worst case bit error probability performance of *M*-ary FH-DPSK for partial-band multitone jamming.

the average bit error probability performance of *M*-ary FH-DPSK for worst case PB multitone jamming.

Before concluding this paper, we wish to alert the reader to a point of pathological behavior that is directly attributable to the assumption of an on-tune tone jammer and is perhaps not obvious from the analytical or graphical results given. In particular, we observe from Fig. 4 that $Q_{\pi/8}$ has a jump discontinuity at $\rho = 1$ and, thus, $P_{B_{max}}$ of (43) will have a similar jump discontinuity at $E_c/N_J = 0.25$ (−6 dB). In fact, since from (39), $Q_{\pi/8} = 0.0625$ as $\rho$ approaches 1 from below, then at $E_c/N_J = -6$ dB, $P_{B_{max}}$ jumps from 0.4375 to 0.5. For other values of $M \geq 4$, a similar jump discontinuity in the worst case tone jammer bit error probability will occur at $\rho = 1$ and $P_H = 1$, or equivalently, from (4) and (5), $E_c/N_J =$

$1/\log_2 M$. Since the range of $E_c/N_J$ in Fig. 6 extends only down to $-2$ dB, these jump discontinuities are not visible on this plot, i.e., the largest value of $E_c/N_J$ at which a discontinuity occurs would correspond to $M = 4$ ($E_c/N_J = 1/2 = -3$ dB).

## IV. CONCLUSIONS

Using a geometric approach, a general relation for the average error probability performance of $M$-ary FH-DPSK in the presence of partial-band multitone jamming has been found, thus enabling a direct comparison for different values of $M$. In particular, when the asymptotic (large bit energy-to-jamming noise spectral density) performances corresponding to the worst case partial band multitone jammer were compared, it was found that $M = 4$ results in the optimum system design. Specifically, FH-DQPSK ($M = 4$) is 2.85, 3.09, and 7.5 dB more antijam resistant than FH-DPSK ($M = 2$), 8-ary FH-DPSK, and 16-ary FH-DPSK, respectively.

## REFERENCES

[1] S. W. Houston, "Modulation techniques for communication, Part I: Tone and noise jamming performance of spread spectrum $M$-ary FSK and 2, 4-ary DPSK waveforms," in *NAECON '75 Rec.*, pp. 51–58.
[2] W. C. Lindsey and M. K. Simon, *Telecommunication Systems Engineering*. Englewood Cliffs, NJ: Prentice-Hall, 1973.

# Tradeoff Between Processing Gain and Interference Immunity in Co-Site Multichannel Spread-Spectrum Communications

IRVING R. SMITH, SENIOR MEMBER, IEEE

*Abstract*—**Multiple single channel spread-spectrum radio communications simultaneously transmitted and received at a small station, aircraft, or ship are subject to mutual interference. Wideband PN and/or FH radios which occupy a large fraction of the available frequency band are more difficult to protect from adjacent channel interference than conventional narrow-band radios. This paper describes the interference susceptibility of various spread-spectrum radios, identifies potential solutions, and determines the tradeoff of processing gain that must be made to achieve immunity from co-site interference.**

## I. INTRODUCTION

IT is desired to determine the amount of spread-spectrum processing gain that must be traded to achieve acceptable adjacent channel interference performance on a platform with multiple independent radio channels. Although antenna isolation, external frequency filtering, and receiver selectivity provide a measure of rejection, they are often collectively unable to reduce transmitter-to-receiver interference below the receiver sensitivity threshold. Emphasis will be placed on the tradeoff between processing gain and interference rejection for given levels of antenna isolation and filter selectivity. Concurrent noise jamming is assumed in all cases.

Reference [1] determined the optimization of processing gain for various direct sequence spread-spectrum modulations. Coherent matched filter reception was employed, and perfect carrier, bit, and chip synchronization were assumed. Similarly, [2] found the optimum processing gain of a noncoherent frequency-hopped (FH) binary FSK system. In both [1] and [2], the systems are stressed by the presence of intentional interference or jamming, and expressions for the probability of error are presented. This paper builds on these results to include the effects of multiple adjacent radio channels.

In the next section, we analytically describe the factors that determine the composite adjacent channel interference spectrum present at the demodulator input. Probability of error expressions presented in [1] and [2] for various spread-spectrum communications systems are modified and numerical results are derived. Methods of interference control are discussed in Section III and a comparison of relative effectiveness is made. Section IV summarizes the results derived and presents the conclusions drawn.

Manuscript received July 1, 1981; revised February 1, 1982.
The author is with the MITRE Corporation, Bedford, MA 01730.

## II. ANALYSIS

For coherent reception of direct sequence PN spread-spectrum modulations, [1, eq. (4) and (5)] give expressions for the in-phase and quadrature demodulator outputs. To include the contribution of adjacent channel interference, these equations can be modified by the addition of the following expressions in [1, eq. (4) and (5)], respectively. Thus, for MSK-type waveforms for which demodulation is performed by integrating over staggered $KT$ time intervals, the in-phase and quadrature interference caused components are

$$I_1 = 2 \int_{-T}^{(K-1)T} \text{PN}\,(2k-1)i(t)\cos w_0 t \, dt,$$

$$k = 1, 2, \cdots, K/2 \qquad (1)$$

$$I_2 = 2 \int_0^{KT} \text{PN}\,(2k)i(t)\sin w_0 t \, dt \qquad (2)$$

where $\text{PN}(k)$ is a binary pseudorandom sequence taking on the values plus and minus one corresponding to the binary states "zero" and "one," respectively. State transitions occur at the rate of one per $T$ seconds. However, as shown in (1) and (2), the odd and even subsequences of $\text{PN}(k)$ are used in the respective in-phase and quadrature channels with transitions at $2T$ second intervals. The interference input signal $i(t)$ is the superposition of up to $N-1$ distinct PN spread-spectrum signals offset in carrier frequency by multiples of the interchannel separation frequency $f_s$. Thus, for $N$ statistically independent sequences $\text{PN}_n(k)$, $n = 1, \cdots, N$, the signal $i(t)$, including data symbols $d_n = \pm 1$, is given by

$$i(t) = \sum C_n d_n$$

$$\cdot \begin{cases} \text{PN}_n\,(2k-1)C(t-2kT)\cos\left[(w_0+nw_s)t+\phi_n\right] \\ \quad -\,\text{PN}_n\,(2k-2)S(t-(2k-2)T) \\ \quad \cdot \sin\left[(w_0+nw_s)t+\phi_n\right], \\ \qquad \text{for } (2k-1)T \leqslant t < 2kT \\ \\ \text{PN}_n(2k-1)C(t-2kT)\cos\left[(w_0+nw_s)t+\phi_n\right] \\ \quad -\,\text{PN}_n(2k)S(t-2kT) \\ \quad \cdot \sin\left[(w_0+nw_s)t+\phi_n\right], \\ \qquad \text{for } 2kT \leqslant t < (2k+1)T \end{cases} \qquad (3)$$

Reprinted from *IEEE Trans. Commun.*, vol. COM-30, pp. 959–966, May 1982.

where $C(t)$ and $S(t)$ are the in-phase and quadrature symbol weighting functions, respectively, each with duration $2T$ and skewed by $T$ seconds with respect to one another. The $C_n$ are determined by the isolation between adjacent radio antennas and the filter attenuation corresponding to the frequency offset between the transmitting radios and the victim receiver. External transmit and receive filters can be used to control co-site interference. A bandpass filter at the output of each spread-spectrum radio transmitter will significantly attenuate the interference induced in the passbands of receivers in adjacent frequency channels. Because it must pass all signals within its passband, a bandpass receive filter will have no effect on co-channel interference. It can, however, protect the receiver from weak signal suppression effects caused by the high power transmitter signal components not attenuated by the bandpass transmit filters.

In order to characterize the demodulator outputs, it is necessary to establish the statistical properties of each input signal component. As in [1], the input jamming signal $j(t)$ will be taken to be a zero-mean stationary Gaussian process which is independent of the other input signal processes. Since $PN(t)$ spreads the product $j(t)\, PN(t)$ over a band much wider than an individual PN channel, then $j(t)$ can be modeled as essentially a white noise process. From this, it follows that the demodulator output components $J_1$ and $J_2$ are zero-mean Gaussian, each with variance equal to the jammer power $J$ within the receiver's passband. Assuming noise jamming with a constant power spectral density $\eta_{0J}$ over the bandwidth $f_B$ of each of the channels jammed, the jammer power per channel is $J = \eta_{0J} f_B$.

As was the case with $j(t)$, the input interference signal $i(t)$ will also be assumed independent of the signal, noise, and jamming input processes. The statistical properties of $I_1$ and $I_2$ depend on the nature of the products $PN(t)\, PN_n(t)$. Unless the PN sequences are synchronous with simultaneous transitions (unlikely for most mobile radio situations), the product of binary NRZ sequences (because it is equivalently a modulo two addition) generates another binary sequence with twice the transition rate. Thus, $PN(t)$ spreads $i(t)$ over a band much wider than a single channel width. Hence, $i(t)$ can be modeled as essentially a white noise process. If the PN sequences are short segments of a long $m$ sequence with balanced statistics (i.e., on the average an equal number of zeros and ones occurring during the symbol interval $KT$), then $I_1$ and $I_2$ are zero-mean Gaussian, each with variance equal to the interference power $I$ within the receiver's passband.

Under worst case conditions where there are $N_1$ channels both above and below the victim receiver channel, all transmitting simultaneously, the interference power incident within the receiver bandwidth $f_B$ is

$$I = \frac{1}{2\pi} \sum_{\substack{n=-N_1 \\ n \neq 0}}^{N_1} C_n \int_{-w_B/2}^{w_B/2} S(w - nw_s)\, dw \tag{4}$$

where $S(w)$ is the power spectral density of the input signal modulation used, e.g., offset QPSK, minimum shift key (MSK), or sinusoidal frequency shift key (SFSK). For large frequency

offsets from the carrier, the transmitter noise floor exceeds $S(w)$.

Assuming that the interference power $I$ is uniformly distributed over the receiver bandwidth $f_B$ with one-sided power spectral density $\eta_{0I} = I/f_B$, then the probability of error analogous to [1, eq. (15)] becomes, for MSK,

$$P_{e_{MSK}} = \text{erfc}\,(\sqrt{R_d/2}) - \tfrac{1}{4}\,\text{erfc}^2\,(\sqrt{R_d/2}) \tag{5}$$

where $\text{erfc}(x)$ is the complementary error function given by

$$\text{erfc}\,(x) = \frac{2}{\sqrt{\pi}} \int_x^\infty e^{-y^2}\, dy \tag{6}$$

and $R_d$ is the effective signal-to-noise ratio expressed as the ratio of the symbol energy $E$ to the sum of noise, jamming, and interference power spectral densities:

$$R_d = \frac{2E}{\eta_0 + \eta_{0I} + \eta_{0J}}$$
$$= \frac{2}{\dfrac{1}{E/\eta_0} + \dfrac{J/S + I/S}{PG}}. \tag{7}$$

In (7), the processing gain $PG$ is taken as $f_B/R$ where $R$ is the information rate $R = 1/KT$. To maintain the error probability (5) (and hence the signal-to-noise ratio) unchanged as it was in the absence of interference, it is necessary to trade off processing gain in accordance with the relation

$$PG_{\text{effective}} = PG/(1 + I/J). \tag{8}$$

Table I illustrates how much processing gain degradation occurs [i.e., $10 \log_{10}\,(PG_{\text{eff}}/PG)$] for various interference-to-jamming power ratios.

To determine the influence of demodulator input $I/S$ ratio on the desired $J/S$ performance and the processing gain degradation, we must evaluate (4) numerically for particular modulations. For MSK, the power spectral density $S(w)$ takes the form

$$S(w) = \frac{8T}{\pi^2} \left\{ \frac{1 + \cos 2wT}{\left[ 1 - \left( \dfrac{2wT}{\pi} \right)^2 \right]^2} \right\}. \tag{9}$$

The power spectral densities of OQPSK and SFSK are given in [3] and [4]. Assuming that external Butterworth filters of order $k$ are used to reject co-site interference, then the coefficients $C_n$ can be described by the composite expression

$$(C_n)_{\text{dB}} = -10 \log_{10}\,[1 + (2nf_s/f_B)^{2k}] + \text{ant}_{\text{dB}} \tag{10}$$

in which a maximum filter stopband attenuation of 80 dB is assumed and the antenna isolation $\text{ant}_{\text{dB}}$ is taken to be $-20$ dB between any pair of antennas independent of frequency. Incorporating (9) in (4) using (10), it is possible to determine

TABLE I
PROCESSING GAIN DEGRADATION IN DECIBELS

| PROCESSING GAIN DEGRAD. IN DB | -.97 | -2.12 | -3 | -4.1 | -7 | -10.4 | -20.04 |
|---|---|---|---|---|---|---|---|
| (I/J) IN DB | -6 | -2 | 0 | +2 | +6 | +10 | +20 |

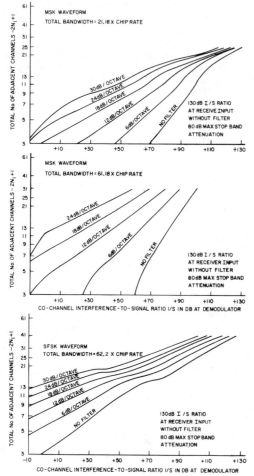

Fig. 1. Adjacent channels allowable with transmit filtering for given I/S ratio.

the number of allowable co-site channels $2N_1 + 1$ as a function of filter attenuation slope $6k$ dB per octave, $I/S$ ratio, and the interchannel separation-to-bandwidth ratio $f_S/f_B$. Fig. 1 illustrates the relative effects of total bandwidth, PN modulation selection, and filter order on the packing density of co-site channels for a given $I/S$ ratio. In each plot, the total bandwidth is fixed. The interchannel separation varies inversely with the number of adjacent channels. From Fig. 1(a), we see that for a bandwidth $f_B$ corresponding to 99 percent of the MSK spectrum power and a design $J/S$ of 22 dB, the maximum number of channels tolerable without more than 1 dB of processing gain degradation is 4, 5, and 7 channels for filter slopes of 18, 24, and 30 dB/octave, respectively. The total bandwidth occupied by all $2N_1 + 1$ channels is 21.2 times the chip rate.

Comparing Fig. 1(b) and (c), it is observed that the superior far sidelobe performance of SFSK allows more co-site channels than MSK for widely spaced channels. However, for densely packed channels, i.e., $f_S T \leq 1.5$, MSK is able to handle more channels than SFSK because of its lower close-in sidelobes.

In contrast to PN systems, the susceptibility of multiple FH radios to interference has been derived in the Appendix. For $N_t$ independent frequency-hopping co-site radios which hop over $N_0$ channels, the probability of at least one of $N_t$ transmitters falling $n$ channels away from a victim receiver channel is given by (A7) in which $\Delta f/\Delta w$ is substituted for $n/N_0$. The effect of channel duty cycle on the probability of error is represented by the product of average call rate $\lambda$ and the mean message duration $\Delta t$.

The interference power is a random variable, a function of $N_t$ and the channel separations $n_1, n_2, \cdots, n_{N_t}$ of the offending transmitters from the victim receiver. For binary FSK/FH, each of the interferers can be transmitting either a MARK or a SPACE during the time the receiver is receiving a MARK. Thus the interference power collected in the MARK filter of a noncoherent FSK demodulator is

$$I(n_1, \cdots, n_{N_t})$$

$$= \sum_{n=n_1}^{n=n_{N_t}} \frac{C_n}{2\pi} \int_{-w_c}^{w_c} S(w - nw_s + e_n w_c)\, dw \qquad (11)$$

where $e_n$ is an RV taking on the values 0 or 1, depending on whether a MARK or SPACE, respectively, was sent by the $n$th transmitter. External filtering and receiver selectivity are assumed to suppress signals within the receiver's linear dynamic range that fall $m$ or more channels away. Co-site interference falling closer than $m$ channels will cause errors with frequency $\frac{1}{2}$. Co-channel interference occurs randomly at a rate (12) that may be tolerable on robust digital voice channels:

$$R_1 = 1 - \sum_{n=m}^{N_0} p_1(n) \qquad (12)$$

where $p_1(n)$ is the probability of an interferer falling $n$ channels away from the victim receiver. For identically distributed independent frequency hops uniformly occurring over the frequency band $p_1$ takes the form

$$p_1(n) = \frac{2}{N_0}\left(1 - \frac{n}{N_0}\right) - \frac{\delta_{n0}}{N_0}, \qquad \text{for } n = 0, 1, \cdots, N_0 \qquad (13)$$

where $\delta_{n0}$ is the Kronecker delta function. From (11), (13), and (A2), the mean interference power due to signals which fall outside the inner $m$ channels can be shown to be

$$I = N X \sum_{n=m}^{N_0} I(n)P_1(n) \tag{14}$$

in which $I(n)$ is the $n$th component of the sum (11) and $N$ is the total number of potential interferers. $X$ is the channel duty cycle. Assuming that the interference power $I$ is uniformly distributed over the receiver's MARK and SPACE filters, then by substituting $\eta_0 + I/2f_c$ for $\eta_0$ in [2, eq. (A4), (A5), (A6), (A10), and (A11)], we are able to determine the combined effect of jamming and adjacent channel interference. The probability of error for a noncoherent FH-binary orthogonal FSK receiver when $K$ of $N_0$ channels are noise jammed is shown by [2] to be

$$P_e = \frac{(N_0 - K)(N_0 - K - 1)}{N_0(N_0 - 1)} P(a_0, b_0)$$

$$+ \frac{2K(N_0 - K)}{N_0(N_0 - 1)} P(a_1, b_1)$$

$$+ \frac{K(K - 1)}{N_0(N_0 - 1)} P(a_2, b_2) \tag{15}$$

where the probabilities $P(a_i, b_i)$, $i = 0, 1, 2$ of jamming neither MARK nor SPACE $i = 0$, either MARK or SPACE filters $i = 1$, and jamming both filters $i = 2$, respectively, are expressed in terms of the Marcum $Q$ function as

$$P(a, b) = \tfrac{1}{2} [1 - Q(\sqrt{b}, \sqrt{a}) + Q(\sqrt{a}, \sqrt{b})]. \tag{16}$$

In the case of nonorthogonal FSK, (15) and (16) become more complex; however, $P_e$ is not a very sensitive function of the degree of orthogonality. With ideal rectangular MARK and SPACE bandpass filters of bandwidth $2f_c$ which are nonoverlapping, and binary detection of MSK frequency shift at the midpoint of each symbol interval, the effective signal-to-noise ratios $a_i$ and $b_i$ take the forms

$$a_i = A \left( \frac{2}{E/\eta_0} + \frac{I}{S} + i\frac{J/S}{2K} \right)^{-1} \tag{17}$$

$$b_i = a_i \frac{B}{A}, \qquad \text{for } i = 0, 1, 2$$

where $A$ and $B$ are represented by the sine integral functions as follows:

$$A = [Si(\pi) - Si(3\pi)]^2/\pi^2 \quad \text{and} \quad B = 2[Si(\pi)/\pi]^2. \tag{18}$$

From the denominator of $a_i$, for full band jamming ($K = N_0$) it is apparent that the effect of interference causes a reduction of the FH processing gain $N_0$ similar to the degradation (8) for PN signals. Thus, the effective processing gain is

$$PG_{\text{effective}} = N_0 \left( 1 + \frac{N_0 I}{J} \right)^{-1} \tag{19}$$

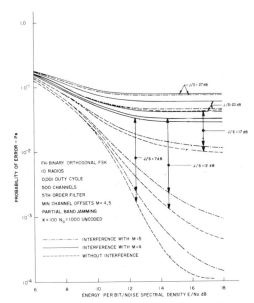

Fig. 2. Error probability for partial band noise jamming.

In contrast to PN radios for the same processing gain degradation, the $I/J$ ratio tolerable is lower by the factor $N_0$. This comparison is based on only one PN channel being jammed, recognizing that it takes a large fraction of the band to jam a single FH channel. If, on the other hand, full band noise jamming is assumed in each case, then $I/J$ for FH is lower by only $N_0/(2N_1 + 1)$.

Receiver susceptibility to interference depends on the frequency offset. The minimum channel offset $m$ is used to characterize receiver selectivity. Signals falling at least $m$ channels away are assumed to cause symbol (bit) errors with probability $P_e$ given in (15). Because of the intense interference caused by co-site transmitters with channel offsets less than $m$, where receiver selectivity is ineffective, the bit error rate for close-in interference is assumed to be $\tfrac{1}{2}$. For $N$ radios, each with duty cycle $X$, the probability $r$ that at least one of $N$ FH signals falls within the offset $m$ is from (12) given by the expression $1 - (1 - R_1 X)^N$. When $X$ and $R_1$, the probability of offsets less than $m$, are both less than 10 percent and $N < 21$, then $r$ is approximately $NXR_1$. Combining the conditional bit error probabilities for "close-in" (offsets less than $m$) and "adjacent" channel (offsets equal or greater than $m$) interference with the corresponding probabilities of channel offset occurrence, it follows that the composite bit error rate is $P_e(1 - r) + r/2$.

The composite bit error rate $P_E$ is plotted in Figs. 2 and 3 for partial and full band noise jamming, respectively, as a function of $J/S$, the jamming-to-signal power ratio, and $E/\eta_0$, the ratio of energy per bit to noise spectral density. As indicated in (17), $P_E$ also depends on the $I/S$ ratio. The average interference power in the numerator of this ratio represents the fraction of the total interference power incident on the victim receiver caused by FH signals with frequency offsets which fall at or outside the minimum channel offset $m$. The complementary "close-in" interference-to-signal ratio, because it can-

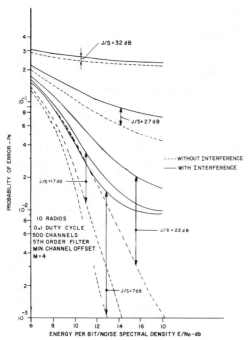

Fig. 3.    Error probability for full band noise jamming.

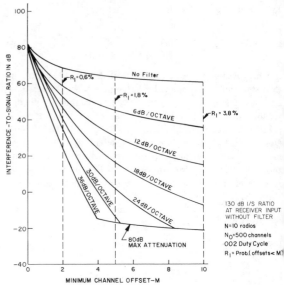

Fig. 4.    Effect of filter selectivity on interference caused by signals offset by at least $M$ channels.

not be significantly attenuated by external bandpass filters, is much larger than the corresponding "adjacent" channel interference-to-signal ratio. This is especially true in the case of multiple co-site radios where the transmitter output level less the attenuation due to antenna isolation frequently exceeds the receiver sensitivity threshold by as much as 130 dB.

Fig. 4 depicts, for MSK signal modulation and Butterworth transmit filtering, the tradeoff between filter slope and the minimum channel offset $m$ necessary to reduce the $I/S$ ratio due to adjacent channel interference to a tolerable level. The 3 dB filter bandwidth is just sufficient to cover both the MARK and SPACE filters corresponding to an instantaneous frequency hop. Electronically tuned filters which hop in synchronism with the radio are assumed. As filter slope increases, the $I/S$ ratio is reduced until it reaches the 80 dB maximum attenuation limit assumed for the filter. Although Fig. 4 is drawn for specific parameter values, (14) indicates that the $I/S$ results can be scaled in direct proportion to decibel changes in the number of radios $N$, the duty cycle $X$, and the receiver input $I/S$ ratio in the absence of external filters.

For each $J/S$ ratio in Fig. 2, there are three curves corresponding to the cases of no interference and interference with minimum offsets of four and five channels, respectively. Note that increasing the offset from 4 to 5 has a marked effect on $P_E$. The $I/S$ ratios corresponding to these offsets can be obtained from Fig. 4. From the $m = 4$ and 5 intercepts on the 30 dB/octave filter slope curve, and adjusting for the parameter differences between Figs. 2 and 4, the $I/S$ ratios are computed to be $-3$ and $-16$ dB, respectively. The former ratio produces significant bit errors, while the latter has a negligible effect. The $m = 4$ curves have at least an order of magnitude lower $P_E$

than the $m = 5$ curves for $E/\eta_0 \geqslant 14$ dB and $J/S \leqslant 17$ dB. For full band noise jamming, Fig. 3 exhibits a similar $P_E$ spread between $J/S$ curves with and without interference for $J/S \leqslant 22$ dB. Since $R_1$ in (12) is approximately proportional to the ratio of the minimum channel offset $m$ to the total number of MARK and SPACE filter channel pairs $N_0$, then the above $P_E$ results can be expressed more universally in terms of the normalized ratio $m/N_0$ rather than simply $m$.

## III. METHODS OF INTERFERENCE CONTROL

Some form of interference reduction is required to achieve multiple co-site spread-spectrum radio operation without degradation. External transmit filtering for PN-MSK signals has been shown in Fig. 1 to be an effective means of reducing adjacent channel interference. Electronically tuned filters can be used with FH radios; however, they may not be fast enough to follow certain high hop rate signals. Adaptive interference cancellation (AIC) has been proposed as an alternative to agile filters. AIC can be used with either FH or PN signals, although not with the same effectiveness. Code division multiplex (CDM) techniques which employ PN codes with low cross-correlation sidelobes trade processing gain for suppression more gracefully than (8). If the rms cross-correlation sidelobes are down by a factor of $M$, then the processing gain degradation is given as

$$PG_{\text{effective}} = PG/(1 + I/MJ). \qquad (20)$$

However, CDM alone is inadequate to protect against co-site interference. A combination of external transmit and receive filtering, CDM, or AIC is required for rejection of adjacent channel interference.

For stationary radios, time division multiplex (TDM) is the most effective single means of interference control. If all potentially interfering signals are assigned their own distinct non-

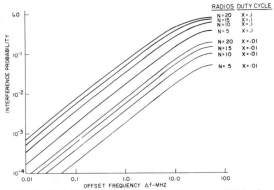

Fig. 5. Probability of interference from at least one of $N$ signals.

overlapping time slots, mutual interference can be avoided. TDM requires all radios to be synchronized to a common net time base. If the individual radios are mobile and/or fall out of sync due to infrequent activity, they must be resynchronized. Radios attempting to transmit before reacquiring sync will interfere with reception elsewhere in the net. Unless adequate time guard bands are provided, mobile TDM radios, although synchronized, will interfere with one another due to propagation delay uncertainty. From the above discussion, it should be clear that a combination of interference control methods is necessary for co-site multiple independent radio operation. The combination of TDM and CDM to provide interference suppression when time slots overlap constitutes an elegant solution to the co-site multichannel problem.

Multiple FH channels which consist of statistically independent sequences in time–frequency space can also be organized into either orthogonal or disjoint frequency sets. Orthogonal frequency sets are constructed such that the cross correlation of any pair of signals vanishes when integrated over a symbol of length $T$, i.e., $(s_i(t), s_j(t))$ equals zero for $i \neq j$. These sets can be identically distributed over the same frequency band (without processing gain loss), ideally causing interference only when two FH channels fall simultaneously on the same frequency. The cross correlations vanish only when there is perfect synchronization. With $T = 40$ $\mu$s, a delay error of just 4 ns at 250 MHz would limit interference reduction to about 80 dB. Realistic tracking bias errors and matched filter dynamic range limit performance.

For $N$ identically distributed statistically independent co-site FH radios, the severity of interference depends on the simultaneity of transmission and reception, the frequency offset between channels, and the relative signal levels. Fig. 5, derived from the Appendix, gives the probability of at least one of 5, 10, 15, or 20 FH radios transmitting coincident with a reception and falling within $\Delta f$ of the victim receiver's RF center frequency. The relative input $I/S$ ratio and receiver selectivity determine the minimum frequency offset $\Delta f$ tolerable. At $\Delta f = 10$ MHz and net duty cycle of $X = 0.1$, the interference probability is found to be, from Fig. 5, unacceptably high for all but the most robust channels. However, for low traffic duty cycles, i.e., $X \leqslant 0.01$, the resultant interference may be infrequent enough to be tolerated without employing suppression control.

FH channels can also be separated into disjoint frequency sets where external filtering and receiver selectivity can be used for interference reduction. This approach by limiting the frequency spread of each FH channel sacrifices processing gain in direct proportion to the number of FH channels required. The desirable features of orthogonality and frequency separation are combined in the $L$-orthogonal signal set. This set is implemented by subdividing the allocated set of FH frequencies into $N$ disjoint subsets, each containing $L$ signals where signals from different subsets are orthogonal. Although $L$-orthogonality allows more FH channels than the simple disjoint case, it also trades processing gain for suppression. Because of orthogonality, adjacent subsets can be spaced closer together. However, both $L$-orthogonal and simple disjoints sets are undesirable for more than four channels due to the $>6$ dB processing gain loss.

## IV. CONCLUSION

The effects of trading off processing gain in both PN–MSK and FH–FSK (MSK) communications to attain immunity from co-site interference have been presented. Comparing (8) and (19) for the same processing gain degradation and $J/S$ performance, it can be seen that the $I/S$ ratio tolerable at the PN demodulator input is much greater than the $I/S$ level the corresponding FM receiver can sustain. When adjacent channels use CDM–PN sequences with low cross correlation, even higher $I/S$ ratios can be tolerated. Although TDM is the only technique that is able to allow $-110$ dBm receivers and 10 W transmitters to coexist, severe interference will occur when a radio with poor timing attempts initial acquisition and transmits in a time slot it is not assigned. A combination of the transmit filtering, waveform modulation, and antenna isolation interference rejection methods is required. Tables II and III summarize the comparison of the effectiveness of these methods in terms of the number of independent PN or FH channels allowable for various frequency allocation bandwidths. These results are parameterized as a function of resultant $J/S$ performance and processing gain degradation. The results of Table II were obtained from Fig. 1(a) and (b). A 2 dB $E/\eta_0$ ratio for MSK detection of 16 kbit/s voice is assumed. Similarly, (19) and Fig. 4 were employed in deriving the results of Table III where a 6.5 dB $E_b/\eta_0$ ratio is needed for noncoherent FSK detection. In comparing these tables, it is clear that, given the same con-

TABLE II
COMPARISON OF NUMBER OF ADJACENT PN CHANNELS ALLOWABLE[1] FOR VARIOUS FILTER SLOPES AND ANTENNA ISOLATIONS

| Co-Channel J/S Ratio | | Frequency Allocation In MHz | Antenna Isolation | | | | | | | |
|---|---|---|---|---|---|---|---|---|---|---|
| | | | 20 dB | | | | 40 dB | | | |
| PG Degrad | I/S Ratio | | MSK Waveform | | | | MSK Waveform | | | |
| | | | No Filter | 18 dB Per Oct. | 24 dB Per Oct. | 30 dB Per Oct. | No Filter | 18 dB Per Oct. | 24 dB Per Oct. | 30 dB Per Oct. |
| 19.5 dB | | 60.6 | 1 | 4 | 5 | 7 | 1 | 5 | 7 | 9 |
| | | 146.4 | 1 | 8 | 12 | 15 | 1 | 13 | 17 | 22 |
| 1 dB | 13.5 dB | 175 | 1 | 10 | 14 | – | 1 | 15 | 20 | – |
| 17.5 dB | | 60.6 | 1 | 4 | 5 | 7 | 1 | 6 | 8 | 9 |
| | | 146.4 | 1 | 9 | 13 | 16 | 1 | 14 | 19 | 23 |
| 3 dB | 17.5 dB | 175 | 1 | 10 | 15 | – | 1 | 16 | 22 | – |

[1] Corresponding to 16 kbit/s CVSD voice.

TABLE III
COMPARISON OF NUMBER OF CO-SITE FH RADIOS ALLOWABLE FOR VARIOUS FILTER SLOPES AND ANTENNA ISOLATIONS

| J/S RATIO FULL-BAND NOISE | | FREQUENCY HOP CHANS/BANDWIDTH | | ANTENNA ISOLATION | | | | | |
|---|---|---|---|---|---|---|---|---|---|
| | | | | 20 DB | | | 30 DB | | |
| PG DEGRAD. | I/S RATIO | DUTY CYCLE | MIN. CHAN. OFFSET | 24 DB /OCT. | 30 DB /OCT. | 36 DB /OCT. | 24 DB /OCT. | 30 DB /OCT. | 36 DB /OCT. |
| J/S = 19.5 DB | | 500/ 18.9MHZ | | | | | | | |
| 1 DB | -13.4 DB | X=0.1 | 4 | 1 | 1 | 1 | 1 | 1 | 1 |
| | | | 5 | 1 | 1 | 1 | 1 | 2 | 3 |
| | | | 6 | 1 | 1 | 1 | 1 | 5 | 5 |
| J/S = 19.5 DB | | 500/ 18.9MHZ | | | | | | | |
| 1 DB | -13.4 DB | X=0.01 | 4 | 1 | 1 | 2 | 1 | 1 | 22 |
| | | | 5 | 1 | 2 | 3 | 1 | 18 | 36 |
| | | | 6 | 1 | 5 | 5 | 3 | 57 | 57 |
| J/S = 17.5 DB | | 500/ 18.9MHZ | | | | | | | |
| 3 DB | -9.5 DB | X=0.1 | 4 | 1 | 1 | 1 | 1 | 1 | 5 |
| | | | 5 | 1 | 1 | 1 | 1 | 4 | 9 |
| | | | 6 | 1 | 1 | 1 | 1 | 14 | 14 |
| J/S = 17.5 DB | | 500/ 18.9MHZ | | | | | | | |
| 3 DB | -9.5 DB | X=0.01 | 4 | 1 | 1 | 5 | 1 | 2 | 56 |
| | | | 5 | 1 | 4 | 9 | 1 | 45 | 90 |
| | | | 6 | 1 | 14 | 14 | 9 | 142 | 142 |

ditions, i.e., antenna isolation, filter slope, and $J/S$ ratio, PN and FH (at low duty cycle) allow a similar number of co-site radio channels. FH, however, is able to pack them in a much smaller frequency band.

## APPENDIX
## CUMULATIVE PROBABILITY OF THE FREQUENCY DIFFERENCE BETWEEN $N$ STATISTICALLY INDEPENDENT FREQUENCY HOPPING RADIOS

It is assumed that there are $N$ co-located simplex radio transceivers, each operating on a different radio net. In any given time interval $(t, t + \Delta t)$, there are $N_t$ radios transmitting (the time interval will be taken short enough, i.e., equal to the reciprocal of the hop rate so that there is at most one message transmitted per radio), $N_r$ radios receiving, and $N - (N_t + N_r)$ radios that are neither transmitting nor receiving. If $\lambda$ is the average call rate on each of the $N$ radio nets, then for Poisson-distributed calls, the probability of having $n$ calls in $\Delta t$ seconds is

$$P(n) = \frac{(\lambda \Delta t)^n}{n!} \exp(-\lambda \Delta t). \tag{A1}$$

For $\lambda \Delta t \ll 1$, there are just two possibilities: either a single call per net occurring with probability $P(1) = \lambda \Delta t$ or no calls per net with probability $P(0) = 1 - \lambda \Delta t$. Thus, with $N$ nets, the probability of having $N_t$ simultaneous transmissions during a hop dwell period $\Delta t$ is given by

$$P(N_t) = \binom{N}{N_t} (\lambda \Delta t)^{N_t} (1 - \lambda \Delta t)^{N - N_t}. \tag{A2}$$

If the $N_t$ hop frequencies $f_n: n = 1, 2, \cdots N_t$ are identically distributed statistically independent random variables whose probability density functions $p(f_n) = 1/(W_2 - W_1)$ are uniform over the $(W_1, W_2)$ frequency band, then the probability of any transmit frequency $f_n$ falling within $\Delta f$ of a co-site receiver center frequency $f_0$ is

$$P(|f_n - f_0| \leq \Delta f) = \frac{\Delta f}{\Delta W} \left(2 - \frac{\Delta f}{\Delta W}\right) \tag{A3}$$

where $\Delta W = W_2 - W_1$.

Combining (A2) and (A3), we can derive the probability of having at least one of $N - 1$ possible transmissions that could occur during the interval $(t, t + \Delta t)$ fall within $\Delta f$ a victim re-

170

ceiver's center frequency. The receiver vulnerability to interference at a given offset frequency is dictated by its selectivity characteristic. For a given ratio of desired to undesired RF signal amplitudes, there is a minimum tolerable offset frequency $\Delta f$ below which significant interference (i.e., BER $> 10$ percent in the case of 16 kbit/s CVSD voice) occurs. Thus, the probability of interference from at least one transmitting radio falling within $\Delta f$ of a receiver's center frequency is

$$P_s = \sum_{N_t=1}^{N-1} \left\{ 1 - \left[ 1 - \frac{\Delta f}{\Delta W} \left( 2 - \frac{\Delta f}{\Delta W} \right) \right]^{N_t} \right\} p(N_t). \quad (A4)$$

On expanding the exponential term in (A4) in a binomial expansion and recognizing that

$$\binom{N_t}{k} \binom{N-1}{N_t} = \binom{N-1}{k} \binom{N-k+1}{N_t-k} \quad \cdot(A5)$$

and

$$\sum_{N_t=k}^{N-1} \binom{N-k+1}{N_t-k} (\lambda \Delta t)^{N_t} (1 - \lambda \Delta t)^{N-1-N_t}$$

$$= (\lambda \Delta t)^k, \quad (A6)$$

then it is possible to show that (A4) reduces to

$$P_s = 1 - \left[ 1 - (\lambda \Delta t) \frac{\Delta f}{\Delta W} \left( 2 - \frac{\Delta f}{\Delta W} \right) \right]^{N-1}. \quad (A7)$$

## REFERENCES

[1] D. L. Schilling, L. B. Milstein, and R. L. Pickholtz, "Optimization of the processing gain of an $M$-ary direct sequence spread spectrum communication system," *IEEE Trans. Commun.*, vol. COM-28, Aug. 1980.

[2] L. B. Milstein, R. L. Pickholtz, and D. L. Schilling, "Optimization of the processing gain of an FSK-FH system," *IEEE Trans. Commun.*, vol. COM-28, July 1980.

[3] S. A. Gronemeyer and A. L. McBride, "MSK and offset QPSK modulation," *IEEE Trans. Commun.*, vol. COM-24, Aug. 1976.

[4] F. Amoroso, "Pulse and spectrum manipulation in the minimum (frequency) shift keying (MSK) format," *IEEE Trans. Commun.*, (Corresp.), vol. COM-24, pp. 381–384, Mar. 1976.

# Part III
# Code Division Multiple Access

THE problem of allowing multiple users to simultaneously access a channel without causing an undue amount of degradation in the performance of any individual user is a classical one in communications. The two most common techniques, frequency division multiple access (FDMA) and time division multiple access (TDMA), attempt to solve the problem by separating the signals in frequency and time, respectively. However, each of these techniques has certain drawbacks associated with it.

For example, in FDMA, any time the channel through which the signals are transmitted is nonlinear, as in a satellite channel with a nonlinear repeater, intermodulation products will be generated. In TDMA, the intermodulation problem does not exist, but accurate synchronization of all the users becomes of paramount importance in system design. Furthermore, if interference such as jamming or multipath is present, large degradations in system performance can result in both FDMA and TDMA systems.

For reasons of these types, code division multiple access (CDMA) has become a competitive multiple accessing scheme in certain situations. It is usually operated in an asynchronous manner so that network timing problems do not exist. There is a penalty for this flexibility, of course, and that penalty is a smaller capacity due to the imperfect orthogonality of the spreading sequences of the various users. The intermodulation problem does not go away with CDMA, but the capability of rejecting external interference is invariably the crucial factor in deciding to implement a CDMA system.

Issues of interest to both researchers and users of CDMA systems are how to design sets of spreading sequences with good correlation properties and how to determine just how many users a given system can support. In this part, "Code Division Multiple Access," there are four papers. The first is by Rowe ("Bounds on the Number of Signals with Restricted Cross Correlation"); it gives an upper bound to the number of users in a synchronous CDMA system in which each user employs either an orthogonal or a biorthogonal alphabet. This bound is presented in terms of the mean-square cross correlation between users, rather than in terms of the maximum cross correlation of the user set.

The next paper is by Pursley, Sarwate, and Stark ("Error Probability for Direct-Sequence Spread-Spectrum Multiple-Access Communications—Part I: Upper and Lower Bounds"). The results of this paper apply to a binary direct-sequence asynchronous CDMA system wherein the peak distortion due to multiple users does not exceed the desired signal level. The limitation of the technique is that the numerical complexity is exponential in the number of simultaneous users; it is therefore most appropriate for a small number of users. This limitation does not exist in the following paper by Geraniotis and Pursley ("Error Probability for Direct-Sequence Spread-Spectrum Multiple-Access Communications—Part II: Approximations"). This latter paper presents an approximate solution which computationally grows only linearly with the number of simultaneous users. In addition, the paper presents results for various quaternary modulation formats.

The final paper in this part, also by Geraniotis and Pursley ("Error Probabilities for Slow-Frequency-Hopped Spread-Spectrum Multiple-Access Communications over Fading Channels"), presents both bounds on, and approximations for, the performance of a slow frequency-hopped noncoherent FSK system. The system is completely asynchronous, and the received signals are further degraded (beyond the multiple access interference) by the effects of a fading channel.

# Bounds on the Number of Signals with Restricted Cross Correlation

HARRISON E. ROWE, FELLOW, IEEE

*Abstract*—We seek the maximum number of users possible for multiuser, spread-spectrum radio systems, as the signal set is varied, under the following ideal conditions: 1) no fading; 2) synchronous system; 3) orthogonal signal set for each user; 4) matched-filter individual receivers; 5) no additive noise.

Welch has given an upper bound on the number of vectors (or equivalently signals) $w$ possible in a $d$-dimensional space, given the maximum cross-correlation magnitude $C_{max}$ between any vector pair. However, the error rate for the systems under study may depend on a suitably defined root-mean-square cross correlation $C_{rms}$, rather than on the maximum cross correlation $C_{max}$ of a vector set.

We extend Welch's results to $d$-dimensional vector sets of size $w$ divided into alphabets of size $A$, the vectors of each alphabet being strictly orthogonal, and give the maximum size of such a vector set as function of $C_{rms}$.

We demonstrate that as Welch's limit is approached, all cross correlations must approach $C_{max}$, and given precise limits on the way in which this must occur. Similarly, close to the present bounds (in terms of $C_{rms}$) all rms cross correlations must be close to $C_{rms}$.

Particular examples are given of signal sets close to the present bounds.

## SYSTEM MODEL

A COMMON channel is shared by $U$ spread-spectrum transmitters. Each has an orthogonal alphabet of $A$ waveforms, each of duration no greater than $T$ seconds and confined to a band of width $B$ Hz. We represent these $A \cdot U$ waveforms by the same number of vectors [1], in a space of $d \equiv 2TB$ dimensions.

$i$th symbol, $k$th user's alphabet $\Leftrightarrow V_i^k$;

$$1 \leqslant k \leqslant U, \quad 1 \leqslant i \leqslant A. \qquad (1)$$

Manuscript received July 9, 1981; revised January 14, 1982. This paper was presented at the National Telecommunications Conference, New Orleans, LA, December 1981.

The author is with Bell Laboratories, Holmdel, NJ 07733.

Reprinted from *IEEE Trans. Commun.*, vol. COM-30, pp. 966–974, May 1982.

All signals have unit energy:

$$|V_i^k|^2 = 1. \tag{2}$$

Each alphabet is strictly orthogonal:

$$V_i^k \cdot V_{i'}^k = 0, \qquad i \neq i'. \tag{3}$$

Define

$$C_{\max} \equiv \max_{i, i', k \neq k'} |V_i^k \cdot V_{i'}^{k'}|. \tag{4}$$

For $C_{\max} \ll 1$ different alphabets are approximately orthogonal.

A related system uses biorthogonal alphabets; each consists of the above $A$ orthogonal signals and their negatives, a total of $2A$ waveforms. We use the convention

$$V_{-i}^k \equiv -V_i^k. \tag{5}$$

In (1) the index $i$ takes on the values $-A \leq i \leq -1$ and $1 \leq i \leq A$, and (3) is modified to

$$V_i^k \cdot V_{i'}^k = 0, \qquad i = \pm i'. \tag{6}$$

Obviously from (5), $V_i^k \cdot V_{-i}^k = -1$.

$U$ receivers correspond to these transmitters. Each consists of $A$ filters matched to the alphabet of its corresponding transmitter. These filters are numbered from 1 to $A$, the impulse response of the $i$th filter corresponding to the vector $V_i^k$. (For the biorthogonal case the $i$th filter gives $+1$ output for $V_i^k$, $-1$ output for $V_{-i}^k$.) Synchronous operation is assumed; for a typical time slot the (common) receiver input is

$$V = \sum_{k=1}^{U} V_{i_k}^k \tag{7}$$

where the $k$th transmitter has sent symbol $i_k$ from its alphabet; either $1 \leq i_k \leq A$ (orthogonal), or $-A \leq i_k \leq -1$ or $1 \leq i_k \leq A$ (biorthogonal). The output of the $i$th matched filter of the $k$th receiver is from (2), (3) or (6), and (7)

$$V_i^k \cdot V = \begin{cases} I_i^k, & i \neq \pm i_k, \\ \pm 1 + I_i^k, & i = \pm i_k, \end{cases} \qquad 1 \leq i \leq A \tag{8}$$

where the $\pm$ applies to the biorthogonal case; the interference parameters $I_i^k$ are given by

$$I_i^k = \sum_{\substack{k'=1 \\ k' \neq k}}^{U} V_i^k \cdot V_{i_{k'}}^{k'}, \qquad 1 \leq i \leq A. \tag{9}$$

$I_i^k$, of course, depends on the set $\{i_{k'} : k' \neq k\}$ of symbols sent by every other transmitter. The $|I_i^k|$ will be small if $C_{\max}$ is sufficiently small.

Each receiver examines its $A$ matched filter outputs, selects the one with the largest magnitude, and declares its corresponding symbol to be the one transmitted. Error will occur for a typical symbol $i$ of receiver $k$ if and only if at least one

other filter has a larger output:

$$|\pm 1 + I_i^k| \leq |I_{i'}^k|, \qquad i' \neq i, \quad 1 \leq i, \quad i' \leq A \tag{10}$$

the $\pm$ again applying to the biorthogonal case.

Rather than characterizing the interference properties of a particular signal set by its $I_i^k$, we consider the statistical model described in the following section.

## STATISTICAL MODEL

Assume each transmitter sends independent, equally likely symbols from its alphabet, independently from all other transmitters. Thus, for orthogonal alphabets the $k$th transmitter sends symbols $V_1^k$, $V_2^k$, $\cdots$, $V_A^k$, each with probability $1/A$; for biorthogonal alphabets it sends $\pm V_1^k$, $\pm V_2^k$, $\cdots$, $\pm V_A^k$, each with probability $1/(2A)$.

The probability of error for each symbol of a given signal set can in principle be determined for this statistical model, by (10) and (9). To do this we would require the complete probability distribution for the interference parameters of each alphabet, i.e., the joint distribution for $\{I_i^k\}$ for each $k$. While this is not possible in general, the means and variances of the $I_i^k$ are readily found.

For this purpose, it is useful to denote the average output of the $k$th transmitter as $\langle V^k \rangle$, where we drop the subscript inside the $\langle \ \rangle$ in accord with a convention used throughout, that parameters averaged over are omitted.

$$\langle V^k \rangle = \begin{cases} \dfrac{1}{A} \sum_{i=1}^{A} V_i^k, & \text{orthogonal} \\ 0, & \text{biorthogonal.} \end{cases} \tag{11}$$

The first and second moments of the output interference quantities $I_i^k$ of (9) are given as follows:

$$\langle I_i^k \rangle = V_i^k \cdot \sum_{\substack{k'=1 \\ k' \neq k}}^{U} \langle V^{k'} \rangle. \tag{12}$$

Define

$$\delta I_i^k \equiv I_i^k - \langle I_i^k \rangle. \tag{13}$$

Then

$$\langle (\delta I_i^k)^2 \rangle = \sum_{\substack{k'=1 \\ k' \neq k}}^{U} \{ \langle (V_i^k \cdot V^{k'})^2 \rangle - (V_i^k \cdot \langle V^{k'} \rangle)^2 \}. \tag{14}$$

Consider first the biorthogonal case, where each alphabet contains $2A$ vectors, consisting of the original $A$ orthogonal vectors and their negatives (5). Substituting (11) into (12) and (14),

$$\langle I_i^k \rangle = 0$$

$$\langle (I_i^k)^2 \rangle = \langle (\delta I_i^k)^2 \rangle = \sum_{\substack{k'=1 \\ k' \neq k}}^{U} \langle (V_i^k \cdot V^{k'})^2 \rangle. \tag{15}$$

175

The mean-square interference to each signal is simply the sum of the mean-square interferences due to the other $U - 1$ users.

This result (15) will not be strictly true for orthogonal alphabets, since $\langle V^k \rangle$ is no longer 0 as in the biorthogonal case (11); rather, for orthogonal alphabets containing $A$ vectors

$$|\langle V^k \rangle| = \frac{1}{\sqrt{A}}. \qquad (16)$$

Nevertheless, in many cases (15) remains approximately true for the orthogonal case as well. To see this, examine the $k'$th term of (14). By the Schwarz inequality this term $\{\ \} \geqslant 0$. The limiting value 0 is attained only when $V_i{}^k \cdot V_{i'}{}^{k'} = $ constant, independent of $i'$; i.e., the interference in the $i$th filter of the $k$th receiver must be identical for every member of the $k'$th alphabet. Now every biorthogonal alphabet has $2^A$ corresponding orthogonal alphabets; only a single one of these could satisfy such a requirement (i.e., interfere with a given filter of another alphabet independently of which symbol of the interfering alphabet was transmitted). This argument suggests that (15) may hold approximately for most signal sets with orthogonal alphabets.

Moreover, frequency-hop signals are frequently used in applications, consisting of contiguous sine waves of different frequencies ("chips"); if there is no phase coherence between signals of different users at the receiver, there can be no dc interference components. A rough analog here would have each transmitter use a biorthogonal alphabet of $2A$ symbols, but have each receiver identify every vector and its negative (5) as a single symbol. Then (15) will hold exactly.

The detector of (10) is optimum for Gaussian interference, i.e., when the $I_i{}^k$ are independent, identically distributed, zero-mean, Gaussian random variables. For a large number of users, $U \gg 1$, the $I_i{}^k$ will be approximately Gaussian. In cases of interest the interference will have zero mean, as discussed above; otherwise the detector must account for the average values of the interference outputs, and if the user set changes, do this adaptively. However, we have no information about the correlation between interference outputs of different filters, i.e., $\langle I_i{}^k I_{i'}{}^k \rangle$, and therefore cannot know for certain that different (Gaussian) $I_i{}^k$ are approximately independent, although this may be so in many cases.

## PERFORMANCE CRITERIA

We adopt the mean-square interference outputs of the receiver filters (15) as a measure of performance. The $U$ active users may be chosen from a possibly much larger set of $M$ potential users, $M \geqslant U$, each possible set $U$ satisfying the conditions of (1)–(3) (and (5), (6) for biorthogonal signals); $C_{max}$ is now defined by (4) with $1 \leqslant k, k' \leqslant M$, i.e., $C_{max}$ is the largest cross-correlation magnitude over the entire signal set of $w \equiv A \cdot M$ vectors. Now define $C_{rms}{}^2$ as the mean-square interference per interferer in the worst case:

$$C_{rms}{}^2 \equiv \max_{i, k, \{k_l\}} \frac{1}{A(U-1)} \sum_{k' \in \{k_l\}} \sum_{i'=1}^{A} (V_i{}^k \cdot V_{i'}{}^{k'})^2 \qquad (17)$$

where

$$\{k_l\} \equiv \{k_1, k_2, \cdots, k_{U-1}\}, \qquad k_l \neq k, \quad 1 \leqslant k_l \leqslant M. \qquad (18)$$

Thus, $C_{rms}{}^2$ is the largest mean-square interference per interferer, for any signal and any of the corresponding $(M - 1)!/[(U - 1)!(M - U)!]$ possible set of interferers. Equations (17) and (18) imply that

$$\frac{1}{A} \sum_{\substack{k'=1 \\ k' \neq k}}^{M} \sum_{i'=1}^{A} (V_i{}^k \cdot V_{i'}{}^{k'})^2 \leqslant (M-1) C_{rms}{}^2, \qquad \text{all } i, k. \qquad (19)$$

We find the maximum number of users (or equivalently signal vectors) as the signal set is varied, for a given $C_{rms}$; the interference powers are then limited by $\langle (I_i{}^k)^2 \rangle \leqslant (U - 1) C_{rms}{}^2$. As noted above, the interference outputs $I_i{}^k$ will be approximately Gaussian. We place no constraint on the correlation between different interference outputs $\langle I_i{}^k I_{i'}{}^k \rangle$. Correlation can only increase the maximum symbol error rate in a biorthogonal system over that corresponding to independent, Gaussian interference. Since, as shown below, most $\langle (I_i{}^k)^2 \rangle$ are close to their upper limit $(U - 1) C_{rms}{}^2$, for correlation to *decrease* the maximum error rate in an orthogonal system, most $\langle I_i{}^k I_{i'}{}^k \rangle$ would have to be positive, a possibility we ignore. Consequently, we assume the maximum symbol error rate will equal or exceed that corresponding to independent Gaussian interference with power $(U - 1) C_{rms}{}^2$ [1, pp. 257-263].

## UPPER BOUNDS—$C_{max}$

Welch has shown for a general set of $w$ unit vectors in $d$ dimensions (i.e., *not* divided into orthogonal or biorthogonal alphabets, as above), denoted by $V_\nu, 1 \leqslant \nu \leqslant w$, that [2]

$$\frac{w^2}{d} \leqslant \sum_{\nu=1}^{w} \sum_{\nu'=1}^{w} (V_\nu \cdot V_{\nu'})^2 \leqslant w(w-1) C_{max}{}^2 + w \qquad (20)[1]$$

where he derived the left-hand member by a sequence of inequalities and the right-hand member by inspection. The outer inequality then yields Welch's result (in different notation and modified form), for the maximum number of vectors with maximum cross correlation $C_{max}$:

$$\frac{w-1}{d-1} \leqslant \frac{1}{1 - C_{max}{}^2 d}. \qquad (21)$$

The equality of (21) is shown as the curve of Fig. 1.

J. E. Mazo has suggested that signal sets close to Welch's limit (21) must have most cross correlations close to $C_{max}$. Now equality in (21) implies equality between the outer members of (20), which in turn requires

$$\sum_{\substack{\nu=1 \\ \nu \neq \nu'}}^{w} \sum_{\nu'=1}^{w} (V_\nu \cdot V_{\nu'})^2 = w(w-1) C_{max}{}^2. \qquad (22)$$

[1] It is interesting to observe that the lower bound is attained by a vector set having $w/d$ vectors along each of the $d$ coordinate axes.

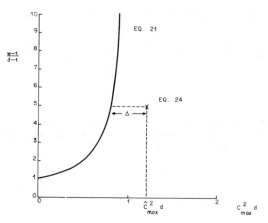

Fig. 1.  Welch's bound, and a typical signal set lying below it.

Since $|V_\nu \cdot V_{\nu'}| \leqslant C_{\max}$ for $\nu \neq \nu'$ and since the $\Sigma\Sigma$ in (22) has $w(w-1)$ terms, equality in (21) therefore implies that

$$|V_\nu \cdot V_{\nu'}| = C_{\max}, \qquad \nu \neq \nu'. \tag{23}$$

Any signal set at Welch's limit must have all cross correlations equal to $C_{\max}$.

More generally, consider possible signal sets close to, but not at, Welch's limit (21). Fig. 1 shows (21), together with a point representing a particular signal set having correlation limit $\hat{C}_{\max}$, lying a distance $\Delta$ to the right of this curve:

$$\frac{w-1}{d-1} = \frac{1}{1 - \hat{C}_{\max}{}^2 d + \Delta}, \qquad 0 \leqslant \Delta. \tag{24}$$

Substituting (24) into (20)

$$\hat{C}_{\max}{}^2 \left(1 - \frac{\Delta}{\hat{C}_{\max}{}^2 d}\right) \leqslant \frac{1}{w(w-1)}$$

$$\cdot \sum_{\substack{\nu=1}}^{w} \sum_{\substack{\nu'=1 \\ \nu \neq \nu'}}^{w} (V_\nu \cdot V_{\nu'})^2 \leqslant \hat{C}_{\max}{}^2. \tag{25}$$

From (25), no more than $w(w-1)/K$ of the $(V_\nu \cdot V_{\nu'})^2$ can be smaller than $\hat{C}_{\max}{}^2 - K\Delta/d$, for $K \geqslant 1$ (see the Appendix). Alternatively expressing this result in terms of the horizontal-axis variable $C_{\max}{}^2 d$ of Fig. 1, no greater fraction than $1/K$ of the $w(w-1)$ quantities $(V_\nu \cdot V_{\nu'})^2 d$ can lie to the left of $\hat{C}_{\max}{}^2 d - K\Delta$. Clearly, this result is useful only if the point (23) representing the particular signal set lies close to the curve (21) representing Welch's limit; however, this point can lie slightly to the right of $C_{\max}{}^2 d = 1$, where (21) has a pole.

A case of particular interest is

$$\hat{C}_{\max}{}^2 d = 1. \tag{26}$$

Here no more than $w(w-1)/K$ of the $w(w-1)$ cross corre-

lations (i.e., no greater fraction than $1/K$) can satisfy

$$(V_\nu \cdot V_{\nu'})^2 d \leqslant 1 - K \frac{d-1}{w-1}. \tag{27}$$

Signal sets close to Welch's limit must have $C_{\mathrm{rms}}$ close to $C_{\max}$. However, it is plausible to suppose that we may modify such a vector set by increasing some cross correlations (and therefore increasing $C_{\max}$) while decreasing others, maintaining $C_{\mathrm{rms}}$ (and perhaps the error rate) unchanged. This is consistent with the fact that certain signal sets proposed for spread-spectrum systems have quite unevenly distributed cross correlations, i.e., $C_{\mathrm{rms}} \ll C_{\max}$ [3], [4]. The appropriate extension to Welch's bound is straightforward and given in the next section.

## UPPER BOUNDS—$C_{\mathrm{rms}}$

We retain the left-hand inequality of (20), given by Welch [2], we modify the central expression ($\Sigma\Sigma$) to a form appropriate for orthogonal alphabets of size $A$, and finally we use the restriction of (19) to give a new right-hand inequality:

$$\frac{w^2}{d} \leqslant \sum_{k,k'=1}^{M} \sum_{i,i'=1}^{A} (V_i{}^k \cdot V_i{}^{k'})^2 \leqslant wA(M-1)C_{\mathrm{rms}}{}^2 + w$$

$$= w(w-A)C_{\mathrm{rms}}{}^2 + w. \tag{28}$$

The outer inequality of (28) yields the appropriate modification of Welch's result (21), for the maximum number of vectors divided into sets of orthogonal alphabets of size $A$, with maximum mean-square interference per interfering alphabet $C_{\mathrm{rms}}{}^2$:

$$\frac{w-A}{d-A} \leqslant \frac{1}{1 - C_{\mathrm{rms}}{}^2 d}. \tag{29}$$

The equality of (29) is shown as the curve of Fig. 2. Note that this curve is identical to the curve of Fig. 1 [the equality of (21)] with redefined axes. For biorthogonal alphabets of size $2A$, the corresponding bound for $w$ is twice as large.

We find analogous results to those of the previous section, for signal sets close to the limit (29). Equality in (29) implies equality in (28), which requires

$$\sum_{k=1}^{M} \sum_{i=1}^{A} \sum_{\substack{k'=1 \\ k' \neq k}}^{M} \sum_{i'=1}^{A} (V_i{}^k \cdot V_{i'}{}^{k'})^2 = wA(M-1)C_{\mathrm{rms}}{}^2.$$

$$\tag{30}$$

The inner pair of $\Sigma\Sigma$ in (30) must be $\leqslant A(M-1)C_{\mathrm{rms}}{}^2$ by (19) for every $i, k$. Therefore, (30) implies equality in (19); this result and (17) require

$$\frac{1}{A} \sum_{i'=1}^{A} (V_i{}^k \cdot V_{i'}{}^{k'})^2 = C_{\mathrm{rms}}{}^2, \quad \text{all } i, k, \quad k' \neq k; \quad M > U.$$

177

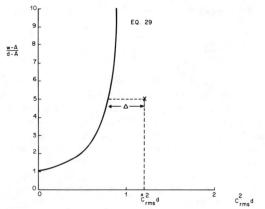

Fig. 2.   The rms bound, and a typical signal set lying below it.

Any signal set at the limit (i.e., with equality) in (29) must possess the property (31); the rms interference to every signal from every other alphabet is equal to $C_{rms}$. Note that (31) is not necessarily satisfied for $M = U$; here

$$\frac{1}{A(U-1)} \sum_{\substack{k'=1 \\ k' \neq k}}^{U} \sum_{i'=1}^{A} (V_i^{\ k} \cdot V_{i'}^{\ k'})^2$$

$$= C_{rms}^2, \qquad \text{all } i, k; \quad M = U. \tag{32}$$

The total interference to every signal is identical (rather than the individual interferences from each of the other alphabets); however, only this result is needed in the $M = U$ case.

More generally, consider possible signal sets close to, but not at, the limit (29), a typical example being illustrated in Fig. 2 for a signal set with $\hat{C}_{rms}^2$ [defined by (17)] lying a distance $\Delta$ to the right of the curve (29). As above, several different cases must be considered (see the Appendix). For $M = U$, no greater fraction than $1/K$ of the $w$ quantities

$$\frac{d}{A(U-1)} \sum_{\substack{k'=1 \\ k' \neq k}}^{U} \sum_{i'=1}^{A} (V_i^{\ k} \cdot V_{i'}^{\ k'})^2$$

can lie to the left of $\hat{C}_{rms}^2\, d - K\Delta$ on Fig. 2. For $M > U$, if the additional condition

$$\frac{1}{A} \sum_{i'=1}^{A} (V_i^{\ k} \cdot V_{i'}^{\ k'})^2 \leqslant \hat{C}_{rms}^2, \qquad \text{all } i, k, \ k' \neq k \tag{33}$$

is imposed, no greater fraction than $1/K$ of the $w(w - A)/A$ quantities $(d/A) \Sigma_{i'=1}^{A} (V_i^{\ k} \cdot V_{i'}^{\ k'})^2$ can lie to the left of $\hat{C}_{rms}^2\, d - K\Delta$ on Fig. 2. If (33) is abandoned, this result is only slightly degraded for $M \gg U$; in general no greater fraction than $1/[K(1 - U/M)]$ of the $w(w - A)/A$ quantities $(d/A) \Sigma_{i'=1}^{A} (V_i^{\ k} \cdot V_{i'}^{\ k'})^2$ can lie to the left of $\hat{C}_{rms}^2\, d - K\Delta$ on Fig. 2.

Any signal set close to the limit (29) must have uniform interference properties, the exact nature of the uniformity

depending on the case (e.g., $M = U$, $M \gg U$). However, such signal sets can differ radically from those close to Welch's limit (21) of the preceding section, having quite unevenly distributed cross correlations, i.e., $C_{rms} \ll C_{max}$.

## LOWER BOUNDS

Consider first a signal set in $d$ dimensions composed of two sets of $d$ vectors each. The first set consists of $d$ unit vectors along the $d$ coordinate axes; the second set consists of $d$ unit vectors with components $\pm 1/\sqrt{d}$, the $\pm$ signs chosen according to a Sylvester-type Hadamard matrix [5].

$$V_1 \quad = \quad [1 \quad 0 \quad 0 \cdots 0 \quad 0]$$

$$V_2 \quad = \quad [0 \quad 1 \quad 0 \cdots 0 \quad 0]$$

$$\vdots$$

$$V_d \quad = \quad [0 \quad 0 \quad 0 \cdots 0 \quad 1]$$

$$V_{d+1} \quad = \frac{1}{\sqrt{d}}[1 \quad 1 \quad 1 \cdots 1 \quad 1] \tag{34}$$

$$V_{d+2} \quad = \frac{1}{\sqrt{d}}[1 \quad -1 \quad 1 \cdots 1 \quad -1]$$

$$\vdots$$

$$V_{2d} \quad = \frac{1}{\sqrt{d}}[1 \quad -1 \quad -1 \cdots \pm 1].$$

By the Hadamard property, $d$ must be a power of 2; we take $M = U$. Since this set of $2d$ vectors contains $U$ alphabets of size $A$, $U \cdot A = 2d$; consequently, $U$ and $A$ must both be powers of 2. From (17) and (18) with $M = U$, and from (34)

$$C_{rms}^2 d = \frac{1}{2\left(1 - \dfrac{A}{2d}\right)}; \qquad C_{max}^2 d = 1. \tag{35}$$

Since $w = 2d$

$$\frac{w - A}{d - A} = \frac{2 - \dfrac{A}{d}}{1 - \dfrac{A}{d}}. \tag{36}$$

If $A/d$ is restricted to the values

$$A/d = 2/d, \cdots, 1/16, 1/8, 1/4, 1/2, 1. \tag{37}$$

The corresponding values for (35) and (36) are

$$C_{rms}^2 d = \frac{1}{2} \frac{1}{1 - \dfrac{1}{d}}, \cdots, 16/31, 8/15, 4/7, 2/3, 1.$$

$$\tag{38}$$

$$\frac{w - A}{d - A} = 2 \frac{1 - \dfrac{1}{d}}{1 - \dfrac{2}{d}}, \cdots, 31/15, 15/7, 7/3, 3, \infty.$$

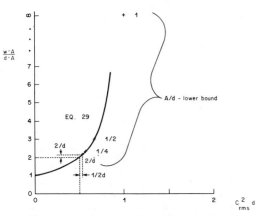

ig. 3. The rms upper bound with semi-Hadamard lower bound. $M = U, d = 2^{\text{integer}}, U = 2^{\text{integer}}, A \equiv 2d/U = 2^{\text{integer}}$

Solving (35) for $A/d$ and substituting into (36)

$$\frac{w - A}{d - A} = \frac{1}{1 - C_{\text{rms}}^2 d} . \tag{39}$$

The values (38) for the signal set (34) lie precisely on the upper bound (29), plotted in Fig. 2. We replot this upper bound, together with the present lower bounds (38) indicated by +'s lying on the upper bound, on Fig. 3. Note that the point $C_{\text{rms}}^2 d = 1, (w - A)/(d - A) = \infty$ corresponds to the case $U = 2, A = d$ for the signal set of (34), and that the point for the case $U = d, A = 2$ (binary alphabets) lies close to the point $C_{\text{rms}}^2 d = 0.5, (w - A)/(d - A) = 1$, when the dimensionality is high, $d \gg 1$, as indicated by the dotted lines in Fig. 3.

Finally, we consider an abstract description of a frequency-hop signal set discussed by Einarsson [6]. Let $\phi_c, c = 1, 2, \cdots, d$ represent unit vectors along the $d$ coordinate axes. Divide this set of unit vectors into $L$ groups, each with $A$ unit vectors:

$$d = AL. \tag{40}$$

Each signal vector is composed of one vector from each group. Using Einarsson's notation, the address of the $k$th user (alphabet) is the $L$-dimensional vector (in a space unrelated to that of the signal vectors $V_i^k$)

$$a_k = [a_{k1} \ a_{k2} \cdots a_{kL}]. \tag{41}$$

The $a_k$ belong to a finite field of $A$ elements $GF(A)$; $A$ is a prime number or a power of a prime number.

$$0 \leqslant a_{k1}, a_{k2}, \cdots, a_{kL} \leqslant A - 1. \tag{42}$$

Then, using the notation of (1), the first member of the $k$th alphabet is

$$V_1^k = \frac{1}{\sqrt{L}} (\phi_{a_{k1}} + \phi_{A + a_{k2}} + \cdots + \phi_{(L-1)A + a_{kL}}). \tag{43}$$

The other members of alphabet $k$ are defined by

$$V_i^k = \frac{1}{\sqrt{L}} (\phi_{y_{k1}} + \phi_{A + y_{k2}} + \cdots + \phi_{(L-1)A + y_{kL}}) \tag{44}$$

where

$$y_k = a_k + (i - 1)\mathbf{1}, \qquad i = 1, 2, \cdots, A \tag{45}$$

and $\mathbf{1}$ is the all-ones vector

$$\mathbf{1} = (1 \quad 1 \cdots 1). \tag{46}$$

Arithmetic operations (addition and multiplication) are mod $A$ if $A$ is prime; if $A$ is a power of a prime these operations are more complicated [6]. The components of $y_k$ are restricted the same as those of $a_k$ in (42).

The vectors of each alphabet are obtained by cyclic permutation of the unit coordinate vectors within each group of $L$. To complete the signal set design we must assign the addresses $a_k$ of (41). Einarsson does this as follows [6].

$$a_k = [(k - 1) \ (k - 1)\beta \ (k - 1)\beta^2 \cdots (k - 1)\beta^{L-1}],$$

$$k = 1, 2, \cdots, A$$

$A$ = prime, or power of prime

$\beta$ = primitive element of $GF(A)$ $\tag{47}$

$2 \leqslant L \leqslant A - 1.$

The final relation of (47) and (40) implies that

$$\sqrt{d + 1/4} + 1/2 \leqslant A \leqslant d/2. \tag{48}$$

There are $A$ alphabets, each with $A$ symbols; hence

$$M = A, \qquad w = MA = A^2. \tag{49}$$

Einarsson shows that vectors from different alphabets (i.e., having different addresses $a_k, a_{k'}$) can have no more than one coordinate axis vector $\phi_c$ in common. Therefore, for any given vector $i$ of alphabet $k$, the cross correlations with the symbols of another alphabet $k'$ are distributed as follows:

$$V_i^k \cdot V_{i'}^{k'} = \begin{cases} \dfrac{1}{L} & \text{for } L \text{ vectors of } k' \\[2mm] 0 & \text{for } A - L \text{ vectors of } k'. \end{cases} \tag{50}$$

Consequently

$$C_{\text{max}}^2 = \frac{1}{L^2} . \tag{51}$$

This signal set satisfies (33), which is more restrictive than (17) or (19), with

$$C_{\text{rms}}^2 = \frac{1}{AL} = \frac{1}{d} . \tag{52}$$

179

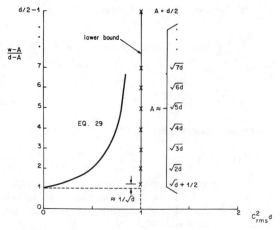

Fig. 4. The rms upper bound, with Einarsson lower bound. $U \leqslant M = A = \text{(prime)}^{\text{integer}}$, $d/A = \text{integer}'$, $\sqrt{d + 1/4} + 1/2 \leqslant A \leqslant d/2$.

Therefore

$$\left(\frac{C_{\text{rms}}}{C_{\max}}\right)^2 = \frac{d}{A^2}. \tag{53}$$

From (48) and (53)

$$\frac{4}{d} \leqslant \left(\frac{C_{\text{rms}}}{C_{\max}}\right)^2 \leqslant \frac{1}{1 + \dfrac{1}{2d} + \sqrt{\dfrac{1}{d} + \dfrac{1}{4d^2}}} \approx 1 - \frac{1}{\sqrt{d}}. \tag{54}$$

For large alphabets the cross correlations are quite unevenly distributed; i.e., $(C_{\text{rms}}/C_{\max})^2 \ll 1$.

By (52) all signal sets described by (40)–(47) must have $C_{\text{rms}}^2 d = 1$. From (49)

$$\frac{w - A}{d - A} = \frac{A(A - 1)}{d - A}. \tag{55}$$

For different $A$'s satisfying (48) we plot the corresponding vertical coordinate (55) at the horizontal coordinate $C_{\text{rms}}^2 d = 1$, in Fig. 4. For minimum and maximum alphabet and signal set size we have respectively

$$A = \sqrt{d + 1/4} + 1/2 \approx \sqrt{d} + 1/2;$$

$$\frac{w - A}{d - A} = \frac{1}{1 - \sqrt{\dfrac{1}{d} - \dfrac{1}{4d^2}} - \dfrac{1}{2d}} \approx 1 + \frac{1}{\sqrt{d}}$$

$$\tag{56}$$

$$A = d/2;$$

$$\frac{w - A}{d - A} = \frac{d}{2} - 1.$$

In general

$$A \approx \sqrt{d}\,\sqrt{\frac{w - A}{d - A}}, \qquad A \ll d. \tag{57}$$

Note that for $U$ active users the interference power to each is $(U - 1)/d$; if all alphabets are used ($U = M$) the interference power is $(A - 1)/d$.

## DISCUSSION

We have given an upper bound, shown in Fig. 2, for the number of users in a multiuser spread-spectrum system with orthogonal (or biorthogonal) alphabets, given the allowable interference power. This interference power will largely determine the error rate for most practical cases.

The descriptive parameter $C_{\text{rms}}$, defined in (17), is the worst rms interference per interfering alphabet. $C_{\text{rms}}$ depends on the particular signal set, with $M$ alphabets; if the number of active users $U$ is less than $M$, $C_{\text{rms}}$ may or may not depend on $U$ as well. In the Hadamard example (34), (35) applies only for $U = M$; e.g., for $U \leqslant M/2 + 1$, $C_{\text{rms}}^2 d = 1$. In contrast, the Einarsson signal set (40)–(47) has $C_{\text{rms}}^2$ given by (52) for every $U \leqslant M = A$ (49).

Our bound [Fig. 2 or (29)] shows the maximum number of vectors $w$ or equivalently alphabets $M = w/A$ possible for a given $C_{\text{rms}}^2$. Any signal set close to this limit must have uniform interference properties, which we have described in a precise way. At the limit:

1) If $U = M$, the total interference to every signal is the same.

2) If $U < M$, the interference to every signal from every other alphabet is the same (however, the interference from different members of the interfering alphabet may vary).

The Hadamard example illustrates case 1 (Fig. 3). However the Einarsson signal set (Fig. 4) does not lie at the limit although it does satisfy the condition (33) that the interference to every signal from every other alphabet is the same. A signal set at Welch's limit [Fig. 1 or (21)] must have $C_{\text{rms}} = C_{\max}$, and the same interference between every pair of signals.

We next inquire how tight is the upper bound of (29) and Fig. 2. The point at $C_{\text{rms}}^2 d = 0$, $(w - A)/(d - A) = 1$ is easily attained by a signal set consisting of $d$ unit vectors along the $d$ coordinate axes; these may be divided into alphabets of any size. Next, the Hadamard signal set of Fig. 3 attains the upper bound for $U = M$ for $C_{\text{rms}}^2 d$ ranging from a little greater than 1/2 to 2/3 for alphabet sizes varying from $A = 2$ to $A = d/2$. This set also has a point at $C_{\text{rms}}^2 d = 1$, $(w - A)/(d - A) = \infty$. Finally, the Einarsson set shown on Fig. 4 is confined to the vertical line $C_{\text{rms}}^2 d = 1$, with $(w - A)/(d - A)$ varying from a little greater than 1 to $d/2 - 1$ for alphabet size varying from $A \approx \sqrt{d}$ to $A = d/2$.

Our upper bounds thus are tight in the sense that signal sets with some alphabet size $A$ can be found that attain the limit for values of $C_{\text{rms}}^2 d$ near either 0 or 1, or in the middle of its range. However, we desire more lower bounds, for signal sets with various restrictions.

1) Can the bound be attained in the mid-range ($C_{rms}^2 d \approx$ 0.5) by signal sets with $U < M$, for large and small alphabet size $A$?

2) At the pole $C_{rms}^2 d = 1$, the Einarsson set (for which we can have $U < M$ as well as $U = M$) attains a maximum value of $(w - A)/(d - A) = d/2 - 1$, for large alphabet size $A = d/2$. How large can $(w - A)/(d - A)$ be for small alphabet size, both for $U = M$ and for $U < M$?

3) How close can we come to the rms bound with signal sets for which $C_{rms} = C_{max}$, i.e., having equal interference between every pair of signals from different alphabets?

The present $C_{rms}$ bound (Fig. 2) was obtained by modifying Welch's analysis for his $C_{max}$ bound (Fig. 1). Welch also presented higher order bounds for $C_{max}^2 d > 1$ [2]; similar results for $C_{rms}^2 d > 1$ are not available.

The present rms bound yields limits on the number of users possible in spread-spectrum systems under various conditions (e.g., [7]).

## APPENDIX

### SIGNALS SETS CLOSE TO WELCH'S AND rms LIMITS

Consider Welch's $C_{max}$ bound (21), and a particular signal set close to it (24), illustrated in Fig. 1. Subtracting $w$ from the outer members of (20), deleting the $\nu = \nu'$ terms from the $\Sigma\Sigma$ [which sum to $w$ by (2)], dividing by $w(w - 1)$, and rearranging the left-hand member, yields

$$\frac{1}{d}\left(1 - \frac{d-1}{w-1}\right) \leqslant \frac{1}{w(w-1)} \sum_{\substack{\nu, \nu'=1 \\ \nu \neq \nu'}}^{w} (V_\nu \cdot V_{\nu'})^2 \leqslant C_{max}^2. \tag{58}$$

The descriptive parameters of the particular signal set under consideration are related by (24); substituting (24) in the left-hand member of (58), and changing the right-hand member to $\hat{C}_{max}^2$, yields (25) after minor rearrangement, where the $\Sigma$ includes $w(w - 1)$ terms.

Now if

$$0 \leqslant x_i \leqslant X \tag{59}$$

then necessarily

$$\frac{1}{N} \sum_{i=1}^{N} x_i \leqslant X. \tag{60}$$

If further

$$X - \Delta \leqslant \frac{1}{N} \sum_{i=1}^{N} x_i \tag{61}$$

then it is obvious that no more than $N/K$ of the $x_i$ can be $\leqslant X - K\Delta$. Note that the limiting case—exactly $N/K$ of the $x_i = X - K\Delta$—is attained when the rest of the $x_i$, $N(1 - (1/K))$ in number, have their maximum value $X$. Applying this

result to (25) yields directly the bounds on cross correlations stated immediately following (25).

Consider finally the $C_{rms}$ bound (29), and a signal set close to it, illustrated in Fig. 2. This signal set has parameters satisfying

$$\frac{w - A}{d - A} = \frac{1}{1 - \hat{C}_{rms}^2 d - \Delta} \tag{62}$$

where $\hat{C}_{rms}^2$ is defined by (17) for the particular signal set under consideration, and, as noted in the Discussion, may depend on $U$ if $U < M$. Following the above analysis for Welch's $C_{max}$ bounds, we observe from (3) that for $k = k'$ in (28) the terms inside the $\Sigma\Sigma$ are 0 for $i \neq i'$, 1 for $i = i'$; therefore, the $k = k'$ terms sum to $w$. Subtracting $w$ from the outer members of (28) and deleting the $k = k'$ terms in the $\Sigma\Sigma$, we have after rearrangement

$$\frac{1}{d}\left(1 - \frac{d - A}{w - A}\right)$$

$$\leqslant \frac{1}{w(w-A)} \sum_{k=1}^{M} \sum_{i=1}^{A} \sum_{\substack{k'=1 \\ k' \neq k}}^{M} \sum_{i'=1}^{A} (V_i^k \cdot V_{i'}^{k'})^2 \leqslant C_{rms}^2. \tag{63}$$

Since the signal set under discussion satisfies (62), substituting (62) in the left-hand member of (63) and changing $C_{rms}^2$ in the right-hand member to $\hat{C}_{rms}^2$ yields

$$\hat{C}_{rms}^2 - \frac{\Delta}{d}$$

$$\leqslant \frac{1}{w(w-A)} \sum_{k=1}^{M} \sum_{i=1}^{A} \sum_{\substack{k'=1 \\ k' \neq k}}^{M} \sum_{i'=1}^{A} (V_i^k \cdot V_{i'}^{k'})^2 \leqslant \hat{C}_{rms}^2 \tag{64}$$

corresponding to (25) for Welch's $C_{max}$ bounds.

The results (31) and (32) suggest that the $M = U$ and the $M > U$ cases must be treated separately. Consider first the $M = U$ case. Rewrite (64) as

$$\hat{C}_{rms}^2 d - \Delta \leqslant \frac{1}{w} \sum_{k=1}^{U} \sum_{i=1}^{A} \frac{d}{U - 1}$$

$$\cdot \left\{ \sum_{\substack{k'=1 \\ k' \neq k}}^{U} \frac{1}{A} \sum_{i'=1}^{A} (V_i^k \cdot V_{i'}^{k'})^2 \right\}$$

$$\leqslant \hat{C}_{rms}^2 d. \tag{65}$$

The quantity inside { } of (65) represents the total interference to $V_i^k$; there are $w = AU$ such quantities. By definition [(17) and (18) with $M = U$] each is $\leqslant (U - 1)\hat{C}_{rms}^2$. By (59)—

(61) no greater fraction than $1/K$ of the $w$ quantities

$$\frac{d}{A(U-1)} \sum_{\substack{k'=1 \\ k' \neq k}}^{U} \sum_{i'=1}^{A} (V_i^k \cdot V_{i'}^{k'})^2$$

can be less than $\hat{C}_{rms}^2 d - K\Delta$.

We treat the $M > U$ case in two steps. First, we impose the condition (33), which places a uniform limit on the rms interference to any signal $V_i^k$ from any other alphabet $k'$. This condition is satisfied by the Einarsson signal set (40)–(47), but not by the Hadamard signal set (34), for example; (33) is stronger than (17) and (18). Rewrite (64) as

$$\hat{C}_{rms}^2 d - \Delta \leqslant \frac{A}{w(w-A)} \sum_{k=1}^{M} \sum_{i=1}^{A} \sum_{\substack{k'=1 \\ k' \neq k}}^{M} d$$

$$\cdot \left\{ \frac{1}{A} \sum_{i'=1}^{A} (V_i^k \cdot V_{i'}^{k'})^2 \right\}$$

$$\leqslant \hat{C}_{rms}^2 d. \tag{66}$$

The quantity inside $\{\ \}$ of (66) is the interference to $V_i^k$ from the $k$th alphabet, limited as noted above by (33). Then (59)–(61) yield directly the result that no greater fraction than $1/K$ of the $w(w-A)/A = MA(M-1)$ quantities $(d/A) \Sigma_{i'=1}^{A} (V_i^k \cdot V_{i'}^{k'})^2$ can be less than $\hat{C}_{rms}^2 d - K\Delta$. However, if we abandon (33), the interference now being limited only by (17) and (18), some of the quantities inside $\{\ \}$ of (66) can now exceed $\hat{C}_{rms}^2$, and so the fraction of $(d/A) \Sigma_{i'=1}^{A} (V_i^k \cdot V_{i'}^{k'})^2$ less than $\hat{C}_{rms}^2 d - K\Delta$ can now be greater than $1/K$.

We therefore consider this problem for $M \geqslant U$, with $\hat{C}_{rms}^2$ of (62) defined by (17) and (18) and subject to no other restrictions. For each $i, k$ reorder the terms inside the $\Sigma_{k'}$, i.e., $d\{\ \}$, such that their magnitudes are nondecreasing; then by (17) and (18) the first $(U-1)$ such terms must sum to $\leqslant (U-1)\hat{C}_{rms}^2 d$. No more than the first $(U-2)$ such terms can exceed $\hat{C}_{rms}^2 d$, and if any do, all successive terms must be less than $\hat{C}_{rms}^2 d$.

First, it is clear that to maximize the number of terms $d\{\ \}$ in (66) that are $\leqslant \hat{C}_{rms}^2 d - K\Delta$, we must replace the $\leqslant$ by $=$. Second, divide such terms into two sets:

1) those belonging to the first $(U-1)$ terms of the reordered $\Sigma_{k'}$ and

2) those belonging to the remaining terms of the reordered $\Sigma_{k'}$.

The largest number in the second category is the same as given following (66) for the case where the additional constraint (33) was imposed, i.e., $MA(M-1)/K$. The number in the first category will be maximized when those in the second category are confined to as few $i, k$ as possible; then for each of the $[MA(M-1)/K]/[K(M-U)]$ such $i, k$ we can have no more than $(U-2)$ additional $d\{\ \}$ at the minimum value $\hat{C}_{rms}^2 d - K\Delta$. Combining these, an upper bound on the number of $d\{\ \} \leqslant \hat{C}_{rms}^2 d - K\Delta$ is $[MA(M-1)/K][1 + (U-2)/(M-U)] < [w(w-A)/A]/[K(1-U/M)]$, as stated following (33).

To summarize the above three paragraphs, if we impose a uniform limit (33) on the rms interference to any signal $V_i^k$ from any other alphabet $k'$, the fraction of the $w(w-A)/A$ quantities $(d/A) \Sigma_{i'=1}^{A} (V_i^k \cdot V_{i'}^{k'})^2$ that can be less than $\hat{C}_{rms}^2 d - K\Delta$ is less than $1/K$. If we abandon this uniform limit (33), this fraction is less than $1/[K(1-U/M)]$, a very small increase for $U \ll M$.

## ACKNOWLEDGMENT

I would like to thank J. E. Mazo for several helpful discussions, and in particular for bringing Welch's bounds to my attention. I am also grateful to D. J. Goodman, B. G. King, W. L. Mammel, V. K. Prabhu, W. R. Young, and O. C. Yue for helpful discussions.

## REFERENCES

[1] J. M. Wozencraft and I. M. Jacobs, *Principles of Communication Engineering.* New York: Wiley, 1965, ch. 4.

[2] L. R. Welch, "Lower bounds on the maximum cross correlation of signals," *IEEE Trans. Inform. Theory,* vol. IT-20, pp. 397–399, May 1974.

[3] A. J. Viterbi, "A processing satellite transponder for multiple access by low-rate mobile users," presented at the *Digital Satellite Commun. Conf.,* Montreal, P.Q., Canada, Oct. 23–25, 1978.

[4] D. J. Goodman, P. S. Henry, and V. K. Prabhu, "Frequency-hopped multilevel FSK for mobile radio," *Bell Syst. Tech. J.,* vol. 59, pp. 1257–1275, Sept. 1980.

[5] M. Harwit and N. J. A. Sloane, *Hadamard Transform Optics.* New York: Academic, 1979, pp. 17,210.

[6] G. Einarsson, "Address assignment for a time-frequency-coded, spread-spectrum system," *Bell Syst. Tech. J.,* vol. 59, pp. 1241–1255, Sept. 1980.

[7] H. E. Rowe, "Bounds on the number of users in spread-spectrum systems," in *Proc. Nat. Telecommun. Conf.,* Nov. 30–Dec. 4, 1980, pp. 69.6.1–69.6.6.

# Error Probability for Direct-Sequence Spread-Spectrum Multiple-Access Communications—Part I: Upper and Lower Bounds

MICHAEL B. PURSLEY, FELLOW, IEEE, DILIP V. SARWATE, SENIOR MEMBER, IEEE, AND
WAYNE E. STARK, STUDENT MEMBER, IEEE

*Abstract*—Upper and lower bounds on the average probability of error are obtained for direct-sequence spread-spectrum multiple-access communications systems with additive white Gaussian noise channels. The bounds, which are developed from convexity properties of the error probability function, are valid for systems in which the maximum multiple-access interference does not exceed the desired signal and the signature sequence period is equal to the duration of the data pulse. The tightness of the bounds is examined for systems with a small number of simultaneously active transmitters. This is accomplished by comparisons of the upper and lower bounds for several values of the system parameters. The bounds are also compared with an approximation based on the signal-to-noise ratio and with the Chernoff upper bound.

## INTRODUCTION

**D**URING the past few years there has been considerable interest in efficient methods for obtaining approximations and bounds for the average probability of error in asynchronous direct-sequence spread-spectrum multiple-access (SSMA) communications systems. Among the published contributions to this problem are the approximation based on the signal-to-noise ratio (SNR) [6, p. 798], the bounds based on moment-space techniques [14], approximations based on series-expansion methods and Gauss quadrature rules ([5] and [13]), approximations based on the integration of the characteristic function [3] (see also [4]), and our preliminary versions of the bounds obtained in the present paper ([1] and [11]). Each of the proposed methods has its advantages and disadvantages, and the choice of method for a given application ultimately depends on the system parameters, the required accuracy, and the available computing equipment. Some of the methods require fairly sophisticated computer software (e.g., [14]), while others are very easy to apply. In particular, the approximation based on the SNR can be evaluated from the tabulated correlation parameters (e.g., [2], [10], and [12]) without the use of a computer.

One of the key issues is whether a *bound* on the error probability is required or an *approximation* will suffice. Generally speaking, it is much easier to obtain an approximation than a bound, but upper and lower bounds together

Manuscript received February 17, 1981; revised January 26, 1982. This work was supported by the Army Research Office under Grant DAAG29-78-G-0114 and Contract DAAG29-81-K-0064.

The authors are with the Coordinated Science Laboratory and the Department of Electrical Engineering, University of Illinois, Urbana, IL 61801.

supply more information. For instance, in order to guarantee that a particular error rate specification is or is not attainable for a given set of system parameters, bounds on the probability of error are required. Moreover, any bound is also an approximation, and an upper and a lower bound together furnish not only an approximation but also a bound on the resulting error in the approximation.

The present paper is devoted to *bounds* on the average probability of error. These bounds are conceptually simpler than the moment-space bounds given in [14] and we have found that they are easier to evaluate for small to moderate values of $K$, the number of simultaneously active transmitters. The relative simplicity of the bounds we present is evident from the fact that the paper gives all of the necessary details to enable the reader to compute these bounds. The evaluation of our bounds does not require the determination of convex hulls, the solution of sets of nonlinear equations, or the computation of high-order moments of the multiple-access interference. Moreover, the numerical results presented in [1] show that our bounds are much tighter than the second-moment bound of [14], and we have found that the improved bounds given in the present paper are also tighter than the single-exponential bounds of [14].

The main disadvantage with the bounds presented in this paper is that the computational requirements increase exponentially in $K$. Thus, these bounds are not suitable for systems with a large number of simultaneously active transmitters. However, they are suitable for packet radio systems and other applications involving bursty data transmission, and they are also suitable for hybrid frequency-hopped direct-sequence SSMA systems. It should be noted that for SSMA systems with relatively few chips per bit (e.g., 31), the number of simultaneously active transmitters must necessarily be small in order to achieve satisfactory performance.

As shown in [8], the bounds developed in this paper can also be modified to give bounds on the probability of error for spread-spectrum communications over certain specular multipath channels. The multipath interference is handled in the same way as the multiple-access interference is handled in the present paper. The main change that must be made is that crosscorrelations are replaced by autocorrelations.

## SYSTEM MODEL

This paper is concerned with bounds on the average probability of error for an asynchronous binary direct-sequence

Reprinted from *IEEE Trans. Commun.*, vol. COM-30, pp. 975–984, May 1982.

spread-spectrum multiple-access (SSMA) communications system with an additive white Gaussian noise channel. The model that is employed in the present paper is taken from [6]. This model has been used in most of the recent performance analyses for asynchronous binary direct-sequence SSMA communications (e.g., [1], [3], [11], [13], and [14]), and it has been generalized [7] to provide a model for asynchronous quaternary direct-sequence SSMA (e.g., [4], [5], and [9]). Although some of the results that we obtain here can be extended in a straightforward manner to quaternary systems, we restrict attention to binary systems throughout the paper.

The received signal in the asynchronous binary direct-sequence SSMA system is the sum of $K$ spread-spectrum signals $s_k(t - \tau_k)$, $1 \leq k \leq K$, plus an additive white Gaussian noise process $n(t)$ which has (two-sided) spectral density $\frac{1}{2}N_0$. For the model of [6], the spread-spectrum signal $s_k(t - \tau_k)$ is given by

$$s_k(t - \tau_k) = \sqrt{2\bar{P}}a_k(t - \tau_k)b_k(t - \tau_k) \cos(\omega_c t + \varphi_k) \quad (1)$$

where $a_k(\cdot)$ is the code waveform, $b_k(\cdot)$ is the data signal, $\tau_k$ is a time-delay parameter which accounts for propagation delay and the lack of synchronism between the signals, and $\varphi_k$ is the phase angle for the $k$th carrier (the time delay for the carrier has been absorbed in $\varphi_k$). The reader is referred to [6] for a detailed description of the code and data waveform. Basically, $a_k(t)$ is a periodic infinite sequence of nonoverlapping rectangular pulses which are called code pulses or chips. Each code pulse has duration $T_c$. The amplitude of the $n$th pulse is $a_n^{(k)}$, where $a_n^{(k)}$ is +1 or −1 for each $n$ and where $(a_n^{(k)}) = \cdots, a_{-1}^{(k)}, a_0^{(k)}, a_1^{(k)}, \cdots$ is a periodic sequence with period $p$. The data signal $b_k(t)$ is a sequence of nonoverlapping rectangular pulses, each of which has duration $T$. The amplitude of the $l$th pulse is denoted by $b_l^{(k)}$. We assume that there are exactly $N$ full code pulses in each data pulse, and therefore it must be that $T = NT_c$. We also assume that $N$ is a integer multiple of $p$. Several of the properties of the code and data waveforms are summarized in a compact form by $a_k(t) = a_j^{(k)}$ for $jT_c \leq t < (j + 1)T_c$ and $b_k(t) = b_l^{(k)}$ for $lT \leq t < (l + 1)T$.

In this paper we are concerned with the *average* probability of error. Consequently, the parameters $b_l^{(k)}$, $\tau_k$, and $\varphi_k$ are treated as random variables. We assume that the collection of all of these parameters (i.e., for $-\infty < l < \infty$ and $1 \leq k \leq K$) is a set of mutually independent random variables and that $P(b_l^{(k)} = +1) = P(b_l^{(k)} = -1) = \frac{1}{2}$ for each $l$ and $k$. From these basic assumptions and certain properties of the SSMA system we can draw the following conclusions. First, because of the symmetry of the problem we may restrict attention to the output of the receiver for signal $s_1(t - \tau_1)$. Second, since only relative time delays and phase angles are important, we may set $\tau_1 = \varphi_1 = 0$. The parameters $\tau_k$ and $\varphi_k$ are then the time delay and phase angle for the $k$th signal relative to the first. Third, the properties of an SSMA system and the stationarity of the noise $n(t)$ permit us to consider only time delays modulo $T$ and phase angles modulo $2\pi$, rather than the absolute values of these parameters.

If $b_0^{(1)}$ is the data bit for the first signal during the interval $[0, T]$, then the output of a correlation receiver matched to the first signal is the random variable

$$Z = \eta + (\tfrac{1}{2}P)^{1/2}T\left\{b_0^{(1)} + \sum_{k=2}^{K} I_{k,1}(\boldsymbol{b}_k, \tau_k, \varphi_k)\right\} \quad (2)$$

where $\boldsymbol{b}_k = (b_{-1}^{(k)}, b_0^{(k)})$ and the channel noise component $\eta$ is given by

$$\eta = \int_0^T n(t)a_1(t) \cos \omega_c t \, dt. \quad (3)$$

The multiple-access interference $I_{k,1}(\boldsymbol{b}_k, \tau_k, \varphi_k)$ which appears in (2) is defined in terms of the continuous-time partial crosscorrelation functions

$$R_{k,i}(\tau) = \int_0^\tau a_k(t - \tau)a_i(t) \, dt, \qquad 0 \leq \tau \leq T \quad (4a)$$

and

$$\hat{R}_{k,i}(\tau) = \int_\tau^T a_k(t - \tau)a_i(t) \, dt, \qquad 0 \leq \tau \leq T \quad (4b)$$

which are defined in [6]. For the present analysis we need consider only $i = 1$ and $k > 1$. It is shown in [6] and [7] that the multiple-access interference component is given by

$$I_{k,1}(\boldsymbol{b}_k, \tau, \varphi) = T^{-1}[b_{-1}^{(k)}R_{k,1}(\tau) + b_0^{(k)}\hat{R}_{k,1}(\tau)] \cos \varphi \quad (5)$$

for $0 \leq \tau < T$ and $0 \leq \varphi < 2\pi$, and that for *rectangular* chip waveforms the continuous-time partial crosscorrelation functions are given by

$$R_{k,i}(\tau) = C_{k,i}(l - N)T_c + [C_{k,i}(l + 1 - N) - C_{k,i}(l - N)](\tau - lT_c) \quad (6a)$$

and

$$\hat{R}_{k,i}(\tau) = C_{k,i}(l)T_c + [C_{k,i}(l + 1) - C_{k,i}(l)](\tau - lT_c) \quad (6b)$$

where $l = \lfloor \tau/T_c \rfloor$ and where $C_{k,i}$ is the aperiodic crosscorrelation function which is defined by

$$C_{k,i}(l) = \begin{cases} \sum_{j=0}^{N-1-l} a_j^{(k)}a_{j+l}^{(i)}, & 0 \leq l \leq N-1, \quad (7a) \\ \sum_{j=0}^{N-1+l} a_{j-l}^{(k)}a_j^{(i)}, & 1-N \leq l < 0 \quad (7b) \end{cases}$$

and $C_{k,i}(l) = 0$ for $|l| \geqslant N$. Notice that from (5)-(7) the multiple-access interference is a linear function of $\tau$ on the interval $[lT_c, (l + 1)T_c]$ provided the chip waveform is a rectangular pulse. Furthermore, it depends upon $\varphi$ only through the term $\cos \varphi$, a property which holds for other types of chip waveforms as well [7], [9].

## AVERAGE PROBABILITY OF ERROR

The receiver which is matched to the first signal produces the output $Z$ at time $T$. This receiver is not optimum for making a decision on the data symbol $b_0{}^{(1)}$ (i.e., $Z$ is not a sufficient statistic), since the total interference is not a white Gaussian noise process. However, this type of receiver is employed in nearly all direct-sequence spread-spectrum systems. It is relatively easy to implement and we believe that its performance is very close to that of the optimal receiver (at least for large $N$).

The actual bit decision is made as follows. If $Z \geqslant 0$ the decision is that $b_0{}^{(1)} = +1$; otherwise, the decision is that $b_0{}^{(1)} = -1$. Thus, an error occurs if $Z \geqslant 0$ when in fact $b_0{}^{(1)} = -1$ or if $Z < 0$ when $b_0{}^{(1)} = +1$. Because of the symmetry of the problem, these two types of errors occur with equal probability. Thus, we may assume in all that follows that $b_0{}^{(1)} = +1$; that is, a positive data pulse is sent by the first transmitter during the time interval $[0, T]$.

The conditional probability of error given that $b_0{}^{(1)} = +1$ is a function of the data symbols $b = (b_2, b_3, \cdots, b_K)$, the delays $\tau = (\tau_2, \tau_3, \cdots, \tau_K)$, and the phase angles $\varphi = (\varphi_2, \varphi_3, \cdots, \varphi_K)$. Since $\eta$ is a zero-mean Gaussian random variable with variance $\frac{1}{4}N_0T$, the conditional probability of error for a given $b, \tau$, and $\varphi$ is

$$P_{e,1}(b,\tau,\varphi) = Q\left(\alpha\left[1 + \sum_{k=2}^{K} I_{k,1}(b_k, \tau_k, \varphi_k)\right]\right) \quad (8)$$

where the function $Q$ is defined by

$$Q(y) = (2\pi)^{-1/2} \int_y^\infty e^{-(1/2)x^2}\, dx$$

and the parameter $\alpha$ is given by $\alpha = (2PT/N_0)^{1/2}$. Notice that if there is no multiple-access interference (e.g., if $K = 1$), then the error probability is just

$$Q(\alpha) = Q([2PT/N_0]^{1/2}) = Q([2E_b/N_0]^{1/2})$$

where $E_b$ is the energy per data bit.

The *average probability of error* $\bar{P}_{e,1}$ is the expected value of $P_{e,1}(b, \tau, \varphi)$. Assuming that the time delays (modulo $T$) are uniformly distributed on $[0, T]$ and the phase angles (modulo $2\pi$) are uniformly distributed on $[0, 2\pi]$, then

$$\bar{P}_{e,1} = (8\pi T)^{1-K} \Sigma_b \iint P_{e,1}(b,\tau,\varphi)\, d\tau\, d\varphi \quad (9)$$

where $\Sigma_b$ denotes the sum over all $b = (b_2, b_3, \cdots, b_K)$ such that $b_k = (b_{-1}{}^{(k)}, b_0{}^{(k)})$ with $b_l{}^{(k)} \in \{-1, +1\}$.

Although $\int d\tau$ is a multidimensional integral over $[0, T]^{K-1}$ and $\int d\varphi$ is a multidimensional integral over $[0, 2\pi]^{K-1}$, each of these integrals can be replaced by a sequence of one-dimensional integrals. This sequence can be bounded recursively to provide bounds on $\bar{P}_{e,1}$.

First, we define

$$G_2(b,\tau,\varphi) = Q\left(\alpha\left[1 + \sum_{k=2}^{K} I_{k,1}(b_k, \tau_k, \varphi_k)\right]\right). \quad (10)$$

Next, for $2 \leqslant n \leqslant K$ let

$$G_{n+1}(b,\tau,\varphi)$$
$$= (8\pi T)^{-1} \Sigma_{b_n} \int_0^{2\pi} \int_0^T G_n(b,\tau,\varphi)\, d\tau_n\, d\varphi_n \quad (11)$$

where $\Sigma_{b_n}$ denotes the sum over all $b_n \in \{-1, +1\}^2$. Notice that $G_n(b, \tau, \varphi)$ depends on $b_k, \tau_k$, and $\varphi_k$ only if $k \geqslant n$. In particular, $G_{K+1}(b, \tau, \varphi)$ does not depend on $b, \tau$, and $\varphi$ at all. Indeed, we see from (9) that

$$G_{K+1}(b,\tau,\varphi) = \bar{P}_{e,1}. \quad (12)$$

The bounds presented in this paper depend primarily on the fact that $Q(x)$ is convex for $x \geqslant 0$. As a result, the bounds are valid for spread-spectrum multiple-access systems for which

$$\sum_{k=2}^{K} I_{k,1}(b_k, \tau_k, \varphi_k) \geqslant -1 \quad (13)$$

for all $b$, $\tau$, and $\varphi$. Because of symmetry properties of the multiple-access interference this condition is equivalent to

$$\left|\sum_{k=2}^{K} I_{k,1}(b_k, \tau_k, \varphi_k)\right| \leqslant 1 \quad (14)$$

for all $b$, $\tau$, and $\varphi$, which is the requirement that the maximum multiple-access interference must be less than the desired signal component at the output of the correlation receiver. This restriction is imposed in all that follows.

The symmetry property mentioned above is due to the following relationships:

$$I_{k,1}(-b_k, \tau_k, \varphi_k) = -I_{k,1}(b_k, \tau_k, \varphi_k), \quad (15a)$$

$$I_{k,1}(b_k, \tau_k, 2\pi - \varphi_k) = I_{k,1}(b_k, \tau_k, \varphi_k), \quad (15b)$$

and

$$I_{k,1}(b_k, \tau_k, \pi - \varphi_k) = -I_{k,1}(b_k, \tau_k, \varphi_k). \quad (15c)$$

Because of these properties we can replace (11) by

$$G_{n+1}(b,\tau,\varphi)$$
$$= (2\pi T)^{-1} \Sigma_{b_n} \int_0^{(1/2)\pi} \int_0^T G_n(b,\tau,\varphi)\, d\tau_n\, d\varphi_n. \quad (16)$$

185

## LOWER BOUNDS FOR THE AVERAGE PROBABILITY OF ERROR

The first step in obtaining a lower bound on $\bar{P}_{e,1}$ is to consider (16) with $n = 2$ and develop a lower bound $G_3^L(b, \tau, \varphi)$ for $G_3(b, \tau, \varphi)$. The bound $G_3^L(b, \tau, \varphi)$ must have the property that it is suitable for use in (16). This limits the types of lower bounds that can be considered. The next step is to obtain a suitable lower bound $G_4^L(b, \tau, \varphi)$ for $G_4(b, \tau, \varphi)$, and so on. The approach that we develop gives a bound $G_n^L(b, \tau, \varphi)$ on $G_n(b, \tau, \varphi)$ with the property that the dependence of $G_n^L(b, \tau, \varphi)$ on $b_n$, $\tau_n$, and $\varphi_n$ is of the same form for each $n$. Hence, the bound on $\bar{P}_{e,1}$ is derived from a sequence of essentially identical bounds on $G_n(b, \tau, \varphi)$ for $n = 3, 4, \cdots, K + 1$.

The sequence is set up as follows. Let $G_n^L(b, \tau, \varphi)$ be given and define

$$\hat{G}_n^L(b, \tau, \varphi) = (4T)^{-1} \Sigma_{b_n} \int_0^T G_n^L(b, \tau, \varphi) \, d\tau_n. \qquad (17)$$

Since $G_n(b, \tau, \varphi) \geqslant G_n^L(b, \tau, \varphi)$ for each $b$, $\tau$, and $\varphi$, then

$$G_{n+1}(b, \tau, \varphi) \geqslant (2/\pi) \int_0^{(1/2)\pi} \hat{G}_n^L(b, \tau, \varphi) \, d\varphi_n. \qquad (18)$$

In order to obtain $G_{n+1}^L(b, \tau, \varphi)$ we make use of the fact that for any positive integer $J$

$$\hat{G}_n^L(b, \tau, \varphi)$$
$$= (4T)^{-1} \Sigma_{b_n} \sum_{l=0}^{N-1} \sum_{j=0}^{J-1} \int_{\Delta(l,j)}^{\Delta(l,j+1)} G_n^L(b, \tau, \varphi) \, d\tau_n$$

$$\qquad (19)$$

where $\Delta(l, i) = (l + J^{-1}i)T_c$ for $0 \leqslant l < N$ and $0 \leqslant i \leqslant J$, and for any positive integer $M$

$$G_{n+1}(b, \tau, \varphi) \geqslant (2/\pi) \sum_{m=0}^{M-1} \int_{\psi(m)}^{\psi(m+1)} \hat{G}_n^L(b, \tau, \varphi) \, d\varphi_n$$

$$\qquad (20)$$

where $\psi(m) = \frac{1}{2} m\pi/M$ for $0 \leqslant m \leqslant M$. We then derive $G_{n+1}^L(b, \tau, \varphi)$ by obtaining lower bounds on the integrals of (19) and (20). These lower bounds are presented in Appendix A.

The lower bound on $\bar{P}_{e,1}$ is in terms of the interference spectrum $S_k^L(i)$ defined as follows. Let $S_k^L(i)$ be the set

$$S_k^L(i) = \{(b_k, l, j): 2JNI_{k,1}(b_k, \bar{\Delta}(l,j), 0) = i\} \qquad (21)$$

where

$$\bar{\Delta}(l, j) = [l + J^{-1}(j + \tfrac{1}{2})] T_c \qquad (22)$$

for $0 \leqslant l < N$ and $0 \leqslant j < J$. Let $S_k^L(i) = |S_k^L(i)|$, the cardi-

nality of the set $S_k^L(i)$. Notice that $S_k^L(i) = 0$ for $|i| > 2JN$, and that the symmetry properties (i.e., (15)) of the interference function $I_{k,1}$ imply that $S_k^L(i) = S_k^L(-i)$. Next, for $0 \leqslant m < M$ define

$$\gamma(m) = M(\pi JN)^{-1} \{\sin [(m + 1)\pi/2M] - \sin [m\pi/2M]\}. \qquad (23)$$

Finally, let $\Sigma_i$ denote the sum over all $i = (i_2, i_3, \cdots, i_K)$ such that $|i_k| \leqslant 2JN$ for $1 < k \leqslant K$, and let $\Sigma_m$ denote the sum over all $m = (m_2, m_3, \cdots, m_K)$ such that $0 \leqslant m_k < M$ for $1 < k \leqslant K$.

The lower bound is then

$$\bar{P}_{e,1} \geqslant (4JNM)^{1-K} \Sigma_i \left\{ \prod_{n=2}^K S_n^L(i_n) \right\}$$

$$\cdot \Sigma_m Q\left( \alpha \left[ 1 + \sum_{k=2}^K i_k \gamma(m_k) \right] \right). \qquad (24)$$

This bound is valid for each choice of the positive integers $J$ and $M$. Larger values of $J$ and $M$ give tighter bounds at the expense of increased computation. As pointed out in Appendix A, the right-hand side of (24) can also be viewed as an approximation to $\bar{P}_{e,1}$, which is obtained from an application of a rectangular integration rule to the integrals of (19) and (20). Hence, the difference between $\bar{P}_{e,1}$ and the lower bound of (24) converges to zero as $J \to \infty$ and $M \to \infty$. Moreover, the approach that we have taken guarantees the convergence is monotonic.

## UPPER BOUNDS FOR THE AVERAGE PROBABILITY OF ERROR

In order to obtain an upper bound on $\bar{P}_{e,1}$ we use a procedure which is analogous to that developed in the previous section. This amounts to finding a sequence of upper bounds $G_n^U(b, \tau, \varphi)$ on the quantities $G_n(b, \tau, \varphi)$ for $n = 3, 4, \cdots, K + 1$. Thus, (17) and (18) are replaced by

$$\hat{G}_n^U(b, \tau, \varphi) = (4T)^{-1} \Sigma_{b_n} \int_0^T G_n^U(b, \tau, \varphi) \, d\tau_n \qquad (25)$$

and

$$G_{n+1}(b, \tau, \varphi) \leqslant (2/\pi) \int_0^{(1/2)\pi} \hat{G}_n^U(b, \tau, \varphi) \, d\varphi_n, \qquad (26)$$

respectively.

Expressions (19) and (20) apply if we simply replace the lower bounds $\hat{G}_n^L(b, \tau, \varphi)$ and $G_n^L(b, \tau, \varphi)$ by the corresponding upper bounds $\hat{G}_n^U(b, \tau, \varphi)$ and $G_n^U(b, \tau, \varphi)$, and reverse the inequality in (20). In Appendix B we obtain upper bounds on the integrals

$$\int_{\Delta'(l,j)}^{\Delta'(l,j+1)} G_n^U(b, \tau, \varphi) \, d\tau_n \qquad (27)$$

where $\Delta'(l, j) = [l + (j/J')] T_c$, and

$$\int_{\psi'(m)}^{\psi'(m+1)} \hat{G}_n{}^U(b, \tau, \varphi) \, d\varphi_n \qquad (28)$$

where $\psi'(m) = \frac{1}{2} m\pi/M'$.

The upper bound on the average probability of error is given in terms of an interference spectrum $S_k{}^U(i)$. This spectrum is defined as $S_k{}^U(i) = |S_k{}^U(i)|$, where $S_k{}^U(i)$ is the set

$$S_k{}^U(i) = \{(b_k, l, j): 2J'NI_{k,1}(b_k, \Delta'(l, j), 0) = i\}. \qquad (29)$$

In this case $J'$ can be of the form $\frac{1}{2}L$ for some integer $L$, but it is sufficient for our purposes to restrict $J'$ to be an integer. Let $\Sigma_i'$ and $\Sigma_m'$ be defined in the same way as $\Sigma_i$ and $\Sigma_m$ with $J$ and $M$ replaced by $J'$ and $M'$. The upper bound is

$$\bar{P}_{e,1} \leqslant (8J'NM')^{1-K} \Sigma_i' \left\{ \prod_{n=2}^{K} S_n{}^U(i_n) \right\}$$

$$\cdot \Sigma_m' Q\left( \alpha \left[ 1 + \sum_{k=2}^{K} (i_k/2J'N) \cos\left(\tfrac{1}{2} i_k \pi/M'\right) \right] \right) \qquad (30)$$

which is valid for all positive integers $J'$ and $M'$. As with the lower bound, the upper bound given by (30) becomes tighter as $J'$ and $M'$ increase. Indeed, if $\bar{P}_{e,1}^U$ denotes the right-hand side of (30), then the error $\bar{P}_{e,1}^U - \bar{P}_{e,1}$ is nonnegative and it decreases monotonically to zero as $J' \to \infty$ and $M' \to \infty$.

## GENERALIZED CHEBYSHEV BOUNDS

We examine a class of upper bounds which includes the Chebyshev and Chernoff bounds. This class is derived from the generalized Chebyshev inequality

$$P\{X \geqslant \beta\} \leqslant \frac{E\{h(X)\}}{h(\beta)} \qquad (31)$$

which holds for any nonnegative, nondecreasing function $h$, any real number $\beta$ such that $h(\beta) > 0$, and any random variable $X$. We consider two such functions: 1) the function $h_1$ defined by $h_1(x) = x^2$ for $x \geqslant 0$ and $h_1(x) = 0$ for $x < 0$, and 2) the function $h_2$ defined by $h_2(x) = \exp(sx)$, where $s$ is an arbitrary nonnegative real number. The latter case is the well-known Chernoff bound, which can be optimized by minimizing the upper bound with respect to $s$.

From (2) and (15) it follows that $\bar{P}_{e,1}$ can be written as

$$\bar{P}_{e,1} = P\left\{ \eta' + \sum_{k=2}^{K} I_{k,1}(b_k, \tau_k, \varphi_k) \geqslant 1 \right\} \qquad (32)$$

where $\eta'$ is a zero-mean Gaussian random variable with variance $\alpha^{-1}$. In (32) $\eta'$, $b$, $\tau$, $\varphi$ are all random variables and $\bar{P}_{e,1}$ is the average probability of error as defined in (9).

Let $\beta = +1$ and define

$$X = \eta' + \sum_{k=2}^{K} I_{k,1}(b_k, \tau_k, \varphi_k). \qquad (33)$$

Application of (31) with $h(x) = h_1(x)$ requires the computation of $E\{h_1(X)\}$. From [6] and the fact that the distribution of $X$ is symmetric about its mean we see that

$$E\{h_1(X)\} = \tfrac{1}{2} E(X^2) = \tfrac{1}{2} (SNR_1)^{-2} \qquad (34)$$

where $SNR_1$ is given in [6] for rectangular pulses and in [7] and [9] for general pulse shapes.

The computation of $E\{h_2(X)\}$ is more complicated so the details are omitted. The result is

$$E\{h_2(X)\} = E\{e^{sX}\}$$

$$= e^{\left(\frac{1}{2} s^2/\alpha\right)} \prod_{k=2}^{K} \left[ \frac{1}{2T} \int_0^T \{I_0(s[R_{k,1}(\tau) + \hat{R}_{k,1}(\tau)]/T) \right.$$

$$\left. + I_0(s[R_{k,1}(\tau) - \hat{R}_{k,1}(\tau)]/T)\} \, d\tau \right] \qquad (35)$$

where $I_0$ is the modified Bessel function of order zero.

The general Chebyshev bounds can also be evaluated for *random* signature sequences which are defined as follows. For each $k$, the sequence $(a_n{}^{(k)})$ is a sequence of independent identically distributed random variables with $P\{a_j{}^{(k)} = +1\} = P\{a_j{}^{(k)} = -1\} = \frac{1}{2}$ for each $j$. Also, for each $i \neq k$, $a_l{}^{(i)}$ and $a_j{}^{(k)}$ are independent. The resulting bound is $\bar{P}_e \leqslant \frac{1}{2} \overline{(SNR)}^{-2}$, where

$$\overline{SNR} = \left\{ \frac{N_0}{2E_b} + \frac{K-1}{NT_c{}^3} \int_0^{T_c} R^2(\tau) \, d\tau \right\}^{-1/2} \qquad (36)$$

The function $R$ is the aperiodic autocorrelation function for the chip waveform [9]. By evaluating $E\{h_2(X)\}$ and applying the result to (31) gives

$$\bar{P}_e \leqslant \min_{s \geqslant 0} \exp\left\{ -s + \tfrac{1}{2} s^2/\alpha \right\}$$

$$\cdot \left[ \frac{1}{T_c} \int_0^{T_c} \frac{2}{\pi} \int_0^{\pi/2} [f(s, \tau, \varphi)]^N \, d\varphi \, d\tau \right]^{K-1} \qquad (37a)$$

where

$$f(s, \tau, \varphi) = \cosh(sT^{-1} R(\tau) \cos\varphi)$$

$$\cdot \cosh(sT^{-1} R(T_c - \tau) \cos\varphi). \qquad (37b)$$

These bounds are specialized to binary PSK direct-sequence systems by letting $R(\tau) = \tau$ for $0 \leqslant \tau \leqslant T_c$, which is the aperiodic autocorrelation function for the rectangular chip waveform.

## NUMERICAL RESULTS

In this section we present some representative numerical results which will give an indication of the tightness of the bounds. Numerical values are given for the bounds for various choices of signature sequences. For $K = 2$ and $K = 3$ the signature sequences are $m$-sequences, and for $K = 4$ the Gold sequences of period 31 are employed since there are only three nonreciprocal $m$-sequences of period 31. The phases of the $m$-sequences employed for our numerical results are the auto-optimal least-sidelobe-energy (AO/LSE) phases given in [10] or the phases which maximize the SNR [2]. The shift-register tap connections and initial loadings are given in [2, Table 5] for the phases which maximize the SNR and in [10, Fig. A.1] for the AO/LSE phases.

Tables of numerical values for the bounds are here for various values of the parameters $K$, $N$, and $\bar{E}_b/N_0$. In order to give an indication of the amount of computation required in each case, we have specified the values of $J$ and $M$ used in (24) to compute the lower bound and the values of $J'$ and $M'$ used in (30) to compute the upper bound.

In Table I we give values for the lower and upper bound [(24) and (30)] denoted by $\bar{P}_{e,1}^L$ and $\bar{P}_{e,1}^U$, respectively, as a function of $J$ and $M$ for $K = 2$ and $N = 31$. For the data in this table, $J' = J$ and $M' = M$. The sequences are the first two AO/LSE sequences given in [10, Fig. A.1(a)]. Since $K$ is only 2, we can approximate the integrals using an algorithm based on Simpson's rule. This calculation gives $2.3975 \times 10^{-5}$ as the approximate value of $\bar{P}_{e,1}$. Notice that the lower bound converges faster than the upper bound. This is due to the type of bounds being used for the integrals in (19), (20), (27), and (28). For example, the integral in (28) is bounded using the rectangular rule which is the weakest bound employed. This accounts for the slower convergence in the upper bound. For $K = 3$ and $N = 31$ the bounds are given in Table II where again $J' = J$ and $M' = M$.

In Table III the upper and lower bounds are given for several values of $K$ and $N$. In Tables IV, V, and VI the bounds are given for $K = 2$, 3, and 4, respectively, and $N = 31$. Also tabulated is the approximation $\tilde{P}_{e,1} = Q(\text{SNR}_1)$ suggested in [6]. The values of $J$, $M$, $J'$, and $M'$ were chosen to give moderate computation for each case. The sequences for Tables IV and V are the first two or three AO/LSE $m$-sequences of [10, Fig. A.1(a)]. For Table VI the sequences are four Gold sequences that have the property that the maximum interference is less than the desired signal. The shift-register tap connection for these sequences is 3551 (in the notation of [10]), and the initial loadings (in octal notation) for sequences 1 through 4 are 1756, 0355, 0432, and 1306, respectively.

The importance of the selection of the phases of signature sequences for direct-sequence SSMA systems is illustrated in Fig. 1. In this figure the upper and lower bounds are shown for two different sets of phases of the *same set* of three $m$-sequences of period 31. These phases, which are given in [2], give the minimum and maximum possible SNR. Notice that at a bit error rate of $10^{-5}$, the difference in performance is about 2 dB. Thus, proper choice of the phases of the signature sequences can significantly improve the efficiency of a direct-sequence SSMA system.

TABLE I
BOUNDS FOR $K = 2$, $N = 31$, AND $\bar{E}_b/N_0 = 10$ dB

| $J$ | $M$ | $\bar{P}_{e,1}^L$ | $\bar{P}_{e,1}^U$ | |
|---|---|---|---|---|
| 20 | 10 | 2.322 | 2.689 | $(\times 10^{-5})$ |
| 40 | 20 | 2.378 | 2.534 | $(\times 10^{-5})$ |
| 60 | 30 | 2.389 | 2.485 | $(\times 10^{-5})$ |
| 80 | 40 | 2.393 | 2.463 | $(\times 10^{-5})$ |
| 160 | 80 | 2.396 | 2.429 | $(\times 10^{-5})$ |
| 300 | 150 | 2.397 | 2.414 | $(\times 10^{-5})$ |

TABLE II
BOUNDS FOR $K = 3$, $N = 31$, AND $\bar{E}_b/N_0 = 12$ dB

| $J$ | $M$ | $\bar{P}_{e,1}^L$ | $\bar{P}_{e,1}^U$ | |
|---|---|---|---|---|
| 8 | 2 | 5.31 | 36.20 | $(\times 10^{-6})$ |
| 8 | 4 | 6.24 | 23.70 | $(\times 10^{-6})$ |
| 16 | 4 | 8.95 | 19.87 | $(\times 10^{-6})$ |
| 16 | 6 | 9.22 | 18.04 | $(\times 10^{-6})$ |
| 32 | 6 | 10.20 | 16.05 | $(\times 10^{-6})$ |
| 32 | 12 | 10.38 | 13.41 | $(\times 10^{-6})$ |
| 64 | 12 | 10.65 | 13.24 | $(\times 10^{-6})$ |
| 64 | 16 | 10.68 | 12.63 | $(\times 10^{-6})$ |

TABLE III
BOUNDS FOR $\bar{E}_b/N_0 = 12$ dB

| $K$ | $N$ | $J$, $M$ | $J'$, $M'$ | $\bar{P}_{e,1}^L$ | $\bar{P}_{e,1}^U$ | |
|---|---|---|---|---|---|---|
| 2 | 31 | 300, 150 | 300, 150 | 4.81 | 4.86 | $(\times 10^{-7})$ |
| 3 | 31 | 64, 16 | 64, 16 | 1.07 | 1.26 | $(\times 10^{-7})$ |
| 4 | 31 | 32, 4 | 8, 16 | 4.01 | 6.16 | $(\times 10^{-5})$ |
| 2 | 127 | 160, 40 | 80, 80 | 3.94 | 3.98 | $(\times 10^{-8})$ |
| 3 | 127 | 12, 16 | 16, 12 | 1.17 | 1.42 | $(\times 10^{-7})$ |
| 2 | 255 | 120, 40 | 40, 80 | 1.98 | 2.00 | $(\times 10^{-8})$ |

TABLE IV
BOUNDS AND APPROXIMATION FOR $K = 2$, $N = 31$ ($J = 160$, $M = 100$; $J' = 80$, $M' = 120$)

| $\bar{E}_b/N_0$ (dB) | $\bar{P}_{e,1}^L$ | $\bar{P}_{e,1}^U$ | $\tilde{P}_{e,1}$ | |
|---|---|---|---|---|
| 4 | 1.44 | 1.45 | 1.44 | $(\times 10^{-2})$ |
| 6 | 3.35 | 3.36 | 3.34 | $(\times 10^{-3})$ |
| 8 | 4.22 | 4.24 | 4.17 | $(\times 10^{-4})$ |
| 10 | 2.40 | 2.42 | 2.34 | $(\times 10^{-5})$ |
| 12 | 4.81 | 4.89 | 5.13 | $(\times 10^{-7})$ |
| 14 | 2.28 | 2.34 | 4.43 | $(\times 10^{-9})$ |

TABLE V
BOUNDS AND APPROXIMATION FOR $K = 3$ AND $N = 31$ ($J = J' = 32$, $M = M' = 15$)

| $\bar{E}_b/N_0$ (dB) | $\bar{P}_{e,1}^L$ | $\bar{P}_{e,1}^U$ | $\tilde{P}_{e,1}$ | |
|---|---|---|---|---|
| 4 | 1.66 | 1.69 | 1.66 | $(\times 10^{-2})$ |
| 6 | 4.58 | 4.77 | 4.56 | $(\times 10^{-3})$ |
| 8 | 8.43 | 9.11 | 8.13 | $(\times 10^{-4})$ |
| 10 | 1.06 | 1.21 | 0.92 | $(\times 10^{-4})$ |
| 12 | 1.04 | 1.29 | 0.70 | $(\times 10^{-5})$ |
| 14 | 9.68 | 13.32 | 4.40 | $(\times 10^{-7})$ |

TABLE VI
BOUNDS AND APPROXIMATION FOR $K = 4$ AND $N = 31$ ($J = 16$, $M = 4; J' = 8, M' = 8$)

| $E_b/N_0$ (dB) | $\bar{P}_{e,1}^L$ | $\bar{P}_{e,1}^U$ | $\tilde{P}_{e,1}$ | |
|---|---|---|---|---|
| 4 | 1.87 | 1.99 | 1.88 | ($\times 10^{-2}$) |
| 6 | 5.85 | 6.67 | 5.94 | ($\times 10^{-3}$) |
| 8 | 1.36 | 1.74 | 1.38 | ($\times 10^{-3}$) |
| 10 | 2.44 | 3.74 | 2.48 | ($\times 10^{-4}$) |
| 12 | 3.74 | 7.29 | 3.84 | ($\times 10^{-5}$) |
| 14 | 5.37 | 14.19 | 6.16 | ($\times 10^{-6}$) |

TABLE VII
BOUNDS, APPROXIMATION, AND CHERNOFF BOUND FOR
$K = 3$ AND $N = 127$ ($J = 12, M = 16; J' = 16, M' = 12$)

| $E_b/N_0$ (dB) | $\bar{P}_{e,1}^L$ | $\bar{P}_{e,1}^U$ | $\tilde{P}_{e,1}$ | $\bar{P}_{e,1}^C$ | |
|---|---|---|---|---|---|
| 4 | 1.35 | 1.36 | 1.35 | 8.67 | ($\times 10^{-2}$) |
| 6 | 2.86 | 2.91 | 2.86 | 22.0 | ($\times 10^{-3}$) |
| 8 | 2.95 | 3.07 | 2.94 | 27.5 | ($\times 10^{-4}$) |
| 10 | 1.11 | 1.22 | 1.08 | 12.6 | ($\times 10^{-5}$) |
| 12 | 1.17 | 1.41 | 0.98 | 16.3 | ($\times 10^{-7}$) |
| 14 | 3.03 | 4.35 | 1.62 | 51.8 | ($\times 10^{-10}$) |

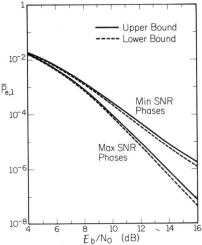

Fig. 1. Upper and lower bounds for two sets of sequences with different phases ($K = 3, N = 31$).

TABLE VIII
APPROXIMATION AND CHERNOFF BOUND FOR RANDOM
SEQUENCES OF LENGTH 127 WITH $K = 3$

| $E_b/N_0$ (dB) | $\tilde{P}_e$ | $\bar{P}_e C$ |
|---|---|---|
| 4 | $1.35 \times 10^{-2}$ | $4.33 \times 10^{-2}$ |
| 6 | $2.85 \times 10^{-3}$ | $1.10 \times 10^{-2}$ |
| 8 | $2.41 \times 10^{-4}$ | $1.37 \times 10^{-3}$ |
| 10 | $1.05 \times 10^{-5}$ | $6.51 \times 10^{-5}$ |
| 12 | $9.29 \times 10^{-8}$ | $1.02 \times 10^{-6}$ |
| 14 | $1.44 \times 10^{-10}$ | $6.69 \times 10^{-9}$ |

Numerical results for $K = 3$ and $N = 127$ are shown in Table VII for the AO/LSE sequences given in [10]. In addition to the bounds of (24) and (30), the approximation $\tilde{P}_{e,1}$ and the Chernoff bound $\bar{P}_{e,1}^C$ are shown. Results on the Chernoff bound $\bar{P}_e^C$ for random sequences of length 127 are presented in Table VIII for $K = 3$. As expected, the Chernoff bound is not very tight. Numerical values for the approximation $\tilde{P}_e = Q(\overline{\text{SNR}})$ for random sequences are also given. We found the Chebyshev bound was far too loose to be of any interest, so detailed numerical results are not presented. As an illustration, the Chebyshev bound is $1.26 \times 10^{-2}$ for random sequences of length 127 with $K = 3$ and $E_b/N_0 = 14$ dB.

We close by giving some comparisons between the bounds presented in this paper and the moment-space bounds of [14]. The numerical results presented in [1] for the second-moment bounds indicate that the bounds presented in this paper are tighter even for small values of $J$. In addition, we have found that our bounds are tighter than the single-exponential moment-space bounds (even for relatively small values of $J$ and $M$). For $K = 2$ and $N = 31$ we computed the single-exponential bounds and the bounds presented in this paper. The two signature sequences are specified by the feedback connections 45 and 67 with loadings 06 and 36, respectively. These are two sequences from the set obtained in [2]. For $E_b/N_0 = 12$ dB the upper and lower single-expo-

nential bounds are $7.830 \times 10^{-7}$ and $5.401 \times 10^{-7}$, respectively. The difference between the upper and lower bounds is $2.429 \times 10^{-7}$. Our bounds for $J' = J = 20$ and $M' = M = 10$ are $9.020 \times 10^{-7}$ and $6.871 \times 10^{-7}$, so the difference is $2.149 \times 10^{-7}$. For $J' = J = 40$ and $M' = M = 80$ we find that our bounds are $7.680 \times 10^{-7}$ and $7.309 \times 10^{-7}$, which is a difference of only $0.371 \times 10^{-7}$.

## APPENDIX A

In this Appendix we present the lower bounds for the integrals that appear in (19) and (20). These are obtained as key steps in the derivation of (24).

The first step is to consider (17)–(20) for $n = 2$. We let $G_2^L(b, \tau, \varphi) = G_2(b, \tau, \varphi)$ and observe from (10) and (13a) that

$$\hat{G}_2^L(b, \tau, \varphi) = (4T)^{-1} \Sigma_{b_2}$$
$$\cdot \int_0^T Q(\alpha[1 + \beta_3 + I_{2,1}(b_2, \tau_2, \varphi_2)]) \, d\tau_2 \tag{A.1}$$

where $\beta_3$ does not depend on $b_2$, $\tau_2$, or $\varphi_2$. The convexity of $Q$ on $[0, \infty)$ implies that

$$T_c^{-1} J \int_{\Delta(l,j)}^{\Delta(l,j+1)} G_2^L(b, \tau, \varphi) \, d\tau_2$$
$$\geqslant Q(\alpha[1 + \beta_3 + \hat{I}_{2,1}(b_2, \varphi_2; l, j)]) \tag{A.2}$$

where

$$\hat{I}_{2,1}(b_2, \varphi_2; l, j)$$
$$= T_c^{-1} J \int_{\Delta(l,j)}^{\Delta(l,j+1)} I_{2,1}(b_2, \tau_2, \varphi_2) \, d\tau_2. \tag{A.3}$$

From a comparison of (10) and (A.1) it is easy to see that the quantity $\beta_3$ is given by

$$\beta_3 = \sum_{k=3}^{K} I_{k,1}(b_k, \tau_k, \varphi_k). \tag{A.4}$$

Furthermore, if we define

$$\Gamma_k(l,j) = \int_{\Delta(l,j)}^{\Delta(l,j+1)} R_{k,1}(\tau)\,d\tau \tag{A.5}$$

and

$$\hat{\Gamma}_k(l,j) = \int_{\Delta(l,j)}^{\Delta(l,j+1)} \hat{R}_{k,1}(\tau)\,d\tau \tag{A.6}$$

for $2 \leqslant k \leqslant K$, then we see from (5) that

$$\hat{I}_{2,1}(b_2, \varphi_2; l, j)$$
$$= J(T_c T)^{-1}\{b_{-1}{}^{(2)}\Gamma_2(l,j) + b_0{}^{(2)}\hat{\Gamma}_2(l,j)\} \cos \varphi_2. \tag{A.7}$$

Notice that we have not made use of (6) in arriving at (A.7). Instead, (13)-(20) and (A.1)-(A.7) are valid for arbitrary time-limited pulses such as the sine pulse considered in [7] and [9]. For the rectangular pulse it follows that

$$\Gamma_k(l,j) = J^{-1}T_c R_{k,1}(\bar{\Delta}(l,j)) \tag{A.8}$$

and

$$\hat{\Gamma}_k(l,j) = J^{-1}T_c \hat{R}_{k,1}(\bar{\Delta}(l,j)) \tag{A.9}$$

because of the linear form of $R_{k,i}(\tau)$ and $\hat{R}_{k,i}(\tau)$ as demonstrated in (6). Alternatively, we can work directly with (A.3) and deduce from (5) and (6) that for the rectangular pulse waveform

$$\hat{I}_{2,1}(b_2, \varphi_2; l, j) = I_{2,1}(b_2, \bar{\Delta}(l,j), \varphi_2)$$
$$= T^{-1}[b_{-1}{}^{(2)}R_{2,1}(\bar{\Delta}(l,j))$$
$$+ b_0{}^{(2)}\hat{R}_{2,1}(\bar{\Delta}(l,j))] \cos \varphi_2. \tag{A.10}$$

If we now combine the above results for the rectangular pulse with (A.2) we have the lower bound

$$T_c^{-1}J \int_{\Delta(l,j)}^{\Delta(l,j+1)} G_2(b, \tau, \varphi)\,d\tau_2$$
$$\geqslant Q(\alpha[1 + \beta_3 + I_{2,1}(b_2, \bar{\Delta}(l,j), \varphi_2)]), \tag{A.11}$$

which can then be employed in (19) to obtain a lower bound for $\hat{G}_2{}^L(b, \tau, \varphi)$. We can simplify the computation of the right-hand side of (19) by replacing the triple sum (over $b_2$, $l$, and $j$) by a single sum as follows. First, observe that

the quantity $T_c^{-1}\{b_{-1}{}^{(2)}R_{2,1}(lT_c) + b_0{}^{(2)}\hat{R}_{2,1}(lT_c)\}$ takes on integer values between $-N$ and $N$ (this follows from (6) and properties of the aperiodic correlation function). Consequently, (6) and (A.10) imply that $(2JT/T_c)I_{2,1}(b_2, \bar{\Delta}(l,j), 0)$ takes on integer values between $-2JN$ and $2JN$. Since $T = NT_c$ we define $S_2{}^L(i)$ as in (21). Let $S_2{}^L(i) = |S_2{}^L(i)|$, the number of elements in $S_2{}^L(i)$. From (19) and (A.11) we have

$$\hat{G}_2{}^L(b, \tau, \varphi) \geqslant (2N')^{-1} \sum_{i=-N'}^{N'} S_2{}^L(i)$$
$$\cdot Q(\alpha[1 + \beta_3 + (i/N')\cos\varphi_2]) \tag{A.12}$$

where $N' = 2JN$. It follows from (20), (23), the convexity of $Q$ on $[0, \infty)$, and the symmetry of the multiple-access interference that if

$$G_3{}^L(b, \tau, \varphi)$$
$$= (2N'M)^{-1} \sum_{i=-N'}^{N'} S_2{}^L(i) \sum_{m=0}^{M-1} Q(\alpha[1 + \beta_3 + i\gamma(m)]) \tag{A.13}$$

then $G_3(b, \tau, \varphi) \geqslant G_3{}^L(b, \tau, \varphi)$, as required.

This completes the derivation of the lower bound for $n = 2$. The next step in obtaining a lower bound for $\bar{P}_{e,1}$ is to employ (A.13) with $\beta_3$ as defined in (A.4) to (17)-(20) with $n = 3$. For example, (17) yields

$$\hat{G}_3{}^L(b, \tau, \varphi) = (2N'M)^{-1}\Sigma_{i_2}\Sigma_{m_2}S_2(i_2)$$
$$\cdot H_3(b, \tau, \varphi; i_2, m_2) \tag{A.14}$$

where

$$H_3(b, \tau, \varphi; i, m)$$
$$= (4T)^{-1}\Sigma_{b_3} \int_0^T Q(\alpha[1 + \beta_4(i,m)$$
$$+ I_{3,1}(b_3, \tau_3, \varphi_3)]\,d\tau_3 \tag{A.15}$$

for $|i| \leqslant N'$ and $0 \leqslant m \leqslant M-1$. We use $\Sigma_{i_n}$ and $\Sigma_{m_n}$ to denote the sums from $i_n = -N'$ to $N'$ and $m_n = 0$ to $M-1$, respectively. The quantity $\beta_4(i, m)$ is defined by

$$\beta_4(i,m) = i\gamma(m) + \sum_{k=4}^{K} I_{k,1}(b_k, \tau_k, \varphi_k). \tag{A.16}$$

Therefore, it does not depend on $b_3$, $\tau_3$, or $\varphi_3$.

A comparison of (A.15) with (A.1) shows that we can obtain a lower bound on $H_3(b, \tau, \varphi; i, m)$ in exactly the same way as we obtained the lower bound on $\hat{G}_2{}^L(b, \tau, \varphi)$; namely, we apply the procedure outlined in (A.2)-(A.12). This gives a lower bound of the form

$$\hat{G}_3{}^L(b,\tau,\varphi) \geqslant M^{-1}(2N')^{-2}\Sigma_{i_2}\Sigma_{i_3}S_2(i_2)S_3(i_3)\Sigma_{m_2}$$

$$\cdot Q(\alpha[1+\beta_4(i_2,m_2)+(i_3/N')\cos\varphi_3]).$$

$$(A.17)$$

We then proceed as in (A.13) to obtain

$$G_4{}^L(b,\tau,\varphi) = (2N'M)^{-2}\Sigma_{i_2}\Sigma_{i_3}S_2(i_2)S_3(i_3)\Sigma_{m_2}\Sigma_{m_3}$$

$$\cdot Q(\alpha[1+\beta_4(i_2,m_2)+i_3\gamma(m_3)]).\quad (A.18)$$

The pattern should be clear at this point, so the reader should be able to verify the final expression (24).

Notice that the lower bound of (A.11) is also the approximation to the integral on the left-hand side of (A.11) that is obtained by applying a rectangular integration rule using the value of the function at the midpoint of the interval. Similarly, the lower bound of (A.13) can be viewed as an application of a rectangular integration rule to the integrals of the right-hand side of (20) using the value of the function at some point (not necessarily the midpoint) of the interval $[\psi(m),$ $\psi(m+1)]$. Thus, the error $G_3(b,\tau,\varphi)-G_3{}^L(b,\tau,\varphi)$ in the bound of (A.13) is nonnegative. Moreover, it converges monotonically to zero as $J \to \infty$ and $M \to \infty$. The same conclusion holds for the error $\bar{P}_{e,1} - \bar{P}^L_{e,1}$ in the lower bound, where $\bar{P}^L_{e,1} = G^L_{K+1}(b,\tau,\varphi)$ is just the right-hand side of (24).

## APPENDIX B

In this Appendix we present upper bounds for the integrals given in (27) and (28). First, the special case $n = 2$ is handled by letting $\hat{G}_2{}^U(b,\tau,\varphi) = \hat{G}_2{}^L(b,\tau,\varphi)$, which is given by (A.1). If the chip waveform is the rectangular pulse, then $\hat{G}_2{}^U(b,\tau,\varphi)$ is a convex function of $\tau_2$ on the interval $[lT_c, (l+1)T_c]$ and so

$$T_c{}^{-1}J \int_{\Delta'(l,j)}^{\Delta'(l,j+1)} G_2{}^U(b,\tau,\varphi)\,d\tau_2$$

$$\leqslant \tfrac{1}{2}Q(a[1+\beta_3+I_{2,1}(b_2,\Delta'(l,j),\varphi_2)])$$

$$+ \tfrac{1}{2}Q(\alpha[1+\beta_3+I_{2,1}(b_2,\Delta'(l,j+1),\varphi_2)]).\quad (B.1)$$

The inequality in (B.1) follows from the fact that the (normalized) integral of a convex function on any subinterval is not greater than the average of the values of the function at the two endpoints of the subinterval. Notice also that the right-hand side of (B.1) is just the trapezoidal approximation to the integral; therefore the difference between the right-hand side and the left-hand side of (B.1) converges monotonically to zero as $J' \to \infty$.

We then proceed as in (A.10)–(A.12) to obtain

$$\hat{G}_2{}^U(b,\tau,\varphi) \leqslant (2N'')^{-1} \sum_{i=-N''}^{N''} S_2{}^U(i)$$

$$\cdot Q(\alpha[1+\beta_3+(i/2J'N)\cos\varphi_2])\quad (B.2)$$

where $N'' = 2J'N$. It is easy to show that the quantity

$$Q(\alpha[1+\beta_3-(i/2J'N)\cos\varphi])$$

$$+ Q(\alpha[1+\beta_3+(i/2J'N)\cos\varphi])$$

is a decreasing function of $\varphi$ on $[0,\tfrac{1}{2}\pi]$. Thus

$$(2M/\pi)\int_{\psi'(m)}^{\psi'(m+1)} \hat{G}_2{}^U(b,\tau,\varphi)\,d\varphi_2$$

$$\leqslant (2N'')^{-1}\sum_{i=-N''}^{N''} S_2{}^U(i)Q(\alpha[1+\beta_3+(i/2J'N)$$

$$\cdot \cos\psi'(m)]).\quad (B.3)$$

Thus we let

$$G_3{}^U(b,\tau,\varphi) = (2N''M')^{-1}\sum_{i=-N''}^{N''} S_2{}^U(i)$$

$$\cdot \sum_{m=0}^{M'-1} Q(\alpha[1+\beta_3+(i/2J'N)\cos\psi'(m)])$$

$$(B.4)$$

which is analogous to (A.13).

Notice that the right-hand side of (B.3) can be obtained from an application of the rectangular integration rule to the left-hand side. For this approximation we use the value of the function at the left endpoint [i.e., at $\psi'(m)$] of the interval $[\psi'(m),\ \psi(m+1)]$. Thus, the error in the resulting approximation converges monotonically to zero as $M' \to \infty$.

The procedure for continuing on to $n = 3$ is the same as outlined in Appendix A, especially (A.14)–(A.18). By following the same pattern for $n = 4, 5, \cdots, K+1$, the reader can verify the bound of (30).

## REFERENCES

[1] D. E. Borth, M. B. Pursley, D. V. Sarwate, and W. E. Stark, "Bounds on error probability for direct-sequence spread-spectrum multiple-access communications," in *Proc. MIDCON Conf.*, Nov. 1979, paper 15/1, pp. 1–14.

[2] F. D. Garber and M. B. Pursley, "Optimal phases of maximal-length sequences for asynchronous spread-spectrum multiplexing," *Electron. Lett.*, vol. 16, pp. 756–757, Sept. 1980.

[3] E. A. Geraniotis and M. B. Pursley, "Error probability for binary PSK spread-spectrum multiple-access communications," in *Proc. Conf. Inform. Sci. Syst.*, Johns Hopkins Univ., Baltimore, MD, Mar. 1981, pp. 238–244.

[4] ——, "Error probability for direct-sequence spread-spectrum multiple-access communications—Part II: Approximations," this issue, pp. 985–995.

[5] D. Laforgia, A. Luvison, and V. Zingarelli, "Exact bit error probability with application to spread-spectrum multiple access communications," in *Proc. IEEE Int. Conf. Commun.*, June 1981, vol. 4, pp. 76.5.1–76.5.5.

[6] M. B. Pursley, "Performance evaluation for phase-coded spread-spectrum multiple-access communications—Part I: System analysis," *IEEE Trans. Commun.*, vol. COM-25, pp. 795–799, Aug. 1977.

[7] ——, "Spread-spectrum multiple-access communications," in *Multi-User Communication Systems*, G. Longo, Ed. New York: Springer-Verlag, 1981, pp. 139–199.

[8]  ——, "Effects of specular multipath fading on spread-spectrum communications," in *New Concepts in Multi-User Communication,* J. K. Skwirzynski, Ed. Alphen aan den Rijn, The Netherlands: Sijtoff and Noordhoff, 1981, pp. 481–505.

[9]  M. B. Pursley, F. D. Garber, and J. S. Lehnert, "Analysis of generalized quadriphase spread-spectrum communications," in *Proc. IEEE Int. Conf. Commun.,* June 1980, vol. 1, pp. 15.3.1–15.3.6.

[10] M. B. Pursley and H. F. A. Roefs, "Numerical evaluation of correlation parameters for optimal phases of binary shift-register sequences," *IEEE Trans. Commun.,* vol. COM-27, pp. 1597–1604, Oct. 1979.

[11] M. B. Pursley, D. V. Sarwate, and W. E. Stark, "On the average probability of error for direct-sequence spread-spectrum multiple-access systems," in *Proc. 1980 Conf. Inform. Sci. Syst.,* Princeton Univ., Princeton, NJ, Mar. 1980, pp. 320–325.

[12] H. F. A. Roefs, "Binary sequences for spread spectrum multiple access communication," Ph.D. dissertation, Dep. Elec. Eng., Univ. of Illinois, Urbana, and Coord. Sci. Lab., Rep. R-785, Aug. 1977.

[13] K. -T. Wu and D. L. Neuhoff, "Average error probability for direct sequence spread spectrum multiple access communication systems," in *Proc. 18th Annu. Allerton Conf. Commun., Contr., and Comput.,* Oct. 1980, pp. 359–368.

[14] K. Yao, "Error probabilty of asynchronous spread spectrum multiple access communication systems," *IEEE Trans. Commun.,* vol. COM-25, pp. 803–809, Aug. 1977.

# Error Probability for Direct-Sequence Spread-Spectrum Multiple-Access Communications—Part II: Approximations

EVAGGELOS A. GERANIOTIS, STUDENT MEMBER, IEEE, AND MICHAEL B. PURSLEY, FELLOW, IEEE

*Abstract*—Approximations are obtained for the average probability of error in an asynchronous direct-sequence spread-spectrum multiple-access communications system. Both binary and quaternary systems are considered, and the chip waveforms are allowed to be arbitrary time-limited waveforms with time duration equal to the inverse of the chip rate. The approximation is based on the integration of the characteristic function of the multiple-access interference. The amount of computation required to evaluate this approximation grows only linearly with the product of the number of simultaneous transmitters and the number of chips per bit. The accuracy of the approximation is extremely good in most cases, but it can be improved, if necessary, by an application of a series expansion. Numerical results are presented for specific chip waveforms and signature sequences.

## I. INTRODUCTION

THIS paper is concerned with the average probability of error for asynchronous binary and quaternary direct-sequence spread-spectrum multiple-access (DS/SSMA) communications over an additive white Gaussian noise channel. The system model, which is described in detail in [8] and [9], is an extension of the model for binary DS/SSMA communications that was introduced in [7] and employed in all of the recent work on the average probability of error for binary DS/SSMA systems (e.g., [10] and [16]-[18]). The spread-spectrum signals are quaternary signals with (possibly offset) in-phase and quadrature components, each of which is a binary direct-sequence spread-spectrum signal. The chip waveforms for these binary signals are pulses of arbitrary shape with duration equal to the inverse of the chip rate. Thus the model is sufficiently general to include the forms of direct-sequence modulation known as quadriphase-shift-keying (QPSK), offset QPSK, and minimum-shift-keying (MSK) modulation.

During the past few years, several approximations and bounds have been developed for the average probability of error in a DS/SSMA system. The approximation suggested in [7] is by far the easiest to compute, due in part to the fact that it is given in terms of the aperiodic *autocorrelation* functions for the signature sequences (crosscorrelation functions are not needed). The upper and lower *bounds* of [11] and [18] give good results for small values of the number $K$ of transmitters. Because the amount of computation required

Manuscript received July 21, 1981; revised February 15, 1982. This work was supported by the Army Research Office under Grant DAAG29-78-G-0114 and DAAG29-81-K-0064.

The authors are with the Coordinated Science Laboratory and the Department of Electrical Engineering, University of Illinois, Urbana, IL 61801.

in the numerical evaluation increases exponentially as $K$ increases, these bounds are not intended for use in systems with more than a few simultaneously active transmitters. All such assessments of computational requirements are based on the amount of computation needed to evaluate the error probability for a given signal once the crosscorrelation functions have been computed. Straightforward evaluation of the crosscorrelation functions for a given signal require on the order of $KN^2$ computations, where $N$ is the signature sequence length (there are algorithms with lower computational requirements for large $K$ and $N$). The most recent contributions to the problem are the methods of approximation given in [4], [6], [16], [17], and in this paper. Taken collectively, the algorithms proposed in these papers cover a wide range of system parameters. Each technique has its advantages and disadvantages, and the ultimate selection of the technique to be used for a given problem depends on the system parameters and the available computer resources (both hardware and software). Some comparisons of the computational requirements of the various methods are given in [16].

In this paper we obtain an approximation with a computational requirement that is linear in the product $NK$. This approximation is based on the integration of the characteristic function of the multiple-access interference component of the output of the correlation receiver. This method, which we refer to as the characteristic-function method, was applied to intersymbol interference problems in [15] (see also [13]). We show that when adapted to DS/SSMA problems, the characteristic-function method provides a very accurate approximation to the average probability of error (it is certainly accurate enough for applications). Moreover, even greater accuracy can be achieved by using this approximation to obtain an expansion point for a Taylor series representation of the actual probability of error. By employing this combination of the characteristic-function method followed by a series-expansion method, we can obtain any prespecified degree of accuracy. Another attractive feature of the characteristic-function method is that it does not require the evaluation of high-order moments. This feature is particularly important for certain applications, such as spread-spectrum communications on channels with selective fading in which high-order moments are difficult to compute.

An overview of the organization of this paper is as follows. Binary DS/SSMA systems are considered in Section II, and the general quaternary DS/SSMA systems are treated in Section III. In each of these sections we first give a brief description of

Reprinted from *IEEE Trans. Commun.*, vol. COM-30, pp. 985–995, May 1982.

the system models. Next, the characteristic-function method is developed for the DS/SSMA application, and the resulting computational requirements are considered. Finally, the moments of the multiple-access interference are evaluated for use in the series-expansion method. In Section IV numerical results are presented for specific waveforms and signature sequences.

## II. BINARY DS/SSMA COMMUNICATIONS

### A. System Model

The model of the asynchronous binary DS/SSMA system that we consider is described in [8, Sect. II] and [9, Sect. II]. This model is a straightforward generalization of the model of [7] to account for chip waveforms $\psi(t)$ of arbitrary shape. The only restriction is that $\psi(t) = 0$ except for $0 \leqslant t < T_c$, where $T_c^{-1}$ is the chip rate for the system. The chip waveform is normalized to have energy equal to $T_c$, that is, $\int_0^{T_c} \psi^2(t)\, dt = T_c$.

As in [11], we consider the output of the correlation receiver which is matched to the first of the $K$ direct-sequence spread-spectrum signals in the DS/SSMA system. If $b_1{}^{(1)}$ is transmitted the output is the random variable

$$Z = \eta + (\tfrac{1}{2}P)^{1/2} T \left\{ b_1{}^{(1)} + \sum_{k=2}^{K} I_{k,1}(\boldsymbol{b}_k, \tau_k, \varphi_k) \right\}. \quad (1)$$

The random variable $\eta$ is Gaussian with mean zero and variance equal to $N_0 T/4$, where $N_0/2$ is the two-sided spectral density of the white Gaussian noise and $T$ is the data bit duration. The vector $\boldsymbol{b}_k = (b_1{}^{(k)}, b_2{}^{(k)})$ represents a pair of consecutive data bits for the $k$th signal ($b_l{}^{(k)}$ is either $+1$ or $-1$) and $\tau_k$ and $\varphi_k$ represent the time delay (modulo $T$) and the phase angle (modulo $2\pi$), respectively, of the $k$th signal relative to the first. Under fairly general conditions $\tau_k$ and $\varphi_k$ are independent and uniformly distributed on the intervals $[0, T]$ and $[0, 2\pi]$, respectively [8, pp. 159–160]. The assumption that each of the $K$ signals has power $P$, which is required for (1), is not necessary for the methods considered in this paper. The equal power assumption is made for notational simplicity and for convenience in presenting numerical results.

The function $I_{k,1}$ which appears in (1) represents the normalized multiple-access interference due to the $k$th signal. This function is defined by

$$I_{k,1}(\boldsymbol{b}_k, \tau, \varphi) = T^{-1}[b_1{}^{(k)} R_{k,1}(\tau)$$
$$+ b_2{}^{(k)} \hat{R}_{k,1}(\tau)] \cos \varphi \quad (2)$$

where the functions $R_{k,m}$ and $\hat{R}_{k,m}$ are given in terms of the aperiodic crosscorrelation function $C_{k,m}$ for the $k$th and $m$th signature sequences by

$$R_{k,m}(\tau) = C_{k,m}(l-N)\hat{R}_\psi(\tau - lT_c)$$
$$+ C_{k,m}(l+1-N)R_\psi(\tau - lT_c) \quad (3)$$

and

$$\hat{R}_{k,m}(\tau) = C_{k,m}(l)\hat{R}_\psi(\tau - lT_c)$$
$$+ C_{k,m}(l+1)R_\psi(\tau - lT_c) \quad (4)$$

where $l = \lfloor \tau/T_c \rfloor$, the integer part of $\tau/T_c$, and $R_\psi$ and $\hat{R}_\psi$ are the partial autocorrelation functions for the chip waveform. That is, for $0 \leqslant s \leqslant T_c$,

$$\hat{R}_\psi(s) = \int_s^{T_c} \psi(t)\psi(t-s)\, dt$$

and $R_\psi(s) = \hat{R}_\psi(T_c - s)$. The expressions (2)–(4) are for a DS/SSMA system in which each of the $K$ spread-spectrum signals has $N$ chips per bit (i.e., $T = NT_c$), and each signature sequence has a period which is a divisor of $N$.

The properties of the asynchronous DS/SSMA system under consideration imply that the collection of variables $\boldsymbol{b}_k$, $\tau_k$, and $\varphi_k$, $2 \leqslant k \leqslant K$, is a set of mutually independent random variables. We assume that $b_l{}^{(k)}$, $-\infty < l < \infty$, is a sequence of independent data bits for each $k$ and that $P(b_l{}^{(k)} = +1) = \frac{1}{2}$ for each $k$ and $l$. The average probability of error $\bar{P}_{e,1}$ for the correlation receiver matched to the first signal is then defined in terms of the mutually independent, uniformly distributed, random variables $\boldsymbol{b}_k$, $\tau_k$, and $\varphi_k$, $2 \leqslant k \leqslant K$. If we let $I$ be the random variable defined by

$$I = \sum_{k=2}^{K} I_{k,1}(\boldsymbol{b}_k, \tau_k, \varphi_k)$$

then $\bar{P}_{e,1}$ may be written as

$$\bar{P}_{e,1} = \tfrac{1}{2} P\{Z \leqslant 0 \,|\, b_1{}^{(1)} = +1\}$$
$$+ \tfrac{1}{2} P\{Z > 0 \,|\, b_1{}^{(1)} = -1\}$$
$$= \tfrac{1}{2} - \tfrac{1}{2} P\{-1 < \eta' + I \leqslant 1\} \quad (5)$$

where $\eta' = (\tfrac{1}{2}PT^2)^{-1/2}\eta$. Notice that $\eta'$ has variance $N_0/2E_b$ where $E_b = PT$ is the energy per data bit.

### B. The Characteristic-Function Method

Let $\phi_1$ denote the characteristic function for the random variable $I$, that is, $\phi_1(u) = E\{e^{ju\,I}\}$ where $u$ is a real variable and $j$ is the complex square root of $-1$. Let $\phi_2$ and $\phi$ be the characteristic functions for $\eta'$ and $\eta' + I$, respectively. Because of the symmetry of the distributions of $\eta'$ and $I$ (and hence of $\eta' + I$), the characteristic functions $\phi_1$, $\phi_2$, and $\phi$ are all real-valued functions (and, therefore, they take values in the interval $[-1, 1]$). The symmetry of the distributions also implies that these characteristic functions are even functions [e.g., $\phi(u) = \phi(-u)$]. Thus, (5) may be written

$$\bar{P}_{e,1} = \tfrac{1}{2} - \pi^{-1} \int_0^1 \int_0^\infty \phi(u) \cos(ux)\, du\, dx$$

$$= \tfrac{1}{2} - \pi^{-1} \int_0^\infty u^{-1} (\sin u)\phi(u)\, du. \quad (6)$$

Since $\eta'$ and $I$ are independent, then $\phi(u) = \phi_1(u)\phi_2(u)$, and since $\eta'$ is a zero-mean Gaussian random variable with variance $N_0/2E_b$ then

$$\phi_2(u) = \exp\left[-(N_0/4E_b)u^2\right]. \tag{7}$$

For computational purposes, it is convenient to replace $\phi(u)$ by $\phi_2(u) - \phi_2(u)[1 - \phi_1(u)]$ in (6). Making this substitution we have

$$\bar{P}_{e,1} = \tfrac{1}{2} - \pi^{-1}\int_0^\infty u^{-1}(\sin u)\phi_2(u)\,du$$

$$+ \pi^{-1}\int_0^\infty u^{-1}(\sin u)\phi_2(u)[1 - \phi_1(u)]\,du. \tag{8}$$

Now if $K = 1$ then $I = 0$ (no multiple-access interference) and hence $\phi_1(u) \equiv 1$. Moreover, if $K = 1$ then $\bar{P}_{e,1} = Q([2E_b/N_0]^{1/2})$ where

$$Q(x) = \int_x^\infty (2\pi)^{-1/2}\exp(-y^2/2)\,dy.$$

Consequently, it must be that (8) reduces to

$$\bar{P}_{e,1} = Q([2E_b/N_0]^{1/2})$$

$$+ \pi^{-1}\int_0^\infty u^{-1}(\sin u)\phi_2(u)[1 - \phi_1(u)]\,du. \tag{9}$$

Since the $K - 1$ multiple-access interference terms $I_{k,1}(b_k, \tau_k, \varphi_k)$, $2 \leq k \leq K$, are mutually independent, then

$$\phi_1(u) = \prod_{k=2}^K \left\{ (8\pi T)^{-1} \right.$$

$$\left. \cdot \Sigma_b \int_0^{2\pi}\int_0^T \exp[juI_{k,1}(b,\tau,\varphi)]\,d\tau\,d\varphi \right\} \tag{10}$$

where $\Sigma_b$ denotes the sum over all $b = (b_1, b_2)$ with $b_l \in \{+1, -1\}$. The expression in (10) can be written as

$$\phi_1(u)$$

$$= \prod_{k=2}^K \left\{ (2N)^{-1}\sum_{l=0}^{N-1} [f(u;l,\theta_{k,1}) + f(u;l,\hat{\theta}_{k,1})] \right\} \tag{11}$$

where

$$f(u;l,g)$$

$$\triangleq \frac{1}{2\pi T_c}\int_0^{2\pi}\int_0^{T_c} \exp\left\{ju[g(l + 1)R_\psi(\tau)\right.$$

$$\left. + g(l)\hat{R}_\psi(\tau)]\frac{\cos\theta}{T}\right\}\,d\tau\,d\theta \tag{12}$$

for an arbitrary function $g$. The functions $\theta_{k,1}$ and $\hat{\theta}_{k,1}$ that appear in (11) are the usual periodic and odd crosscorrelation functions for the binary signature sequences ([7] or [12]). It is easy to see that (12) implies

$$f(u;l,g) = \frac{2}{\pi T_c}\int_0^{(1/2)\pi}\int_0^{T_c} \cos\left\{u[g(l + 1)R_\psi(\tau)\right.$$

$$\left. + g(l)\hat{R}_\psi(\tau)]\frac{\cos\theta}{T}\right\}\,d\tau\,d\theta. \tag{13}$$

For a binary phase-shift-keyed (PSK) system the chip waveform $\psi(t)$ is the rectangular pulse of duration $T_c$ (i.e., $\psi(t) = 1$ for $0 \leq t < T_c$ and $\psi(t) = 0$ otherwise). For this waveform, $R_\psi(\tau) = \tau$ and $\hat{R}_\psi(\tau) = T_c - \tau$ for $0 \leq \tau \leq T_c$. Thus, for the binary PSK system, (13) reduces the following expression which was obtained in [4]:

$$f(u;l,g) = \frac{2}{\pi}\int_0^{(1/2)\pi} \mathrm{sinc}\left\{u[g(l + 1) - g(l)]\frac{\cos\theta}{2\pi N}\right\}$$

$$\cdot \cos\left\{u[g(l + 1) + g(l)]\frac{\cos\theta}{2N}\right\}\,d\theta \tag{14}$$

where $\mathrm{sinc}(x) = (\pi x)^{-1}\sin(\pi x)$ for $x \neq 0$ and $\mathrm{sinc}(0) = 1$.

Next, we consider a binary system with a chip waveform given by $\psi(t) = \sqrt{2}\sin(\pi t/T_c)$ for $0 \leq t < T_c$ and $\psi(t) = 0$ otherwise. This waveform, which we refer to as the sine pulse, is the basic waveform employed in MSK. The partial autocorrelation functions $R_\psi$ and $\hat{R}_\psi$ for the sine pulse are given in [8] and [9]. The results of [9] imply that the sine pulse is superior to the rectangular pulse for DS/SSMA systems if the performance criterion is the signal-to-noise ratio as defined in [7]–[9]. The techniques described in the present paper are employed to see if this conclusion is true if the average probability of error is the performance criterion.

In order to evaluate $\bar{P}_{e,1}$ via the characteristic-function method, we first evaluate $\phi_1(u)$ from (11) and (13). In general, for each value of $u$, $l$, and $k$, a numerical evaluation of the double integral of (13) is required. In certain special cases it is possible to obtain simpler expressions which lead to reduced computational requirements [e.g., for PSK we only need to evaluate the single integral of (14)]. Let $n_\tau$ and $n_\varphi$ be the number of points required for numerical evaluation (e.g., via Simpson's rule) of the integrals over $[0, T_c]$ and $[0, \tfrac{1}{2}\pi]$, respectively. The amount of computation required for the numerical evaluation of the double integral in (13) is proportional to the product $n_\tau n_\varphi$. Notice that this evaluation is independent of $E_b/N_0$. Also notice that for PSK the use of (14) leads to a numerical evaluation with $n_\tau = 1$.

From (11) we see that for each value of $u$ the number of times that the double integral of (13) must be evaluated is at most $2N(K - 1)$. This fact combined with the conclusions above implies that for each value of $u$, the computational effort grows no faster in $N$ and $K$ than the product $2n_\tau n_\varphi N(K - 1)$. Moreover, since the double integral of (13) does not depend on $E_b/N_0$, the computation of $\phi_1(u)$ need not be repeated for each value of $E_b/N_0$ of interest. The computational requirement for direct integration is proportional to $(4Nn_\tau n_\varphi)^{K-1}$, and the evaluation must be repeated for each value of $E_b/N_0$.

Finally, we consider the evaluation of the average probability of error from (9). Notice that for each integer $L$

$$\frac{1}{\pi} \int_{L\pi}^{\infty} u^{-1}(\sin u)\phi_2(u)[1 - \phi_1(u)] \, du$$

$$\leqslant \frac{2}{\pi} \int_{L\pi}^{\infty} \phi_2(u) \, du$$

$$= \left(\frac{8}{\pi}\right)^{1/2} \alpha Q(L\pi/\alpha) \tag{15}$$

where $\alpha \triangleq (2E_b/N_0)^{1/2}$. Thus an approximation of the integral in (9) by an integral from 0 to $L\pi$ gives an error less than the arbitrary positive number $\epsilon$ provided that

$$L > \left(\frac{\alpha}{\pi}\right) Q^{-1}(\sqrt{\pi/8}\,\epsilon/\alpha). \tag{16}$$

For instance, if $E_b/N_0$ is not greater than 14 dB, then $L = 20$ guarantees an error of less than $\epsilon = 10^{-14}$. Suppose that the numerical integration technique employs $n_u$ points; then the total computational effort is proportional to $2n_\tau n_\varphi n_u N(K - 1)$. We actually evaluate the integral from 0 to $L\pi$ by summing the $n$-point Simpson's-rule approximations to the integrals from $\lambda\pi$ to $(\lambda + 1)\pi$ for $\lambda = 0, 1, \cdots, L - 1$. This gives $n_u = nL$. We have found that $n_\varphi \geqslant 10$, $n_\tau \geqslant 10$, and $n \geqslant 20$ are sufficient to obtain very good approximations for the integrals in question (see [4, Tables 2a and 2b]).

### C. The Series-Expansion Method

The average probability of error $\bar{P}_{e,1}$ can be expressed as

$$\bar{P}_{e,1} = E\{Q(\alpha[1 + I])\}, \tag{17}$$

which can be written as a Taylor series expansion about a nominal point $x$ in the range $0 < x \leqslant \alpha$. This gives

$$\bar{P}_{e,1} = Q(x) + \sum_{n=1}^{\infty} \frac{\alpha^n}{n!} \mu_n Q^{(n)}(x) \tag{18}$$

where

$$\mu_n = E\{(1 - \alpha^{-1}x + I)^n\} \tag{19}$$

and $Q^{(n)}$ is the $n$th derivative of $Q$. We can write (18) in terms of the Hermite polynomials $H_n(x)$, which are defined by $H_0(x) = 1, H_1(x) = 2x,$ and

$$H_{n+1}(x) = 2xH_n(x) - 2nH_{n-1}(x)$$

for $n \geqslant 1$. The resulting expression is

$$\bar{P}_{e,1} = Q(x) + \frac{e^{-1/2x^2}}{\sqrt{\pi}} \sum_{n=1}^{\infty} \frac{\mu_n}{n!} \left(\frac{-\alpha}{\sqrt{2}}\right)^n H_{n-1}(x/\sqrt{2}). \tag{20}$$

In the Appendix it is shown that the infinite series in (20) converges, and an upper bound is obtained on the truncation error that results when the series is approximated by the sum of a finite number of terms. Unlike previous asymptotic results, the bound that we obtain is valid for the full ranges of the parameters involved, and it does not depend on the moments of the interference. Let $M$ be defined in terms of the maxima $M_{k,1}$ and $\hat{M}_{k,1}$ of the periodic and odd crosscorrelation functions [10, p. 1598] by the expression

$$M = \left(1 - \frac{x}{\alpha}\right) + \frac{1}{N} \sum_{k=2}^{K} \max\{M_{k,1}, \hat{M}_{k,1}\}. \tag{21}$$

If $J > (\alpha M)^2 - 1$, then the truncation error $E_J$ that results when only the first $J$ terms are retained is bounded by

$$E_J < \frac{ce^{-(x/2)^2}(\alpha M)^{J+1}}{\sqrt{2\pi(J+1)(J+1)!}} \left(1 - \frac{\alpha M}{\sqrt{J+1}}\right)^{-1} \tag{22}$$

where $c \approx 1.086435$.

In order to evaluate the moments $\mu_n$ we consider the expansion

$$\mu_n = \sum_{s=0}^{n'} \binom{n}{2s} E\{I^{2s}\}(1 - \alpha^{-1}x)^{n-2s} \tag{23}$$

where $n' = [\frac{1}{2}n]$. Because of the symmetry of the multiple-access interference [see (2)], $E\{I^{2s+1}\} = 0$ for each nonnegative integer $s$. The moments $E\{I^{2s}\}$ are evaluated by means of a recursion ([17] and [5, eq. (36)]) which is presented in [4] for rectangular pulses. The only change that must be made for arbitrary chip waveforms is that (24) of [4] must be replaced by

$$f_n(l, g) = T^{-2n} \sum_{m=0}^{2n} \binom{2n}{m} M_\psi(m, n)[g(l+1)]^m [g(l)]^{2n-m} \tag{24}$$

where

$$M_\psi(m, n) = \frac{1}{T_c} \int_0^{T_c} [R_\psi(\tau)]^m [\hat{R}_\psi(\tau)]^{2n-m} \, d\tau. \tag{25}$$

As discussed in [4], the main difficulty with the series-expansion method is the large number of terms that must be retained in order to obtain a good approximation. The infinite series of (18) converges slowly for large values of $M$ and $E_b/N_0$. If the ratio $\mu_{n+1}/\mu_n$ is nonincreasing in $n$ and upper bounded by 1 (for sufficiently large $n$), then the convergence is somewhat faster [4]. Unfortunately, for many systems of interest these conditions on the moments are not met (e.g., they are usually not satisfied for $K > N/10$).

The choice of $x$ greatly influences the speed of convergence of the series in (18). For good convergence the nominal point $x$ must be selected such that $Q(x)$ is close to $\bar{P}_{e,1}$. We found that if $x$ is taken to be the signal-to-noise ratio $SNR_1$ of [7], the convergence is better than for the choice $x = \sqrt{2E_b/N_0}$ that is employed in [17]. However, neither of these two values is satisfactory. We obtain good convergence by choosing $x$ such that $Q(x) = \tilde{P}_{e,1}$ where $\tilde{P}_{e,1}$ is the approximation obtained from the characteristic-function method. If $\tilde{P}_{e,1}$ is accurate to two significant figures, then $\bar{P}_{e,1}$ can be approximated as accurately as desired using (18) with only a moderate number of terms (see [4, Tables 3a and 3b]).

## III. QUATERNARY DS/SSMA COMMUNICATIONS

### A. System Model

A detailed description of the model that we employ for the asynchronous quaternary DS/SSMA communications system is given in [8] and [9]. The chip waveform $\psi(t)$ is a time-limited pulse of arbitrary shape as discussed in Section II. The energy in the pulse is $T_c$.

For offset quaternary systems we allow a time offset of the form $t_0 = (\nu + \frac{1}{2})T_c$ between the in-phase and quadrature components of the quaternary signal ($\nu$ is a positive integer). In addition we consider quaternary systems which have no offset (i.e., $t_0 = 0$). The primary example of the latter type of system is quadriphase-shift-keying with orthogonal biphase-coded carriers (QPSK). Examples of offset quaternary spread-spectrum systems are offset QPSK (OQPSK) and minimum-shift-keying (MSK). These spread-spectrum systems are discussed in [3], [8], and [9].

The outputs of the quadrature and in-phase branches of the correlation receiver matched to the first signal are denoted by $Z_1$ and $Z_2$, respectively. As shown in [8, pp. 183–189], these random variables can be expressed in terms of the aperiodic correlation functions for the signature sequences, and the resulting expressions are analogous to (1) and (2) of Section II. The outputs are given by

$$Z_m = \eta_m + (\tfrac{1}{2}P)^{1/2} T[b_1^{(m)} + I_m] \tag{26}$$

where

$$I_m = \sum_{k=2}^{K} \{I_{2k-1,m}(b_{2k-1}, \tau_k, \varphi_k)$$

$$+ I_{2k,m}(b_{2k}, \tau_k, \varphi_k)\} \tag{27}$$

where $m = 1$ denotes the quadrature branch and $m = 2$ denotes the in-phase branch. The random variables $\eta_1$ and $\eta_2$ are independent and Gaussian with mean zero and variance $\frac{1}{4}N_0 T$. The vectors $b_n \in \{+1, -1\}^2$, $1 \leq n \leq 2K$, represent pairs of successive data bits where $n = 2k - 1$ and $n = 2k$ correspond to the quadrature and in-phase components,

respectively, of the $k$th spread-spectrum signal. Analogous to the model for binary systems in Section II, the variables $b_{2k-1}$, $b_{2k}$, $\tau_k$, and $\varphi_k$, $1 \leq k \leq K$, are random variables which are independent and uniformly distributed on appropriate sets. The functions $I_{n,m}$ which appear in (27) are defined as follows. Let $b = (b_1, b_2)$ with $b_l \in \{+1, -1\}$. Then,

$$I_{n,1}(b, \tau, \varphi) = T^{-1}\{b_1 R_{n,1}(\tau) + b_2 \hat{R}_{n,1}(\tau)\} \cos \varphi \tag{28a}$$

for $n = 2k - 1$ and

$$I_{n,1}(b, \tau, \varphi) = T^{-1}\{b_1 R_{n,1}(\tau + t_0)$$
$$+ b_2 \hat{R}_{n,1}(\tau + t_0)\} \sin(-\varphi) \tag{28b}$$

for $n = 2k$ where $\tau + t_0$ is to be interpreted modulo $[0, T]$. Similarly,

$$I_{n,2}(b, \tau, \varphi) = T^{-1}\{b_1 R_{n,2}(\tau - t_0) + b_2 \hat{R}_{n,2}(\tau - t_0)\} \sin \varphi \tag{28c}$$

for $n = 2k - 1$ and

$$I_{n,2}(b, \tau, \varphi) = T^{-1}\{b_1 R_{n,2}(\tau) + b_2 \hat{R}_{n,2}(\tau)\} \cos \varphi \tag{28d}$$

for $n = 2k$. In (28c) the quantity $\tau - t_0$ is to be interpreted modulo $[0, T]$. The functions $R_{n,m}$ and $\hat{R}_{n,m}$ are defined in (3) and (4). Contrary to an assertion made in [6], the random variables $I_{2k-1,m}(b_{2k-1}, \tau_k, \varphi_k)$ and $I_{2k,m}(b_{2k}, \tau_k, \varphi_k)$ are statistically dependent.

The average probability of error $\bar{P}_{e,1}$ for the first receiver is given by

$$\bar{P}_{e,1} = \frac{1}{4} \sum_{m=1}^{2} \{P(Z_m \leq 0 \mid b_1^{(m)} = +1)$$

$$+ P(Z_m > 0 \mid b_1^{(m)} = -1)\}$$

$$= \frac{1}{2} - \frac{1}{2} \sum_{m=1}^{2} P(-1 < \eta_m' + I \leq 1) \tag{29}$$

where $\eta_m' = (\frac{1}{2}PT^2)^{-1/2}\eta_m$ is a zero-mean Gaussian random variable with variance $(2E_b/N_0)^{-1}$ for $m = 1$ and 2.

### B. The Characteristic-Function Method

As in Section II, $\bar{P}_{e,1}$ can be expressed in terms of the characteristic functions of the multiple-access interference and the channel noise. Let $\phi_1^{(m)}$ be the characteristic function for $I_m$ and let $\phi_2$ be as in (7). Then we can write [cf. (8)]

$$\bar{P}_{e,1} = Q([2E_b/N_0]^{1/2}) + \pi^{-1} \int_0^\infty u^{-1}(\sin u)\phi_2(u)$$

$$\cdot \left[1 - \frac{1}{2} \sum_{m=1}^{2} \phi_1^{(m)}(u)\right] du. \tag{30}$$

In general, the characteristic functions $\phi_1^{(1)}$ and $\phi_1^{(2)}$ are not the same. However, in one important special case these two functions are identical. This is the special case in which each of the $K$ spread-spectrum signals is such that the signature sequence for the quadrature component is the *reverse* of the signature sequence for the in-phase component. The use of reverse sequences for the quadrature and in-phase components of a quaternary spread-spectrum signal was proposed in [8] (see also [3, p. 310] or [9, p. 15.3.5]). Correlation properties for pairs of reverse sequences are given in [12]. The key result that we need here is that if the quadrature and in-phase sequences are reverse sequences, then for $n = 2k$

$$R_{n,2}(\tau) = \hat{R}_{n-1,1}(T - \tau) \tag{31a}$$

and

$$R_{n-1,2}(\tau) = \hat{R}_{n,1}(T - \tau) \tag{31b}$$

for $0 \leqslant \tau \leqslant T$. This follows from (3) and (4), [12, eq. (5.19)], and the fact that $R_\psi(s) = \hat{R}_\psi(T_c - s)$. Then, from (26)–(28) and (31), it follows that $I_1$ and $I_2$ have the same distribution and hence the same characteristic function.

The next step in the analysis is to obtain appropriate expressions for $\phi_1^{(m)}(u)$. Proceeding as in Section II, we find that

$$\phi_1^{(m)}(u) = \prod_{k=2}^{K} \left\{ (4N)^{-1} \sum_{l=0}^{N-1} H_k^{(m)}(u; l, \nu) \right\} \tag{32}$$

for $m = 1$ and 2, where the functions $H_k^{(m)}$ depend on the signature sequences for the $k$th and first signals and on the chip waveform $\psi(t)$. In order to specify the functions $H_k^{(m)}$, it is convenient to consider separately the offset systems and the systems without an offset.

*Quaternary Systems With Offset:* For offset quaternary systems we introduce two functions $f$ and $\hat{f}$ which are analogous to the function defined in (12). Let $N = \{0, 1, \cdots, N-1\}$ represent the integers modulo $N$. For an arbitrary real-valued function $g$ defined on $N$, let

$$\Gamma_g(l, \tau) = [g(l+1)R_\psi(\tau) + g(l)\hat{R}_\psi(\tau)]/T$$

for each $l \in N$ and $\tau \in [0, T_c]$. Then, for arbitrary functions $g$ and $h$, let

$$f(u; l, g; n, h)$$
$$\triangleq \frac{1}{2\pi T_c} \int_0^{2\pi} \int_0^{(1/2)T_c} \exp\{ju[\Gamma_g(l, \tau) \cos\theta$$
$$- \Gamma_h(n, \tau + \tfrac{1}{2}T_c) \sin\theta]\} \, d\tau \, d\theta \tag{33a}$$

and

$$\hat{f}(u; l, g; n, h)$$
$$\triangleq \frac{1}{2\pi T_c} \int_0^{2\pi} \int_{(1/2)T_c}^{T_c} \exp\{ju[\Gamma_g(l, \tau) \cos\theta$$
$$- \Gamma_h(n, \tau - \tfrac{1}{2}T_c) \sin\theta]\} \, d\tau \, d\theta. \tag{33b}$$

The need for two functions $f$ and $\hat{f}$, rather than one as in Section II, is because of the offset. However, since $\hat{R}_\psi(\tau) = R_\psi(T_c - \tau)$, then

$$\hat{f}(u; l, g; n, h) = f(u; n, h; l, g)$$

so our expressions can be written in terms of the function $f$ alone.

Next we define the functions $H_k^{(m)}$ by

$$H_k^{(m)}(u; l, i; n, j)$$
$$= f(u; l, \theta_{i,m}; n, \theta_{j,m}) + f(u; l, \theta_{i,m}; n, \hat{\theta}_{j,m})$$
$$+ f(u; l, \hat{\theta}_{i,m}; n, \theta_{j,m}) + f(u; l, \hat{\theta}_{i,m}; n, \hat{\theta}_{j,m}) \tag{34}$$

for $m = 1$ and 2, $1 < k \leqslant K$, $1 \leqslant l \leqslant 2K$ and $1 \leqslant j \leqslant 2K$. Finally, we define the functions $H_k^{(1)}$ and $H_k^{(2)}$ of (32) by

$$H_k^{(1)}(u; l, \nu) = H_k^{(1)}(u; l, 2k-1; l+\nu, 2k)$$
$$+ H_k^{(1)}(u; l+\nu+1, 2k; l, 2k-1) \tag{35a}$$

and

$$H_k^{(2)}(u; l, \nu) = H_k^{(2)}(u; l, 2k; l-\nu-1, 2k-1)$$
$$+ H_k^{(2)}(u; l-\nu, 2k-1; l, 2k) \tag{35b}$$

where the expressions $l + \nu$, $l + \nu + 1$, $l - \nu$, and $l - \nu - 1$ are interpreted modulo $N$.

For OQPSK systems the chip waveform $\psi(t)$ is a rectangular pulse and $R_\psi(\tau) = \tau$ for $0 \leqslant \tau \leqslant T_c$. In this case (33a) reduces to

$$f(u; l, g; n, h)$$
$$= \frac{1}{4\pi} \int_0^{2\pi} \text{sinc} \left[ \frac{u}{4\pi N} \{[g(l+1) - g(l)] \cos\theta \right.$$
$$- [h(n+1) - h(n)] \sin\theta\} \right]$$
$$\cdot \cos \left[ \frac{u}{4N} \{[g(l+1) + 3g(l)] \cos\theta \right.$$
$$\left. - [3h(n+1) + h(n)] \sin\theta\} \right] d\theta. \tag{36}$$

For more general systems (including MSK) in which the chip waveform is arbitrary, we cannot reduce (33a) to a single integral. However, we can write (33a) in the following form which is somewhat simpler:

$$f(u; l, g; n, h)$$
$$= \frac{2}{\pi T_c} \int_0^{(1/2)\pi} \int_0^{(1/2)T_c} \cos [u\{[\Gamma_g(l, \tau)]^2$$
$$+ [\Gamma_h(n, \tau + \tfrac{1}{2}T_c)]^2\}^{1/2} \cos\theta] \, d\tau \, d\theta. \tag{37}$$

This is the form that we employ in the evaluation of the error probability for MSK.

*Quaternary Systems With No Offset:* In the absence of an offset between the in-phase and quadrature components of the quaternary spread-spectrum signals, the characteristic functions $\phi_1^{(m)}$ are somewhat simpler to evaluate. First, we define a function $\tilde{f}$ by letting $\tilde{f}(u; l, g; n, h)$ be equal to the expression on the right-hand side of (37) with $\Gamma_h(n, \tau + \frac{1}{2}T_c)$ being replaced by $\Gamma_h(n, \tau)$ and the upper limit on the inner integral changed from $\frac{1}{2}T_c$ to $T_c$. Then let $\tilde{H}_k^{(m)}$ be defined by (34) with $H_k^{(m)}$ and $f$ replaced by $\tilde{H}_k^{(m)}$ and $\tilde{f}$. The functions $\tilde{H}_k^{(m)}$ are then defined by [cf. (35)]

$$\tilde{H}_k^{(1)}(u; l) = \tilde{H}_k^{(1)}(u; l, 2k-1; l, 2k) \tag{38a}$$

and

$$\tilde{H}_k^{(2)}(u; l) = \tilde{H}_k^{(2)}(u; l, 2k; l, 2k-1), \tag{38b}$$

and, finally, we have

$$\phi_1^{(m)} = \prod_{k=2}^K \left\{ (4N)^{-1} \sum_{l=0}^{N-1} \tilde{H}_k^{(m)}(u; l) \right\}. \tag{39}$$

For QPSK, which is by far the system of greatest interest among the non-offset quaternary systems, the expressions are much simpler than for general quaternary systems. Since the chip waveform is a rectangular pulse for QPSK, the function $\tilde{f}$ can be written in the following form [cf. (36)]:

$$\tilde{f}(u; l, g; n, h)$$
$$= \frac{1}{2\pi} \int_0^{2\pi} \mathrm{sinc}\left[ \frac{u}{2\pi N} \{[g(l+1)-g(l)] \cos\theta \right.$$
$$\left. - [h(n+1)-h(n)] \sin\theta \} \right]$$
$$\cdot \cos\left[ \frac{u}{2N} \{[g(l+1)+g(l)] \cos\theta \right.$$
$$\left. - [h(n+1)+[h(n)] \sin\theta \} \right] d\theta. \tag{40}$$

Notice from (38) that we only need to evaluate (40) for the case $n = l$.

### C. The Series-Expansion Method

This section follows very closely the presentation of Section II-C and [4], so many of the details are omitted. The error probability $\bar{P}_{e,1}$ defined in (29) can be expressed as [cf. (17)]

$$\bar{P}_{e,1} = \frac{1}{2} \sum_{m=1}^2 Q(\alpha[1 + I_m]). \tag{41}$$

This can be expanded about nominal points $x_1$ and $x_2$ to give

$$\bar{P}_{e,1} = \frac{1}{2} \sum_{m=1}^2 \left\{ Q(x_m) + \sum_{n=1}^\infty \frac{\alpha^n}{n!} \mu_n^{(m)} Q^{(n)}(x_m) \right\} \tag{42}$$

where the moments $\mu_n^{(m)}$ are given by

$$\mu_n^{(m)} = E\{(1 - \alpha^{-1}x_m + I_m)^n\}. \tag{43}$$

Instead of the parameter $M$ of (21) we consider the parameters $M_1$ and $M_2$ which are defined by

$$M_m = (1 - \alpha^{-1}x_m) + \frac{1}{N} \sum_{k=2}^K [\max \{M_{2k-1,m}, \dot{M}_{2k-1,m}\} + \max \{M_{2k,m}, \dot{M}_{2k,m}\}]. \tag{44}$$

Then, $E_J^{(m)}$ is given by (22) with $M$ replaced by $M_m$, and the truncation error is given by $E_J = \frac{1}{2}(E_J^{(1)} + E_J^{(2)})$, where $J$ is the number of terms retained in the infinite series of (42).

The expansions for the moments $\mu_n^{(m)}$ are given by (23) with $I$ and $x$ replaced by $I_m$ and $x_m$, respectively. Following [4] we define

$$W_{2k-1,m} = I_{2k-1,m}(b_{2k-1}, \tau_k, \Psi_k^{(m)}) \tag{45a}$$

and

$$W_{2k,m} = I_{2k,m}(b_{2k}, \tau_k, 3(\Psi_k^{(m)} - \tfrac{1}{2}\pi)) \tag{45b}$$

where $\Psi_k^{(m)}$ is 0 for $m = 1$ and $\frac{1}{2}\pi$ for $m = 2$. From (27) and (28) we see that the multiple-access interference can be written as

$$I_m = \sum_{k=2}^K W_k^{(m)}$$

where

$$W_k^{(1)} = W_{2k-1,1} \cos\varphi_k - W_{2k,1} \sin\varphi_k \tag{46a}$$

and

$$W_k^{(2)} = W_{2k-1,2} \sin\varphi_k + W_{2k,2} \cos\varphi_k. \tag{46b}$$

Next, we consider the partial sums $I_j^{(m)}$ for $2 \leq j \leq K$ and $m = 1$ or 2. Let $I_2^{(m)} = W_2^{(m)}$ and for $3 \leq j \leq K$, let

$$I_j^{(m)} \triangleq \sum_{k=2}^j W_k^{(m)} = I_{j-1}^{(m)} + W_j^{(m)}. \tag{47}$$

Notice that $I_K^{(m)} = I_m$. Because of the symmetry of the distribution of the multiple-access interference, the odd-order moments of $I_j^{(m)}$ are zero.

For the even-order moments of $I_j^{(m)}$ we must consider the correlations

$$w_k^{(m)}(r, s) \triangleq E\{[W_{2k-1,m}]^{2r}[W_{2k,m}]^{2s}\} \qquad (48)$$

for $2 \leqslant k \leqslant K$. Although it may be tempting to assume that $W_{2j-1,m}$ and $W_{2j,m}$ are independent (such an assumption is made in [6]), they are in general statistically dependent. In fact, it is not hard to find examples where they are very strongly dependent.

Define the even-order moments $\nu_{j,p}^{(m)}$ for $p = 0, 1, 2, \cdots$ by $\nu_{1,p}^{(m)} = 0$ and $\nu_{j,p}^{(m)} = E\{[I_j^{(m)}]^{2p}\}$ for $j \geqslant 2$. These moments are given by the recursion

$$\nu_{j,p}^{(m)} = \nu_{j-1,p}^{(m)} + \sum_{n=0}^{p-1}\binom{2p}{2n}\nu_{j,n}^{(m)}$$

$$\cdot \sum_{i=0}^{p-n} B(p-n, i)w_j^{(m)}(i, p-n-i) \qquad (49)$$

where

$$B(q, i) = [(2q)!]\binom{q}{i}[q!2^q]^{-2} \qquad (50)$$

for $0 \leqslant i \leqslant q$.

The correlations $w_k^{(m)}(r, s)$ can be expressed in terms of the periodic and odd crosscorrelation functions for the signature sequences. First, let

$$f_{r,s}(l, g; n, h)$$

$$= \frac{1}{T_c}\int_0^{(1/2)T_c}[\Gamma_g(l, \tau)]^{2r}[\Gamma_h(n, \tau + \tfrac{1}{2}T_c)]^{2s}\,d\tau \qquad (51)$$

and then define [cf. (34)]

$$H_{k,r,s}^{(m)}(l, i; n, j)$$

$$= f_{r,s}(l, \theta_{i,m}; n, \theta_{j,m}) + f_{r,s}(l, \theta_{i,m}; n, \hat{\theta}_{j,m})$$

$$+ f_{r,s}(l, \hat{\theta}_{i,m}; n, \theta_{j,m}) + f_{r,s}(l, \hat{\theta}_{i,m}; n, \hat{\theta}_{j,m}). \qquad (52)$$

Just as (52) is analogous to (34), we define $H_{k,r,s}^{(m)}$ in terms of $H_{k,r,s}^{(m)}$ by equations which are analogous to (35). Roughly speaking, we simply omit the variable $u$ in (35) and add the additional subscripts $r$, $s$ to $H_k^{(m)}$ and $H_k^{(m)}$. Finally, we have

$$w_k^{(m)}(r, s) = (4N)^{-1}\sum_{l=0}^{N-1} H_{k,r,s}^{(m)}(l, \nu)$$

which is similar in some respects to (32).

In the absence of an offset (i.e., if $t_o = 0$), then the above expressions are somewhat simpler [for example, we only need to consider $n = l$ in (51) and (52)]. Moreover, if reverse sequences are employed on the in-phase and quadrature com-

ponents, then for $x_1 = x_2$ we have $\mu_n^{(1)} = \mu_n^{(2)}$ and $w_k^{(1)}(r, s) = w_k^{(2)}(r, s)$.

## IV. NUMERICAL RESULTS

In this section we apply the characteristic-function method described in Sections II-B and III-B to the evaluation of the average probability of error for several direct-sequence SSMA systems. For all of the numerical results presented in this paper, the characteristic-function method is employed with parameters $n = L = 20$ and $n_\varphi = n_\tau = 10$ (these parameters are defined in Section II-B).

First, we consider binary direct-sequence SSMA systems with a rectangular chip waveform (i.e., binary PSK/SSMA). The numerical results presented here complement those that we presented in [4]. In Fig. 1(a) the probability of error is shown for $N = 255$. The five different curves correspond to five different values of $K$, the number of transmitted signals. The $K$ signature sequences for the data presented in Fig. 1(a) correspond to the first $K$ entries of Table A.1(d) of [10]. These are auto-optimal least-sidelobe-energy (AO/LSE) phases of maximal-length sequences ($m$-sequences) of period 255. In Fig. 1(b) the probability of error is shown for $N = 511$ and $K = 3, 6, 12$, and 24. For $K = 3$, the three signature sequences are the AO/LSE Gold sequences specified (in the notation of [10]) by the shift-register tap connections 1371, 1365, and 1555 and initial loadings 530, 101, and 400, respectively (these loadings are octal representations of the binary loadings given in [14]). For the remaining values of $K$, we employ the first $K$ sequences and loadings listed in the first two columns of Table 1 of [14]. These are AO/LSE Gold sequences.

A comparison of the probability of error for two different sets of binary signature sequences is presented in Table I. Each set has six sequences ($K = 6$) of period 31. The six $m$-sequences are the AO/LSE phases listed in Figure A.1(a) of [10]. The six Gold sequences are also in their AO/LSE phases; they are generated by a shift register having tap connection 3551 with initial loadings 1014, 1570, 0610, 1742, 1476, and 1734. The fact that the Gold sequences give better performance than $m$-sequences is of no particular importance and not likely to be true in general. In this particular case, it may be due to the fact that some of these $m$-sequences are reciprocals of others (for period 31 there are only three nonreciprocal $m$-sequences [12]).

The performance of binary direct-sequence SSMA systems for two different chip waveforms is compared in Table II for $N = 31$. The two chip waveforms are the rectangular pulse and the sine pulse. The AO/LSE $m$-sequences [10, Table A.1(a)] are used in Table II(a) and (b), and the left-hand columns of Table II(c). The sequences for the right-hand columns of Table II(c) are the Gold sequences generated by shift register 3551 with loadings 1756, 0355, 0432 and 1306. Notice that the *relative* performance for the two waveforms depends on the value of $E_b/N_0$, a result which is not predicted by the analysis of signal-to-noise ratio [8]. For AO/LSE sequences, the sine pulse gives better performance than the rectangular pulse for small values of $E_b/N_0$, but the reverse is true for large values of $E_b/N_0$.

For the next results the signature sequences are the three

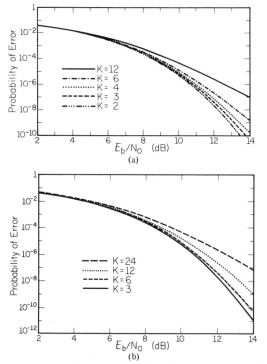

Fig. 1. (a) Probability of error for binary PSK direct-sequence SSMA ($N = 255$). (b) Probability of error for binary PSK direct-sequence SSMA ($N = 511$).

TABLE I
PROBABILITY OF ERROR FOR BINARY PSK SSMA SYSTEM
($K = 6, N = 31$)

| $E_b/N_0$ (dB) | $m$-sequences | Gold sequences | |
|---|---|---|---|
| 8 | 3.523 | 3.397 | ($\times 10^{-3}$) |
| 9 | 2.117 | 1.984 | ($\times 10^{-3}$) |
| 10 | 1.297 | 1.166 | ($\times 10^{-3}$) |
| 11 | 8.218 | 6.986 | ($\times 10^{-4}$) |
| 12 | 5.437 | 4.318 | ($\times 10^{-4}$) |
| 13 | 3.776 | 2.778 | ($\times 10^{-4}$) |
| 14 | 2.756 | 1.869 | ($\times 10^{-4}$) |

TABLE II
PROBABILITY OF ERROR FOR BINARY DS/SSMA SYSTEMS
WITH TWO DIFFERENT CHIP WAVEFORMS ($N = 31$)

(a) $K = 2$

| $E_b/N_0$ (dB) | Rectangular | Sine Pulse | |
|---|---|---|---|
| 8 | 4.216 | 4.066 | ($\times 10^{-4}$) |
| 9 | 1.125 | 1.088 | ($\times 10^{-4}$) |
| 10 | 2.397 | 2.370 | ($\times 10^{-5}$) |
| 11 | 3.952 | 4.110 | ($\times 10^{-6}$) |
| 12 | 4.815 | 5.464 | ($\times 10^{-7}$) |
| 13 | 4.101 | 5.262 | ($\times 10^{-8}$) |
| 14 | 2.342 | 3.447 | ($\times 10^{-9}$) |

(b) $K = 3$

| $E_b/N_0$ (dB) | Rectangular | Sine Pulse | |
|---|---|---|---|
| 8 | 8.487 | 7.833 | ($\times 10^{-4}$) |
| 9 | 3.147 | 2.881 | ($\times 10^{-4}$) |
| 10 | 1.076 | 0.933 | ($\times 10^{-4}$) |
| 11 | 3.468 | 3.306 | ($\times 10^{-5}$) |
| 12 | 1.081 | 1.099 | ($\times 10^{-5}$) |
| 13 | 3.347 | 3.736 | ($\times 10^{-6}$) |
| 14 | 1.048 | 1.317 | ($\times 10^{-6}$) |

(c) $K = 4$

| $E_b/N_0$ (dB) | $m$-sequences | | Gold sequences | | |
|---|---|---|---|---|---|
| | Rectangular | Sine Pulse | Rectangular | Sine Pulse | |
| 8 | 1.626 | 1.412 | 1.416 | 1.210 | ($\times 10^{-3}$) |
| 9 | 7.732 | 6.484 | 6.273 | 5.132 | ($\times 10^{-4}$) |
| 10 | 3.623 | 2.938 | 2.648 | 2.071 | ($\times 10^{-4}$) |
| 11 | 1.714 | 1.348 | 1.081 | 0.810 | ($\times 10^{-4}$) |
| 12 | 8.363 | 6.412 | 4.330 | 3.131 | ($\times 10^{-5}$) |
| 13 | 4.281 | 3.217 | 1.719 | 1.213 | ($\times 10^{-5}$) |
| 14 | 2.322 | 1.721 | 0.682 | 0.476 | ($\times 10^{-5}$) |

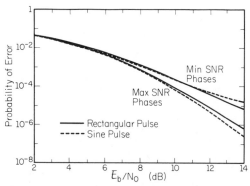

Fig. 2. Probability of error for rectangular and sine pulse chip waveforms for two different sets of signature sequence phases ($K = 3$, $N = 31$).

nonreciprocal $m$-sequences of period 31. For each of two chip waveforms—the rectangular pulse and the sine pulse—the probability of error is presented for two different sets of phases of the signature sequences. One set is the set of phases which maximizes the signal-to-noise ratio (SNR) for the chip waveform in question, and the other is the set of phases which minimizes the signal-to-noise ratio. The shift-register connections and initial loadings for these phases are given in [2, Table 5]. The error probabilities for both sets of phases and both chip waveforms are shown in Fig. 2. Notice that if the maximum SNR phases of the signature sequences are employed, the sine pulse gives a lower probability of error than the rectangular pulse for all values of $E_b/N_0$. This is as predicted by the signal-to-noise ratio analysis of [8]. The curves of Fig. 2 suggest that the selection of phases for the signature sequences is more important for the sine

pulse than for the rectangular pulse. Even for the rectangular pulse, however, there is a 2 dB performance difference for the two sets of phases at a bit error rate of $10^{-5}$.

The remaining results are for quaternary direct-sequence SSMA systems. Contrary to a statement made in [6], QPSK systems and OQPSK systems do not have the same probability of error, even if identical sets of signature sequences are employed for the two types of systems. Consequently, we give results for both QPSK and OQPSK systems in addition to MSK systems. All numerical results for OQPSK and MSK systems are for an offset of $t_0 = \frac{1}{2}T_c$.

The probability of error for a QPSK direct-sequence SSMA

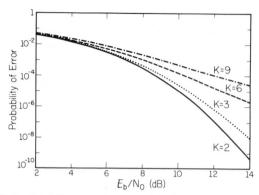

Fig. 3. Probability of error for QPSK direct-sequence SSMA ($N = 127$).

TABLE III
PROBABILITY OF ERROR FOR QUATERNARY DS/SSMA
SYSTEMS ($N = 31$)

(a) $K = 2$

| $E_b/N_0$ (dB) | QPSK | OQPSK | MSK | |
|---|---|---|---|---|
| 8 | 7.364 | 7.422 | 6.841 | ($\times 10^{-4}$) |
| 9 | 2.420 | 2.458 | 2.217 | ($\times 10^{-4}$) |
| 10 | 6.741 | 6.935 | 6.135 | ($\times 10^{-5}$) |
| 11 | 1.572 | 1.651 | 1.446 | ($\times 10^{-5}$) |
| 12 | 3.062 | 3.323 | 2.945 | ($\times 10^{-6}$) |
| 13 | 5.005 | 5.727 | 5.300 | ($\times 10^{-7}$) |
| 14 | 6.837 | 8.476 | 8.299 | ($\times 10^{-8}$) |

(b) $K = 3$

| $E_b/N_0$ (dB) | QPSK | OQPSK | MSK | |
|---|---|---|---|---|
| 8 | 2.082 | 2.090 | 1.824 | ($\times 10^{-3}$) |
| 9 | 1.032 | 1.039 | 0.882 | ($\times 10^{-3}$) |
| 10 | 4.971 | 5.031 | 4.177 | ($\times 10^{-4}$) |
| 11 | 2.369 | 2.414 | 1.987 | ($\times 10^{-4}$) |
| 12 | 1.140 | 1.172 | 0.974 | ($\times 10^{-4}$) |
| 13 | 5.659 | 5.869 | 5.037 | ($\times 10^{-5}$) |
| 14 | 2.949 | 3.082 | 2.775 | ($\times 10^{-6}$) |

system is shown in Fig. 3 for signature sequences of period $N = 127$. The four different curves are for four different values of $K$, the number of transmitters in the system. The signature sequences are the AO/LSE $m$-sequences of Fig. A.1(c) of [10], and they are employed in pairs with reverse sequences for the in-phase and quadrature components of the signals as discussed in Section II-B.

Three different quaternary direct-sequence SSMA systems (QPSK, OQPSK, and MSK) are compared in Table III for signature sequences of period $N = 31$. The signature sequences are the first $K$ pairs of reverse sequences from Fig. A.1(a) of [10]: $K = 2$ for Table III(a) and $K = 3$ for Table III(b). As we observed in Table II, the relative performance for two different chip waveforms may depend on the value of $E_b/N_0$. A comparison of the data for QPSK and MSK in Table III(a) shows that the relative performance of these two systems is also dependent on $E_b/N_0$ for the particular signature sequences that were employed. The MSK system gives better performance than the QPSK system for values of $E_b/N_0$ less than 12 dB, but the reverse is true for larger values of $E_b/N_0$. Notice that for these particular sequences, QPSK gives better

TABLE IV
PROBABILITY OF ERROR FOR QPSK AND OQPSK DIRECT-
SEQUENCE SSMA SYSTEMS ($N = 127$)

(a) $K = 2$

| $E_b/N_0$ (dB) | QPSK | OQPSK | |
|---|---|---|---|
| 8 | 2.968 | 2.964 | ($\times 10^{-4}$) |
| 9 | 6.654 | 6.632 | ($\times 10^{-5}$) |
| 10 | 1.116 | 1.107 | ($\times 10^{-5}$) |
| 11 | 1.365 | 1.339 | ($\times 10^{-6}$) |
| 12 | 1.220 | 1.170 | ($\times 10^{-7}$) |
| 13 | 8.308 | 7.705 | ($\times 10^{-9}$) |
| 14 | 4.365 | 3.942 | ($\times 10^{-10}$) |

(b) $K = 3$

| $E_b/N_0$ (dB) | QPSK | OQPSK | |
|---|---|---|---|
| 8 | 4.299 | 4.299 | ($\times 10^{-4}$) |
| 9 | 1.156 | 1.154 | ($\times 10^{-4}$) |
| 10 | 2.521 | 2.505 | ($\times 10^{-5}$) |
| 11 | 4.445 | 4.372 | ($\times 10^{-6}$) |
| 12 | 6.465 | 6.245 | ($\times 10^{-7}$) |
| 13 | 8.174 | 7.692 | ($\times 10^{-8}$) |
| 14 | 9.705 | 8.824 | ($\times 10^{-9}$) |

performance than OQPSK for both $K = 2$ and $K = 3$ over the entire range of $E_b/N_0$ considered in Table III. Also notice that these results illustrate the fact that QPSK and OQPSK do not have the same probability of error.

Comparisons of QPSK and OQPSK are given in Table IV for signature sequences of period $N = 127$. The AO/LSE $m$-sequences from Fig. A.1(c) of [10] are employed with reverse sequences on the in-phase and quadrature components of each signal. In this case we see that OQPSK gives better performance than QPSK for both $K = 2$ and $K = 3$ over the entire range of $E_b/N_0$ considered in Table IV. From the data of Tables III and IV, it appears that the relative performance of QPSK and OQPSK is strongly dependent on the signature sequences that are used.

## APPENDIX

We first give a proof of the convergence of the infinite series in (20). A bound on the Hermite polynomials $H_n(x)$ which is valid for all $n$ is [1]

$$|H_n(x)| < ce^{(1/2)x^2}[(n!)2^n]^{1/2} \tag{A1}$$

where $c \approx 1.086435$. We can write (20) as

$$\bar{P}_{e,1} = Q(x) + \sum_{n=1}^{\infty} h_n(x) \tag{A2}$$

where

$$h_n(x) = \frac{e^{-(1/2)x^2}}{\sqrt{\pi}} \left(\frac{\mu_n}{n!}\right)\left(\frac{-\alpha}{\sqrt{2}}\right)^n H_{n-1}(x/\sqrt{2}). \tag{A3}$$

From (A1) and (A3) we see that

$$|h_n(x)| < \frac{e^{-(1/4)x^2}}{\sqrt{2\pi}} \left(\frac{\alpha^n \mu_n}{\sqrt{(n!)n}}\right). \tag{A4}$$

From (19) and (21) we see that for a fixed value of $x$, $0 \leqslant \mu_n \leqslant M^n$ so that (A4) implies

$$|h_n(x)| < \frac{e^{-(1/4)x^2}}{\sqrt{2\pi}} \left[ \frac{(\alpha M)^n}{\sqrt{(n!)n}} \right]. \tag{A5}$$

Applying the ratio test for convergence we see that the series in (A2) is absolutely convergent provided that

$$\frac{n^{1/2}}{n+1} (\alpha M) < 1, \tag{A6}$$

which is true for sufficiently large $n$ (e.g., $n > (\alpha M)^2$). Therefore, the series in (20) is absolutely convergent.

Next, we consider the truncation error $E_J$ that results when we retain only the first $J$ terms of the infinite series in (20). The above analysis shows that $E_J$ is bounded by

$$E_J < \frac{c}{\sqrt{2\pi}} e^{-(1/4)x^2} \sum_{n=J+1}^{\infty} \frac{(\alpha M)^n}{\sqrt{(n!)n}}. \tag{A7}$$

The series which appears in (A7) can be bounded as follows:

$$\sum_{n=J+1}^{\infty} \frac{(\alpha M)^n}{\sqrt{(n!)n}} = \frac{(\alpha M)^{J+1}}{\sqrt{(J+1)!(J+1)}} \sum_{j=0}^{\infty} (\alpha M)^j$$

$$\cdot \sqrt{\frac{(J+1)!(J+1)}{(j+J+1)!(j+J+1)}}$$

$$< \frac{(\alpha M)^{J+1}}{\sqrt{(J+1)!(J+1)}} \sum_{j=0}^{\infty} (\alpha M)^j$$

$$\cdot \sqrt{(J+1)^{-j}}$$

$$= \frac{(\alpha M)^{J+1}}{\sqrt{(J+1)!(J+1)}} \{1 - \alpha M(J+1)^{-1/2}\}^{-1}. \tag{A8}$$

The desired result (22) follows from (A7) and (A8). Notice that the last step requires that we choose $J$ to be greater than $(\alpha M)^2 - 1$ to guarantee convergence of the series.

## REFERENCES

[1] M. Abramowitz and I. A. Stegun, Eds., *Handbook of Mathematical Functions,* National Bureau of Standards, Washington, DC, 1964.,

[2] F. D. Garber and M. B. Pursley, "Optimal phases of maximal-length sequences for asynchronous spread-spectrum multiplexing," *Electron. Lett.*, vol. 16, pp. 756–757, Sept. 1980.

[3] ——, "Performance of offset quadriphase spread-spectrum multiple-access communication," *IEEE Trans. Commun.*, vol. COM-29, pp. 305–314, Mar. 1981.

[4] E. A. Geraniotis and M. B. Pursley, "Error probability for binary PSK spread-spectrum multiple-access communications," in *Proc. 1981 Conf. Inform. Sci. Syst.*, Johns Hopkins Univ., Baltimore, MD, Mar. 1981, pp. 238–244.

[5] E. Y. Ho and Y. S. Yeh, "A new approach for evaluating the error probability in the presence of intersymbol interference and additive Gaussian noise," *Bell Syst. Tech. J.*, vol. 49, pp. 2249–2266, Nov. 1970.

[6] D. Laforgia, A. Luvison, and V. Zingarelli, "Exact bit error probability with application to spread-spectrum multiple access communications," in *Conf. Rec. 1981 IEEE Int. Conf. Commun.*, vol. 4, June 1981, pp. 76.5.1–76.5.5.

[7] M. B. Pursley, "Performance evaluation for phase-coded spread-spectrum multiple-access communications—Part I: System analysis," *IEEE Trans. Commun.*, vol. COM-25, pp. 795–799, Aug. 1977.

[8] ——, "Spread-spectrum multiple-access communications," in *Multi-User Communication Systems*, G. Longo, Ed. Vienna and New York: Springer-Verlag, 1981, pp. 139–199.

[9] M. B. Pursley, F. D. Garber, and J. S. Lehnert, "Analysis of generalized quadriphase spread-spectrum communications," in *Conf. Rec. 1980 IEEE Int. Conf. Commun.*, vol. 1, June 1980, pp. 15.3.1–15.3.6.

[10] M. B. Pursley and H. F. A. Roefs, "Numerical evaluation of correlation parameters for optimal phases of binary shift-register sequences," *IEEE Trans. Commun.*, vol. COM-27, pp. 1597–1604, Oct. 1979.

[11] M. B. Pursley, D. V. Sarwate, and W. E. Stark, "Error probability for direct-sequence spread-spectrum multiple-access communications—Part I: Upper and lower bounds," this issue, pp. 975–984.

[12] D. V. Sarwate and M. B. Pursley, "Crosscorrelation properties of pseudorandom and related sequences," *Proc. IEEE*, vol. 68, pp. 593–619, May 1980.

[13] O. Shimbo and M. I. Celebiler, "The probability of error due to intersymbol interference and Gaussian noise in digital communication systems," *IEEE Trans. Commun.*, vol. COM-19, pp. 113–119, Apr. 1971.

[14] R. Skaug, "Numerical evaluation of the nonperiodic autocorrelation parameter for optimal phases of maximal length sequences," *Inst. Elec. Eng. Proc.*, vol. 127, pt. F, pp. 230–237, June 1980.

[15] J. C. Vanelli and N. M. Shehadeh, "Computation of bit-error probability using the trapezoidal integration rule," *IEEE Trans. Commun.*, vol. COM-22, pp. 331–334, Mar. 1974.

[16] K.-T. Wu, "Average error probability for DS-SSMA communications: The Gram–Charlier expansion approach," in *Proc. 19th Annu. Allerton Conf. Commun., Contr., Comput.*, Sept. 1981, pp. 237–246; see also, "Direct sequence spread spectrum communications: Applications to multiple access and jamming resistance," Ph.D. dissertation, Univ. Michigan, Ann Arbor, 1981.

[17] K.-T. Wu and D. L. Neuhoff, "Average error probability for direct sequence spread-spectrum multiple access communication systems," in *Proc. 18th Annu. Allerton Conf. Commun., Contr., Comput.*, Oct. 1980, pp. 359–368.

[18] K. Yao, "Error probability of asynchrnous spread spectrum multiple access communication systems," *IEEE Trans. Commun.*, vol. COM-25, pp. 803–809, Aug. 1977.

# Error Probabilities for Slow-Frequency-Hopped Spread-Spectrum Multiple-Access Communications Over Fading Channels

EVAGGELOS A. GERANIOTIS, STUDENT MEMBER, IEEE, AND MICHAEL B. PURSLEY, FELLOW, IEEE

*Abstract*—Bounds and approximations are obtained for the average probability of error in an asynchronous slow-frequency-hopped spread-spectrum multiple-access communications system with non-coherent binary frequency-shift-keyed (FSK) data transmission. Both nonselective fading and wide-sense-stationary uncorrelated-scattering fading are considered.

## I. INTRODUCTION

SEVERAL communications systems currently being developed have the following common features. Frequency-hopped spread-spectrum modulation is employed with a hopping rate not greater than the data rate. Multiple-access capability is required, because with high probability two or more terminals will be transmitting simultaneously. During transmission the spread-spectrum signals encounter severe fading, which causes reduced signal strength and may produce intersymbol interference or other dispersive effects. These systems are described in current terminology as slow-frequency-hopped (SFH) spread-spectrum multiple-access (SSMA) communications systems with fading channels.

In this paper we present bounds and approximations for the average probability of error for SFH/SSMA communications over fading channels. Two important classes of fading models are considered: the class of nonselective Rician fading channels—which includes the additive white Gaussian noise channel and the nonselective Rayleigh fading channel as special cases—and the selective wide-sense-stationary uncorrelated-scattering fading channel. The data modulation is binary frequency-shift-keying (FSK), but many of the results apply to differential phase-shift-keying (DPSK) as well. Noncoherent demodulation of the data is employed, partly because we do not require coherent frequency hopping and dehopping. The communications network is assumed to be asynchronous; that is, a given terminal makes no attempt to coordinate its transmissions with those of other terminals. This may be due to the lack of an accurate timing reference or because of the variation in propagation times among the different communication paths in the net-

work. The point here is that even if the transmitters have a common clock they cannot adjust their transmission times to provide coordinated arrival times at *all* of the receivers in the network.

In the analysis of SFH/SSMA systems there are two approaches to the modeling of the frequency hopping patterns: general random-process models may be employed or specific (deterministic) sets of hopping patterns may be considered. The random-process models are often used in an attempt to match certain characteristics of extremely complex hopping patterns which have very long periods. Also, random-process models serve as substitutes for deterministic models when the communications engineer is given little or no information about the structure of the hopping patterns to be used in the system. Both random patterns and a special class of deterministic patterns (based on Reed–Solomon codes) are considered in this paper.

The results obtained in this paper are bounds and approximations for the bit error probability. These results are useful for both uncoded FH/SSMA systems and fully-interleaved coded FH/SSMA systems. For coded systems which employ random-error-correcting codes, full interleaving is usually necessary for satisfactory performance. We have also obtained results (similar to those presented in Section III) on the probability of error for FH/SSMA systems which employ certain burst-error-correcting codes, but this topic is beyond the scope of the present paper.

A brief outline of the paper is as follows. The model for the SFH/SSMA system is presented in Section II where our models for the various subsystems and signals are described. The effect of nonselective fading on the probability of error in an SFH/SSMA system is considered in Section III. A more precise analysis is given in Section IV for the special case in which the channel exhibits nonselective Rayleigh fading. Finally, selective fading is considered in Section V.

## II. SYSTEM MODEL

The transmitter for the slow-frequency-hopped spread-spectrum signal is shown in Fig. 1. There are $K$ such transmitters in the spread-spectrum multiple-access system. The $k$th data signal $b_k(t)$ is a sequence of positive and negative rectangular pulses of duration $T$. The amplitude of the $l$th pulse for the $k$th signal is denoted by $b_l^{(k)}$ (i.e., $b_k(t) = b_l^{(k)}$ for $lT \leqslant t < (l + 1)T$), and $b_l^{(k)}$ is either $+1$ or $-1$ for each $k$ and $l$. The data signal $b_k(t)$ is the input to an FSK modulator, and

Manuscript received August 28, 1981; revised February 11, 1982. This paper was presented in part at the IEEE International Conference on Communications, Denver, CO, June 1981. This work was supported by the Naval Research Laboratory under Contract N00014-80-C-0802 and by the Joint Services Electronics Program under Contract N00014-79-C-0424.

The authors are with the Coordinated Science Laboratory and the Department of Electrical Engineering, University of Illinois, Urbana, IL 61801.

Reprinted from *IEEE Trans. Commun.*, vol. COM-30, pp. 996–1009, May 1982.

204

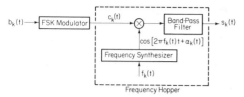

Fig. 1.   Transmitter model.

the corresponding output is

$$c_k(t) = \cos\{2\pi[f_c + b_k(t)\Delta]t + \theta_k(t)\} \tag{1}$$

where $\Delta$ is one-half the spacing between the two FSK tones. The signal $\theta_k(t)$ is the phase signal introduced by the FSK modulator; that is, if $b_l^{(k)} = m$ then $\theta_k(t) = \theta_{k,m}$ for $lT \leqslant t < (l+1)T$ where $\theta_{k,m}$ is the phase of the tone at frequency $f_c + m\Delta$ for $m = +1$ or $m = -1$.

The FSK signal is then frequency hopped according to the $k$th hopping pattern $f_k(t)$ which is derived from a sequence $(f_j^{(k)}) = \cdots, f_{-1}^{(k)}, f_0^{(k)}, f_1^{(k)}, \cdots$ according to

$$f_k(t) = f_j^{(k)}, \qquad jT_h \leqslant t < (j+1)T_h. \tag{2}$$

The parameter $T_h$ is the time between hops (also the dwell time for this system). For slow frequency-hopping $T_h$ is an integer multiple of $T$. The frequencies $f_j^{(k)}$ are all from the set $S = \{\nu_n: 1 \leqslant n \leqslant q\}$ which is ordered such that $\nu_n < \nu_{n+1}$ for each $n$. Let $\Delta'$ be the minimum spacing between the frequencies in the set $S$, and let $N_b \triangleq T_h/T$ be the number of data bits per hop.

The bandpass filter shown in Fig. 1 removes unwanted frequency components present at the output of the multiplier. The signal at the output of the filter is

$$s_k(t) = \sqrt{2P} \cos[2\pi\tilde{f}_k(t)t + \tilde{\varphi}_k(t)] \tag{3}$$

where

$$\tilde{f}_k(t) = f_c + b_k(t)\Delta + f_k(t) \tag{4}$$

and

$$\tilde{\varphi}_k(t) = \theta_k(t) + \alpha_k(t). \tag{5}$$

The signal $\alpha_k(t)$ represents the phase shifts introduced by the frequency hopper as it switches from one frequency to another. Accordingly, $\alpha_k(t)$ is constant on the time intervals that $f_k(t)$ is constant. Let $\alpha_j^{(k)}$ denote the value of $\alpha_k(t)$ on the interval $[jT_h, (j+1)T_h)$.

The quantity $P$ that appears in (3) is the power of the $k$th signal at the receiver in the absence of fading. In order to account for fading, we will multiply $P$ by a suitable factor to obtain the average power in the received signal. For simplicity we have assumed that the signals $s_k(t)$ all have the same power. However, as we will point out later, we obtain error probability bounds that are valid even if the power levels are not equal.

Since we are considering an asynchronous multiple-access system, we allow an arbitrary time delay $\tau_k$ for the $k$th communication link ($1 \leqslant k \leqslant K$). Thus, the received signals are $s_k(t - \tau_k)$, $1 \leqslant k \leqslant K$. For the random hopping patterns that will be considered in subsequent sections, it is sufficient to consider time delays modulo $T_h$. In order to allow for the possibility of deterministic periodic hopping patterns, we consider time delays modulo $NT_h$ where $N$ is the period of the patterns for deterministic hopping patterns or $N = 1$ for random hopping patterns. Thus, we may restrict attention to time delays in the range $0 \leqslant \tau_k < NT_h$. Similarly, we are only concerned with phase angles modulo $2\pi$, so we may restrict attention to phase angles in the interval $[0, 2\pi]$.

The analysis presented in this paper does not account for adjacent channel interference in the frequency-hopping system or for interference between the two FSK tones of a given signal. Instead we are primarily concerned with multiple-access interference and the effects of fading such as intersymbol interference and reduced signal strength. In order to focus on multiple-access interference and fading, we made certain simplifying assumptions concerning the frequency spacings $\Delta$ and $\Delta'$. It is enough for our purposes to have

$$\Delta' \gg \Delta + T^{-1} \tag{6a}$$

and

$$\Delta \gg T^{-1}. \tag{6b}$$

However, it is possible to relax these conditions somewhat, expecially for nonselective fading. For example, if the fading is nonselective then it is sufficient to replace the constraint $\Delta \gg T^{-1}$ by the condition $\Delta = m/2T$ for any positive integer $m$ (the case $m = 1$ is of greatest interest). In the absence of time-selective fading our results are valid if $\Delta'$ is about $3(\Delta + T^{-1})$ or larger, and they are likely to be fairly good approximations even if $\Delta' \approx 2(\Delta + T^{-1})$. However, frequency dispersion can expand the signal bandwidths so that $\Delta' \gg \Delta + T^{-1}$ is needed for time-selective fading.

Under our assumptions, the frequency band that contains the signals $s_k(t)$ is approximately the band from $f_c + \nu_1 - \Delta$ to $f_c + \nu_q + \Delta$. The center of this band is at frequency $f_c' = f_c + \frac{1}{2}(\nu_q + \nu_1)$. The (one-sided) bandwidth $W$ is approximately $\nu_q - \nu_1 + 2\Delta$. Under our assumptions $W \approx \nu_q - \nu_1 \geqslant (q-1)\Delta'$.

In the absence of fading and noise the received signal is given by

$$s(t) = \sum_{k=1}^{K} s_k(t - \tau_k). \tag{7}$$

We focus our attention on the receiver for the $i$th signal, and in doing so we may select the time reference such that $\tau_i = 0$. The variables $\tau_k$ are then delays (modulo $NT_h$) relative to this time reference.

The receiver for the $i$th signal is shown in Fig. 2. The received signal $\tilde{s}(t)$, which is a faded version of $s(t)$, is the input to the first bandpass filter. This filter has center frequency ap-

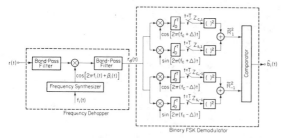

Fig. 2. Receiver model.

proximately $f_c'$ and bandwidth approximately $W$ so $\tilde{s}(t)$ is passed without distortion. This filter is followed by the $i$th dehopper which is synchronized in frequency and time to the $i$th frequency-hopping signal $f_i(t)$. The dehopper introduces a phase signal $\beta_i(t)$ which is analogous to the phase signal $\alpha_i(t)$ introduced by the frequency hopper. The phase signal $\beta_i(t)$ is constant during the time intervals between hops (i.e., when $f_i(t)$ is constant). The constant value of $\beta_i(t)$ for $jT_h \leqslant t < (j+1)T_h$ is denoted by $\beta_j^{(i)}$.

The time delays, phase angles, and data symbols are modeled as mutually independent random variables, each of which is uniformly distributed on the appropriate set (cf. [4] or [6]). The random time delays are the random variables $\tau_k$. The random phase angles that are of primary interest are $\theta_{k,m}$, $\alpha_j^{(k)}$, and $\beta_j^{(i)}$. An important feature of our model for asynchronous spread-spectrum multiple-access systems is that addition of phase angles is modulo-$2\pi$ addition. This feature is critical to our assertions concerning the distributions and the statistical independence of the phase angles (the basis for these assertions is given on pp. 159–160 of [6]).

The output of the dehopper is then passed through a bandpass filter which is designed to remove certain unwanted signals such as the double-frequency components of the $i$th signal itself, the sum and difference frequency components due to the other $K-1$ signals (except, of course, those that happen to be at the same frequency as the $i$th signal), and the thermal noise that is outside the frequency band occupied by the $i$th signal. The bandwidth $B$ of this bandpass filter is less than $\Delta'$ but usually larger than $2(\Delta + T^{-1})$. If $(\Delta + T^{-1}) \ll B < \Delta'$ then the thermal noise present at the output of the bandpass filter which follows the dehopper has a bandwidth larger than that of the FSK demodulator. This simplifies the analysis of the demodulator.

As shown in Fig. 2 the FSK demodulator has two branches. Each branch forms a statistic $\tilde{R}_m^2$ where $m = 1$ corresponds to the upper branch and $m = -1$ corresponds to the lower branch. Each of these two branches has two components. In the in-phase component the signal is multiplied by $\cos[2\pi(f_c + m\Delta)t]$, and the quadrature component is multiplied by $\sin[2\pi(f_c + m\Delta)t]$.

Consider the reception of the data bit $b_l^{(i)}$. The outputs of the in-phase components of the two branches are given by

$$Z_{c,m} = \int_{lT}^{(l+1)T} r_d(t) \cos[2\pi(f_c + m\Delta)t]\, dt \qquad (8)$$

for $m = \pm 1$, where $r_d(t)$ is the output of the bandpass filter which follows the $i$th dehopper (i.e., $r_d(t)$ is the input to the $i$th FSK demodulator). Notice that in general $Z_{c,m}$ depends on both $l$ and $i$. However, if the random hopping patterns are stationary and identically distributed and the fading process is stationary and not frequency selective, then the distribution of the random variable $Z_{c,m}$ will not depend on either $l$ or $i$. In case the hopping patterns are deterministic or the fading is frequency selective, then we provide upper bounds on the probability of error which are independent of $l$ and $i$. The outputs of the quadrature components of the two branches are denoted by $Z_{s,m}$ for $m = \pm 1$. The random variables $Z_{s,m}$, which are defined by (8) with $\cos[\cdot]$ replaced by $\sin[\cdot]$, have the same properties as $Z_{c,m}$.

## III. PERFORMANCE OF FH/SSMA SYSTEM WITH NONSELECTIVE FADING

The channels considered in this section are the nonselective slow-fading channels. For the frequency-hopped spread-spectrum system described in the previous section, this means that the signal at the input to the first bandpass filter in the $i$th receiver (see Fig. 2) is

$$r(t) = n(t) + \sum_{k=1}^{K} y_k(t - \tau_k) \qquad (9)$$

where for $lT \leqslant t < (l+1)T$ the signal $y_k(t)$ is given by

$$y_k(t) = \sqrt{2P} A_l^{(k)} \cos[2\pi \tilde{f}_k(t)t + \tilde{\varphi}_k(t) + \Theta_l^{(k)}]. \qquad (10)$$

The thermal noise $n(t)$ is white Gaussian noise with spectral density $\frac{1}{2}N_0$. Notice from comparisons of (9) and (10) with (7) and (3), respectively, that $y_k(t)$ is a faded version of $s_k(t)$ and, in the absence of noise, $r(t)$ is $\tilde{s}(t) = \sum_{k=1}^{K} y_k(t - \tau_k)$ which is a faded version of $s(t)$.

The amplitude of the fading signal $y_k(t)$ during the time interval $lT \leqslant t < (l+1)T$ is represented by a nonnegative random variable $A_l^{(k)}$, and the phase shift due to the fading is denoted by $\Theta_l^{(k)}$. In this section the only assumption that we make concerning the signal amplitudes is that they are constant during the data bit interval. The sequence of amplitudes $(A_l^{(k)}) = \cdots, A_{-1}^{(k)}, A_0^{(k)}, A_1^{(k)}, \cdots$ may be any stationary random sequence. In particular, we place no restrictions on the statistical dependence of amplitudes in different data bit intervals. Consider the set $\{A_l^{(k)}: jN_b \leqslant l < (j+1)N_b\}$ of amplitudes for the data bits that are transmitted during the $j$th hopping interval $[jT_h, (j+1)T_h]$. This interval contains the data bits $b_l^{(k)}$ for $jN_b \leqslant l < (j+1)N_b$. Among the cases of interest are the two extreme cases (i) $A_l^{(k)} = A_m^{(k)}$ for all $l$ and $m$ in the same hopping interval and (ii) $A_l^{(k)}$ and $A_m^{(k)}$ are independent if $l \neq m$ but $l$ and $m$ are in the same hopping interval. Case (i) corresponds to a system with no interleaving and a channel with fading which is slow relative to the hopping rate. An example of case (ii) arises in a system which is fully interleaved (e.g., if a random-error-correcting code is to be employed). Although these are the two specific cases of greatest interest, there is no need to restrict attention to such special cases in this section. Similarly, we impose no restrictions on the phase sequence $(\Theta_l^{(k)})$; all that is required is a constant value for the phase during the data bit intervals. Notice from (1)–(5) that for $lT \leqslant t < (l+1)T$ the phase of the signal in (10) is given by

$$\Psi_l^{(k)} = \theta_{k,m} + \alpha_j^{(k)} + \Theta_l^{(k)} \tag{11}$$

where $j$ is the integer part of $l/N_b$. Under quite general conditions the phase $\Psi_l^{(k)}$ is uniformly distributed on $[0, 2\pi]$ because the addition in (11) is modulo-$2\pi$. For example, it is enough to assume that *one* of the phase angles which appears on the right-hand side of (11) is uniformly distributed and that they are mutually independent (see [6, pp. 159–160] for the rationale for this statement).

There are two different phenomena which contribute to errors in the system under consideration. First, even in the absence of noise and fading, errors may occur in a frequency-hopped spread-spectrum multiple-access system when a signal is hopped to a frequency slot that is occupied by another signal. Whenever two different signals simultaneously occupy one frequency slot we say a *hit* occurs. Second, even in the absence of hits, errors may occur due to the fading and additive noise. The first step in analyzing the overall probability of error is to evaluate the probability of a hit for various types of hopping patterns.

*A. Probability of a Hit*

Consider as before the $i$th receiver during reception of the $l$th data bit. For a nonselective fading channel we say that a *hit* from the $k$th signal occurs during the $l$th data bit if

$$f_k(t - \tau_k) = f_i(t) \tag{12}$$

for at least one value of $t$ in the $l$th data bit interval $[lT, (l+1)T)$. As pointed out in Section II, we can let $N = 1$ in considering stationary random hopping patterns. It follows that the probability $P_l^{(k)}$ of a hit from the $k$th signal during the $l$th data bit interval does not depend on $l$ for such patterns. If the $K$ hopping patterns $\{(f_j^{(k)}): 1 \leqslant k \leqslant K\}$ are also mutually independent and identically distributed, then $P_l^{(k)}$ does not depend on $k$ either, and hence we denote it by $P$ for such patterns. We first consider two different models for stationary random hopping patterns and give the value of $P$ for each case.

Suppose the random process $(f_j^{(k)})$ is a stationary Markov process with transition probabilities given by

$$P(f_{j+1}^{(k)} = v_n \mid f_j^{(k)} = v_r) = (q-1)^{-1} \tag{13}$$

for $1 \leqslant n \leqslant q$, $1 \leqslant r \leqslant q$, and $n \neq r$. It follows that for these patterns

$$P(f_{j+1}^{(k)} = f_j^{(k)}) = 0 \tag{14}$$

and hence

$$P = \frac{1}{q}\left(1 + \frac{1}{N_b}\right). \tag{15}$$

Because of (13) the process $(f_j^{(k)})$ is a random process with first-order distribution given by

$$P(f_j^{(k)} = v_n) = q^{-1}, \qquad 1 \leqslant n \leqslant q. \tag{16}$$

If instead of (13) we consider random hopping patterns for which $f_{j+1}^{(k)}$ is independent of $f_j^{(k)}$ and the distribution of $f_j^{(k)}$ is given by (16) for each $j$, then (14) should be replaced by

$$P(f_{j+1}^{(k)} = f_j^{(k)}) = q^{-1} \tag{17}$$

and thus the probability of a hit is

$$P = \frac{1}{q}\left[1 + \frac{1}{N_b}\left(1 - \frac{1}{q}\right)\right]. \tag{18}$$

Notice that

$$P \leqslant \frac{1}{q}\left(1 + \frac{1}{N_b}\right) \tag{19}$$

and if $q$ is large then

$$P \approx \frac{1}{q}\left(1 + \frac{1}{N_b}\right) \tag{20}$$

[cf. (15)]. Thus, for large $q$ these *memoryless* hopping patterns give approximately the same probability of a hit as the first-order Markov patterns.

In general, for a set of deterministic hopping patterns the probability $P_l^{(k)}$ depends on both $k$ and $l$. One set of deterministic hopping patterns that has very good properties is derived from a Reed-Solomon code, so we refer to it as a set of Reed-Solomon (RS) patterns [7]. Given a prime number $q$ of frequency slots, the particular set of RS patterns of interest here consists of $N = q - 1$ sequences of period $N$ (of course, we can always choose a subset if fewer patterns are needed). Each sequence is *nonrepeating*: $f_j^{(k)} \neq f_{j+n}^{(k)}$ for $1 \leqslant n \leqslant N - 1$. Let the function $\delta(\cdot, \cdot)$ be defined by

$$\delta(u, v) = \begin{cases} 1 & u = v \\ 0 & u \neq v. \end{cases} \tag{21}$$

The property of RS patterns that is of primary importance here is that for any two patterns $(f_j^{(k)})$ and $(f_j^{(i)})$

$$\sum_{n=0}^{N-1} \delta(f_n^{(k)}, f_j^{(i)}) \leqslant 1 \tag{22}$$

for each $j$. Property (22) is actually valid for any set of nonrepeating patterns.

Since $\tau_k$ is uniform on $[0, NT_h]$ for $k \neq i$, then it follows from (22) that

$$P_l^{(k)} \leqslant P \triangleq \frac{T_h + T}{NT_h} = \frac{1}{N}\left(1 + \frac{1}{N_b}\right) = \frac{1}{q-1}\left(1 + \frac{1}{N_b}\right). \tag{23}$$

Actually (22) implies the stronger statement that either $P_l^{(k)} = 0$ or $P_l^{(k)} = P$. Since the number of frequency slots $q$ is larger than the period $N = q - 1$, then it is possible to choose the hopping pattern $(f_j^{(k)})$ such that $P_l^{(k)} = 0$ for $N_b$ different values of $l$ in the range $0 \leqslant l \leqslant NN_b$. Notice that for large $q$

$$P \approx \frac{1}{q}\left(1 + \frac{1}{N_b}\right) \tag{24}$$

is a good approximation for the upper bound [cf. (15) and (20)].

Of primary interest for our subsequent analysis is the probability $\hat{P}_l$ of one or more hits from the $K - 1$ signals (corresponding to $k \neq i$) during the $l$th data bit interval. For stationary *random* patterns $\hat{P}_l$ does not depend on $l$ so we denote it by $\hat{P}$. If the patterns are also mutually independent and identically distributed then

$$\hat{P} = 1 - (1 - P)^{K-1} \tag{25}$$

where $P$ is the probability of a hit from a given signal. For the first-order Markov patterns (25) and (15) imply

$$\hat{P} = 1 - \left\{1 - \frac{1}{q}\left(1 + \frac{1}{N_b}\right)\right\}^{K-1}. \tag{26}$$

If the patterns are sequences of independent random variables (i.e., memoryless patterns) satisfying (16) then

$$\hat{P} = 1 - \left\{1 - \frac{1}{q}\left[1 + \frac{1}{N_b}\left(1 - \frac{1}{q}\right)\right]\right\}^{K-1}. \tag{27}$$

Next we consider the probability $\hat{P}_l$ of one or more hits in the $l$th data bit interval for *deterministic* patterns. Since the random variables $\tau_k, k \neq i$, are mutually independent, then for any deterministic hopping patterns

$$\hat{P}_l = 1 - \left\{\prod_{\substack{k=1 \\ k \neq i}}^{K} [1 - P_l^{(k)}]\right\}. \tag{28}$$

For RS patterns (23) implies

$$\begin{aligned}
\hat{P}_l &\leqslant \hat{P} \\
&\triangleq 1 - [1 - P]^{K-1} \\
&= 1 - \left\{1 - \frac{1}{q-1}\left(1 + \frac{1}{N_b}\right)\right\}^{K-1}
\end{aligned} \tag{29}$$

where the symbol $\hat{P}$ is used to denote an upper bound. Notice from (27) and (29) that for large $q$

$$\hat{P} \approx 1 - \left\{1 - \frac{1}{q}\left(1 + \frac{1}{N_b}\right)\right\}^{K-1} \tag{30}$$

for the sequences of independent random elements *and* the RS sequences. Notice from (26) that the expression given in (30) is the exact value of $\hat{P}$ for the first-order Markov patterns.

### B. Bounds and Approximations for the Probability of Error

For a nonselective fading channel the bit error probability $P_{e,l}$ in a slow-frequency-hopped spread-spectrum multiple-access communications systems can be written as

$$P_{e,l} = P_0(1 - \hat{P}_l) + P_{1,l}\hat{P}_l \tag{31}$$

where $P_0$ is the conditional probability of error for the $l$th bit given that there are no hits and $P_{1,l}$ is the conditional probability of error for the $l$th data bit given there is at least one hit. Noitce that $P_0$ does not depend on $l$. In general $P_{1,l}$ depends on $l$ but, as will be seen from the numerical results, it is sufficient for many purposes to use the bounds $0 \leqslant P_{1,l} \leqslant \frac{1}{2}$.

Recall that for stationary random hopping patterns $\hat{P}_l$ does not depend on $l$ (and hence it is denoted by $\hat{P}$). For RS patterns $\hat{P}_l$ depends on $l$ but its upper bound $\hat{P}$, given by (29), does not. Hence, for all of these patterns we have the lower bound

$$P_{e,l} \geqslant P_L \triangleq P_0(1 - \hat{P}) \tag{32}$$

and the upper bound

$$P_{e,l} \leqslant P_U \triangleq P_L + \frac{1}{2}\hat{P} = \hat{P}_0 + (\frac{1}{2} - P_0)\hat{P} \tag{33}$$

where $\hat{P}$ is given by (26), (27), or (29), depending on which type of hopping patterns are employed. The lower bound is the same as we previously presented in [5], but the upper bound of (33) is a slight improvement of the upper bound presented in [5].

It is tempting to use $P_{e,l} \geqslant P_0$ in place of (32), and we certainly believe this tighter lower bound to be valid for independent time delays, data streams, and hopping patterns. Under these conditions it is intuitively clear that multiple-access interference cannot decrease the average probability of error. However, the lower bound of (32) has the advantage that it holds under more general conditions (such as for dependent time delays, data streams, and hopping patterns).

The bounds given in (32) and (33) are valid even if the power levels are not the same for the various signals or the hopping patterns are statistically dependent. As might be expected, the imposition of additional restrictions on the system leads to more precise results. In Section IV we present such results for a more restrictive channel model. However, even with the full generality of the nonselective fading channel model considered in this section, several improvements can be made. In what follows we give a tighter lower bound and a useful approximation for systems with equal power signals and certain constraints on the hopping patterns and the binary data streams. Specifically, the hopping patterns are assumed to be stationary, mutually independent, identically distributed random patterns, and the data sequences are stationary, memoryless, independent random sequences with distribution given by $P(b_n^{(k)} = m) = \frac{1}{2}$ for $m = +1$ and $m = -1$.

The lower bound can be improved for such systems by providing a nonzero lower bound for the term $P_{1,l}\hat{P}_l$ of (31). One such bound is obtained as follows. Consider the conditional probability of error in the $l$th data bit given a "full" hit from the $k$th signal [i.e., given that (12) holds for *all* $t$ in $[lT, (l + 1)T)]$ and given the $k$th data signal is equal to $-b_l^{(i)}$ for the *two* consecutive bit intervals of interest. This conditional probability of error is equal to ½. The conditional probability of a "full" hit (given a hit has occurred) is not smaller than $(T_h - T)/(T_h + T)$, and the probability of two consecutive transmissions of a particular tone is ¼. Finally, we use the fact that (25) implies

$$\hat{P}_l \geqslant (K - 1)P(1 - P)^{K-2}$$

which is just the statement that the probability of one or more hits is not less than the probability of exactly one hit. From the above we conclude that

$$P_{1,l}\hat{P}_l \geqslant \frac{(N_b - 1)}{8(N_b + 1)} (K - 1)P(1 - P)^{K-2}$$

so that the improved lower bound is

$$P_{e,l} \geqslant \tilde{P}_L \triangleq P_L + \frac{(N_b - 1)}{8(N_b + 1)} (K - 1)P(1 - P)^{K-2}. \quad (34)$$

We use a tilde ($\sim$) to denote bounds and approximations which are valid for the restricted class of systems only (i.e., equal power signals, memoryless independent data sequences, independent hopping patterns).

An approximation which is valid under the same conditions is

$$P_{e,l} \approx \tilde{P}_A \triangleq P_L + \frac{1}{2}(\frac{1}{2} + P_0)(K - 1)P(1 - P)^{K-2}. \quad (35)$$

This approximation is very accurate whenever $q/K$ is large because it is based on the assumption that the probability of a multiple hit (i.e., hits from two or more signals in a given data bit interval) is negligibly small in comparison to the probability of a hit from only one signal.

Comparisons of the bounds and the approximation are given in Table I for various values of $P_0$, $K$, $q$, and $N_b$. The hopping patterns are the first-order Markov patterns for the data in Table I, but in view of (30) the results would not be significantly different for the other patterns described above.

### C. The Nonselective Rician Fading Channel

The bounds and approximation given in (32)–(35) can be applied to any particular nonselective fading channel by substituting the appropriate expression for $P_0$ in these results. In this section we consider the Rician nonselective fading model in which each transmitted signal results in a received signal that is the sum of a nonfaded version of the transmitted signal and a (nonselective) Rayleigh faded version of the trans-

### TABLE I
#### LOWER BOUNDS, APPROXIMATION, AND UPPER BOUND ON THE PROBABILITY OF ERROR FOR AN FH/SSMA SYSTEM

| $P_0$ | $P_L$ | $\tilde{P}_L$ | $\tilde{P}_A$ | $P_U$ |
|---|---|---|---|---|
| (a) $K = 15$, $q = 1000$, and $N_b = 10$ | | | | |
| 0.100 | 0.098 | 0.100 | 0.103 | 0.106 |
| 0.050 | 0.049 | 0.051 | 0.053 | 0.057 |
| 0.030 | 0.030 | 0.031 | 0.034 | 0.037 |
| 0.020 | 0.020 | 0.021 | 0.024 | 0.027 |
| 0.010 | 0.010 | 0.011 | 0.014 | 0.017 |
| 0.005 | 0.005 | 0.006 | 0.009 | 0.013 |

| $P_0$ | $P_L$ | $\tilde{P}_L$ | $\tilde{P}_A$ | $P_U$ |
|---|---|---|---|---|
| (b) $K = 15$, $q = 100$, and $N_b = 5$ | | | | |
| 0.100 | 0.084 | 0.096 | 0.128 | 0.162 |
| 0.050 | 0.042 | 0.054 | 0.081 | 0.120 |
| 0.030 | 0.025 | 0.037 | 0.063 | 0.103 |
| 0.020 | 0.017 | 0.029 | 0.054 | 0.095 |
| 0.010 | 0.008 | 0.020 | 0.045 | 0.086 |
| 0.005 | 0.004 | 0.016 | 0.041 | 0.082 |

| $P_0$ | $P_L$ | $\tilde{P}_L$ | $\tilde{P}_A$ | $P_A$ |
|---|---|---|---|---|
| (c) $K = 25$, $q = 250$, and $N_b = 20$ | | | | |
| 0.100 | 0.090 | 0.100 | 0.118 | 0.138 |
| 0.050 | 0.045 | 0.056 | 0.070 | 0.093 |
| 0.030 | 0.027 | 0.037 | 0.051 | 0.075 |
| 0.020 | 0.018 | 0.028 | 0.042 | 0.066 |
| 0.010 | 0.009 | 0.019 | 0.032 | 0.057 |
| 0.005 | 0.005 | 0.015 | 0.027 | 0.053 |

mitted signal. The difference in the propagation times for these two components is sufficiently small compared with the data bit duration $T$ that the overall channel is nonselective. This model is discussed in [9] where the nonfaded component is called the *fixed* or *specular* component and the Rayleigh-faded component is called the *random* or *scatter* component. In some applications the nonfaded component arises from a direct path between the transmitter and the receiver, and the faded component arises from a reflection.

The amplitude $S$ of the sum of the two components of the received signal is a random variable with a Rician distribution (see [8] or [9]). Since we are interested in the conditional probability of error given there are no hits, we can assume in all that follows that only the components of the $i$th signal are present at the $i$th receiver (during the data bit interval under consideration). Let $\rho$ be the normalized bit energy to noise density ratio, so that $S^2\rho$ is the actual received energy to noise density ratio. Hence, for noncoherent FSK the probability of error given $S = a$ is ½ exp $(-\frac{1}{2} a^2 \rho)$. For the Rician channel the density function $f_S$ for the amplitude $S$ is

$$f_S(a) = (a/\sigma^2) \exp\{-\frac{1}{2}(a^2 + \alpha^2)/\sigma^2\}I_0(\alpha a/\sigma^2) \quad (36)$$

for $a > 0$, where $\alpha^2$ represents the strength of the nonfaded component, $2\sigma^2$ is the expected value of the strength of the faded component, and $I_0$ is the zeroth-order modified Bessel function. The average probability of error for noncoherent

FSK[1] is [9]

$$P_0 = \int_0^\infty \tfrac{1}{2} \exp\left(-\tfrac{1}{2}\,a^2\rho\right) f_S(a)\,da$$

$$= \frac{\exp\{-\tfrac{1}{2}\,\alpha^2\rho/(\sigma^2\rho+1)\}}{2(\sigma^2\rho+1)}. \tag{37}$$

If $\bar{E}$ denotes the average energy per bit in the received signal then

$$\Lambda \triangleq \bar{E}/N_0 = (\alpha^2 + 2\sigma^2)\rho. \tag{38}$$

Let $\gamma^2$ denote the ratio of the power in the faded component to the power in the unfaded component; that is, $\gamma^2 = 2\sigma^2/\alpha^2$. Then we can write

$$P_0 = \frac{(\gamma^2+1)}{\gamma^2\Lambda + 2(\gamma^2+1)} \exp\{-\Lambda/[\gamma^2\Lambda + 2(\gamma^2+1)]\}. \tag{39}$$

Two limiting cases of interest are $\sigma^2 = 0$ and $\alpha^2 = 0$. If $\sigma^2 = 0$ ($\gamma^2 = 0$) then there is no faded component, and the channel is just an additive white Gaussian noise channel. In this case $\Lambda = \alpha^2\rho$, and the probability of error reduces to

$$P_0 = \tfrac{1}{2}\exp\{-\tfrac{1}{2}\Lambda\}. \tag{40}$$

If $\alpha^2 = 0$ the channel is a nonselective Rayleigh fading channel, $\Lambda = 2\sigma^2\rho$, and the probability of error is

$$P_0 = \frac{1}{\Lambda+2}. \tag{41}$$

An examination of (39) as a function of $\gamma^2$ shows that for $\gamma^2 = 10$ the probability of error for Rician fading is nearly the same as for Rayleigh fading. For example, if $\Lambda$ is 12 dB then $P_0$ is $1.81 \times 10^{-2}$ for $\gamma^2 = 0$, $4.41 \times 10^{-2}$ for $\gamma^2 = 0.1$, $4.53 \times 10^{-2}$ for $\gamma^2 = 1.0$, and $5.58 \times 10^{-2}$ for $\gamma^2 = 10.0$. The value of $P_0$ for Rayleigh fading ($\gamma^2 = \infty$) is $5.60 \times 10^{-2}$.

In order to apply (36)–(41) to the slow-frequency-hopped spread-spectrum multiple-access system, consider first the expressions (9) and (10) for the received signal. The amplitudes $A_l^{(k)}$ are random variables with a density function of the form given in (36). In general, the parameters $\alpha$ and $\sigma$ may depend on $i$, in which case $\Lambda$ and $\gamma$ also depend on $i$. The probability $P_0$ then depends on $i$ and is given by (37) with $\alpha$ and $\sigma$ replaced by $\alpha_i$ and $\sigma_i$ or by (39) with $\gamma$ and $\Lambda$ replaced by $\gamma_i$ and $\Lambda_i$. It follows from (9) and (10) that the parameter $\rho$ is given by $\rho = PT/N_0$.

The next step is to substitute for $P_0$ in (32)–(35) using the expressions (37) or (39). If $P_0$ depends on $i$ the bounds of (32) are valid, but, of course, they will also depend on $i$. Notice that if $\alpha$ and $\sigma$ depend on $i$, then the average power in the received signal also depends on $i$. That is, the signals are not required to have equal power. The *approximation* given in (35) is also valid even if $P_0$ depends on $i$, provided that $\alpha_k \approx \alpha_i$ and $\sigma_k \approx \sigma_i$ for all $k$.

In Fig. 3 the approximation $\tilde{P}_A$, which is given by (35) with $P_0$ replaced by the expression in (39), is shown as a function of $\Lambda = \bar{E}/N_0$ for various values of $\gamma^2$. For the data presented in Fig. 3, the values of $\alpha$ and $\sigma$ (and, hence, $\gamma$ and $\Lambda$) do not depend on $i$. Additional numerical data can be obtained from Table I by evaluating $P_0$ from (37) or (39). Notice that for Rayleigh fading with $\bar{E}/N_0$ less than 20 dB the value of $P_0$ is less than 0.01. From Table I we see that for $P_0 \leqslant 0.01$, the value of $P_U$ is always less than $2\tilde{P}_A$ and the value of $\tilde{P}_A$ is always less than $2\tilde{P}_L$ for the values of $K, q$, and $N_b$ considered in Table I. For $K = 15$, $q = 1000$, and $N_b = 10$, we see that for $P_0 \leqslant 0.01$ we always have $\tilde{P}_A \leqslant 1.2\,\tilde{P}_L$ and $P_U \leqslant 1.25\,\tilde{P}_A$. Thus, for Rayleigh fading or Rician fading with $\gamma^2 \geqslant 1$, the bounds and approximations given in this section are sufficiently accurate for the design of slow-frequency-hopped spread-spectrum multiple-access systems. Further evidence of this is given in the next section.

## IV. NONSELECTIVE RAYLEIGH FADING

In this section we present a more precise analysis of the effects of multiple-access interference and nonselective fading for the special case in which the fading is Rayleigh. This analysis provides a more accurate approximation and a tighter upper bound than can be obtained by specializing the results of Section III to Rayleigh fading. The system and channel models are as presented in Section III, and the received signal is as given in (9) and (10).

Since we are considering only Rayleigh fading in the present section, the random amplitudes $A_l^{(k)}$ have a Rayleigh distribution. The density function for $A_l^{(k)}$ is given by (36) with $\alpha = 0$ and $\sigma = \sigma_k$. In general the second moments $\mu_k \triangleq 2\sigma_k^2$ are different for different signals. For the analysis presented in this section we assume that the fading for different signals is statistically independent. Stated precisely, the requirement is that $A_{l_1}^{(1)}, A_{l_2}^{(2)}, \cdots, A_{l_K}^{(K)}$ are mutually independent for any choice of $l_1, l_2, \cdots, l_K$.

The starting point for the analysis of the receiver is (8). Since in practice $f_c \gg T^{-1}$ for a spread-spectrum system, the high frequency terms in the integrand of (8) may be ignored. First consider the in-phase component of the receiver of Fig. 2. The output of the integrator at the sampling instant

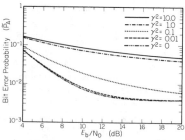

Fig. 3. Bit error probability for FH/SSMA communications with Rician fading ($K = 15$, $q = 1000$, and $N_b = 10$).

---

[1] Corresponding results for binary DPSK are obtained by replacing $\rho$ by $2\rho$ in (37).

is

$$Z_{c,m} = D_{c,m} + (P/8)^{1/2} T \sum_{k \neq i} F_{c,m}^{(k,i)} + N_{c,m}. \tag{42}$$

The first term $D_{c,m}$ is the component due to the signal $s_i(t)$. If the transmitted data bit is $b_\lambda^{(i)}$ for $\lambda = jN_b + p$ then

$$D_{c,m} = (P/8)^{1/2} T A_\lambda^{(i)} \delta(b_\lambda^{(i)}, m)$$
$$\cdot \cos [\theta_{i,m} + \alpha_j^{(i)} - \beta_j^{(i)} + \Theta_\lambda^{(i)}] \tag{43}$$

where $\delta(\cdot, \cdot)$ is as defined in (21). Since $D_{c,m}$ is the output of the integrator in the absence of multiple-access and channel noise, it is called the desired signal component.

The multiple-access interference $F_{c,m}^{(k,i)}$ from the $k$th signal depends upon the delay $\tau_k$. For convenience let $l_k = \lfloor \tau_k/T_h \rfloor$ and $n_k = \lfloor (\tau_k - l_k T_h)/T \rfloor$ where $\lfloor u \rfloor$ denotes the integer part of the real number $u$. Then $F_{c,m}^{(k,i)}$ can be expressed as

$$F_{c,m}^{(k,i)} = d_1(l_k) \hat{A}_1^{(k)}(\tau_k) \cos \varphi_1^{(k)}(l_k, \tau_k)$$
$$+ d_2(l_k) \hat{A}_2^{(k)}(\tau_k) \cos \varphi_2^{(k)}(l_k, \tau_k) \tag{44}$$

for $0 \leq n_k < p$. The following expressions define the terms in (44). First, the functions $d_1$ and $d_2$ are defined by

$$d_1(l) = d_2(l) = \delta(f_{j-l}^{(k)}, f_j^{(i)}) \tag{45}$$

for $0 \leq l < N$; that is, $d_1(l)$ and $d_2(l)$ are indicator functions for hits between the two hopping patterns $(f_j^{(k)})$ and $(f_j^{(i)})$ for an offset of $l$ hops. Next, for fixed $n_k$ and $l_k$ let

$$L = (j - l_k)N_b + p - n_k \tag{46}$$

and denote the data symbols $b_{L-1}^{(k)}$ and $b_L^{(k)}$ by $m'$ and $m''$, respectively. Let

$$\hat{A}_\nu^{(k)}(\tau_k) = A_{L-2+\nu}^{(k)} e_\nu^{(k)}(\tau_k) \tag{47}$$

for $\nu = 1$ and $\nu = 2$ where

$$e_1^{(k)}(\tau_k) = \delta(m', m)[\tau_k - l_k T_h - n_k T]/T \tag{48}$$

and

$$e_2^{(k)}(\tau_k) = \delta(m'', m)[l_k T_h + (n_k + 1)T - \tau_k]/T. \tag{49}$$

Also, for $0 \leq l < N$, define

$$\varphi_1^{(k)}(l, \tau_k) = \theta_{k,m'} + \alpha_{j-l}^{(k)} - \beta_j^{(i)}$$
$$- 2\pi[f_c + m'\Delta + f_{j-l}^{(k)}]\tau_k + \Theta_{L-1}^{(k)} \tag{50}$$

and let $\varphi_2^{(k)}(l, \tau_k)$ be defined as the right-hand side of (50) with $m'$ replaced by $m''$ and $L - 1$ replaced by $L$.

The multiple-access interference $F_{c,m}^{(k,i)}$ for $p < n_k \leq N_b - 1$ is given by (44) with $l_k$ replaced by $l_k + 1$ in the arguments of the functions $d_1$, $d_2$, $\varphi_1^{(k)}$, and $\varphi_2^{(k)}$. The only remaining case is $n_k = p$ for which $F_{c,m}^{(k,i)}$ is defined by (44) with $l_k$ replaced by $l_k + 1$ in the arguments of the functions $d_1$ and $\varphi_1^{(k)}$ only.

The expressions for the multiple-access interference are based on the following considerations. First, recall that the data pulse of interest is the $p$th pulse of the $j$th dwell interval for the $i$th signal. Since $\tau_i = 0$, this corresponds to the time interval from $(jN_b + p)T$ to $(jN_b + p + 1)T$, where $N_b$ is the number of data pulses in each dwell interval and $T$ is the pulse duration. Since (with probability 1) the time delay $\tau_k$ is not an integer multiple of $T$, two consecutive data pulses from the $k$th signal must be considered as possible interference for the pulse under consideration for the $i$th signal. The integer $L$ given by (46) specifies the two possible interfering pulses from the $k$th signal. These two pulses are in the time intervals determined by $l_k$ and $n_k$, and they correspond to the data symbols $b_{L-1}^{(k)}$ and $b_L^{(k)}$. Whether or not these pulses actually interfere with the pulse under consideration depends on the values of $d_1(l_k)$, $d_2(l_k)$, $d_1(l_k + 1)$, and $d_2(l_k + 1)$. A value of 1 for one or more of these functions indicates interference from the $k$th signal.

If $n_k$ is in the range $0 \leq n_k < p$, both of the potentially interfering pulses belong to the $(j - l_k)$th dwell interval of the $k$th signal. If $p < n_k \leq N_b - 1$ both of these pulses belong to the $(j - 1 - l_k)$th dwell interval. If $n_k = p$ one pulse is in the $(j - l_k)$th dwell interval and the other is in the $(j - 1 - l_k)$th. This is the reason that the three cases $0 \leq n_k < p$, $p < n_k \leq N_b - 1$, and $n_k = p$ are considered separately in defining the multiple-access interference $F_{c,m}^{(k,i)}$.

The terms $\delta(m', m)$ and $\delta(m'', m)$ indicate whether the potential interference is in the upper branch ($m = +1$) or the lower branch ($m = -1$) of the receiver for the $i$th signal (see Fig. 2). The functions $\varphi_1^{(k)}$ and $\varphi_2^{(k)}$ represent the phase of the interference relative to the phase of the $i$th signal.

If we set $A_l^{(k)} = 1$ and $\Theta_l^{(k)} = 0$ for all $l$ in (43), (47), and (50), the resulting expressions give the desired signal component and the multiple-access interference component for a system with a (nonfaded) additive white Gaussian noise channel. The remaining component of $Z_{c,m}$ is $N_{c,m}$ which is due to the channel noise process $n(t)$. It is easy to show that $N_{c,m}$ is a zero-mean Gaussian random variable with variance $N_0 T/16$.

The quadrature components are defined by expressions which are analogous to (42)–(44). In fact, $Z_{s,m}$ and $N_{s,m}$ are defined in the same way as above, and the only change that must be made in the definitions of $D_{s,m}$ and $F_{s,m}^{(k,i)}$ is that $\cos (\cdot)$ should be replaced by $-\sin (\cdot)$ in (43) and (44).

The error probability of interest is the average probability of a bit error for the $\lambda$th bit of the $i$th signal (as before, $\lambda = jN_b + p$). This average is with respect to the phase angles, time delays, and data sequences. It is convenient to first condition on the time delays and data symbols. We evaluate the conditional error probability $P_e(b, \tau)$, where $b$ is the vector of all the data symbols $\{b_l^{(k)}: 1 \leq k \leq K, -\infty < l < \infty\}$ and $\tau = (\tau_1, \tau_2, \cdots, \tau_K)$. Of course, for each value of $\tau$, $P_e(b, \tau)$

211

depends on only a finite number of the data symbols. In principle, the average probability of error $\bar{P}_e$ can be obtained by averaging $P_e(\boldsymbol{b}, \tau)$ with respect to $\boldsymbol{b}$ and $\tau$.

The analysis leading up to the expression for $P_e(\boldsymbol{b}, \tau)$ is in terms of conditional variances, conditional expectations, and conditional distributions. In all cases these are conditioned on $\boldsymbol{b}$ and $\tau$, even though for notational convenience we have not explicitly displayed the dependence on $\boldsymbol{b}$ and $\tau$. The phases, on the other hand, are treated as random variables throughout, and all expectations involve averaging with respect to the appropriate phase angles. Because of the symmetry of the problem, we can restrict attention to the case $b_\lambda^{(i)} = +1$ in all that follows. As a result $D_{c,-1} = 0$ in (42).

Consider the outputs $Z_{c,m}$ and $Z_{s,m}$ for $m = +1$ and $m = -1$. These random variables are conditionally Gaussian given $\boldsymbol{b}$ and $\tau$. Consequently, the conditional probability of error for noncoherent detection and slow nonselective Rayleigh fading is [8, p. 587]

$$P_e(\boldsymbol{b}, \tau) = \sigma_{c,-1}^2 (\sigma_{c,1}^2 + \sigma_{c,-1}^2)^{-1} \qquad (51)$$

where $\sigma_{c,m}^2$ is the conditional variance of $Z_{c,m}$. In writing (51) we have used the fact that $\sigma_{c,m}^2$ is also the conditional variance of $Z_{s,m}$ (by symmetry).

From (42) we see that $Z_{c,m}$ is the sum of several random variables which, as we discuss below, are conditionally independent given $\boldsymbol{b}$ and $\tau$. As a result, the variance $\sigma_{c,m}^2$ of $Z_{c,m}$ is the sum of the variance of these random variables. To compute $\sigma_{c,m}^2$ we must evaluate these variances.

First, consider the desired signal term $D_{c,m}$. As pointed out above, $D_{c,-1} = 0$ since $b_\lambda^{(i)} = +1$. For the fading model under consideration, $D_{c,1}$ is a zero-mean, conditionally Gaussian random variable. Its conditional variance is $PT^2 \mu_i / 16$.

Next we consider the multiple-access interference. To do so, it is necessary to consider the nature of the statistical dependence between $A_l^{(k)}$ and $A_{l+1}^{(k)}$ and between $\Theta_l^{(k)}$ and $\Theta_{l+1}^{(k)}$ for $l$ and $l+1$ in the same hopping interval. These are the random variables which describe the fading during adjacent data bits. We consider the two extreme cases described in Section III: (i) the fading is constant in the sense that $A_l^{(k)} = A_{l+1}^{(k)}$ and $\Theta_l^{(k)} = \Theta_{l+1}^{(k)}$ whenever $l$ and $l+1$ are in the same hopping interval and (ii) the fading is independent for adjacent data bits in the same hopping interval.

Under our assumptions, the multiple-access interference component $F_{c,m}^{(k,i)}$ is a zero-mean conditionally Gaussian random variable with conditional variance $\frac{1}{2} \mu_k \sigma_m^2(k, i)$. For constant fading, as described by case (i) above, we have

$$\sigma_m^2(k, i) = d_1(l_k)[e_1^{(k)}(\tau_k) + e_2^{(k)}(\tau_k)]^2 \qquad (52)$$

for $0 \leqslant n_k < p$,

$$\sigma_m^2(k, i) = d_1(l_k + 1)[e_1^{(k)}(\tau_k)]^2$$
$$+ d_1(l_k)[e_2^{(k)}(\tau_k)]^2 \qquad (53)$$

for $n_k = p$, and

$$\sigma_m^2(k, i) = d_1(l_k + 1)[e_1^{(k)}(\tau_k) + e_2^{(k)}(\tau_k)]^2 \qquad (54)$$

for $p < n_k < N_b$. For independent fading, as described by case (ii) above, we have

$$\sigma_m^2(k, i) = d_1(l_k)\{[e_1^{(k)}(\tau_k)]^2 + [e_2^{(k)}(\tau_k)]^2\} \qquad (55)$$

for $0 \leqslant n_k < p$. If $n_k = p$, $\sigma_m^2(k, i)$ is given by (53), and for $p < n_k < N_b$

$$\sigma_m^2(k, i) = d_1(l_k + 1)\{[e_1^{(k)}(\tau_k)]^2 + [e_2^{(k)}(\tau_k)]^2\}. \qquad (56)$$

Because of the independence of the fading for different signals, the random variables $F_{c,m}^{(k,i)}$ are independent when conditioned on the data bits and time delays. As a result

$$\sigma_{c,m}^2 = (PT^2/16)\left\{\delta(1, m)\mu_i + \sum_{k \neq i} \mu_k \sigma_m^2(k, i)\right\} + N_0 T/16. \qquad (57)$$

Thus, (51) can be written as

$$P_e(\boldsymbol{b}, \tau)$$
$$= \frac{(\bar{E}/N_0)^{-1} + \sum_{k \neq i} \mu_k \mu_i^{-1} \sigma_{-1}^2(k, i)}{1 + 2(\bar{E}/N_0)^{-1} + \sum_{k \neq i} \mu_k \mu_i^{-1} [\sigma_{-1}^2(k, i) + \sigma_1^2(k, i)]} \qquad (58)$$

where $\bar{E} = \mu_i PT$ is the energy per bit for the received signal (in the absence of multiple-access interference).

In order to evaluate the average probability of error $\bar{P}_e$, we must average the expression in (58) with respect to the time delays and data symbols. This is, of course, a difficult computation since it involves the evaluation of $K - 1$ dimensional integrals. However, we can obtain an approximation $\bar{P}_A$ and an upper bound $\bar{P}_U$ which are relatively easy to compute. This is accomplished by observing that $P_e(\boldsymbol{b}, \tau)$ depends on $\tau_k$ only through $t_k$, $l_k$, and $n_k$ where $t_k = (\tau_k - l_k T_h - n_k T)/T$. We can thus obtain a discrete approximation to the integral with respect to $t_k$ by approximating the uniform distribution on $[0, 1]$ by the discrete distribution with probability mass $J^{-1}$ at points $J^{-1}, 2J^{-1}, \cdots, (J - 1)J^{-1}$ and probability mass $(2J)^{-1}$ at the endpoints 0 and 1. We find that for the first-order Markov patterns and constant fading [case (i)]

$$P\{\sigma_m(k, i) = jJ^{-1}\} = p_j, \qquad 0 \leqslant j \leqslant J \qquad (59)$$

where the quantities $p_j$ are defined completely by the fact that their sum is 1 and

$$p_j = \begin{cases} (2Jq)^{-1}(1 + N_b^{-1}), & j = 1, 2, \cdots, J-1 \\ (4Jq)^{-1}(1 + N_b^{-1}) + (4q)^{-1}(1 - N_b^{-1}), & j = J. \end{cases}$$

(60)

For independent fading [case (ii)]

$$P\{\sigma_m(k,i) = jJ^{-1}\} = p_j, \qquad 0 \leqslant j \leqslant J \tag{61a}$$

$$P\{\sigma_m(k,i) = [j^2 + (J-j)^2]^{1/2} J^{-1}\} = q_j, \qquad 0 \leqslant j \leqslant \tfrac{1}{2} J \tag{61b}$$

where $p_j$ and $q_j$ are defined by

$$p_j = \begin{cases} (2Jq)^{-1}(1 + N_b^{-1}), & j = 1, 2, \cdots, J-1 \\ (2Jq)^{-1}, & j = J \end{cases} \tag{62a}$$

$$q_j = \begin{cases} (2Jq)^{-1}(1 - N_b^{-1}), & j = 1, 2, \cdots, J/2 - 1 \\ (4Jq)^{-1}(1 - N_b^{-1}), & j = J/2 \end{cases} \tag{62b}$$

and

$$p_0 = 1 - \sum_{j=1}^{J} p_j - \sum_{j=1}^{J/2} q_j. \tag{62c}$$

In (61) and (62) we assume $J$ is an even integer.

An *upper bound* can be obtained as follows. The conditional probability of error $P_e(\boldsymbol{b}, \boldsymbol{\tau})$ given by (58) is not convex in $t_k$ $(1 \leqslant k \leqslant K, k \neq i)$. However, if we upper bound the sum of squares of (53) for case (i) or of (55), (53), and (56) for case (ii), by the square of the sum (all terms are nonnegative) the upper bound on $P_e$ becomes a convex function of the $t_k$'s. We then obtain a discrete approximation to the integral with respect to $t_k$ as for $\overline{P}_A$. The upper bound $\overline{P}_U$ is the same for cases (i) and (ii) and the distribution of $\sigma_m(k,i)$ is given by (59), where the $p_j$ are defined by

$$p_j = \begin{cases} (4Jq)^{-1}(1 + N_b^{-1}), & j = 1, 2, \cdots, J-1 \\ (1 + J^{-1})(4q)^{-1}(1 + N_b^{-1}), & j = J. \end{cases} \tag{63}$$

A similar approximation and upper bound can be obtained for the sequences of independent random elements.

Finally, we note that in order that the approximation and bound presented in this section be tight and computationally efficient, we need to assume that $\mu_k = \mu_i$ for all $k \neq i$. If this is not the case, we can still work with $\mu_k' = \max_k\{\mu_k\}$, but the approximation and the upper bound obtained above are not expected to be very tight, so that it might be preferable to work with the bounds suggested in Section III which are not affected by the different power levels.

In Table II the approximation obtained in this section is compared with the improved lower bound, the approximation, and the upper bound of Section III-B for the first-order Markov hopping patterns and $\mu_k = \mu_i$ for all $k$. The approximation $\overline{P}_A$ [for both cases (i) and (ii)] and the bound $\overline{P}_U$ are evaluated for $J = 4$. It turns out that they are rather insensitive to increases in $J$ as long as $J \geqslant 4$. Values for $\overline{P}_A$ are given in Table II (a) for i) constant fading and ii) independent fading. The notations $\overline{P}_A^{(i)}$ and $\overline{P}_A^{(ii)}$, respectively, are used for these two cases. Independent fading turns out to be the most favorable case although the difference is less than 10 percent. Also notice that the bound $\overline{P}_U$ (common for both cases) differs from $\overline{P}_A$ of (i) or (ii) by at most 20 percent; therefore, in Table II (b) and (c) we present data on $\overline{P}_U$ only (not on $\overline{P}_A$). The purpose is for comparison with $\widetilde{P}_L, \widetilde{P}_A$, and $P_U$. In comparing $\overline{P}_U$ and $\widetilde{P}_A$ we note that $\widetilde{P}_A$ appears to be an upper bound for the nonselective Rayleigh case. Also for $q = 1000$, $K = 15$, $N_b = 10$, and $E/N_0 \leqslant 20$ dB the results of Table II (c) show that $\overline{P}_U \leqslant 1.17 \, \widetilde{P}_L, \widetilde{P}_A \leqslant 1.3 \, \overline{P}_U$, and $P_U \leqslant 1.52 \, \overline{P}_U$. Similar observations can be made for the data provided in Table II (a) and (b). As a final comment we point out that since the approximations $\overline{P}_A$ and the bound $\overline{P}_U$ are expected to be very close to the true probability of error, their favorable comparison with the simpler bounds $\widetilde{P}_L$ and $P_U$ and the approximation $\widetilde{P}_A$ strongly suggests the use of the latter for the design of SFH/SSMA systems.

TABLE II
BIT ERROR PROBABILITY FOR NONSELECTIVE RAYLEIGH FADING

| $\overline{E}/N_0$ (dB) | $\widetilde{P}_L$ | $\overline{P}_A^{(ii)}$ | $\overline{P}_A^{(i)}$ | $\overline{P}_U$ | $\widetilde{P}_A$ | $P_U$ | |
|---|---|---|---|---|---|---|---|
| (a) $K = 5$, $q = 100$, and $N_b = 10$ | | | | | | | |
| 6 | 1.64 | 1.17 | 1.72 | 1.72 | 1.74 | 1.82 | $(\times 10^{-1})$ |
| 8 | 1.19 | 1.25 | 1.26 | 1.27 | 1.28 | 1.37 | $(\times 10^{-1})$ |
| 10 | 8.41 | 8.95 | 9.02 | 9.09 | 9.21 | 10.14 | $(\times 10^{-2})$ |
| 12 | 5.80 | 6.29 | 6.36 | 6.44 | 6.54 | 7.52 | $(\times 10^{-2})$ |
| 15 | 3.28 | 3.73 | 3.81 | 3.90 | 3.97 | 5.01 | $(\times 10^{-2})$ |
| 20 | 1.37 | 1.78 | 1.87 | 1.97 | 2.02 | 3.10 | $(\times 10^{-2})$ |
| $\infty$ | 0.44 | 0.83 | 0.92 | 1.02 | 1.06 | 2.16 | $(\times 10^{-2})$ |

| $\overline{E}/N_0$ (dB) | $\widetilde{P}_L$ | | $\overline{P}_U$ | | $\widetilde{P}_A$ | $P_U$ | |
|---|---|---|---|---|---|---|---|
| (b) $K = 10$, $q = 1000$, and $N_b = 10$ | | | | | | | |
| 6 | 1.67 | | 1.68 | | 1.69 | 1.70 | $(\times 10^{-1})$ |
| 8 | 1.20 | | 1.22 | | 1.22 | 1.24 | $(\times 10^{-1})$ |
| 10 | 8.35 | | 8.51 | | 8.54 | 8.74 | $(\times 10^{-2})$ |
| 12 | 5.65 | | 5.79 | | 5.82 | 6.04 | $(\times 10^{-2})$ |
| 15 | 3.05 | | 3.18 | | 3.20 | 3.44 | $(\times 10^{-2})$ |
| 20 | 1.07 | | 1.20 | | 1.22 | 1.46 | $(\times 10^{-2})$ |
| $\infty$ | 0.10 | | 0.23 | | 0.24 | 0.49 | $(\times 10^{-2})$ |

| $\overline{E}/N_0$ (dB) | $\widetilde{P}_L$ | | $\overline{P}_U$ | | $\widetilde{P}_A$ | $P_U$ | |
|---|---|---|---|---|---|---|---|
| (c) $K = 15$, $q = 1000$, and $N_b = 10$ | | | | | | | |
| 6 | 1.66 | | 1.69 | | 1.70 | 1.72 | $(\times 10^{-1})$ |
| 8 | 1.20 | | 1.23 | | 1.23 | 1.26 | $(\times 10^{-1})$ |
| 10 | 8.36 | | 8.60 | | 8.65 | 8.97 | $(\times 10^{-2})$ |
| 12 | 5.67 | | 5.90 | | 5.94 | 6.28 | $(\times 10^{-2})$ |
| 15 | 3.08 | | 3.30 | | 3.33 | 3.69 | $(\times 10^{-2})$ |
| 20 | 1.12 | | 1.33 | | 1.35 | 1.73 | $(\times 10^{-2})$ |
| $\infty$ | 0.16 | | 0.36 | | 0.38 | 0.76 | $(\times 10^{-2})$ |

## V. SELECTIVE FADING

In this section we consider a general wide-sense stationary uncorrelated-scattering (WSSUS) fading channel. This model is described in detail in [1] and [8, ch. 9] and is employed in the analysis of direct-sequence SSMA communications over fading channels in [4]. We assume that $f_c \gg q\Delta'$, so that narrow-band signal models can be employed. The input to the $k$th channel is $s_k(t - \tau_k)$ where

$$s_k(t) = \text{Re} \{x_k(t) \exp (j2\pi f_c t)\} \tag{64}$$

and

$$x_k(t) = \sqrt{2P} \exp \{j(2\pi[b_k(t)\Delta + f_k(t)]t + \theta_k(t) + \alpha_k(t))\}. \tag{65}$$

The corresponding output is $y_k(t - \tau_k)$ where

$$y_k(t) = \gamma_0 s_k(t) + \text{Re} \{u_k(t) \exp (j2\pi f_c t)\} \tag{66}$$

and

$$u_k(t) = \gamma_k \int_{-\infty}^{\infty} h_k(t, \tau)x_k(t - \bar{\tau}) \, d\tau \tag{67}$$

so that the received signal for this channel is given by (9).

If $\gamma_0 = 1$ then there is a (nonfaded) specular component present in the output of the channel, and the channel is a Rician fading channel (as in [4]). In this case $\gamma_k^2$ plays the same role as the parameter $\gamma^2$ of Section III. If $\gamma_0 = 0$ there is no specular component, and the channel is a Rayleigh fading channel. In this case $\gamma_k^2$ plays the same role as the parameter $\mu_k$ of Section IV.

The fading process $h_k(t, \tau)$ (which can be thought of as the time-varying impulse response of a low-pass filter) is a zero-mean complex Gaussian random process with auto-covariance

$$E\{h_k(t, \tau)h_k^*(s, \sigma)\} = \rho_k(t - s, \tau)\delta(\tau - \sigma) \tag{68}$$

where $\delta(\cdot)$ is the Dirac delta function and

$$\int_{-\infty}^{\infty} \rho_k(0, \tau) \, d\tau = 1.$$

Two special cases of the model considered in [1] and [4] are the purely time-selective and purely frequency-selective WSSUS fading channels (see also [2] and [3]).

In the present paper we consider a somewhat more general model which is both time and frequency selective. This is a special doubly-dispersive model that is characterized by

$$\rho_k(t - s, \tau) = r_k(t - s)g_k(\tau). \tag{69}$$

If $g_k(\tau) \equiv \delta(\tau)$ the channel is not frequency selective. If $r_k(\xi) \equiv 1$ it is not time selective.

As usual [1]-[4], some limitations are imposed on the selectivity of the channel. First it is assumed that

$$g_k(\tau) \approx 0 \qquad \text{for } |\tau| > T \tag{70}$$

which is a constraint on the frequency selectivity of the channel that allows us to restrict attention to the intersymbol interference from the two adjacent data bits. This assumption can be relaxed, but the error probability computations become more difficult. The second assumption is that two signals which are transmitted at different frequencies have nonoverlapping spectra at the receiver. This is primarily a limitation on the time selectivity of the channel, but it also is related to the spacing $\Delta$ [i.e., (6b)].

The analysis of the receiver follows that of Section IV, so many of the details are omitted. The output of the in-phase component of each of the two branches of the $i$th receiver is

$$Z_{c,m} = \gamma_0 D_{c,m} + (P/8)^{1/2}T$$
$$\cdot \left[ \gamma_i F_{c,m} + \sum_{k \neq i} (\gamma_0 I_{c,m}^{(k,i)} + \gamma_k F_{c,m}^{(k,i)}) \right]$$
$$+ N_{c,m}. \tag{71}$$

The terms $D_{c,m}$ and $I_{c,m}^{(k,i)}$ are the nonfaded components of the desired signal and multiple-access interference, respectively. These terms are defined by (43)-(50) with $A_l^{(k)} = 1$ and $\Theta_l^{(k)} = 0$. The terms $F_{c,m}$ and $F_{c,m}^{(k,i)}$ are (normalized) faded versions of the desired signal and the multiple-access interference due to the $k$th signal. These terms are defined for the $\lambda$th data pulse ($\lambda = jN_b + p$) by

$$F_{c,m} = \text{Re} (\tilde{F}_m) \tag{72}$$

and

$$F_{c,m}^{(k,i)} = \text{Re} (\tilde{F}_m^{(k,i)}) \tag{73}$$

where

$$\tilde{F}_m = T^{-1} \int_{\lambda T}^{(\lambda+1)T} \int_{-\infty}^{\infty} h_i(t, \tau)\Gamma_{i,m}(t, \tau)$$
$$\cdot \exp [j\Phi_i(t, \tau)] \, d\tau \, dt \tag{74}$$

and

$$\tilde{F}_m^{(k,i)} = T^{-1} \int_{\lambda T}^{(\lambda+1)T} \int_{-\infty}^{\infty} h_i(t - \tau_k, \tau)$$
$$\cdot \Gamma_{k,i,m}(t, \tau_k + \tau) \exp [j\Phi_{k,i}(t, \tau_k + \tau)] \, d\tau \, dt. \tag{75}$$

In (74) and (75) the functions $\Gamma_{k,i,m}$ and $\Phi_{k,i}$ are given by

$$\Gamma_{k,i,m}(t, \tau) = \delta[f_k(t - \tau), f_i(t)]\delta[b_k(t - \tau), m] \tag{76}$$

and

$$\Phi_{k,i}(t,\tau) = -2\pi[f_c + b_k(t-\tau)\Delta + f_k(t-\tau)]\tau$$
$$+ \theta_k(t-\tau) + \alpha_k(t-\tau) - \beta_i(t). \quad (77)$$

The functions $\Gamma_{i,i,m}$ and $\Phi_{i,i}$ and denoted by $\Gamma_{i,m}$ and $\Phi_i$, respectively. Notice that $\tilde{F}_m$ is nonzero if and only if both $f_i(t-\tau) = f_i(t)$ and $b_i(t-\tau) = m$ for some $t$ and $\tau$ (similarly for $\tilde{F}_m^{(k,i)}$). This is a result of our assumption for the time-selectivity of the channel and the size of $\Delta$.

In the analysis below we are dealing with the conditional probability of error given there are no hits (i.e., conditioned on no multiple-access interference for the $\lambda$th bit of the $i$th signal). This probability does not depend on $\tau_k$ nor on $b_l^{(k)}$ for $k \neq i$. It may depend on the data bits $b_{\lambda-1}^{(i)}$ and $b_{\lambda+1}^{(i)}$, however, because of possible intersymbol interference. Thus, it is convenient to first consider the conditional probability of error given no hits and given the data bits $\boldsymbol{b}_i = (b_{\lambda-1}^{(i)}, b_{\lambda+1}^{(i)})$. Let this conditional probability be denoted by $P_0(\boldsymbol{b}_i)$, and let $P_0$ denote the average of $P_0(\boldsymbol{b}_i)$ with respect to the data symbols $b_{\lambda-1}^{(i)}$ and $b_{\lambda+1}^{(i)}$.

### A. WSSUS Rayleigh Fading Model ($\gamma_0 = 0$)

The bounds of (32)–(34) and the approximation of (35) are employed except that fading must be accounted for in $P_0$ and $P$. For $\gamma_0 = 0$ and $K = 1$, $Z_{c,m}$ is the sum of two random variables $((P/8)^{1/2}T F_{c,m}$ and $N_{c,m})$ of which the first is conditionally Gaussian and the second is Gaussian. Furthermore, it is not hard to see that $Z_{c,1}$ and $Z_{c,-1}$ are conditionally independent, and so are $Z_{s,1}$ and $Z_{s,-1}$. Since $\sigma_{c,m} = \sigma_{s,m}$ then [8, p. 587]

$$P_0(\boldsymbol{b}_i) = \sigma_{c,-1}^2(\sigma_{c,1}^2 + \sigma_{c,-1}^2)^{-1} \quad (78)$$

is the conditional probability of error, given there are no hits where

$$\sigma_{c,m}^2 = (PT^2/8)\gamma_i^2 \operatorname{var}\{F_{c,m}\} + N_0 T/16 \quad (79)$$

is the conditional variance of $Z_{c,m}$ given no hits and given $\boldsymbol{b}_i$. It is convenient to normalize $\sigma_{c,m}^2$ and write (78) as

$$P_0(\boldsymbol{b}_i) = v_{-1}(v_1 + v_{-1})^{-1} \quad (80)$$

where $v_m$ is given in terms of $\bar{\bar{E}} = \gamma_i^2 PT$ by

$$v_m = \operatorname{var}\{F_{c,m}\} + (2\bar{\bar{E}}/N_0)^{-1}. \quad (81)$$

The expression for $\operatorname{var}\{F_{c,m}\}$ depends on the position of the data bit within the interval $[jT_h, (j+1)T_h]$. For the $p$th bit of the $j$th hop ($\lambda = jN_b + p$) define $\delta_m = \delta(1, m)$, $\delta_m' = \delta(b_{\lambda-1}^{(i)}, m)$, and $\delta_m'' = \delta(b_{\lambda+1}^{(i)}, m)$. Let

$$H_i(v) = 2T^{-2} \int_0^v (v-u)r_i(u)\,du \quad (82)$$

and define

$$E_i = \int_0^T g_i(\tau)H_i(\tau)\,d\tau \quad (83a)$$

$$\hat{E}_i = \int_0^T g_i(\tau)H_i(T-\tau)\,d\tau \quad (83b)$$

and

$$G_i = T^{-2} \int_0^T g_i(\tau) \int_0^\tau \int_0^{T-\tau} r_i(t-s)\,dt\,ds. \quad (83c)$$

The following expressions for $\operatorname{var}\{F_{c,m}\}$ are derived in the Appendix. First, for $p = 0$ we find

$$\operatorname{var}\{F_{c,m}\} = \tfrac{1}{2}[(\delta_m'' + \delta_q)E_i + 2\delta_m\hat{E}_i + 2\delta_m(\delta_m'' + \delta_q)G_i]. \quad (84)$$

For $p = N_b - 1$ (84) is valid, provided we replace $\delta_m''$ by $\delta_m'$. Finally, for $0 < p < N_b - 1$, the expression is

$$\operatorname{var}\{F_{c,m}\} = \tfrac{1}{2}[(\delta_m' + \delta_m'')E_i + 2\delta_m\hat{E}_i + 2\delta_m(\delta_m' + \delta_m'')G_i]. \quad (85)$$

For the first-order Markov hopping patterns and the RS hopping patterns the quantity $\delta_q$ that appears in (84) is identically 0. For the sequences of independent random elements $\delta_q$ is a random variable with $P\{\delta_q = 1\} = q^{-1}$ and $P\{\delta_q = 0\} = 1 - q^{-1}$. Finally, to obtain $P_0$ we should average $P_0(\boldsymbol{b}_i)$ with respect to $\boldsymbol{b}_i$ and $\delta_q$.

Notice that for $0 < p < N_b - 1$ (i.e., for the internal bits of each dwell interval), $\operatorname{var}\{F_{c,m}\}$ does not depend on the hopping pattern. It turns out that the average probability of error for these bits ($0 < p < N_b - 1$) is larger than that of the first and last bits ($p = 0$ and $p = N_b - 1$). Thus, we use (85), and not (84), in order to obtain an upper bound on $P_0$ which applies for all values of $p$. As a consequence of using (85), we obtain a bound on $P_0$ which does not depend on the hopping pattern.

In order to obtain the limiting error probability (as the channel becomes nonselective) it suffices to let $g_i(\tau) = \delta(\tau)$ and $r_i(u) = 1$. We then have $E_i = G_i = 0$ and $\hat{E}_i = \frac{1}{2}$ so that $P_0$ is given by (41) with $\Lambda = \bar{\bar{E}}/N_0 = \gamma_i^2 \rho$. Similarly, to obtain the irreducible error probability (as $\rho \to \infty$) we simple disregard the second term in the right-hand side of (81).

For the WSSUS Rayleigh fading model we say that a hit occurs from the $k$th signal whenever $\tau_k$, $b_k(t)$, and $f_k(t)$ are such that $\operatorname{var}\{F_{c,m}^{(k,i)}\} \neq 0$. The probability $P$ of such a hit depends upon $N_b$ and $q$. For the first-order Markov hopping patterns we have

$$P \leqslant P_u \triangleq \frac{1}{q}\left(1 + \frac{3}{N_b}\right). \quad (86)$$

215

In deriving (86) we used the fact that for the selective fading model used in this section, as many as four adjacent bits from the $k$th signal may interfere with each bit of the $i$th signal. The expressions (25) and (32)–(35) apply with $P$ replaced by $P_u$ and $P_0$ evaluated as explained above. For memoryless hopping patterns the corresponding result is

$$P \leqslant P_u \triangleq \frac{1}{q}\left[1 + \frac{3}{N_b}\left(1 - \frac{1}{q}\right)\right].$$  (87)

Both bounds in (86) and (87) are tight for $N_b \geqslant 3$. For the RS hopping patterns the corresponding result is

$$P \leqslant P_u = \frac{1}{q-1}\left(1 + \frac{1}{N_b}\right).$$  (88)

Notice that the bound in (88) is the same as in (23) which was obtained under nonselective fading conditions. This is due to the fact that the RS hopping patterns do not repeat within a period.

### B. WSSUS Rician Fading Model ($\gamma_0 = 1$)

In this case the conditional error probability given there are no hits is [8, p. 587]

$$P_0(b_i) = \sigma_{c,-1}^2(\sigma_{c,1}^2 + \sigma_{c,-1}^2)^{-1}$$
$$\cdot \exp\left[-\tfrac{1}{2}(D_{c,1}^2 + D_{s,1}^2)(\sigma_{c,1}^2 + \sigma_{c,-1}^2)^{-1}\right].$$  (89)

Upon normalization, (89) reduces to

$$P_0(b_i) = v_{-1}(v_1 + v_{-1})^{-1}\exp\{-\tfrac{1}{2}(v_1 + v_{-1})^{-1}\}$$  (90)

where $v_m$ is now defined by

$$v_m = \gamma_i^2 \, \text{var}\, \{F_{c,m}\} + (1 + \gamma_i^2)(2\bar{E}/N_0)^{-1},$$  (91)

$\bar{E}/N_0 = (1 + \gamma_i^2)\rho$, and var $\{F_{c,m}\}$ is given by (82)–(85). Finally, in order to obtain $P_0$ we have to average $P_0(b_i)$ with respect to the data bits $(b_{\lambda+1}^{(i)}, b_{\lambda-1}^{(i)})$.

For Rician fading the hits from the $k$th signal may occur from either the direct-path component or the faded component. The probability of a hit from the $k$th signal due to the direct-path component is the same as for nonselective Rayleigh fading (this was evaluated in Section III). The probability of a hit due to the faded component is evaluated above (for Rayleigh fading). The union bound provides a simple upper bound on the probability of a hit. This is given by

$$P \leqslant P_u' = \frac{2}{q}\left(1 + \frac{2}{N_b}\right)$$  (92)

for first-order Markov hopping patterns and

$$P \leqslant P_u' = \frac{2}{q}\left[1 + \frac{2}{N_b}\left(1 - \frac{1}{q}\right)\right]$$  (93)

TABLE III
BIT ERROR PROBABILITY FOR RAYLEIGH FREQUENCY-SELECTIVE FADING ($K = 15$, $q = 1000$, AND $N_b = 10$)

| $\bar{E}/N_0$ (dB) | $\sigma = 0.05T$ | $\sigma = 0.1T$ | $\sigma = 0.15T$ | $\sigma = 0.2T$ | |
|---|---|---|---|---|---|
| 6 | 1.75 | 1.82 | 1.91 | 2.01 | ($\times 10^{-1}$) |
| 8 | 1.28 | 1.35 | 1.44 | 1.54 | ($\times 10^{-1}$) |
| 10 | 0.91 | 0.97 | 1.06 | 1.17 | ($\times 10^{-1}$) |
| 12 | 6.31 | 6.88 | 7.71 | 8.84 | ($\times 10^{-2}$) |
| 15 | 3.63 | 4.13 | 4.94 | 6.08 | ($\times 10^{-2}$) |
| 20 | 1.59 | 2.04 | 2.82 | 3.95 | ($\times 10^{-2}$) |
| $\infty$ | 0.58 | 1.00 | 1.76 | 2.89 | ($\times 10^{-2}$) |

TABLE IV
BIT ERROR PROBABILITY FOR RICIAN FREQUENCY-SELECTIVE FADING ($K = 15$, $q = 1000$, $N_b = 10$, AND $\sigma = 0.05T$)

| $\bar{E}/N_0$ (dB) | $\gamma^2 = 0.1$ | $\gamma^2 = 0.5$ | $\gamma^2 = 1$ | $\gamma^2 = 10$ | $\gamma^2 = 1000$ | |
|---|---|---|---|---|---|---|
| 6 | 0.98 | 1.42 | 1.60 | 1.77 | 1.78 | ($\times 10^{-1}$) |
| 8 | 0.49 | 0.94 | 1.13 | 1.30 | 1.31 | ($\times 10^{-1}$) |
| 10 | 0.23 | 0.61 | 0.78 | 0.93 | 0.94 | ($\times 10^{-1}$) |
| 12 | 1.23 | 3.96 | 5.34 | 6.58 | 6.63 | ($\times 10^{-2}$) |
| 15 | 0.86 | 2.26 | 3.14 | 3.94 | 3.97 | ($\times 10^{-2}$) |
| 20 | 0.82 | 1.24 | 1.58 | 1.92 | 1.94 | ($\times 10^{-2}$) |
| $\infty$ | 0.81 | 0.83 | 0.86 | 0.93 | 0.94 | ($\times 10^{-2}$) |

for memoryless random hopping patterns. For RS hopping patterns $P$ is still bounded as in (88); that is,

$$P \leqslant P_u' = P_u.$$  (94)

By substituting for $P_0$ in (32)–(35) and replacing $P$ by $P_u'$ in (25) we have lower bounds, an approximation, and an upper bound on the average probability of error.

In Tables III and IV the approximation $\tilde{P}_A$ given in (35) is obtained for purely frequency-selective Rayleigh and Rician fading channels, respectively. The system parameters are $K = 15$, $q = 1000$, and $N_b = 10$. First-order Markov hopping patterns are employed. The covariance function of the frequency-selective channel is triangular, so that the rms multipath spread $\sigma$ defined by $\sigma^2 = \int_{-\infty}^{\infty} \tau^2 \, g(\tau)d\tau$ is related to the parameter $d$ of [3] by $d = 2.22 \, \sigma/T$. We let $\gamma_k = \gamma$ for all $k$. In Table III, $\tilde{P}_A$ is given as a function of $\bar{E}/N_0 = \gamma^2\rho$ for four values of $\sigma/T$. In Table IV, $\tilde{P}_A$ is given as a function of $\bar{E}/N_0 = (1 + \gamma^2)\rho$ for $\sigma = 0.05 \, T$ and for five different values of $\gamma^2$. Notice that as $\gamma^2 \to \infty$ the probability $\tilde{P}_A$ is *not* the same as the second column of Table III. Although $P_0$ is the same in this limiting case, the fact that $P_u < P_u'$ [compare (86) to (92)] implies that the two cases give different values of the bit error probability.

Finally, we compare $\tilde{P}_A$ for nonselective and frequency-selective Rayleigh fading for $K = 15$, $q = 1000$, and $N_b = 10$ (first-order Markov hopping patterns are employed). From Tables II (c) and III we see that the probability of error for the frequency-selective case is, for $\bar{E}/N_0 = 12$ dB and $\sigma = 0.05$, 1.1 times that for nonselective fading, and it becomes 1.5 times the corresponding probability for nonselective fading as $\sigma$ increases to 0.2 $T$. Similarly for $\bar{E}/N_0 = 20$ dB, the ratio of the two probabilities ranges from 1.2 for $\sigma = 0.05 \, T$ to 2.9 for $\sigma = 0.2 \, T$.

## APPENDIX

In this Appendix we develop the expressions for $\text{var}\,\{F_{c,m}\}$. As in [4] we can write $\text{var}\,\{F_{c,m}\}$ as

$$\text{var}\,\{F_{c,m}\} = E\,[\text{Re}\,\{\tilde{F}_m\}]^2 = \tfrac{1}{2}E[\tilde{F}_m\tilde{F}_m^*] \qquad \text{(A1)}$$

where we used the fact that [1] $E\{h_i(t,\tau)h_i(s,\sigma)\} = 0$. Upon substitution for (74), (68), and (69) in (A1) we find

$$\text{var}\,\{F_{c,m}\}$$

$$= \tfrac{1}{2}\,T^{-2}\int_{-\infty}^{\infty}g_i(\tau)\int_{\lambda T}^{(\lambda+1)}\int_{\lambda T}^{(\lambda+1)T}r_i(t-s)\Gamma_{i,m}(t,\tau)$$

$$\cdot\,\Gamma_{i,m}(s,\tau)\cdot\exp\,\{j[\Phi_i(t,\tau)-\Phi_i(s,\tau)]\}\,dt\,ds\,d\tau.$$

$$\text{(A2)}$$

Notice that $\Gamma_{i,m}(t,\tau)\Gamma_{i,m}(s,\tau)\neq 0$ only for those $t$, $s$, and $\tau$ for which the following three conditions hold: $f_i(t-\tau) = f_i(t)$, $f_i(s-\tau) = f_i(s)$, and $b_i(t-\tau) = b_i(s-\tau) = m$. But these three conditions imply $\alpha_i(t-\tau) = \alpha_i(t)$, $\alpha_i(s-\tau) = \alpha_i(s)$, $\theta_i(t-\tau) = \theta_i(s-\tau)$, respectively. Also $\alpha_i(t) = \alpha_i(s) = \alpha_j^{(i)}$ and $\beta_i(t) = \beta_i(s) = \beta_j^{(i)}$ for $t$ and $s$ in $[\lambda T, (\lambda+1)T)$. Consequently, $\Phi_i(t,\tau) = \Phi_i(s,\tau)$ for these values of $t, s,$ and $\tau$. As a result we may let

$$\exp\,\{j[\Phi_i(t,\tau)-\Phi_i(s,\tau)]\} = 1$$

in (A2).

The next step is to write (A2) as

$$\text{var}\,\{F_{c,m}\}$$

$$= \frac{1}{2}\sum_{l=-\infty}^{\infty}\left[d(l)\sum_{n=0}^{p-1}B_m(l,n) + B_m(l,p)\right.$$

$$\left. + d(l+1)\sum_{n=p+1}^{N_b-1}B_m(l,n)\right] \qquad \text{(A3)}$$

where for $n\neq p$

$$B_m(l,n)$$

$$= \int_0^T g_i(\tau+lT_h+nT)[\Delta_m(l,n+1)E_i(\tau)$$

$$+\,\Delta_m(l,n)\hat{E}_i(\tau) + 2\Delta_m(l,n+1)$$

$$\cdot\,\Delta_m(l,n)G_i(\tau)]\,d\tau \qquad \text{(A4a)}$$

and for $n = p$

$$B_m(l,p)$$

$$= \int_0^T g_i(\tau+lT_h+pT)[d(l+1)\Delta_m(l,p+1)E_i(\tau)$$

$$+\,d(l)\Delta_m(l,p)\hat{E}_i(\tau) + 2d(l+1)d(l)$$

$$\cdot\,\Delta_m(l,p+1)\Delta_m(l,p)G_i(\tau)]\,d\tau. \qquad \text{(A4b)}$$

In (A3) and (A4) we also need the definitions

$$d(l) = \delta(f_{j-1}^{(i)}, f_j^{(i)}) \qquad \text{(A5)}$$

$$\Delta_m(l,n) = \delta(b_{\lambda-lN_b-n}^{(i)}, m) \qquad \text{(A6)}$$

and [cf. (82), (83)]

$$E_i(\tau) = T^{-2}\int_{\lambda T}^{\lambda T+\tau}\int_{\lambda T}^{\lambda T+\tau}r_i(t-s)\,dt\,ds = H_i(\tau) \qquad \text{(A7a)}$$

$$\hat{E}_i(\tau) = T^{-2}\int_{\lambda T+\tau}^{(\lambda+1)T}\int_{\lambda T+\tau}^{(\lambda+1)T}r_i(t-s)\,dt\,ds = H_i(T-\tau) \qquad \text{(A7b)}$$

$$G_i(\tau) = T^{-2}\int_{\lambda T}^{\lambda T+\tau}\int_{\lambda T+\tau}^{(\lambda+1)T}r_i(t-s)\,dt\,ds$$

$$= T^{-2}\int_0^{\tau}\int_{\tau}^{T}r_i(t-s)\,dt\,ds. \qquad \text{(A7c)}$$

Notice that the result of (A3) is quite general and it accounts for the intersymbol interference due to many data bits. However, because of the assumption (70) only the terms $l = 0$, $n = 0$ and $l = -1$, $n = N_b - 1$ of (A3) give nonzero contributions, and thus (A3) reduces to (84) and (85).

## REFERENCES

[1] P. A. Bello, "Characterization of randomly time-variant linear channels," *IEEE Trans. Commun. Syst.*, vol. CS-11, pp. 360–393, Dec. 1963.

[2] P. A. Bello and B. D. Nelin, "The influence of fading spectrum on the binary error probabilities of incoherent and differentially coherent matched filter receivers," *IRE Trans. Commun. Syst.*, vol. CS-10, pp. 160–168, June 1962.

[3] ——, "The effect of frequency selective fading on the binary error probabilities of incoherent and differentially coherent matched filter receivers," *IEEE Trans. Commun. Syst.*, vol. CS-11, pp. 170–186, June 1963.

[4] D. E. Borth and M. B. Pursley, "Analysis of direct-sequence spread-spectrum multiple-access communication over Rician fading channels," *IEEE Trans. Commun.*, vol. COM-27, pp. 1566–1577, Oct. 1979.

[5] E. A. Geraniotis and M. B. Pursley, "Error probability bounds for slow-frequency-hopped spread-spectrum multiple-access communications over fading channels," in *Proc. IEEE Int. Conf. Commun.*, June 1971, vol. 4, pp. 76.3.1–76.3.7.

[6] M. B. Pursley, "Spread-spectrum multiple-access communications," in *Multi-User Communication Systems*, G. Longo, Ed. Vienna and New York: Springer-Verlag, 1981, pp. 139–199.

[7] G. Solomon, "Optimal frequency-hopping sequences for multiple-access," in *Proc. Symp. Spread-Spectrum Commun.*, 1973, vol. 1, AD-915 852, pp. 33–35.

[8] M. Schwartz, W. R. Bennett, and S. Stein, *Communication Systems and Techniques*. New York: McGraw-Hill, 1966, part III.

[9] G. L. Turin, "Error probabilities for binary symmetric ideal reception through nonselective slow fading and noise," *Proc. IRE*, vol. 46, pp. 1603–1619, Sept. 1958.

# Part IV
# Nonlinear Effects

THE effect of a nonlinear channel on the performance of a communications system has received a good deal of attention for many years. Since devices such as traveling-wave tubes (TWT's) operate most efficiently when driven into saturation and since TWT's have for many years been a standard component in a satellite repeater, many articles have appeared in the literature describing the effect of such devices on communications systems. These articles ranged from analyses of intermodulation products in FDMA systems to probability of error analyses for digital modulation schemes.

With the increased use of spread-spectrum techniques, the same types of considerations are relevant; hence the need for accurate performance analysis when spread signals are transmitted through nonlinear devices is evident. In this section, there are two papers that deal with the effects of a limiter on the performance of a spread-spectrum communication system. The first is by Baer ("Interference Effects of Hard Limiting in PN Spread-Spectrum Systems"); it considers a biphase PN system operating over a hard-limited channel in the presence of strong interference. Baer illustrates that the resulting intermodulation products can negate the antijam capability of systems designed with even a large "processing gain." The second paper, coauthored by Aein and Pickholtz ("A Simple Unified Phasor Analysis of PN Multiple Access to Limiting Repeaters"), presents a justification for using a relatively straightforward phasor analysis technique to determine the probability of error of a quadriphase system operating over a nonlinear channel in the presence of strong interference. The analysis is quite general, and even allows for the possibility of AM/PM conversion. In both papers, the key application is satellite communications.

# Interference Effects of Hard Limiting in PN Spread-Spectrum Systems

HANS P. BAER, MEMBER, IEEE

*Abstract*—It is shown that the use of a hard limiter in PN spread-spectrum systems leads to the generation of intermodulation products which become narrow-band signals after the despreading process in the PN receiver. The analysis treats the situation where a bandpass limiter is inserted into the transmission path of the PN signal prior to the PN receiver. The input to the bandpass limiter is the superposition of the PN signal with white Gaussian noise and one interferer. This interferer may be either a narrow-band signal or another wide-band PN signal. It is found that the amplitudes of the resulting narrow-band intermodulation products are independent of the spread-spectrum processing gain and of the type of interference. A complete loss of the antijamming capability may thus result even with extremely large "processing gain."

## I. INTRODUCTION

THE various effects of system components with nonlinear transfer characteristics on signal quality in a pseudonoise-code-division-multiple-access (PN-CDMA) system have received considerable attention for many years [1]-[4]. It has been demonstrated that the use of a nonlinear device in a PN-CDMA system, for example, a nonlinear power amplifier in a satellite repeater, leads to comparable degradation in system performance as in frequency-division-multiple-access (FDMA) systems, where the power ratios of the different input signals of a nonlinear device may substantially change at the output favoring the stronger input signal, and where intermodulation products may fall inside signal bands.

In addition to the theories presented in [1]-[4] the present investigation provides an explicit analysis of the intermodulation terms which are generated when the input signal of the nonlinear device consists of a superposition of several deterministic and stochastic signals. The nonlinear device investigated in this paper is a bandpass limiter which can be considered to model a satellite repeater using a traveling wave tube (TWT) as a power amplifier. The PN spread signal and intermodulation products are analyzed after the despreading process in the receiver, i.e., after the multiplication of the received signal with a synchronized replica of the PN transmitter code signal.

## II. ANALYSIS OF THE BANDPASS LIMITER

Fig. 1 shows a simplified functional block diagram of a PN spread-spectrum receiver and a bandpass limiter. The bandpass limiter consists of the input bandpass filter $BP_i$, the hard limi-

Manuscript received July 1, 1981; revised January 19, 1982. This paper was presented at the 4th Symposium on Electromagnetic Compatibility, Zurich, Switzerland, March 1981.

The author was with the Institute for Communication Technology, Swiss Federal Institute of Technology, CH-8092 Zurich, Switzerland. He is now with Zellweger Uster Ltd., CH-8610 Uster, Switzerland.

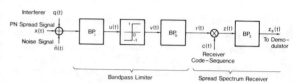

Fig. 1. Functional block diagram of bandpass limiter and spread-spectrum receiver.

ter, and the output bandpass filter $BP_o$. The bandpass filter $BP_i$ is assumed to be an ideal filter with bandwidth equal to the bandwidth $B_x$ of the PN spread signal $x(t)$. The hard limiter is followed by the bandpass filter $BP_o$ which confines the output signal from the nonlinearity to the fundamental band which is centered at the carrier frequency $\omega_0$.

The input to the bandpass limiter consists of the superposition of the PN spread signal $x(t)$ with white Gaussian noise $\tilde{n}(t)$ and an interference signal $q(t)$. This interferer $q(t)$ may be either a narrow-band signal or another PN spread-spectrum signal from a different PN code sequence than that used for $x(t)$.

The PN spread signal $x(t)$ may be represented as

$$x(t) = Ac(t) \cos \{\omega_0 t + \phi_x(t)\} \tag{1}$$

where $c(t)$ denotes the PN code signal, $\omega_0$ is the carrier frequency, and $\phi_x(t)$ is the phase defined by the data signal to be transmitted. The code chips of the normalized bipolar PN code signal attain only the amplitude values $+1$ or $-1$.

The interferer $q(t)$ is also assumed to be angle modulated with amplitude $\alpha$ and carrier frequency $\omega_1(\omega_1 \simeq \omega_0)$ and thus may be written

$$q(t) = \alpha p(t) \cos \{\omega_1 t + \phi_q(t)\}. \tag{2}$$

The function $p(t)$ determines the type of interfering signal. A narrow-band interferer results if the value of $p(t)$ is chosen to be identically 1. If, however, $p(t)$ is defined to be a PN code signal of the same frequency as the PN code signal $c(t)$, then the interferer is also a PN spread signal, for example, the signal of another subscriber in a code-division-multiple-access (CDMA) system.

The purpose of the input filter $BP_i$ in Fig. 1 is to reject as much as possible of the noise signal $\tilde{n}(t)$ which is assumed to be Gaussian and white with one-sided power spectral density $\eta$. It is also assumed that $x(t)$ and $q(t)$ are only negligibly distorted by the filter $BP_i$.

The first step of the demodulation process in the PN re-

Reprinted from *IEEE Trans. Commun.*, vol. COM-30, pp. 1010–1017, May 1982.

ceiver of Fig. 1 is the despreading operation of the input signal. The large transmission bandwidth of the PN spread signal $x(t)$ is compressed by the multiplication of the input signal with a synchronized replica of the transmitter PN code signal $c(t)$. The resulting output signal is then fed to the narrow-band bandpass filter $BP_z$ which eliminates most of the wide-band noise. The bandwidth $B_z$ of $BP_z$ is determined by the bandwidth of the angle-modulated information signal as it appears prior to the bandspreading process in the PN transmitter.

### A. Calculation of the Received Signal r(t)

Several methods have been proposed for the calculation of the effect of memoryless nonlinearities on the sum of input signals. Of these, the "transform method" [5] appears to be the most widely applicable. This method yields the autocorrelation function of the signal at the output of the nonlinearity from knowledge of the joint characteristic function of the input signals. For spread-spectrum signals, however, this method cannot give a complete analysis because it does not supply phase information about the output signals. These phase values play a major role in the despreading process in the PN receiver. Consequently, the transform method must be modified to enable a useful analysis of PN spread-spectrum systems. We carry out this modification as follows. The transfer characteristic of the nonlinear device is first characterized by its transform as suggested in [5]. Thereafter, however, a time-domain approach is applied [6].

The output $v$ as a function of the input $u$ of a memoryless nonlinearity with a transfer characteristic $g(u)$ can be represented by means of the inverse Fourier transform of the transfer function $G(jw)$. For the hard limiter

$$v = g(u)$$
$$= \frac{1}{2\pi} \int_{-\infty}^{\infty} G(jw) \exp\{jwu\}\, dw$$
$$= \begin{cases} 1, & u > 0 \\ 0, & u = 0 \\ -1, & u < 0. \end{cases} \tag{3}$$

The transfer characteristic $g(u)$ of a hard limiter is thus the signum function. It is well known that the Fourier transform of the signum function is given by [7]

$$G(jw) = \int_{-\infty}^{\infty} g(u) \exp\{-juw\}\, du = \frac{2}{jw}. \tag{4}$$

We now make a major simplifying assumption, namely that the phases $\phi_x(t)$ and $\phi_q(t)$ are constant. The input signal to the hard limiter then becomes just

$$u(t) = Ac(t) \cos \omega_0 t + \alpha p(t) \cos \omega_1 t + n(t). \tag{5}$$

The transmitted information signal is thus represented by a dc level as is also the interfering signal. Our approach hereafter is to identify the signal term, interference term, noise term, and intermodulation terms in the receiver output so that the power in these terms can be computed. These comparisons made for constant information and interfering signals will be good approximations whenever the actual phases $\phi_x(t)$ and $\phi_q(t)$ are low-frequency signals compared to the chip rate of the PN sequences and the carrier frequencies $\omega_0$ and $\omega_1$. The following theory, therefore, is applicable to most practical PN spread-spectrum systems.

The term $n(t)$ in (5) denotes Gaussian noise of power

$$\sigma^2 = \eta B_x. \tag{6}$$

Inserting (5) into (3) yields

$$v(t) = \frac{1}{2\pi} \int_{-\infty}^{\infty} G(jw) \exp\{jw(Ac(t) \cos \omega_0 t$$
$$+ \alpha p(t) \cos \omega_1 t + n(t))\}\, dw$$
$$= \frac{1}{2\pi} \int_{-\infty}^{\infty} G(jw) \exp\{jwAc(t) \cos \omega_0 t\}$$
$$\cdot \exp\{jw\alpha p(t) \cos \omega_1 t\} \exp\{jwn(t)\}\, dw. \tag{7}$$

The exponential functions appearing in (7) may be written

$$\exp\{jwAc(t) \cos \omega_0 t\}$$
$$= \cos\{Awc(t) \cos \omega_0 t\} + j \sin\{Awc(t) \cos \omega_0 t\}. \tag{8}$$

As the PN code signal $c(t)$ attains only the values $+1$ or $-1$, (8) simplifies to

$$\exp\{jwAc(t) \cos \omega_0 t\}$$
$$= \cos\{Aw \cos \omega_0 t\} + jc(t) \sin\{Aw \cos \omega_0 t\}. \tag{9}$$

Applying standard identities [7, p. 974] we find that the cosine and sine terms in (9) may be written

$$\cos\{Aw \cos \omega_0 t\}$$
$$= J_0(Aw) + 2 \sum_{k=1}^{\infty} (-1)^k J_{2k}(Aw) \cos 2k\omega_0 t,$$

$$\sin\{Aw \cos \omega_0 t\}$$
$$= 2 \sum_{k=0}^{\infty} (-1)^k J_{2k+1}(Aw) \cos (2k+1)\omega_0 t \tag{10}$$

where $J_i(Aw)$ denotes the Bessel function of the first kind, order $i$ and argument $Aw$.

The exponential function in (9) thus may be written as

$$\exp\{jwAc(t) \cos \omega_0 t\}$$
$$= J_0(Aw) + 2 \sum_{k=1}^{\infty} (j)^k c^k(t) J_k(Aw) \cos k\omega_0 t. \tag{11}$$

221

We introduce the notation

$$n_j(w, t) = \exp\{jwn(t)\} \tag{12}$$

for the complex valued random process appearing in (7) and due to the Gaussian noise $n(t)$. The output signal $v(t)$ of the hard limiter of Fig. 1 as given by (7) can be written with the aid of (4), (11), and (12) as

$$v(t) = \frac{1}{j\pi} \int_{-\infty}^{\infty} \frac{1}{w} n_j(w, t)\{J_0(Aw)J_0(\alpha w)$$

$$+ 2J_0(\alpha w) \sum_{i=1}^{\infty} (j)^i c^i(t) J_i(Aw) \cos i\omega_0 t$$

$$+ 2J_0(Aw) \sum_{k=1}^{\infty} (j)^k p^k(t) J_k(\alpha w) \cos k\omega_1 t$$

$$+ 4 \sum_{i=1}^{\infty} \sum_{k=1}^{\infty} (j)^{i+k} c^i(t)p^k(t) J_i(Aw)J_k(\alpha w)$$

$$\cdot \cos i\omega_0 t \cos k\omega_1 t\} dw. \tag{13}$$

The bandpass filter $BP_o$ removes higher harmonics from $v(t)$ to produce $r(t)$. The received signal $r(t)$ thus contains only signal components in the fundamental band centered at $\omega_0$.

With $\omega_0 \simeq \omega_1$ and

$$\Delta\omega = \omega_1 - \omega_0 \tag{14}$$

we see from (13) that $r(t)$ can be written

$$r(t) = \frac{1}{\pi} \int_{-\infty}^{\infty} \frac{1}{w} n_j(w, t)\left\{ \frac{1}{j} J_0(Aw)J_0(\alpha w) \right.$$

$$+ 2J_0(\alpha w)J_1(Aw)c(t) \cos \omega_0 t$$

$$+ 2J_0(Aw)J_1(\alpha w)p(t) \cos \omega_1 t$$

$$+ 2 \sum_{i=1}^{\infty} (-1)^i c^{i+1}(t)p^i(t) J_{i+1}(Aw)$$

$$\cdot J_i(\alpha w) \cos\{(\omega_0 - i\Delta\omega)t\}$$

$$+ 2 \sum_{k=1}^{\infty} (-1)^k c^k(t)p^{k+1}(t) J_k(Aw)$$

$$\left. \cdot J_{k+1}(\alpha w) \cos\{(\omega_0 + (k+1)\Delta\omega)t\} \right\} dw. \tag{15}$$

Only those intermodulation signal components for which the condition

$$|i - k| = 1 \tag{16}$$

for the summation variables $i$ and $k$ in the last summation of (13) is fulfilled appear at the output of bandpass filter $BP_o$ since only these terms give contributions whose frequency is close to $\omega_0$ as follows from the identity

$$\cos i\omega_0 t \cos k\omega_1 t$$

$$= \frac{1}{2} \cos(i\omega_0 t - k\omega_1 t) + \frac{1}{2} \cos(i\omega_0 t + k\omega_1 t)$$

$$= \frac{1}{2} \cos\{(i - k)\omega_0 t - k\Delta\omega t\}$$

$$+ \frac{1}{2} \cos\{(i + k)\omega_0 t + k\Delta\omega t\}. \tag{17}$$

The signal $r(t)$ as given by (15) is the wide-band signal that enters the PN receiver of Fig. 1.

## III. ANALYSIS OF THE SPREAD-SPECTRUM RECEIVER

The despread signal $z(t)$ in Fig. 1 can, with the aid of (15), be written as

$$z(t)$$

$$= r(t)c(t)$$

$$= \frac{1}{j\pi} \int_{-\infty}^{\infty} \frac{1}{w} n_j(w, t)J_0(Aw)J_0(\alpha w)c(t) dw$$

$$+ \frac{2}{\pi} \int_{-\infty}^{\infty} \frac{1}{w} n_j(w, t)J_0(\alpha w)J_1(Aw) dw \cos \omega_0 t$$

$$+ \frac{2}{\pi} \int_{-\infty}^{\infty} \frac{1}{w} n_j(w, t)J_0(Aw)J_1(\alpha w) dw \, c(t)p(t) \cos \omega_1 t$$

$$+ \frac{2}{\pi} \sum_{i=1}^{\infty} (-1)^i c^i(t)p^i(t) \cos\{(\omega_0 - i\Delta\omega)t\}$$

$$\cdot \int_{-\infty}^{\infty} \frac{1}{w} n_j(w, t)J_{i+1}(Aw)J_i(\alpha w) dw$$

$$+ \frac{2}{\pi} \sum_{k=1}^{\infty} (-1)^k c^{k+1}(t)p^{k+1}(t)$$

$$\cdot \cos\{(\omega_0 + (k+1)\Delta\omega)t\}$$

$$\cdot \int_{-\infty}^{\infty} \frac{1}{w} n_j(w, t)J_k(Aw)J_{k+1}(\alpha w) dw. \tag{18}$$

All signal components in (18) contain the noise contribution $n_j(w, t)$ as a factor; the signal $z(t)$, therefore, is of a stochastic nature. To make signal level comparisons, therefore, we must calculate ensemble averages (with respect to the noise signal) of the terms in $z(t)$ for every instant $t$. The representation of $z(t)$ according to (18) enables the identification of the resulting information signal and interfering term averages.

### A. Analysis of the Information Signal

The signal

$$z_{nx}(t) = \frac{2}{\pi} \int_{-\infty}^{\infty} \frac{1}{w} n_j(w, t)J_0(\alpha w)J_1(Aw) dw \cos \omega_0 t$$

$$\tag{19}$$

appearing in (18) is identified as direct contribution of the modulated information signal of frequency $\omega_0$ to $z(t)$. The separation of $n_j(w, t)$ in (12) into sine and cosine components as

$$n_j(w, t) = \exp\{jwn(t)\}$$

$$= \cos\{wn(t)\} + j\sin\{wn(t)\} \tag{20}$$

yields [using the symmetry properties of the integrand of (19)]

$$z_{nx}(t) = \frac{4}{\pi} \int_0^\infty \frac{1}{w} \cos\{wn(t)\} J_0(\alpha w)$$

$$\cdot J_1(Aw)\, dw \cos \omega_0 t. \tag{21}$$

The noise signal $n(t)$ at any time $t$ is a zero-mean Gaussian random variable with variance $\sigma^2 = \eta B_x$. Thus, for each $t$

$$E[\sin\{wn(t)\}] = 0 \tag{22a}$$

$$E[\cos\{wn(t)\}] = \exp\left\{-\frac{w^2\sigma^2}{2}\right\}. \tag{22b}$$

Using (22b) in (21), we find that the ensemble average of $z_{nx}(t)$ for every instant $t$ becomes

$$E[z_{nx}(t)] = A_x \cos \omega_0 t \tag{23}$$

where

$$A_x = \frac{4}{\pi} \int_0^\infty \frac{1}{w} J_0(\alpha w) J_1(Aw) \exp\left\{-\frac{w^2\sigma^2}{2}\right\} dw. \tag{24}$$

The quantity $A_x$ is the expected value of the amplitude of the resulting information signal.

### B. Analysis of the Interferer q(t)

The method of Section III-A can also be applied to the signal

$$z_{nq}(t) = \frac{2}{\pi} \int_{-\infty}^\infty \frac{1}{w} n_j(w, t) J_0(Aw) J_1(\alpha w)\, dw$$

$$\cdot c(t)p(t) \cos \omega_1 t \tag{25}$$

appearing in (18) which is the direct contribution of the interferer $q(t)$ to the despread signal $z(t)$. The result is

$$E[z_{nq}(t)] = A_q c(t)p(t) \cos \omega_1 t \tag{26}$$

where

$$A_q = \frac{4}{\pi} \int_0^\infty \frac{1}{w} J_0(Aw) J_1(\alpha w) \exp\left\{-\frac{w^2\sigma^2}{2}\right\} dw \tag{27}$$

is the average amplitude of the direct interfering term.

### C. Analysis of the Intermodulation Products

The last two summations in (18) represent the intermodulation products of $x(t)$, $q(t)$, and $n(t)$ in the frequency range near $\omega_0$ that appear in the despread signal $z(t)$. Calculation of the expected values of these signal terms for every instant $t$ yields

$$E[z_{nxq}(t)]$$

$$= E\left[\frac{2}{\pi} \sum_{i=1}^\infty (-1)^i c^i(t)p^i(t) \cos\{(\omega_0 - i\Delta\omega)t\}\right.$$

$$\cdot \int_{-\infty}^\infty \frac{1}{w} n_j(w, t) J_{i+1}(Aw) J_i(\alpha w)\, dw$$

$$+ \frac{2}{\pi} \sum_{k=1}^\infty (-1)^k c^{k+1}(t)p^{k+1}(t)$$

$$\cdot \cos\{(\omega_0 + (k+1)\Delta\omega)t\}$$

$$\left. \cdot \int_{-\infty}^\infty \frac{1}{w} n_j(w, t) J_k(Aw) J_{k+1}(\alpha w)\, dw\right]$$

$$= \sum_{i=1}^\infty c^i(t)p^i(t) A_{xqi} \cos\{(\omega_0 - i\Delta\omega)t\}$$

$$+ \sum_{k=1}^\infty c^{k+1}(t)p^{k+1}(t) A_{xqk} \cos\{(\omega_0 + (k+1)\Delta\omega)t\} \tag{28}$$

where

$$A_{xqi} = (-1)^i \frac{4}{\pi} \int_0^\infty \frac{1}{w} J_{i+1}(Aw) J_i(\alpha w)$$

$$\cdot \exp\left\{-\frac{w^2\sigma^2}{2}\right\} dw \tag{29}$$

and

$$A_{xqk} = (-1)^k \frac{4}{\pi} \int_0^\infty \frac{1}{w} J_k(Aw) J_{k+1}(\alpha w)$$

$$\cdot \exp\left\{-\frac{w^2\sigma^2}{2}\right\} dw \tag{30}$$

are the average amplitudes of the intermodulation terms between the signal and the interferer.

### D. Analysis of the Noise Signal

We now make another simplification by assuming that the information signal, the interfering signal, and their intermodulation products in (18) can be replaced by their average values. This will be a good approximation when the signal-to-noise ratio is large. We can then use (23), (26), and (28) to write

$$z(t)$$

$$= A_x \cos \omega_0 t + A_q c(t) p(t) \cos \omega_1 t$$

$$+ \sum_{i=1}^{\infty} c^i(t) p^i(t) A_{xqi} \cos \{(\omega_0 - i\Delta\omega) t\}$$

$$+ \sum_{k=1}^{\infty} c^{k+1}(t) p^{k+1}(t) A_{xqk} \cos \{(\omega_0 + (k+1)\Delta\omega) t\}$$

$$+ z_n(t). \tag{31}$$

The noise signal $z_n(t)$ in (31) denotes the first integral in (18); this is the stochastic part of $z(t)$, namely

$$z_n(t) = \frac{1}{j\pi} \int_{-\infty}^{\infty} \frac{1}{w} n_j(w, t) J_0(Aw) J_0(\alpha w) c(t) \, dw. \tag{32}$$

This noise signal seems to be intractable to analysis to determine its power within the passband of the filter $BP_z$ as we require. Thus we make now another simplifying assumption by neglecting the multiplication by the PN code signal $c(t)$, i.e., we assume that

$$z_n(t) = \frac{1}{j\pi} \int_{-\infty}^{\infty} \frac{1}{w} n_j(w, t) J_0(Aw) J_0(\alpha w) \, dw. \tag{33}$$

In the following, the resulting noise power of $z_n(t)$ at the output of the narrow-band bandpass filter $BP_z$ is calculated. As the multiplication of the received noise signal with the receiver PN code signal $c(t)$, which has been neglected, would actually spread more of the noise power outside the band of the filter $BP_z$, our results will be rather conservative estimates of the true values.

The signal $z_n(t)$ of (33) can be rewritten

$$z_n(t) = \int_0^{\infty} \beta(w) \sin \{wn(t)\} \, dw \tag{34}$$

where

$$\beta(w) = \frac{2}{\pi} \frac{1}{w} J_0(Aw) J_0(\alpha w). \tag{35}$$

The autocorrelation function of $z(t)$ is

$$R_{zn}(\tau) = E[z_n(t) z_n(t + \tau)]$$

$$= \int_0^{\infty} \int_0^{\infty} \beta(w_1) \beta(w_2) E \left[ \sin \{w_1 n(t)\} \right.$$

$$\left. \cdot \sin \{w_2 n(t + \tau)\} \right] \, dw_1 \, dw_2. \tag{36}$$

The expectation may be rewritten

$$E \left[ \sin \{w_1 n(t)\} \sin \{w_2 n(t + \tau)\} \right]$$

$$= \frac{1}{2} E \left[ \cos \{w_1 n(t) - w_2 n(t + \tau)\} \right]$$

$$- \frac{1}{2} E \left[ \cos \{w_1 n(t) + w_2 n(t + \tau)\} \right]. \tag{37}$$

Since $n(t)$ is a stationary Gaussian process, so also are

$$n_1(t, \tau) = w_1 n(t) - w_2 n(t + \tau),$$

$$n_2(t, \tau) = w_1 n(t) + w_2 n(t + \tau) \tag{38}$$

for any fixed value of $\tau$. The variances of $n_1(t, \tau)$ and $n_2(t, \tau)$ thus become [11]

$$\sigma_1^2(\tau) = \sigma^2 \left\{ w_1^2 + w_2^2 - 2w_1 w_2 \frac{R_n(\tau)}{\sigma^2} \right\} \tag{39}$$

and

$$\sigma_2^2(\tau) = \sigma^2 \left\{ w_1^2 + w_2^2 + 2w_1 w_2 \frac{R_n(\tau)}{\sigma^2} \right\} \tag{40}$$

where $R_n(\tau)$ is the autocorrelation of the bandpass noise $n(t)$. The expectation in (37) is now found to be

$$E \left[ \sin \{w_1 n(t)\} \sin \{w_2 n(t + \tau)\} \right]$$

$$= \frac{1}{2} \exp \left\{ -\frac{\sigma_1^2(\tau)}{2} \right\} - \frac{1}{2} \exp \left\{ -\frac{\sigma_2^2(\tau)}{2} \right\}. \tag{41}$$

Thus, (36) becomes

$$R_{zn}(\tau) = \int_0^{\infty} \int_0^{\infty} \beta(w_1) \beta(w_2) \left( \frac{1}{2} \exp \left\{ -\frac{\sigma_1^2(\tau)}{2} \right\} \right.$$

$$\left. - \frac{1}{2} \exp \left\{ -\frac{\sigma_2^2(\tau)}{2} \right\} \right) dw_1 \, dw_2. \tag{42}$$

The power density of $z_n(t)$ at the frequency $\omega = \omega_0$ can now be calculated as the Fourier transform of the autocorrelation function evaluated at frequency $\omega_0$, namely

$$S_n(\omega_0) = 2 \int_0^{\infty} R_{zn}(\tau) \cos \omega_0 \tau \, d\tau$$

$$= \int_0^{\infty} \int_0^{\infty} \beta(w_1) \beta(w_2) \int_0^{\infty} \exp \left\{ -\frac{\sigma_1^2(\tau)}{2} \right\}$$

$$\cdot \cos \omega_0 \tau \, d\tau \, dw_1 \, dw_2$$

$$- \int_0^{\infty} \int_0^{\infty} \beta(w_1) \beta(w_2) \int_0^{\infty} \exp \left\{ -\frac{\sigma_2^2(\tau)}{2} \right\}$$

$$\cdot \cos \omega_0 \tau \, d\tau \, dw_1 \, dw_2. \tag{43}$$

The exponential functions in the integrand may be rewritten as

$$\exp\left\{-\frac{\sigma_1^2(\tau)}{2}\right\} = \exp\left\{-\frac{w_1^2\sigma^2}{2}\right\} \exp\left\{-\frac{w_2^2\sigma^2}{2}\right\}$$

$$\cdot \exp\left\{w_1 w_2 R_n(\tau)\right\},$$

$$\exp\left\{-\frac{\sigma_2^2(\tau)}{2}\right\} = \exp\left\{-\frac{w_1^2\sigma^2}{2}\right\} \exp\left\{-\frac{w_2^2\sigma^2}{2}\right\}$$

$$\cdot \exp\left\{-w_1 w_2 R_n(\tau)\right\}. \tag{44}$$

The term $\exp\left\{w_1 w_2 R_n(\tau)\right\}$ can be expanded into a series:

$$\exp\left\{w_1 w_2 R_n(\tau)\right\}$$

$$= 1 + \frac{w_1 w_2 R_n(\tau)}{1!} + \frac{\left\{w_1 w_2 R_n(\tau)\right\}^2}{2!} + \cdots. \tag{45}$$

The autocorrelation of the bandpass noise $n(t)$ is

$$R_n(\tau) = \eta B_x \frac{\sin \pi B_x \tau}{\pi B_x \tau} \cos \omega_0 \tau \tag{46}$$

where we have assumed that the input bandpass filter $BP_i$ (see Fig. 1) is ideal with bandwidth $B_x$.

If we analyze (43)–(46) and consider that $\omega_0 \gg 2\pi B_x$ for most practical PN spread-spectrum systems, then we see that only those summands in the series representation of $\exp\left\{-\sigma_1^2(\tau)/2\right\}$ and $\exp\left\{-\sigma_2^2(\tau)/2\right\}$ which contain $\cos \omega_0 \tau$ as a factor yield a nonvanishing contribution to the integral over $\tau$ in (43). For the cosine terms raised to an odd power we can apply the following identity:

$$\cos^{2m-1} \omega_0 \tau = \frac{1}{2^{2m-2}} \left[ \binom{2m-1}{m-1} \cos \omega_0 \tau \right.$$

$$+ \sum_{i=0}^{m-2} \binom{2m-1}{i}$$

$$\left. \cdot \cos\left\{(2m-2i-1)\omega_0\tau\right\} \right]. \tag{47}$$

Inserting (46) into (45) and applying (47) gives with (44) and (43)

$$S_n(\omega_0) = \eta \frac{4}{\pi^3} \sum_{i=1}^{\infty} \frac{1}{i!(i-1)!} \left(\frac{\sigma^2}{2}\right)^{2i-2}$$

$$\cdot \int_0^{\infty} \left(\frac{\sin \lambda}{\lambda}\right)^{2i-1} d\lambda$$

$$\cdot \left(\int_0^{\infty} w^{2i-2} \exp\left\{-\frac{w^2\sigma^2}{2}\right\} J_0(Aw)\right.$$

$$\left. \cdot J_0(\alpha w)\, dw\right)^2. \tag{48}$$

We now make our final simplifying assumption by assuming that the spectrum $S_n(\omega)$ is flat within the passband of the narrow-band filter $BP_z$. The resulting noise power $P_n$ of $z_n(t)$ at the output of $BP_z$ is then simply

$$P_n = 2B_z S_n(\omega_0) \tag{49}$$

where $B_z$ denotes the bandwidth of the filter $BP_z$.

## IV. NUMERICAL RESULTS AND INTERPRETATION

### A. Intermodulation Products

The signal $z_s(t)$ obviously contains the wanted information signal of frequency $\omega_0$. The corresponding amplitude value is $A_x$. This signal is narrow-band as it depends no longer on the wide-band PN code signal $c(t)$ [see (23)]. The remaining part of the interferer $q(t)$ in the frequency range of interest depends both on the signals $c(t)$ and $p(t)$ [see (26)]. This signal of amplitude $A_q$, therefore, is wide-band. Consequently, its effect on the information signal is reduced by the processing gain of the PN system.

Of special interest are, of course, amplitude value and bandwidth of the intermodulation products. It has been assumed that the signals $c(t)$ and $p(t)$ attain only the values $+1$ and $-1$. The result of the product $c(t)p(t)$ raised to an even power thus becomes 1. Consequently, the terms in the last two summations of (28) do not depend upon $c(t)$ or $p(t)$ when $i$ is even and $k$ is odd. These intermodulation products are, therefore, narrow-band signals. The despreading process in the receiver which normally spreads the spectrum of interfering signals leads in this case to a compression of the bandwidth of the intermodulation products. It is important to notice that *the power of these intermodulation components* at the output of $BP_z$ thus *is independent of the processing gain.*

It is somewhat surprising that the mentioned narrow-band intermodulation products do not depend upon the type of interfering signal $q(t)$. The results are identical for a narrow-band interferer ($p(t) = 1$) or a wide-band PN spread signal where $p(t)$ denotes a PN code signal.

The bandwidth of a narrow-band intermodulation product generally exceeds the bandwidth $B_z$ of the bandfilter $BP_z$ (see the examples in the following paragraph). The resulting power at the output of $BP_z$ thus depends on both bandwidth and center frequency of the intermodulation product. The following representation of numerical results leaves this dependence unconsidered. If $\omega_0$, $\omega_1$ and bandwidth $B_z$ are known, then the relevant power of the intermodulation product at the output of $BP_z$ can easily be calculated by means of an estimation of its power density spectrum.

In order to present more general results, the effect of the intermodulation products is described by a comparison of the resulting power of the restored information signal and the power of the strongest narrow-band intermodulation product. The corresponding ratio $(X/I)_{out}$ depends, of course, on both the signal-to-noise ratio $(X/N)_{in}$ of $x(t)$ and $n(t)$ at the input of the hard limiter and on the power ratio $(X/Q)_{in}$ of $x(t)$ and $q(t)$. Examination of Fig. 2 and (31) indicates that if $(X/Q)_{in}$ is less than 0 dB, then the strongest product is always

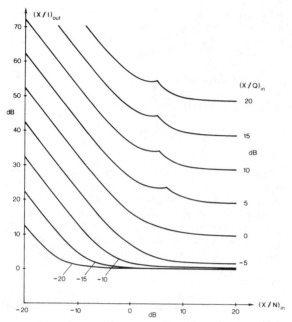

Fig. 2.   Power ratio $(X/I)_{out}$ of information signal and strongest narrow-band intermodulation product after despreading as a function of the input signal-to-noise ratio $(X/N)_{in}$; parameter: power ratio $(X/Q)_{in}$ of information signal $x(t)$ and interferer $q(t)$.

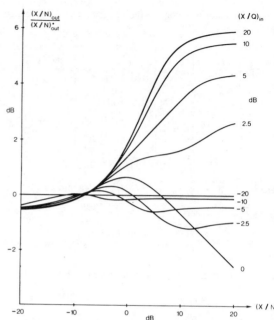

Fig. 3.   Quotient of the output signal-to-noise ratio $(X/N)_{out}$ with hard limiting and signal-to-noise ratio $(X/N)_{out}*$ without hard limiting as a function of the input signal-to-noise ratio $(X/N)_{in}$; parameter: power ratio $(X/Q)_{in}$ of information signal $x(t)$ and interferer $q(t)$.

the third-order cross product of frequency $\omega_0 + 2\Delta\omega$ which is represented by the value $k = 1$ in the second summation of (31). The bandwidth of this interfering signal is equal to $B_z$ if $q(t)$ is not phase modulated ($\phi_q(t)$ constant). For values of $(X/Q)_{in}$ greater than 0 dB and high signal-to-noise ratios $(X/N)_{in}$, the fifth-order term with the center frequency $\omega_0 - 2\Delta\omega$ [$i = 2$ in (31)] and with a bandwidth of about $3B_z$ (if $\phi_q(t)$ is constant) becomes predominant. The transition from the third-order to the fifth-order product leads to the sharp bends in the curves of Fig. 2.

The results demonstrate that if $(X/Q)_{in}$ is less than 5 dB and for large signal-to-noise ratios the transmission of PN signals may be heavily impaired. This is only the case, of course, if the important signal components of center frequency $\omega_0 \pm 2\Delta\omega$ fall at least partially within the passband of $BP_z$.

The power of an investigated intermodulation product is the same as in a comparable narrow-band system [8]. It is obvious that the power of the different signal components at the output of a symmetrical limiter remains unchanged if the input signal is additionally multiplied by a PN code signal which attains only the values +1 and −1. Nevertheless, it is only the present analysis which enables exact identification of the spectral characteristics of the different signal components.

### B. The Calculation of the Noise Power

The bandwidth of the noise power density spectrum in the frequency band of interest is at least $B_x$. As the despread signal is band limited by the narrow-band filter $BP_z$, only a

small fraction of bandwidth $B_z$ and center frequency $\omega_0$ will, however, appear at the output of $BP_z$. As mentioned in Section III-D, the noise term is investigated by an analysis of the predominant direct feedthrough component $z_n(t)$ and the multiplication with $c(t)$ in the receiver is neglected. If the value of the power density $S_n(\omega_0)$ for frequency $\omega_0$ is known and if a flat spectrum within the passband of $BP_z$ is assumed, then the power at the output of $BP_z$ can easily be calculated [see (49)]. In Fig. 3 the signal-to-noise ratio $(X/N)_{out}$ in the investigated PN system is compared to the corresponding value $(X/N)_{out}*$ in an identical system without a hard limiter. A comparison of these results with the corresponding values for a conventional narrow-band system [8] leads to the conclusion that both systems exhibit similar performance with respect to noise power.

### V. CONCLUSIONS

It has been shown that a hard limiter in the transmission path may lead to the generation of strong narrow-band intermodulation products at the output of a spread-spectrum receiver. The power of these products is independent of the spread-spectrum processing gain and, in the worst case, it can be equal to the power of the information signal (see Fig. 2). The transmission of a PN signal thus can be impeded even if a sufficient suppression of interference due to the processing gain is expected.

This effect reduces to a large extent the expense of a jammer which intends to interrupt the transmission of PN signals. With the mere knowledge of the center frequency $\omega_0$, i.e., by means of an unmodulated carrier, it is possible to contaminate the despread signal by strong narrow-band cross products.

The same type of interfering signals occur, for example, in PN-CDMA systems with a hard-limiting satellite repeater. Simultaneous reception of two or more PN signals leads to intermodulation products at the repeater output. After despreading in the receiver, some of these intermodulation products appear as narrow-band signals at the output of $BP_z$. This may heavily degrade overall system performance.

The equality of the powers of comparable intermodulation products in PN and narrow-band systems enables the application of the well-known results in case of the superposition of three and more narrow-band signals at the input of a hard-limiting device [9]. In particular, we can expect that the interfering effect decreases with an increasing number of equally strong PN signals.

The results of the present analysis confirm the hypotheses of [10] which were based on measurements with a hard-limiting satellite repeater.

## ACKNOWLEDGMENT

The author is grateful to Prof. Dr. P. Leuthold of the Swiss Federal Institute of Technology, Zurich, Switzerland, for his encouragement and his support of this research, and to Prof. Dr. J. L. Massey of the Swiss Federal Institute of Technology, for his helpful comments on this paper.

## REFERENCES

[1] J. M. Aein, "Multiple access to a hard-limiting communication-satellite repeater," *IEEE Trans. Space Electron. Telem.*, vol. SET-10, pp. 159–167, 1964.

[2] J. W. Schwartz, J. M. Aein, and J. Kaiser, "Modulation techniques for multiple access to a hard-limiting satellite repeater," *Proc. IEEE*, vol. 54, pp. 763–777, 1966.

[3] D. R. Anderson and P. A. Wintz, "Analysis of a spread-spectrum multiple-access system with a hard limiter," *IEEE Trans. Commun. Technol.*, vol. COM-17, pp. 285–290, 1969.

[4] D. S. Arnstein, "Power division in spread spectrum systems with limiting," *IEEE Trans. Commun.*, vol. COM-27, pp. 574–582, 1979.

[5] W. B. Davenport and W. L. Root, *Random Signals and Noise*. New York: McGraw-Hill, 1958.

[6] P. C. Jain, "Limiting of signals in random noise," *IEEE Trans. Inform. Theory*, vol. IT-18, pp. 332–340, 1972.

[7] I. S. Gradshteyn and I. M. Ryshik, *Tables of Integrals, Series, and Products*. New York: Academic, 1980.

[8] J. J. Jones, "Hard limiting of two signals in random noise," *IEEE Trans. Inform. Theory*, vol. IT-9, pp. 34–42, 1963.

[9] P. D. Shaft, "Limiting of several signals and its effect on communication system performance," *IEEE Trans. Commun. Technol.*, vol. COM-13, pp. 504–512, 1965.

[10] H. J. Kochevar, "Spread spectrum multiple access communication experiment through a satellite," *IEEE Trans. Commun.*, vol. COM-25, pp. 853–856, 1977.

[11] A. Papoulis, *Probability, Random Variables, and Stochastic Processes*. Tokyo, Japan: McGraw-Hill Kogakusha, 1965.

# A Simple Unified Phasor Analysis for PN Multiple Access to Limiting Repeaters

JOSEPH M. AEIN, SENIOR MEMBER, IEEE, RAYMOND L. PICKHOLTZ, FELLOW, IEEE

*Abstract*—This paper presents a simple phasor model for analyzing coherent direct-sequence, pseudonoise (PN), spread-spectrum carriers accessing limiting repeaters in the presence of strong interference. Limiting repeaters are found in several applications, one being satellite relays. The model used provides a rigorous upper bound on system bit error rate depending only on an equivalent $E_b/N_0$. Presented for the first time is an application of the model to RF limiters having an AM-to-PM conversion effect.

For quadriphase PN, receiver $E_b/N_0$ is given by the heuristic approach of replacing the limiter with an average power-limited linear device plus an additional loss $L$ due to limiting. As long as there is no AM-to-PM conversion in the limiter, $L$ is the same as that obtained by passing multitone CW plus noise through the limiter. For quadriphase PN carriers passing through a limiter with AM-to-PM conversion, $L$ can depend on the type of coherent phase tracker used. $E_b/N_0$ for biphase PN, in contrast to quadriphase PN, need not have the usual heuristic form given by quadriphase PN.

## I. INTRODUCTION

THE channel to be considered is depicted in Fig. 1. $s(t)$ denotes the desired PN carrier; $I(t)$ can represent an interferer, a sum of other PN carriers, thermal noise, or a combination of all of these. $x(t)$ denotes the composite input to the repeater, while $y(t)$ denotes the limiter output of unity[1] envelope and phase equal to that of $x(t)$. This results in a repeater output power level of 1/2 W. $n(t)$ represents an additive white Gaussian noise whose single-sided noise power density is taken to be $kT_r/2P_r$ where $k$ is Boltzmann's constant, $T_r$ is system receiving noise temperature, and $P_r$ is the net received (total) repeater power. The $2P_r$ value normalizes the postlimiter noise power density to the 1/2 W power level of the unity envelope repeater.

The receiver local PN reference sequence is assumed to be in track with the transmitted signal. Moreover, the receiver RF reference frequency and phase are assumed to be coherent with the desired transmitted signal. (PN carrier locking performance is itself a profound subject, addressed, for example, in [1] and [2].) The correlator multiplies the received signal coherently (at RF) by the PN phase-coded sequence and then integrates over a data bit of $T$ seconds duration. Binary data at the transmitter are encoded by multiplying the PN carrier by a plus or minus data bit, $a = \pm 1$.

Manuscript received May 15, 1981; revised September 30, 1981.

J. M. Aein is with the Institute for Defense Analyses, Alexandria, VA 22311.

R. L. Pickholtz is with the Department of Electrical Engineering and Computer Science, George Washington University, Washington, DC 20006.

[1] We choose unity for analytic convenience.

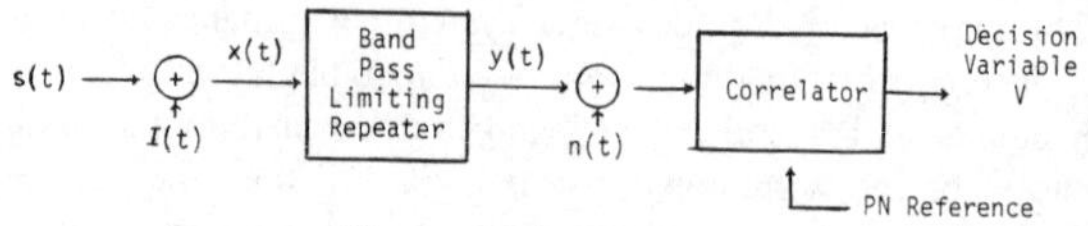

Fig. 1. PN limiting channel.

At the receiver the correlator decision variable $V$ is then tested for its sign to detect the received bit.

It has been asserted that the correlator integrates a sufficiently large number $n$ of PN chips per data bit that the decision variable $V$ is described by a normal distribution. The error probability is then approximated by

$$P_e \doteq \int_d^\infty e^{-u^2/2} \frac{du}{\sqrt{2\pi}} \doteq \frac{e^{-d^2/2}}{\sqrt{2\pi}d}, \qquad \text{for } P_e < 10^{-2}. \quad (1)$$

Heuristically, consider the limiter to be an average-power-limited, ideal linear amplifier. Let $d^2$ denote twice the correlator output energy-to-noise density ratio. The signal energy part of $V$ would then be $(S/I)P_rT$, where $T$ is the bit duration. The noise power density $N_0$ would be $P_r/W + kT_r$ where $W$ is taken[2] to be $1/\Delta T$, the reciprocal of the chip duration. Thus $d_{\text{linear}}^2$ would be $2(S/I)W_0[1 + W_0/W]^{-1}$; $W_0 \triangleq P_r/kT_r$. Note that $W_0$ has the dimensions of bandwidth and is the Shannon capacity of the (linear) channel in the case of Gaussian $s(t)$, $I(t)$, and infinite $W$. If $W_0 \ll W$ the system is said to be power limited, $[1 + W_0/W]^{-1} \approx 1$, and $d_{\text{linear}}^2 \approx 2(S/I)TW_0$; whereas, if $W_0 \gg W$, the system is said to be bandwidth limited and $d_{\text{linear}}^2 \approx 2(S/I)TW$.

Next, to account for the effects of the limiter, a power loss factor $L$ is introduced and $d^2$ is taken to be $Ld_{\text{linear}}^2$ where $L$ depends on the statistics of $I(t)$ and $s(t)$. In the past $L$ has been obtained by at least the following techniques.

1) Experimentally, insert two tones into the bandpass limiter and measure the relative output powers at the frequencies of the input tones to determine $L$.

2) Analytically, via the classic spectral methods of [5]-[7] ([8] presents this calculation for two tones and noise).

3) Analytically, via the transform method of [9] and as reproduced in the text by Spilker [10].

4) Analytically, via the second moment method of [11].

(The list of authors who have contributed to limiter analysis is quite extensive. We offer our apologies to all those contributors whose efforts we have not listed.) This calcula-

[2] By this definition, it follows that $n$, the number of PN chips integrated by the correlator during a bit, is the product of $TW$.

Reprinted from *IEEE Trans. Commun.*, vol. COM-30, pp. 1018–1026, May 1982.

tion for $d^2$ only applies to small signal-to-total-interference power ratios $S/I$.

The applicability of PN systems to satellite communications was recognized early in the evolution of PN technology by several investigators. The capability of PN carriers for providing multiple access was described in [3] and [4]. The first person to our knowledge to predict satellite system performance was Cahn, who later published in [2].

This approach has been reasonably successful (with the exception of biphase PN) in predicting performance. There are at least three theoretical problems that must be addressed in order to have full confidence in the method.

1) Find a direct analytic method for calculating limiter suppression $L$ for PN signals and relate the results to those obtained from CW tone testing.

2) The central limit theorem provides *no theoretical basis* for justifying the use of (1). Calculating $P_e$ involves integrating the *tail* of the probability density function of the correlator output $V$. Physically we are asking whether the number of chips per bit, $n = TW$, is sufficiently large to overcome a small $(S/I)$ $P_r$ so that we are sufficiently "up on" the tail to justify use of the normal distribution.

3) Since $S/I$ has to be small, quantitatively how small must $S/I$ be to justify (1)?

As will be shown, biphase PN essentially fails to produce the desired results. Why, then, does quadriphase PN succeed?

Using Cramér-Chernoff bounds [12], [13], it has been shown that for quadriphase PN and sufficiently small $S/I$, (1) is correct where $L$ is found using first and second moments of the limiter output phasor, and $n$ has to be sufficiently large to obtain an adequate $d^2$ that $P_e < e^{-d^2/2}$ plus negligible correction terms. In all cases the correction terms to $d^2$ [13] go to zero as $S/I$ goes to zero. In the remaining portions of this paper we provide the phasor model followed by pertinent examples. The first class of examples (Section III) is well known. In the last example (Section IV), the analysis of an RF limiter with an AM-to-PM mechanism is new. In every case we use *simple* mathematical formulations.

## II. PHASOR REPRESENTATION AND AN ERROR BOUND

Referring to Fig. 1, $x(t)$ at the input to the limiter is the sum of the desired signal $s(t)$ plus interference $I(t)$.

$$x(t) = \sum_{l=0}^{n-1} a\sqrt{2S}\ \text{rect}_{\Delta T}\ (t - l\Delta T)\ \cos\left[\omega_0 t + \mu_l \frac{\pi}{2}\right]$$

$$+ |I(t)| \cos\left[\omega_0 t + \theta_I(t)\right] \qquad (2)$$

where

| | |
|---|---|
| $a$ | $= +1$ or $-1$, the sign of the binary data bit |
| $S$ | $=$ power in desired PN carrier |
| $\text{rect}_{\Delta T}(t)$ | $= 1,\qquad 0 \leqslant t < \Delta T$ |
| | $= 0,\qquad$ otherwise |

$$\Delta T \triangleq \frac{1}{W}$$ $=$ duration of the phase coding, $\mu_l \dfrac{\pi}{2}$, otherwise known as a "chip"

$n = TW$ $=$ number of PN chips per data bit (processing gain)

$\mu_l$ $=$ pseudorandom phase code sequence

$= 0$ or 2 for biphase PN coding

$= 0, 1, 2,$ or 3 for quadriphase PN coding

$|I(t)|$ $=$ envelope of $I(t)$, the total interference

$I$ $=$ average power in $I(t)$

$\theta_I(t) \equiv \theta(t) =$ phase angle[3] of $I(t)$.

First, a digression on the PN phase coding sequence $\mu_l$. How the $\mu_l$ are generated and stored in the terminals is essential to the PN system. For our purposes we will assume that the $\mu_l$ are obtained by a purely random coin toss (Bernoulli trials) for biphase PN. For quadriphase PN the $\mu_l$ are taken from combining adjacent pairs of the toss sequence into $\mu_l$ events with four equiprobable values. A very close approximation to Bernoulli trails are the codes generated by maximal-length sequence generators [19].

For quadriphase PN, we will be able to show that the four equiprobable phase positions $(\mu_l \pi/2)$ can be made equivalent[4] to an angle, say $\psi$, which is uniformly distributed over $[0, 2\pi)$. Replacing $\mu_l \pi/2$ by a uniformly distributed $\psi$ is not necessarily correct for calculating higher order moments.

The interference term in (2), can be expanded over each $l$th $\Delta T$ chip interval into quadrature components $I_{cl}(t)$ and $I_{sl}(t)$ with respect to the desired signal and then combined with the desired signal to obtain the quadrature components of the limiter input, $x$, relative to the PN *phase-coded* carrier $\omega_0 t + \mu_l \pi/2$. The phasor diagram for $x(t)$ over the $l$th chip interval is shown below in Figs. 2 and 3, where we assume a minus bit, $a = -1$, is being transmitted.

The $u$, $v$ axes are the quadrature components with respect to the transmitted carrier $\omega_0 t + \mu_l \pi/2$. In Fig. 3, biphase PN produces for $I$ either the solid or dotted phasors, each occurring with probability 1/2. Quadriphase PN produces either the solid dotted or dashed phasors, each occurring with probability 1/4. Each interference phasor position changes in a pseudorandom way from chip to chip, by virtue of the reference being relative to the PN sequence $\mu_l$. The resultant $x$ is a phasor from the origin to the vector sum of $\sqrt{2S}$ + $I$ at one of its random positions. For the idealized case of uniform $\mu_l$ phase coding,[5] the tip of $I$ would fall purely at random on a circle of radius $|I|$ and center $-\sqrt{2S}$.

The limiter output $y \equiv y_l$ is obtained by limiting $x \equiv x_l$ to the intersection of $x_l$ with the unit circle centered on the origin as shown in Fig. 4.

---

[3] For typographical convenience we shall henceforward delete the interference subscript of $\theta$.

[4] In the sense that first and second moments of $z$ will be equal in the limit of small $S/I$.

[5] That is to say, $\mu_l$ randomly takes on one of $2^{-n}$ values such that $\mu_l(\pi/2) \to \psi_l$ a uniform $[0, 2\pi)$ r.v. as $n \to \infty$.

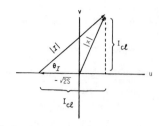

Fig. 2.   Limiter input phasor diagram, $\mu_l = 0$.

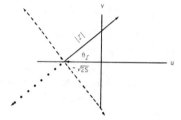

Fig. 3.   Limiter input phasor diagram, all values $\mu_l$.

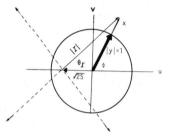

Fig. 4.   Limiter output at the $l$th chip.

The limiter output $y$ equals $|y| \angle \Phi$ with $|y| \equiv 1$, and can be expressed as

$$y(t) = \sum_{l=0}^{n-1} \text{rect}_{\Delta T} (t - l\Delta T) \cos \left[ \omega_0 t + \mu_l \frac{\pi}{2} + \Phi(t) \right] \quad (3)$$

where

$$\Phi(t) = \tan^{-1} \frac{I_{sl}(t)}{a\sqrt{2S} + I_{cl}(t)}.$$

Because the correlator in Fig. 1 is a linear device we can separate $V$ into the sum of two components where $V_1$ is the correlator output due to $y(t)$ while $V_2$ is the correlator output due to $n(t)$. By definition of the correlating operation, it can be easily shown that

$$V_1 = \sum_{l=0}^{n-1} \frac{1}{\Delta T} \int_{l\Delta T}^{(l+1)\Delta T} \cos \Phi(t) \, dt. \quad (4)$$

A correlator gain factor $2/\Delta T$ was introduced for convenience in normalizing the mean of $V_1$. Assume that $\Phi(t)$ is constant

over the chip interval $[l\Delta T, (l + 1) \Delta T]$ and takes the value $\Phi_l$. This then reduces $V_1$ to a stochastic sum

$$V_1 = \sum_{l=0}^{n-1} \cos \Phi_l \triangleq \sum_{l=0}^{n-1} z_l. \quad (5)$$

Referring to Fig. 4, the correlator output is the projection of $y_l$ on the horizontal $u$ axis. Since $|y_l| \equiv 1$, $z_l \equiv \cos \Phi_l$. Later we will return to the phasor diagram of Fig. 4 to express $z_l$ in terms of $\sqrt{2S}$ and $I$.

We complete the description of $V$ by characterizing $V_2$ and then invoke the Cramér–Chernoff bound. Since $n(t)$ is Gaussian (with noise power density $N_0 = kT_r/2P_r$) and the correlator is linear, it follows that $V_2$ is a zero-mean normal random variable with a variance $\sigma^2 V_2 = n N_0 W$ (with the correlator gain factor equal to $2/\Delta T$). Also note that $V_2$ is statistically independent of $V_1$. Because $V_2$ is normally distributed, it can be subdivided into the sum of $n$ independent, identically distributed, normal random variables $\eta_l$, each with variance $\sigma \eta^2 = N_0 W$.

Finally, we have

$$V = \sum_{l=0}^{n-1} (z_l + \eta_l) \quad (6)$$

$$z_l = \cos \Phi_l$$

$$\eta_l \sim N(0, N_0 W).$$

Referring to Fig. 4 it should be clear that since $\mu_l \pi/2$ is chosen at random from chip to chip, $y_l$, hence, $\cos \Phi_l = z_l$ will be identically distributed from chip to chip, provided the following hold:

1) $|I|_l$ is identically distributed and independent chip-to-chip or $|I|_l = \text{constant}$;
2) $\theta$ is a constant from chip to chip; or
3) $\theta$ is independently and uniformly distributed over $[0, 2\pi)$ chip-to-chip.

For $|I|_l$ constant, $I(t)$ becomes a CW tone on the PN frequency when $\theta$ is constant. In the off-frequency case, $\theta$ can be taken randon, provided the difference between the interference frequency and PN frequency is greater than $1/T$.

On assuming that the above conditions hold, the properties required in [13] are satisfied for upper bounding $P_e$ for small $S/I$. $P_e \triangleq \text{prob} \,(V > 0 \,|\, a = -1)$ is upper bounded by

$$P_e < e^{-nu} \quad (7)$$

$$u \triangleq \frac{\alpha(1 - \alpha)}{2} - \epsilon$$

$$\alpha \triangleq (Ez)^2/(\text{var } z + \sigma_\eta^2)$$

$\epsilon$ = a term smaller than $\alpha$ for small $\alpha$ and which approaches zero faster than $\alpha$ as $\alpha \to 0$.

In [13] an exact expression for $\epsilon$ is given and the largest

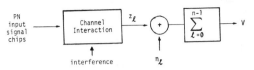

Fig. 5.  General channel model.

allowed value of $\alpha$ is provided. For small $S/I$, $\epsilon$ is negligible and $1 - \alpha \doteq 1$. Then if we define $d^2 \triangleq n\alpha$

$$P_e < e^{-d^2/2} \qquad (8)$$

which is exponentially comparable to (1).

The channel of Fig. 1 is a special case of Fig. 5.[6]

## III. PREVIOUS RESULTS

### A. Biphase PN and I(t) is a Strong CW Tone

Here $I(t) = \sqrt{2I} \cos[\omega_0 t + \theta]$. The CW interferer is on tune to the desired PN frequency and maintains a relatively constant phase angle over a data bit. Denote $z$ for $\mu_l \pi/2 = 0$ as $z_0$ and $z$ for $\mu_l \pi/2 = 180°$ by $z_{180}$. Refer to Fig. 4 and note that when $\theta \to \theta + 180°$, $z_{180}$ can be obtained from $z_0$ by replacing $\cos\theta$ by $-\cos\theta$. Then, since $z_0$ and $z_{180}$ are equally likely

$$Ez = (z_0 + z_{180})/2$$

$$Ez^2 = (z_0^2 + z_{180}^2)/2$$

which yields, upon expanding $z = \cos\Phi$ in a Taylor series and retaining only terms in $\sqrt{S/I}$:

$$Ez \doteq \sqrt{S/I}\, \sin^2\theta$$

$$\text{var } z \doteq \cos^2\theta.$$

Then $d^2$ for small $S/I$ is given by

$$d^2 = n\alpha$$

$$= 2(\sin^4\theta)(S/I)W_0\left[1 + 2(\cos^2\theta)\frac{W_0}{W}\right]^{-1}. \qquad (9)$$

Now note that formally (9) *deviates* from the desired form $d^2 = Ld_{\text{linear}}^2$ in that the bandwidth spread factor $W_0/W$ is multiplied by $2\cos^2\theta$ and is therefore dependent on $\theta$. Of greater importance is the fact that the limiter loss $L$ is equal to $\sin^4\theta$. The bandwidth-limited case cannot be defined because, for $\theta$ in the neighborhood $\pi/2$, $W_0/W$ does not dominate 1. However, because $\cos^2\theta < 1$, the power-limited case is well defined. The limiter loss factor $L$ can be very drastic for small values of $|\theta|$. From Fig. 4, when $\theta = 0$, $|y| = +1$ or $-1$ independent of $\sqrt{2S}$ according to whether $\mu_l = 0$ or 2, respectively. Since $\mu_l = 0$ or 2 with equiprob-

ability, $Ez = 0$. The deleterious effects of tone interference on biphase signals were exhibited in [21].

If the CW interferer is off frequency by more than $1/T$, then $\theta$ can be treated as if it were uniformly distributed chip-to-chip rather than a constant. In this special case the first and second moments in $z$ can be averaged over $\theta$ to yield $(Ez)^2 \doteq (1/4)(S/I)$ and var $z \doteq 1/2$. Then $d^2 = Ld_{\text{linear}}^2$ with $L = 1/4$ or $-6$ dB, the well-known captured limiter loss. If the interferer is malicious he can slowly (relative to $1/T$) walk his center frequency through the PN carrier frequency creating long error bursts. How often he can do this will depend on how well he can narrow his uncertainty in the PN carrier frequency.

The heuristic approach to $d_{\text{linear}}^2$ does not account for possible RF coherence between the desired biphase PN signal and the interfering tone. On the other hand, as will be seen next, quadriphase PN is "equivalent" to uniform phase coding in $(\mu_l \pi/2)$ for small $S/I$ and is therefore independent of the coherence between desired and interfering RF carriers.

### B. Quadriphase PN and I(t) is SSMA[7] Plus a CW Interference Term

In this case, $I(t)$ is composed of the sum of two components, $I(t) = I_1(t) + I_2(t)$.

1) A thermal noise component $I_1(t)$ flat over the band $W$ and having power equal to $\Sigma = \Sigma S_i$. This signal is used to model $i = 1, 2, \cdots, M$ independent PN multiple access carriers, each with power $S_i$.

2) A tone component $I_2(t) = \sqrt{2I} \cos[\omega_0 t + \theta_{I_2}]$.

$I_2(t)$ is an interfering constant envelope tone which can be reduced in average power to $I$ by some form of spatial antenna processing (nulling) at the repeater input. Without antenna processing, $I$ would be much larger than $\Sigma$ and therefore $S/(I + \Sigma) \to (S/I)$. After processing, $I$ is still large compared to $S$ but is no longer large compared to $\Sigma$. We will assume that $S/(I + \Sigma) \ll 1$.

Referring to Fig. 6, $I$ is composed of the sum of a "fixed" vector of length $\sqrt{2I}$ at angle $\theta$ and a thermal noise vector $I_1$, added to the tip of the "fixed" vector with a length that is Rayleigh distributed with rms length $\sqrt{\Sigma}$ and an angle that is uniformly distributed over $[0, 2\pi)$. We assume[8] that with high probability the instantaneous envelope of $I(t)$ over the $l$th chip, $r_l$, given by the length of the sum $I$ of the fixed plus thermal noise phasor, $I_1 + I_2$, is large compared to $\sqrt{2S}$. Thus, $z = \cos\Phi$ and $z^2$ can be expanded in Taylor series retaining terms to $\sqrt{2S}/r$ where the $l$ subscript is suppressed. The net conditional resultant angle $\theta$ for each $I = I_1 + I_2$ phasor, when averaged over the four PN induced angles, can be shown by direct calculation as equivalent to averaging with a uniform distribution, provided $S/E(r^2/2) = S/(I + \Sigma) \ll 1$. Thus, the resultant interference vector $I$ can be treated as having an independent uniform $[0, 2\pi)$ phase distribution and Rician distribution $r$ with "specular" power

---

[6] In fact, multiple interferers could be distributed throughout a cascade of channels. This can happen with multihop satellite links or limiting in the ground station transmitter as well as at the satellite.

[7] Spread spectrum multiple access (see, e.g., [3]).

[8] As we will shortly show, the limiting subcases of (a) $\Sigma \equiv 0$, $S/I \ll 1$, and (b) $I \equiv 0$ and $S/\Sigma \ll 1$ do not require this assumption.

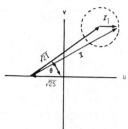

Fig. 6. Noise plus tone phasor diagram.

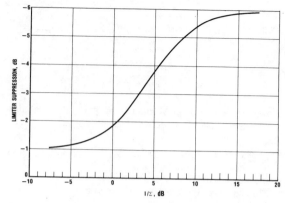

Fig. 7. Limiter suppression $L$ versus ratio of CW interference power to summed power of PN accesses.

component $I$ and noise power component $\Sigma$. The probability density function for $r$ is given by [15]

$$p(r) = \frac{r}{\Sigma} e^{-(r^2 + 2I)/2\Sigma} I_0\left(\frac{\sqrt{2I}r}{\Sigma}\right) \qquad (10)$$

where $I_0(\cdot)$ is the modified Bessel function of zero order. Using Fig. 4, $z = \cos \Phi$ can be expanded to produce

$$z \cong \cos \theta + \frac{\sqrt{2S}}{r}(\cos^2 \theta - 1)$$

$$z^2 \cong \cos^2 \theta + 2\frac{\sqrt{2S}}{r}(\cos^3 \theta - \cos \theta). \qquad (11)$$

Averaging (11) with respect to $\theta$ and then averaging with respect to $r$ produces

$$Ez \doteq -\frac{\sqrt{2S}}{2\Sigma} e^{-I/\Sigma} \int_0^\infty e^{-r^2/2\Sigma} I_0\left(\frac{\sqrt{2I}r}{\Sigma}\right) dr$$

$$Ez^2 \doteq \frac{1}{2}.$$

The value of the above integral is given in [16, 11.4.31]. Forming $d^2 = n\alpha$, collecting terms, and rearranging them produces

$$d^2 = 2L\left(\frac{I}{\Sigma}\right)(S'/I + \Sigma)W_0\left[1 + \frac{W_0}{W}\right]^{-1} \qquad (12)$$

in which

$$L\left(\frac{I}{\Sigma}\right) = \frac{\pi}{4}\left(\frac{I}{\Sigma} + 1\right)\left[e^{-I/2\Sigma} I_0\left(\frac{I}{2\Sigma}\right)\right]^2.$$

As a direct consequence of using quadriphase PN, $d^2 = L d_{\text{linear}}^2$ where $L(I/\Sigma)$ is plotted in Fig. 7.

To our knowledge, Cahn in 1961 [11] was the first to obtain the expression for $L$ given above in (12). The calculation was made for a desired CW tone plus an interfering tone (offset in frequency from the desired tone) plus Gaussian noise.[9] Equation (12) demonstrates that, for quadriphase

[9] Using the methods of Rice, Davenport, and Root, Jones in [8] (1963) derived the autocorrelation function of the output of a hard limiter with two CW tones, offset in frequency plus Gaussian noise. The result is an infinite series involving confluent hypergeometric functions. Limiting results for strong and weak noise are then obtained but a closed-form expression for $L(\cdot)$ was not found.

PN, $L$ can be obtained from multitone CW analysis. Reference [10] also obtains the same expression for $L$ for two CW tones plus Gaussian noise using the Chebyshev transform method due to [9]. When $I$ dominates $\Sigma$ for fixed $S$, i.e., as $I/\Sigma$ gets large, $L$ is asymptotic to $-6$ dB $(1/4)$. For fixed $S$ and $\Sigma$, $S \ll \Sigma$ and $I$ going to zero, we find $L$ approaching $-1.05$ dB $(\pi/4)$.

The effect of an antenna beam shaping processor preceding the limiting repeater not only improves $S/I$, but also reduces limiter loss. The limiter loss can be reduced from $-6$ dB to $-1.8$ dB for $I \approx \Sigma$. This is only 0.7 dB more loss than for *perfect* cancellation of the CW tone, where $L = -1.05$ dB.

As a final note we mention that the two extreme cases of $I \gg \Sigma$ and $I = 0$ ($L = 1/4$ and $\pi/4$) can be derived directly. For $I \gg \Sigma$, set $\Sigma = 0$, $\sqrt{2I} = r$ and expand $z = \cos \Phi$ and $z^2$ to terms in $\sqrt{S/I}$. Directly average over four phase positions $\mu_I \pi/2$ and get the same result as if $\mu_I \pi/2 \rightarrow \psi_I$ uniform and $L = 1/4$. For $I \equiv 0$ and $S/\Sigma \ll 1$, note that $P(\Phi) \doteq (1/2\pi)$ $(1 + \sqrt{\pi S/\Sigma} \cos \Phi)$ and directly average $z = \cos \Phi$ and $z^2 = \cos^2 \Phi$.

## IV. QUADRIPHASE PN, $I(t)$ IS A STRONG CW TONE, AND THE LIMITER HAS AN AM-TO-PM CONVERSION MECHANISM

Up to now the limiter has been idealized as producing at its output unity envelope and preserving the phase of the input. Practical RF limiting transponders can exhibit an output phase shift which is a function of the input envelope [17]. The limiter still produces a unity output envelope.

The output phase angle of the limiter is modeled to be the linear sum of the input angle, $\Phi$ in the previous examples, plus a function $f$ of the input envelope $|R|$ (see [18]). We further assume that $f(|R|)$ is frequency independent, i.e., constant, over the passband of the channel. Thus

$$y = 1 \cdot \cos\left[\omega_0 t + \Phi + f(|R|)\right] \qquad (13)$$

$f(|R|) = $ additional excess phase angle produced by AM-to-PM mechanism in the limiter.

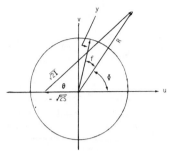

Fig. 8. Phasor diagram of overdriven limiter.

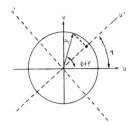

Fig. 9. Phasor diagram for desired PN receiver.

The phasor diagram is shown in Fig. 8.[10] Note that even though $I$ is a constant, the limiter input $x(t)$, as expressed in either (2) or Fig. 8, would still have amplitude modulation on its envelope $|R|$. Hence, $f(|R|)$ is excited for all practical interferers.

For Section III, it can be shown that the PN receiver local oscillator will stay coherent with no phase offset with respect to the desired transmitted PN carrier once in track. Thus, $z$ is the projection of the phasor $y$ onto the $u$ axis. Because of the AM-to-PM conversion in the limiter, we cannot assume that the receiver local oscillator phase angle is in perfect alignment with the transmitter phase angle.

Assume that the PN receiver local oscillator becomes slewed by an angle $\gamma$, with respect to the transmitted phase. Due to the AM-to-PM phase characteristic, $\gamma$ can depend on the particular realization of the phase tracking circuits in the receiver and possibly on slow interference envelope modulation if present. For the moment, treat $\gamma$ as a fixed parameter.

The effect of $\gamma$ is to rotate the $u$, $v$ axes into $u'$, $v'$ by the angle $\gamma$ as shown in Fig. 9. The receiver now takes the projection of $y$ onto $u'$.

Thus, $z$ for each chip is given by

$$z = \cos\,[\Phi + f(|R|) - \gamma]. \qquad (14)$$

First expand $f\,(|R|)$ in a Taylor's series, keeping only terms to order $\sqrt{S/I}$.

Note from Fig. 8 that $|R|$ is given, to terms in $\sqrt{S/I}$, by

$$|R| \doteq \sqrt{2I}\,[1 - \sqrt{S/I}\,\cos\theta].$$

Expanding $f(|R|)$ to order $\sqrt{S/I}$ produces

$$f(|R|) \doteq \bar{f} - G\sqrt{S/I}\,\cos\theta \qquad (15)$$

$$G \doteq \sqrt{2I}f'(\sqrt{2I})$$

$$\bar{f} = f(\sqrt{2I})$$

and $f'(\cdot)$ is the first derivative of $f(\cdot)$. Substituting (15) into (14) produces

$$z = \cos\,[\Phi + (\bar{f} - \gamma) - G\sqrt{S/I}\,\cos\theta]. \qquad (16)$$

[10] For mathematical convenience we have defined $f(\cdot)$ as positive with respect to $\omega_0 t$. In most limiters the actual angle will lag so that $f(\cdot)$ will be negative valued.

Next expand $z$ about the angles $\Phi$, $(\bar{f} - \gamma)$, and $G\sqrt{S/I}$ $\cos\theta$ retaining only terms of order $\sqrt{S/I}$. Then average over the angle $\theta$. From Fig. 8, to first order in $\sqrt{S/I}$, $\cos\Phi$ and $\sin\Phi$ are given by

$$\cos\Phi \doteq \cos\theta + \sqrt{S/I}\,(\cos^2\theta - 1)$$

$$\sin\Phi \doteq \sin\theta + \sqrt{S/I}\,\sin\theta\,\cos\theta. \qquad (17)$$

Utilizing (17) in the trigonometric expansion of (16) and averaging over $\theta$ produces

$$Ez = -\tfrac{1}{2}\,[\cos\,(\bar{f} - \gamma) - G\sin\,(\bar{f} - \gamma)]\sqrt{S/I}. \qquad (18)$$

Once again because we are using quadriphase PN, we can average[11] as if $\theta$ were uniformly distributed over $[0, 2\pi)$.

Repeating the process for $z^2$ produces to order $\sqrt{S/I}$

$$\text{var}\,z = \tfrac{1}{2}. \qquad (19)$$

Note that only $Ez$ depends on the limiter excess phase through $\bar{f} \equiv f(\sqrt{2I})$, its derivative $f'(\sqrt{2I})$, and the phase-tracker reference phase $\gamma$. The variance of $z$ is $1/2$ independent of the limiter AM-to-PM characteristic.

Thus, from (18) and (19) we can conclude that $d^2 = Ld_{\text{linear}}{}^2$ with $L$ given by

$$L = \tfrac{1}{4}\,[\cos\,(\bar{f} - \gamma) - G\sin\,(\bar{f} - \gamma)]^2. \qquad (20)$$

Since limiter loss can depend on the relative receiver RF phase tracking angle $\gamma$, $L$ therefore need not be the same as that measured by multitone frequency analysis or experiment.

We now hypothesize a receive carrier phase-tracking law which lends itself to simple analysis and directly relates to multitone analysis. (This tracker is not known by us to be representative of any real modem implementation.)

Let the receiver phase tracker measure the quadrature components $u$ and $v$ given below.

$$u = \cos\,[\Phi + f(|R|)]$$

$$v = \sin\,[\Phi + f(|R|)] \qquad (21)$$

where $\Phi$ and $f(|R|)$ are shown in Fig. 8. The phase tracker then averages in time over $n$ ($= TW$) PN chips, the value of $u$ and $v$. Assuming ergodicity, the time average over a data

[11] This was checked with a $z_{0^\circ}$, $z_{90^\circ}$, $z_{180^\circ}$, and $z_{270^\circ}$ average.

bit duration of $u$, $v$ is equal to the expectation of $u$ and $v$ values over the phasor angles $\theta + \mu_l \pi/2$. Once again, since we are using quadriphase PN, the discrete average over the four phase positions is equal to the average over an equivalent uniform $\theta$, $[0, 2\pi)$, for small $S/I$.

For small $S/I$, using the expansion of $f(|R|)$ in (15) and the expansions for cos $\Phi$, sin $\Phi$, (17), one obtains $Eu$ and $Ev$ (over one bit interval) accurate to terms in $\sqrt{S/I}$.

$$(Eu)_{\text{bit}} = -\tfrac{1}{2} \sqrt{S/I} \, (\cos \bar{f} - G \sin \bar{f}) \tag{22}$$

$$(Ev)_{\text{bit}} = -\tfrac{1}{2} \sqrt{S/I} \, (\sin \bar{f} + G \cos \bar{f}).$$

The above averages of $u$, $v$ do not account for the effect of the additive local receiver thermal noise. Averaging only over one bit duration will produce too much variance in the tracker estimate of $\gamma$. Thus, a longer, many-data-bit time average is required to lower the variance in $\gamma$ to a negligible level. Since $Eu$ and $Ev$ contain the sign of the data bit, long-term averaging $Eu$, $Ev$ would produce a value of zero if the data bits are equilikely.

Instead, form at the end of each data bit interval, say $t$,

$$\gamma_t \triangleq \tan^{-1} \frac{(Eu)_t}{(Ev)_t} = \tan^{-1} \left( \frac{\sin \bar{f} + G \cos \bar{f}}{\cos \bar{f} - G \sin \bar{f}} \right)_t . \tag{23}$$

$\gamma_t$ is now independent of the data bits and can be serially averaged in time.

Rather than average $\gamma_t$ directly, we see from (18) and (20) that we need to average the following:

$$\cos \gamma_t = \left[ \frac{\cos \bar{f} - G \sin \bar{f}}{\sqrt{1 + G^2}} \right]_t \tag{24}$$

$$\sin \gamma_t = \left[ \frac{\sin \bar{f} + G \cos \bar{f}}{\sqrt{1 + G^2}} \right]_t .$$

In order to reduce the resulting variance in the estimates, time average over many data bits cos $\gamma_t$ and sin $\gamma_t$. Assume the time average is taken over a very long duration relative to fluctuations in $I(t)$. Designate the result of this time average with an overbar and use $\overline{\cos \gamma_t}$ and $\overline{\sin \gamma_t}$ in place of cos $\gamma_t$, sin $\gamma_t$, respectively, in a trigonometric expansion of (18).[12]

The reason for the care in setting up the phase tracker time averages is that we want to introduce a small amount of slow envelope (amplitude) modulation in the interference in order to see an example[13] of the effect of AM-to-PM on a quadriphase PN carrier. In particular, let the interference signal contain a small amount of amplitude modulation

---

[12] Relative to Section III, note that if there is no AM-to-PM $f(\cdot) \equiv 0$, $f' = 0$, $G = 0$, $\bar{f} = 0$, and $\overline{\cos \gamma_t} = 1$. Thus, $\gamma = 0$ and the receiver tracks the transmitter phase with zero slow angle as previously claimed.
[13] Remember that this effect is *dependent* on the particulars of the phase tracker implementation.

represented by

$$I(t) = \sqrt{2I_0} \left( 1 - \sqrt{\frac{\Delta I}{I_0}} \cos \omega_e t \right) \cos (\omega_0 t + \theta). \tag{25}$$

Here $\sqrt{2I} = \sqrt{2I_0} \, (1 - \sqrt{\Delta I/I_0} \cos \omega_e t)$ is assumed to be a slowly varying function of time and $\Delta I/I_0$ is small. In particular, we assume $2\pi/\omega_e$ is long compared to the duration of a data bit but short compared to the time average ($t$-average) over which $\overline{\cos \gamma_t}$, $\overline{\sin \gamma_t}$ were performed. In this case

$$\bar{f} \equiv f(\sqrt{2I})$$

$$G \equiv \sqrt{2I} f'(\sqrt{2I})$$

$$\sqrt{2I} = \sqrt{2I_0} \left( 1 - \sqrt{\frac{\Delta I}{I_0}} \cos \omega_e t \right)$$

and hence $Ez$, given by (18), is a slowly varying function of time.

We now 1) expand the appropriate quantities to order $\sqrt{\Delta I/I_0} < 1$, 2) determine the time averages of cos $\gamma_t$, sin $\gamma_t$, and 3) evaluate $Ez$.

For small $\Delta I/I_0$ the following can be shown.

$$\bar{f} \doteq \bar{f}_0 - G_0 \sqrt{\frac{\Delta I}{I_0}} \cos \omega_e t$$

$$\bar{f}_0 \triangleq f(\sqrt{2I_0})$$

$$G_0 \triangleq G(\sqrt{2I_0}) \equiv \sqrt{2I_0} f'(\sqrt{2I_0})$$

$$G \doteq G_0 \left( 1 - \frac{\Delta I}{I_0} \cos \omega_e t \right) - H_0 \sqrt{\frac{\Delta I}{I_0}} \cos \omega_e t$$

$$H_0 \triangleq (\sqrt{2I_0})^2 f''(\sqrt{2I_0});$$

$$f'' \text{ is second derivative of } f$$

$$\frac{1}{\sqrt{1 + G^2}} \doteq \frac{1}{\sqrt{1 + G_0^2}} \left[ 1 + \frac{G_0}{1 + G_0^2} \right.$$

$$\left. \cdot (G_0 + H_0) \sqrt{\frac{\Delta I}{I_0}} \cos \omega_e t \right]. \tag{26}$$

In (18), cos $(\bar{f} - \gamma)$ and sin $(\bar{f} - \gamma)$ can be expanded in terms of cos $\bar{f}$, sin $\bar{f}$, $\overline{\cos \gamma_t}$ and $\overline{\sin \gamma_t}$. These terms in turn can be expanded for small $\Delta I/I_0$. In determining cos $\gamma_t$, sin $\gamma_t$, in (24) note that to order $\sqrt{\Delta I/I_0}$

$$\cos \bar{f} = \cos \bar{f}_0 + (\sin \bar{f}_0) G_0 \sqrt{\frac{\Delta I}{I_0}} \cos \omega_e t \tag{27}$$

$$\sin \bar{f} = \sin \bar{f}_0 - (\cos \bar{f}_0) G_0 \sqrt{\frac{\Delta I}{I_0}} \cos \omega_e t.$$

Then using (26) and (27) in (24), retaining only terms in

$\sqrt{\Delta I/I_0}$, we obtain

$$\overline{\cos \gamma_t} = \frac{\cos \bar{f}_0 - G_0 \sin \bar{f}_0}{\sqrt{1 + G_0^2}} \tag{28}$$

$$\overline{\sin \gamma_t} = \frac{\sin \bar{f}_0 + G_0 \cos \bar{f}_0}{\sqrt{1 + G_0^2}} .$$

We note from (28) that the interference AM envelope does not affect our *particular* method for phase estimating $\gamma$ even though there is AM-to-PM conversion in the limiter.

Now returning to the expansion of (18), replace $\cos \gamma$, $\sin \gamma$ with $\cos \gamma_t$, $\sin \gamma_t$, respectively, as given by (28). This produces the following:

$$\cos (\bar{f} - \gamma) \doteq \frac{1 - G_0^2 \sqrt{\dfrac{\Delta I}{I_0}} \cos \omega_e t}{\sqrt{1 + G_0^2}}$$

$$G \sin (\bar{f} - \gamma) \doteq - \frac{\left[ G_0^2 - G_0 H_0 \sqrt{\dfrac{\Delta I}{I_0}} \cos \omega_e t \right]}{\sqrt{1 + G_0^2}} . \tag{29}$$

Inserting (29) into (18), expanding $(\sqrt{I})^{-1}$, squaring $Ez$, and retaining terms of order $\sqrt{\Delta I/I_0}$ yields the result

$$(Ez)^2 = \frac{1}{4} \left( \frac{S}{I_0} \right) \left[ (1 + G_0^2) \left( 1 - 2 \left[ \frac{G_0^2 - G_0 H_0}{(1 + G_0^2)} \right] \right. \right.$$

$$\left. \left. \cdot \sqrt{\frac{\Delta I}{I_0}} \cos \omega_e t \right) \right] \tag{30}$$

$$G_0 \triangleq \sqrt{2I_0} f'(\sqrt{2I_0})$$

$$H_0 \triangleq (\sqrt{2I_0})^2 f''(\sqrt{2I_0}).$$

Limiter loss $L$ for our chosen phase tracker and slowly varying jammer envelope with small AM index is then given by

$$L = \tfrac{1}{4} (1 + G_0^2) \left( 1 - 2 \left[ \frac{G_0^2 - G_0 H_0}{(1 + G_0^2)} \right] \sqrt{\frac{\Delta I}{I_0}} \cos \omega_e t \right). \tag{31}$$

Note that $L$ varies slowly in synchronism with the interference envelope.

Suppose $\Delta I = 0$; then $L$ would be a constant $(1/4)(1 + G_0^2)$. This result is in complete agreement with the results of [18, eq. 9], which analyzed a large (limiter-capturing) CW tone plus small frequency offset tones passed through a limiter with AM-to-PM. With the particular receiver phase tracker used here and no AM in the interference envelope, we thus obtain the same limiter suppression as that produced by multifrequency tone analysis.

Since $G_0^2$ must be positive, $(1/4)(1 + G_0^2)$ is smaller loss than the $-6$ dB of an ideal limiter. An erroneous conclusion is that AM-to-PM conversion is desirable and therefore more

$G_0^2$ should be built into the limiter. The limiter loss is sensitive to interference envelope modulation. In particular, the term $2[G_0^2 - G_0 H_0][1 + G_0^2]^{-1}$ is the limiter AM-to-PM amplification factor on the percent interference envelope swing in (31). When the phase function $f$ has a large second derivative, the amplification factor could bring the oscillatory term in the limiter loss, (31), close to 1. This would result in a limiter loss "peak" greatly in excess of $-6$ dB.

An excess phase function $f(\cdot)$, which frequently occurs in practice, is $f(|R|) = \theta_0 - c \ln |R|$ where $\theta_0$ is a fixed reference phase relative to $|R|$ equal to one volt. The AM-to-PM conversion coefficient $c$ is device dependent. If given in degrees per decibel, $\delta$, then $c = (2\pi/360)(20/\ln 10) \delta = 0.1516 \delta$. Usually the excess phase $f(\cdot)$ increasingly lags with increasing $|R|$.

With logarithmic $f(\cdot)$, $G_0$ and $H_0$ are easily evaluated as $G_0 = -c$ and $H_0 = c$ and the amplification factor $A$ is then $A \triangleq 2[G_0^2 - G_0 H_0] \cdot [1 + G_0^2]^{-1} = 2[c^2 + c^2][1 + c^2]^{-1} = 4c^2/(1 + c^2) < 4$. Since the amplification factor is always less than 4 (and usually considerably less than 4, e.g., 2.4 when $\delta$ has the large value of $10°/dB$), a quadriphase PN receiver whose phase tracker behaves as postulated herein should be robust to logarithmic AM-to-PM effects. This conclusion cannot be made when $f(\cdot)$ has steeper slopes than the logarithmic function.

In frequency division multiple access (FDMA), the AM-to-PM conversion effect is well known [20]. Although the desired-carrier output is larger than expected, by the factor $1 + G_0^2$, intelligible crosstalk is coherently introduced into the desired carrier baseband from the large limiter capturing signal. This effect is extremely deleterious and FDMA systems should therefore be designed with a minimum of AM-to-PM conversion. The "equivalent" AM-to-PM effect on PN carriers is the manifestation of a slowly varying limiter loss with potentially large swings as given by (31) when the limiter has steeper slopes in the excess phase function $f(\cdot)$.

## V. CONCLUSIONS

This paper presents a unified phasor model of considerable generality for analyzing the bit error rate performance of coherent PN carrier systems accessing a limiting RF repeater in the presence of strong interference in the form of multiple access noise, a CW tone, or a combination of both. The bit error rate is strictly upper bounded by an exponential which for practical purposes is equal to the bit error rate given by a linear channel with an additional limiter loss factor $L$. A specific formula, involving only first and second moments of the channel output phasors, is provided for calculating $L$ in the case of strong interference. In quadriphase PN with no AM-to-PM conversion in the limiter, $L$ is the same as that obtained by passing multitone CW plus noise through the limiter. This fact is of considerable utility to the repeater developer who may not have access to a PN modem.

For strong interference, use of quadriphase PN has the equivalent effect of randomly distributing the interference RF phase angle uniformly over $[0, 2\pi]$. From this, we can conclude that the jammer need not be on the same frequency as the PN carrier(s) as long as the frequency offset is not so great that channel filters will produce amplitude modulation on the interference tone. The offset tone in frequency merely produces an additional phase angle $\Delta W \Delta T$ (where $\Delta W$ is the frequency offset) in the interference phasor over any chip, $\Delta t$. The phase angle is then randomized over $[0, 2\pi)$ by the PN sequence.

For quadriphase PN carriers passing through a limiter with AM-to-PM conversion the limiter loss factor $L$ will depend on the type of coherent phase tracker implemented in the PN receiver. $L$ will also vary with any slow envelope fluctuation in the interferer.

Quadriphase PN systems accessing limiters with logarithmic AM-to-PM excess phase functions $f(\cdot)$ should be relatively insensitive to the AM-to-PM effect, provided adequate phase tracker performance is achieved. For the phase tracker postulated here, measuring $L$ with multiple frequency tones is justified only when $f(\cdot)$ is logarithmic. We have not justified this conclusion for other receiver phase trackers, nor can we conclude that multitone CW testing, in general, is adequate to characterize PN system performance with limiting channels having an AM-to-PM mechanism. Moreover, for limiters operated in a region where the phase functions $f(\cdot)$ are more steeply sloped than logarithmic on input envelope (linear on input dB), PN systems can become unduly sensitive to interference envelope fluctuation.

## ACKNOWLEDGMENT

The authors wish to acknowledge the helpful discussions and assistance rendered by R. D. Turner and C. R. Cahn. Turner critically reviewed the manuscript. Dr. Cahn provided key insights into the analysis of limiters with AM-to-PM mechanism.

## REFERENCES

[1] Special Issue, *IEEE Trans. Commun.*, vol. COM-25, Aug. 1977.

[2] C. R. Cahn, "Spread spectrum applications and state-of-the art equipments," *Magnavox Commun. Navigation*, Nov. 22, 1972.

[3] J. M. Aein, "Multiple access to a hard-limiting communication-satellite repeater," *IEEE Trans. Space Electron. Telem.*, vol. SET-10, Dec. 1964.

[4] J. W. Schwartz, J. M. Aein, and J. Kaiser, "Modulation techniques for multiple access to a hard-limiting satellite repeater," *Proc. IEEE*, vol. 54, May 1966.

[5] S. O. Rice, "Mathematical analysis of random noise," *Bell Syst. Tech. J.*, vol. 23, pp. 282–332, 1944; also vol. 24, pp. 46–156, 1945.

[6] W. R. Bennett, "Methods of solving noise problems," *Proc. IRE*, vol. 44, pp. 609–638, May 1956.

[7] W. B. Davenport, Jr., "Signal-to-noise ratios in band-pass limiters," *J. Appl. Phys.*, vol. 24, pp. 720–727, June 1953.

[8] J. J. Jones, "Hard-limiting of two signals in random noise," *IEEE Trans. Inform. Theory*, vol. IT-9, Jan. 1963.

[9] N. M. Blachman, "Detectors, band-pass nonlinearities, and their optimization: Inversion of the Chebyshev transform," *IEEE Trans. Inform. Theory*, vol. IT-17, pp. 398–404, July 1971.

[10] J. J. Spilker, Jr., *Digital Communications by Satellite* (Information and System Sciences Series). Englewood Cliffs, NJ: Prentice-Hall, 1977.

[11] C. R. Cahn, "A note on signal-to-noise ratio in band-pass limiters," *IRE Trans. Inform. Theory*, vol. IT-7, Jan. 1961.

[12] J. M. Aein, "Normal approximations to the error rate for hard-limited coherent correlators," *IEEE Trans. Commun. Technol.*, vol. COM-15, Feb. 1967.

[13] ——, "Error rate for peak-limited coherent binary channels," *IEEE Trans. Commun. Technol.*, vol. COM-16, Feb. 1968.

[14] J. M. Aein and W. Doyle, "A note on cascaded limiters," *IEEE Trans. Space Electron. Telem.*, vol. SET-11, Mar. 1965.

[15] W. B. Davenport, Jr. and W. L. Root, *An Introduction to the Theory of Random Signals and Noise*. New York: McGraw-Hill, 1958.

[16] *Handbook of Mathematical Functions With Formulas, Graphs, and Mathematical Tables*, U.S. Dep. Commerce, Nat. Bur. Stand., June 1964.

[17] J. C. Fuenzalida, O. Shimbo, and W. L. Cook, "Time-domain analysis of intermodulation effects caused by nonlinear amplifiers," *COMSAT Tech. Rev.*, vol. 3, Spring 1973.

[18] P. Y. Chang and O. Shimbo, "Effects of one large signal on small signals in a memoryless nonlinear bandpass amplifier," *IEEE Trans. Commun.*, vol. COM-28, May 1980.

[19] S. W. Golomb, *Shift Register Sequences*. San Francisco, CA: Holden-Day, 1967.

[20] R. C. Chapman and J. B. Millard, "Intelligible cross-talk between frequency modulated carriers through AM-to-PM conversion," *IEEE Trans. Commun. Syst.*, June 1964.

[21] J. L. Sevy, "The effect of limiting a biphase or quadriphase signal plus interference," *IEEE Trans. Aerosp. Electron. Syst.*, vol. AES-5, pp. 387–395, May 1969.

# Part V
# New Systems, Technology, and Uses

THIS last part describes system applications of spread-spectrum communications and some new technology which will allow us to use spectrum spreading more efficiently.

Much of what we would like to do is currently being delayed by technology limitations such as linearity and range of SAW devices, Bragg cells, speed of microprocessors, memory size, and speed, size, and cost of frequency synthesizers; these are only a few examples. Since new spread-spectrum systems and systems used in conjunction with them are being built at a rapid rate, the impact of technology limitations is obvious.

In addition to affecting overall system design, technology plays a crucial role in the design and performance of synchronization systems. Synchronization is often the major problem in spread-spectrum communications. Since a code is employed for transmission, the receiver must lock its code to the incoming code. Techniques that minimize this time to synchronize in the presence of noise and jamming are often complex, and the resulting time is often longer than desired. Furthermore, once synchronized, the system is still vulnerable to a jamming signal that may, if placed properly, cause the system to either become unsynchronized or operate with a high error rate.

This section contains five papers that either illustrate new ways of accomplishing various functions necessary to the successful operation of a spread-spectrum system (e.g., synchronization) or present new uses of spread-spectrum communications. There are two papers dealing with synchronization. The first one, by Yost and Boyd ("A Modified PN Code Tracking Loop: Its Performance Analysis and Comparative Evaluation"), shows that by somewhat modifying a standard delay-locked loop it is possible to obtain, for example, tracking performance equivalent to that of a standard loop, but with the simplicity of implementation of a dithering loop. The price one pays for this type of performance is reduced pull-in range and higher probability of false lock. Both synchronization and acquisition are considered in the paper by Baier, Dostert, and Pandit ("A Novel Spread-Spectrum Receiver Synchronization Scheme Using a Saw-Tapped Delay Line"). This paper contains both experimental and analytical results, and, as its title implies, makes heavy use of the matched-filtering capabilities of a surface-acoustic-wave tapped delay line.

Another interesting application of SAW technology to spread-spectrum communications is presented in the paper by LaRosa, Marynowski, and Henrich ("A Simple Modulator for Sinusoidal Frequency Shift Keying"). Here, a SAW filter is used to generate a sequence of phase-coded rectangular chips. This is potentially advantageous in a system where a sharp spectral rolloff is desired.

The fourth paper in this section is by Mizuno ("Randomization Effect of Errors by Means of Frequency-Hopping Techniques in a Fading Channel"). Mizuno shows that the use of a large enough number of frequency slots over which to hop produces an effect equivalent to using time interleaving of data sequences in order to randomize the errors that occur in a signal transmitted over a fading channel.

Finally, the fifth paper, written by Giordano, Sunkenberg, dePedro, Stynes, Brown, and Lee ("A Spread-Spectrum Simulcast MF Radio Network"), describes a network designed to operate reliably over a channel that includes both interference and fading. It is shown that the system is robust with respect to individual links and, hence, exhibits both good connectivity and low vulnerability to jamming.

---

# A Modified PN Code Tracking Loop: Its Performance Analysis and Comparative Evaluation

RICHARD A. YOST, MEMBER, IEEE, AND ROBERT W. BOYD, MEMBER, IEEE

*Abstract*—A modified noncoherent PN code tracking loop (MCTL) has been previously presented by the authors [3]. The loop was shown to have the hardware simplicity of the tau-dither loop and a tracking performance superior to the traditional delay-locked code tracking loop (TCTL). Further work conducted herein expands on these previous results by considering the band-limiting effects of the bandpass arm filters on the performance of the MCTL. It will be shown that an even greater improvement in tracking performance is experienced by the MCTL over the TCTL when such effects are included. Furthermore, it was mentioned in [3] that a specific *S/N* existed below which the MCTL acquisition behavior was superior to the TCTL and above which its behavior was inferior. The band-limiting effects will be shown to have a more deleterious effect on the TCTL and, as such, the crossover *S/N* will increase, thus enhancing the comparable capabilities in favor of the MCTL. This paper also discusses the implementation sensitivities to hardware gain and phase imbalance by expanding on the noise-free work in [3] to include the noisy case.

## I. INTRODUCTION

A TRADITIONAL noncoherent PN code tracking loop (TCTL) [1] based on the delay-lock loop principle is shown in Fig. 1(a). Disregarding for the time being the inserted "amplifier" (see Section III), the basic principle behind this loop is the generation of two signals, one of which is derived by correlating the received waveform with an early version (advanced one-half chip) of the PN code and the other by correlating the received waveform with a late version (retarded by one-half chip) of the PN code.[1] These two signals are then individually filtered, squared, and compared to form the loop error signal.

A major disadvantage with the TCTL as configured in Fig. 1(a) is the difficulty in maintaining gain balance through each of the loop branches (filtering, amplification, and squaring). To circumvent this problem, a different configuration for the TCTL is possible and is shown in Fig. 1(b). Here the sum ($\Sigma$) and difference ($\Delta$) of the early and late signals are first generated, then individually filtered, and finally mixed to form the same error signal as in Fig. 1(a). Any gain imbalance between the sum and difference channels has no biasing effect on the performance since the error signal is now formed by a final mixing operation rather than the original comparison (or subtraction). Furthermore, since the error signal generated in both

Manuscript received July 6, 1981; revised January 13, 1982. This paper was presented at the National Telecommunications Conference, Houston, TX, December 1980.

The authors are with the Government Systems Group, Harris Corporation, Melbourne, FL 32901.

[1] The one-half chip offset is not mandatory; other offsets are also useful, but for the work herein only one-half chip offsets will be considered.

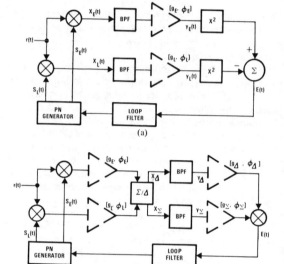

Fig. 1.   Traditional code tracking loops (TCTL). (a) Classified square-law loop. (b) Reconfigured square-law loop.

of these configurations is identical, overall performance is unchanged.

Each of these configurations requires bandpass filtering to establish a good signal-to-noise ratio prior to generating the error signal. The bandwidths of these filters must be wide enough to pass the data signal and are therefore dependent on the data rate. In multirate applications, the complexity of the code tracking loop becomes even greater since a filter matched to each data rate must be used to minimize the tracking jitter at each data rate [2]. In order to reduce the hardware required for such applications, a third loop configuration is presented and shown in Fig. 2. Instead of using the sum channel as the reference, the on-time or data channel is utilized. Because the error signal is no longer identical to the error signal generated by the first two configurations, this loop is referred to as a modified code tracking loop (MCTL).

The MCTL was initially presented by the authors in [3]. Performance measures such as rms tracking jitter, acquisition, and the sensitivity of the MCTL to gain and phase imbalances were evaluated for a simplified model of the MCTL and then a comparison with the analogous parameters for the TCTL was made. With the elimination of an entire IF processing channel, it was pointed out that the MCTL complexity was nearly equivalent to the well-known dithering loop [4], whose

Reprinted from *IEEE Trans. Commun.*, vol. COM-30, pp. 1027–1036, May 1982.

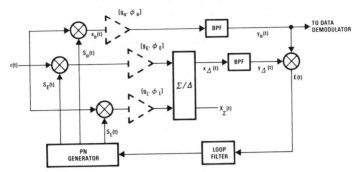

Fig. 2.   Modified code tracking loop (MCTL).

primary advantage was its simplicity. However, it was shown that the MCTL did not suffer the loss in tracking performance (with respect to the TCTL) that the dithering loop experienced. In fact, for a given IF signal-to-noise ratio, the MCTL produced a lower rms tracking jitter than that achievable by a TCTL with the same loop bandwidth. Furthermore, it was also shown that over a wide range of values, the sensitivity of the MCTL to hardware gain and phase imbalances was nearly equivalent to that of the TCTL shown in Fig. 1(b), which was configured to minimize the imbalance effects.

Just as the MCTL was shown to have certain advantages over the TCTL, it also contained some disadvantages. First, the pull-in range of the MCTL was reduced by 33 percent from the TCTL capabilities (see Fig. 3 in this paper). Second, for a given tracking offset, the error signal voltage was greater for the traditional loops, causing the TCTL to respond more quickly than the MCTL. Furthermore, the relative flattening of the MCTL discriminator characteristic as the time offset increases implies a greater probability to loop loss of lock. And finally, the maximum search rate of the MCTL was shown to be smaller (except at low IF $S/N$) than that for the TCTL. With respect to the latter comparison, it was further shown in [3] that the actual search rate reduction was strongly dependent on the IF $S/N$ since, for the same tracking jitter, the MCTL can operate under a lower $S/N$ than the TCTL and thus have a wider bandwidth.

Accuracy of these performance results in [3] was limited by the preciseness of the mathematical model used. In particular, two model deficiencies were the lack of filtering effects on the desired signal, and the absence of noise in developing the sensitivity expressions. The work herein removes these deficiencies and readdresses the performance measures described above. In addition, the analysis provides a means of assessing the practical impact of IF filter types and bandwidths, and of baseband data formats, on the behavior of the MCTL.

## II. ANALYSIS OF MCTL

Ultimately, the primary objective of this analysis is to determine the rms tracking jitter of the MCTL and then compare it to that for the TCTL. In addition to achieving this objective, the analysis will simultaneously culminate in an expression for the loop squaring loss; such a loss takes into con-

sideration the bandpass filtering effects on the desired signal and establishes a criterion upon which the optimum bandpass filter bandwidth can be chosen. The procedure used here is similar to the analytical approach established by Simon [2]. Finally, the comparison of the maximum search rate for the MCTL and TCTL is revisited by considering the newly evaluated squaring loss.

With reference to Fig. 2, the following definitions are presented:

$$S(t) \triangleq \sqrt{2P_s}\, d(t - \tau) S_{\text{PN}}(t - \tau) \cos{(\omega t + \phi)}$$

$$S_0(t) \triangleq \sqrt{2} S_{\text{PN}}(t - \hat{\tau}) \cos{[(\omega + \omega_1)t + \theta]}$$

$$S_E(t) \triangleq \sqrt{2} S_{\text{PN}}\left(t - \hat{\tau} + \frac{\Delta}{2}\right) \cos{[(\omega + \omega_1)t + \theta]}$$

$$S_L(t) \triangleq \sqrt{2} S_{\text{PN}}\left(t - \hat{\tau} - \frac{\Delta}{2}\right) \cos{[(\omega + \omega_1)t + \theta]}$$

$$n(t) \triangleq \sqrt{2}\, \{N_c(t) \cos{(\omega t + \phi)} - N_s(t) \sin{(\omega t + \phi)}\}$$

(1)

where $d(t)$ is the baseband data waveform with spectral density $S_d(f)$, $P_S$ is the total signal power, $S_{\text{PN}}(t)$ is the pseudorandom waveform, $\Delta$ is the PN code chip duration in seconds, and $\tau - \hat{\tau}$ is the tracking error. In addition, $N_c(t)$ and $N_s(t)$ are independent, low-pass filtered white Gaussian noise processes with a spectral height of $N_0/2$.

Consider the on-time channel signal flow first. Ignoring the double frequency terms, the correlator output becomes

$$x_0(t) = \sqrt{P_s}\, d(t - \tau) \overline{S_{\text{PN}}(t - \tau) S_{\text{PN}}(t - \hat{\tau})} \cos{(\omega_1 t + \beta)}$$
$$+ \sqrt{P_s}\, d(t - \tau)\{S_{\text{PN}}(t - \tau) S_{\text{PN}}(t - \hat{\tau})$$
$$- \overline{S_{\text{PN}}(t - \tau) S_{\text{PN}}(t - \hat{\tau})}\} \cos{(\omega_1 t + \beta)} + n_0(t)$$

(2)

where the overhead bar denotes statistical expectation, $\beta \triangleq \theta - \phi$, and

$$n_0(t) = \sqrt{2} S_{\text{PN}}(t - \hat{\tau}) \cos{[(\omega + \omega_1)t + \theta]}n(t).$$

Contributing to this correlator output is the desired signal, a correlation self-noise, and a noise term. When the code rate $(1/\Delta)$ is much larger than the loop bandwidth $(B_L)$, the self-noise can be neglected [5]. Furthermore, for long pseudo-random codes with $\epsilon \triangleq \tau - \hat{\tau}$

$$R_{PN}(\epsilon) \triangleq \overline{S_{PN}(t)S_{PN}(t+\epsilon)}$$

$$= \begin{cases} 1 - \dfrac{|\epsilon|}{\Delta} & |\epsilon| \leqslant \Delta \\ 0 & |\epsilon| > \Delta \end{cases} \tag{3}$$

Therefore, (2) becomes

$$x_0(t) = \sqrt{P_s}\,d(t-\tau)R_{PN}(\epsilon)\cos(\omega_1 t + \beta) + n_0(t). \tag{4}$$

Now consider the effects of the bandpass filter. With the filter impulse response denoted as $h(t)$ and its equivalent low-pass filter impulse response as $h_l(t)$, the filter output can be written as

$$y_0(t) = \sqrt{P_s}\,d_f(t-\tau)R_{PN}(\epsilon)\cos(\omega_1 t + \beta) + n_{0f}(t) \tag{5}$$

where

$$d_f(t) = \int_{-\infty}^{\infty} d(\lambda)h_l(t-\lambda)\,d\lambda$$

$$n_{0f}(t) = N_{c0}(t)\cos(\omega_1 t + \beta) - N_{s0}(t)\sin(\omega_1 t + \beta)$$

$$N_{c0}(t) = \int_{-\infty}^{\infty} N_c(\lambda)S_{PN}(\lambda - \hat{\tau})h_l(t-\lambda)\,d\lambda$$

$$N_{s0}(t) = \int_{-\infty}^{\infty} N_s(\lambda)S_{PN}(\lambda - \hat{\tau})h_l(t-\lambda)\,d\lambda. \tag{6}$$

In a similar fashion the filtered difference channel is determined to be

$$y_\Delta(t) = \sqrt{P_s}\,d_f(t-\tau)\{R_{PN}(\epsilon - \Delta/2) - R_{PN}(\epsilon + \Delta/2)\}$$

$$\cdot \cos(\omega_1 t + \beta) + n_{\Delta f}(t) \tag{7}$$

where

$$n_{\Delta f}(t) = N_{c\Delta}(t)\cos(\omega_1 t + \beta) - N_{s\Delta}(t)\sin(\omega_1 t + \beta)$$

$$N_{c\Delta}(t) = \int_{-\infty}^{\infty} N_c(\lambda)\{S_{PN}(\lambda - \hat{\tau} + \Delta/2)$$

$$- S_{PN}(\lambda - \hat{\tau} - \Delta/2)\}h_l(t-\lambda)\,d\lambda$$

$$N_{s\Delta}(t) = \int_{-\infty}^{\infty} N_s(\lambda)\{S_{PN}(\lambda - \hat{\tau} + \Delta/2)$$

$$- S_{PN}(\lambda - \hat{\tau} - \Delta/2)\}h_l(t-\lambda)\,d\lambda. \tag{8}$$

The error signal is now found by mixing the on-time signal with the difference signal; from (5) and (7) and ignoring the double frequency terms, the error can be written as

$$E(t) = y_0(t)y_\Delta(t)$$

$$= \tfrac{1}{2}P_s R_{PN}(\epsilon)\{R_{PN}(\epsilon - \Delta/2) - R_{PN}(\epsilon + \Delta/2)\}$$

$$\cdot d_f^2(t-\tau) + N(t) \tag{9}$$

where

$$N(t) = \sqrt{P_s}\,d_f(t-\tau)\{\tfrac{1}{2}R_{PN}(\epsilon)N_{c\Delta}(t)$$

$$+ \tfrac{1}{2}[R_{PN}(\epsilon - \Delta/2) - R_{PN}(\epsilon + \Delta/2)]N_{c0}(t)\}$$

$$+ \tfrac{1}{2}N_{c0}(t)N_{c\Delta}(t) + \tfrac{1}{2}N_{s0}(t)N_{s\Delta}(t). \tag{10}$$

With the discriminator characteristic defined as

$$g(\epsilon) \triangleq \tfrac{1}{2}R_{PN}(\epsilon)\{R_{PN}(\epsilon - \Delta/2) - R_{PN}(\epsilon + \Delta/2)\} \tag{11}$$

and comparatively illustrated in Fig. 3, the error signal can be written simply as

$$E(t) = P_s d_f^2(t-\tau)g(\epsilon) + N(t). \tag{12}$$

Furthermore, consider the following equality:

$$d_f^2(t-\tau)g(\epsilon) = \langle d_f^2(t-\tau)\rangle g(\epsilon)$$

$$+ \{d_f^2(t-\tau) - \langle d_f^2(t-\tau)\rangle\}g(\epsilon)$$

where $d_f^2(t-\tau)$ has been decomposed into its mean value and a self-noise term. In [6] it has been shown that the self-noise term can be neglected if the loop bandwidth is small compared to the data rate. Therefore

$$\overline{\langle d_f^2(t-\tau)\rangle}g(\epsilon) = D_m g(\epsilon)$$

where

$$D_m \triangleq \int_{-\infty}^{\infty} S_d(f)|H_l(j2\pi f)|^2\,df$$

is the notation introduced in [2] and used here for consistency.

By normalizing any loop gains, the estimate $\hat{\tau}$ of the PN code delay $\tau$ can be written in terms of $E(t)$ as

$$\frac{\hat{\tau}}{\Delta} = \frac{F(p)}{p}\{E(t)\} \tag{13}$$

where $F(p)$ is the loop filter transfer function and the operator $p$ signifies $d/dt$. Using (12) and carrying out a portion of the

240

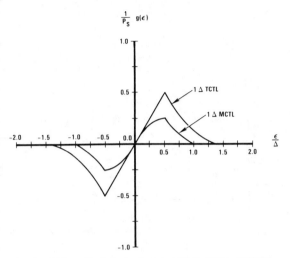

Fig. 3.  1Δ loop discriminator characteristics for TCTL and MCTL.

differentiation explicitly, (13) becomes

$$\frac{1}{\Delta}\frac{d\epsilon}{dt} = \frac{1}{\Delta}\frac{d\tau}{dt} - F(p)[2P_sD_m]\left\{\frac{g(\epsilon)}{2} + \frac{N(t)}{2P_sD_m}\right\}. \quad (14)$$

If consideration now is focused on large loop signal-to-noise ratios whereby $g(\epsilon) = \epsilon/\Delta$, and if it is further assumed that $d\tau/dt = 0$, the rms tracking jitter can be directly written from (14) as

$$\frac{\sigma}{\Delta} \triangleq \sqrt{\frac{\overline{\epsilon^2}}{\Delta^2}} = \frac{1}{P_sD_m}\sqrt{\overline{P_N(\epsilon)}} \quad (15)$$

where $P_N(\epsilon)$ is the portion of the power of $N(t)$ in the loop bandwidth. The final objective then is to determine $P_N(\epsilon)$; this will be accomplished by first evaluating the autocorrelation function of $N(t)$.

The autocorrelation function of $N(t)$ is defined as

$$R_N(\delta, \epsilon) \triangleq E\{N(t)N(t+\delta)\}. $$

With

$$R_{d_f}(\delta) \triangleq \int_{-\infty}^{\infty} S_d(f)|H_l(j2\pi f)|^2 e^{j2\pi f\delta}\, df$$

$$R_{N_{c0}}(\delta) \triangleq \frac{N_0}{2}\int_{-\infty}^{\infty}|H_l(j2\pi f)|^2 e^{j2\pi f\delta}\, df$$

$$= \frac{1}{2}R_{N_{c\Delta}}(\delta) \quad (16)$$

the autocorrelation of $N(t)$ can be simplified to

$$R_N(\delta, \epsilon) = \frac{1}{4}P_sR_{d_f}(\delta)\{2R_{PN}^2(\epsilon)$$
$$+ [R_{PN}(\epsilon + \Delta/2) - R_{PN}(\epsilon - \Delta/2)]^2\}R_{N_{c0}}(\delta)$$
$$+ R_{N_{c0}}^2(\delta). \quad (17)$$

Now, the process $N(t)$ can be approximated to be a white noise process because the MCTL bandwidth is usually much narrower than the bandwidth of $N(t)$. The equivalent single-sided spectral density of this white noise process is

$$N_e(\epsilon) = 2\int_{-\infty}^{\infty} R_N(\delta, \epsilon)\, d\delta$$

$$= \frac{1}{2}P_s\{R_{PN}^2(\epsilon) + [R_{PN}(\epsilon + \Delta/2)$$
$$- R_{PN}(\epsilon - \Delta/2)]^2\} \cdot \{S_{d_f}(f)*S_{N_{c0}}(f)|_{f=0}\}$$
$$+ S_{N_{c0}}(f)*S_{N_{c0}}(f)|_{f=0} \quad (18)$$

where $S_{d_f}(f)$ is the spectral density of the filtered data wave-form, $S_{N_{c0}}(f)$ is the spectral density of $N_{c0}(t)$, and "$*$" corresponds to traditional convolution. From (16) and recalling that autocorrelation and spectral density form a Fourier transform pair, the equalities

$$S_{d_f}(f)*S_{N_{c0}}(f)|_{f=0} = \frac{N_0}{2}\int_{-\infty}^{\infty} S_d(x)|H_l(j2\pi x)|^4\, dx \quad (19)$$

$$S_{N_{c0}}(f)*S_{N_{c0}}(f)|_{f=0} = \frac{N_0^2}{4}\int_{-\infty}^{\infty}|H_l(j2\pi x)|^4\, dx$$

can be written directly. Furthermore, with the definitions [2]

$$K_L \triangleq \frac{\displaystyle\int_{-\infty}^{\infty}|H_l(j2\pi f)|^4\, df}{\displaystyle\int_{-\infty}^{\infty}|H_l(j2\pi f)|^2\, df} \quad (20)$$

$$K_D \triangleq \frac{\displaystyle\int_{-\infty}^{\infty} S_d(f)|H_l(j2\pi f)|^4\, df}{\displaystyle\int_{-\infty}^{\infty} S_d(f)|H_l(j2\pi f)|^2\, df}$$

(18) becomes

$$N_e(\epsilon) = 2P_sN_0D_m^2 \cdot \left\{\frac{\frac{1}{8}[2R_{PN}^2(\epsilon) + \{R_{PN}(\epsilon + \Delta/2) - R_{PN}(\epsilon - \Delta/2)\}^2]K_D}{D_m} + \frac{\frac{1}{4}K_L\dfrac{1}{D_m}\dfrac{1}{S/N}}{D_m}\right\} \quad (21)$$

where $S/N$ is the IF signal-to-noise ratio and defined as

$$S/N \triangleq P_s/(N_0B_{IF})$$

with

$$B_{IF} \triangleq \int_{-\infty}^{\infty}|H_l(j2\pi f)|^2\, df.$$

Now, for large loop signal-to-noise ratios $N_e(\epsilon)$ can be simplified so that $\overline{P_N(\epsilon)}$ becomes

$$
\overline{P_N(\epsilon)} = \overline{N_e(\epsilon)} B_L
$$

$$
= 2P_s N_0 D_m{}^2 \left\{ \frac{\frac{1}{4}\left[2\frac{\sigma^2}{\Delta^2}+1\right]K_D + \frac{1}{4}K_L \frac{1}{D_m S/N}}{D_m} \right\} B_L.
$$

(22)

Using (22) in (15) and recombining terms, the rms jitter becomes

$$
\frac{\sigma}{\Delta} = \sqrt{\frac{\frac{1}{2}\frac{N_0 B_L}{P_s D_m}\left\{K_D + \frac{K_L}{D_m S/N}\right\}}{1 - \frac{N_0 B_L}{P_s D_m}K_D}}
$$

$$
\approx \sqrt{\frac{K_D + \frac{K_L}{D_m S/N}}{2 D_m \rho_L}}
$$

(23)

where $\rho_L = P_s/(N_0 B_L)$. In more compact and typical form (23) can be written as

$$
\frac{\sigma}{\Delta} = \sqrt{\frac{1}{2\rho_L S_L}}
$$

(24)

where $S_L$ is the squaring loss and defined as

$$
S_L \triangleq \frac{D_m}{K_D + \frac{K_L}{D_m S/N}}
$$

$$
= \frac{D_m}{K_D + K_L \frac{B_{IF}/R_s}{R_d/D_m}}.
$$

(25)

In the latter part of (25), $R_s$ is the data symbol rate and $R_d$ is the signal-to-noise ratio in the data symbol bandwidth.

Fig. 4 illustrates the squaring loss for the MCTL (25) as a function of the ratio of IF bandwidth to symbol bandwidth and with $R_d$ as the parameter. The loss curves are for NRZ data where

$$
S_d(f) = \frac{1}{R_s}\frac{\sin^2(\pi f/R_s)}{(\pi f/R_s)^2}
$$

and a two-pole Butterworth filter of the form

$$
|H_l(j2\pi f)|^2 = \frac{1}{1+(f/f_c)^4}
$$

where $f_c$, the 3 dB bandwidth, is related to $B_{IF}$ by

$$
f_c = \frac{2}{\pi}\sin\left(\frac{\pi}{4}\right)B_{IF}.
$$

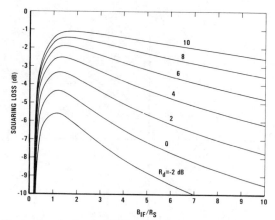

Fig. 4. Squaring loss for MCTL with two-pole Butterworth filters and NRZ data modulation.

Furthermore, typical closed-form expressions for $K_L$ may be found in [7] for various filter types, but for the filter considered here $K_L = 0.75$. Now, in order to minimize the tracking jitter and thus optimally choose the IF bandwidth for a given $R_d$, the squaring loss needs to be minimized. From Fig. 4 and in most cases for NRZ data, the squaring loss is minimized by choosing the IF bandwidth of a two-pole Butterworth filter to be typically 1 to 2 times the symbol rate.

Note that when no filtering effects are considered $D_m = K_D = K_L = 1$, (24) becomes

$$
\frac{\sigma}{\Delta} = \sqrt{\left(\frac{1}{2(S/N)}+\frac{1}{2(S/N)^2}\right)\left(\frac{B_L}{B_{IF}}\right)}
$$

(26)

which was determined in [3]. It was further shown in [3] that upon comparing (26) for the MCTL to the corresponding formula for the TCTL, a smaller rms jitter for identical loop signal-to-noise ratios was obtainable with the MCTL. However, in order to make a more accurate comparison and take into consideration the above filtering effects, it is now appropriate to compare the corresponding squaring losses for the two loops.

For the TCTL, the squaring loss can be written as [2]

$$
S_L = \frac{D_m}{K_D + K_L\left(\frac{2B_{IF}/R_s}{R_d D_m}\right)}.
$$

(27)

Fig. 5 compares the squaring loss of the MCTL to the TCTL. Here NRZ data and a two-pole Butterworth filter are again assumed. Furthermore, the ratio $B_{IF}/R_s$ is fixed at that ratio which minimizes the squaring loss for a given $R_d$. For low values of $R_d$ and an equivalent rms jitter, the MCTL can now utilize a loop bandwidth approximately two times the loop bandwidth for the TCTL. Or alternatively, for equivalent loop bandwidths, the tracking jitter for the MCTL is a factor of 1.35 smaller than the jitter for the TCTL. Thus, by considering these realistic filtering effects, the MCTL has shown a tracking

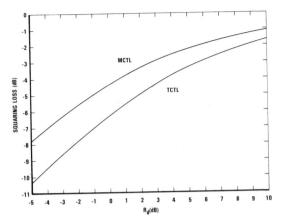

Fig. 5.  Squaring loss comparison for NRZ data modulation.

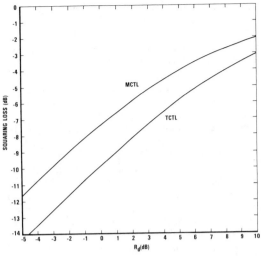

Fig. 6.  Squaring loss comparison for Manchester data modulation.

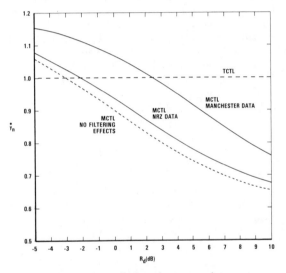

Fig. 7.  Normalized search rate comparison.

improvement over the TCTL greater than that improvement predicted in [3]. As expected, such an advantage diminishes as $R_d$ increases. An even greater improvement is achieved if Manchester coded data are considered. Fig. 6 forms a comparison similar to Fig. 5 but for Manchester coded data. Again, the squaring loss and, hence, tracking jitter is less for the MCTL than the TCTL. The importance here is not so much the actual improvement but rather, for Manchester coded data, the improvement of the MCTL over the TCTL does not diminish as rapidly as it does for NRZ data.

Reference [3] included a discussion of the acquisition behavior of the MCTL and showed initially that a 40 percent decrease in search rate capability, as compared to the TCTL, would be experienced with the MCTL. However, it was further pointed out that such a superficial comparison was biased against the MCTL since the MCTL can maintain the same rms tracking jitter as the TCTL with a wider loop bandwidth. Consequently, it was shown that a more significant comparison of the search rate capability could be made by requiring $\sigma_{MCTL} = \sigma_{TCTL}$. For the squaring loss considered herein, the normalized search rate [3] becomes

$$\dot{y}_n = \begin{cases} 1.0 & \text{TCTL} \\ 0.6\dfrac{S_{LM}}{S_{LT}} & \text{MCTL} \end{cases} \quad (28)$$

where $S_{LM}$ and $S_{LT}$ are the squaring losses for the MCTL and TCTL, respectively. With the data from Figs. 5 and 6, (28) can be evaluated and is shown in Fig. 7 as a function of $R_d$. Also illustrated here is the normalized search rate for the MCTL with no filtering effects and $B_{IF} = R_s$ where

$$S_{LM} = \frac{R_d}{R_d + 1}$$

and

$$S_{LT} = \frac{R_d}{R_d + 2}.$$

A number of observations are available from Fig. 7. First, in all cases there exists a value of $R_d$ below which the MCTL has a higher search rate capability than the TCTL and above which the opposite is true. Second, the TCTL degrades more rapidly when filter effects are considered since the crossover point moves further to the right, thus indicating an improvement of the MCTL over the TCTL. And finally, the acquisition performance tradeoff between the MCTL and TCTL must consider the type of data format since the range of applicable $R_d$ values becomes larger for Manchester coded data.

### III. SENSITIVITY TO GAIN/PHASE IMBALANCE

One concern in implementing PN code tracking loops is how they respond to differences in component values caused by temperature, aging, mismatch, etc. Hardware gain and/or phase imbalances were shown in [3] to produce two effects with noise-free conditions: a change in discriminator gain and a constant bias. The tracking bias is usually the most significant since it causes a loss in signal power (by the square of the PN code's autocorrelation function) in addition to a ranging error, while the gain variation may be somewhat removed by an automatic gain control (AGC). This analysis has been expanded to include the effect of noise when gain and/or phase imbalances exist, resulting in an additional bias term and slightly increased tracking jitter variance. Results for the TCTL will be included since they are not readily available in the literature and are necessary for comparison purposes.

Returning to the loop configurations of Figs. 1 and 2, we call attention to the (dashed) amplifiers that have been heretofore neglected. An amplifier has been shown in each independent hardware path having a gain and phase term (in brackets) associated with each. With this representation, all circuit functions (e.g., mixers, filters, combiner) may be taken to be ideal (i.e., unity gain, no delay) with the combined gain and phase delay of these functions assigned to the appropriate amplifiers. Using this convention and making the usual assumptions of broad-band stationary Gaussian interference and spread bandwidth ≫ data bandwidth ≫ loop bandwidth allows evaluation of the loop's error term through mundane but very tedious analysis as presented in the Appendix.

For the usual square-law loop of Fig. 1(a), one obtains

$$E(t) = P_s \frac{g_E{}^2 + g_L{}^2}{2} \left\{ \frac{\epsilon}{\Delta} + \left( \frac{\rho^2 - 1}{\rho^2 + 1} \right) \right.$$

$$\left. \cdot \left[ \frac{1}{4} + \frac{\epsilon^2}{\Delta^2} + \frac{N_0 B_{IF}}{P_s} \right] + N_E \right\} \tag{29}$$

where $\epsilon \triangleq \tau - \hat{\tau}$ is the PN code tracking error, $N_E$ is a zero-mean random variable, $B_{IF}$ is the noise equivalent bandwidth of the arm filters, and $\rho \triangleq g_E/g_L$ is a measure of gain imbalance. The variance of the tracking jitter is given by

$$\sigma_{\epsilon/\Delta}{}^2 = \frac{B_L}{B_{IF}} \frac{2(\rho^4 + 1)}{(\rho^2 + 1)^2} \left\{ \frac{1}{2} \left[ 1 + \frac{4(\rho^4 - 1)}{\rho^4 + 1} \frac{\bar{\epsilon}}{\Delta} + \frac{4\bar{\epsilon}^2}{\Delta^2} \right] \right.$$

$$\left. \cdot \frac{N_0 B_{IF}}{P_s} + \left( \frac{N_0 B_{IF}}{P_s} \right)^2 \right\} \tag{30}$$

where $B_L$ is the (one-sided) loop bandwidth. As one should expect, the phase delays $\phi_E$ and $\phi_L$ have no effect on this loop due to the presence of the square-law devices.

The loop will act to make the expected value of $E(t)$ go to zero, hence, the steady state bias error $(\bar{\epsilon})$ must satisfy the expression

$$\frac{\bar{\epsilon}}{\Delta} + \frac{\rho^2 - 1}{\rho^2 + 1} \left[ \frac{1}{4} + \frac{\bar{\epsilon}^2}{\Delta^2} + \frac{N_0 B_{IF}}{P_s} \right] = 0. \tag{31}$$

Fig. 8 shows the value of the normalized bias error $(\bar{\epsilon}/\Delta)$ as

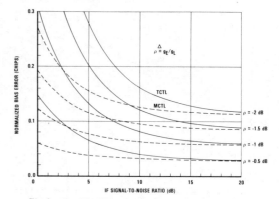

Fig. 8. Tracking bias resulting from arm gain imbalances.

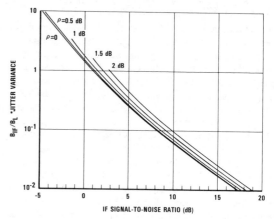

Fig. 9. Tracking jitter for classical square-law loop with arm gain imbalance.

a function of IF signal-to-noise ratio for several values of arm gain imbalance $\rho$. A code tracking loop operating at a 5 dB SNR having $\rho = 0.5$ dB (12 percent arm gain imbalance) would experience a bias error of 0.07 chip, corresponding to a 0.65 dB loss of signal power. Fig. 9 illustrates the effect of gain imbalance on the variance of the tracking jitter as given by (31). Clearly the most significant effect of gain imbalance in a square-law loop is the bias error. Mixers, amplifiers, filters, and square-law devices all contribute to gain variations in this loop, with the major problems being filters (which may be switchable to accommodate varying data rates) and the square-law devices. The biases caused by gain imbalance are so significant that the sum–difference loop of Fig. 1(b) is often employed despite its increased hardware complexity.

The sum–difference loop reduces the gain sensitivity problem by forming the sum (early + late) and difference (early − late) immediately after mixing. Since the error term is formed by multiplication of the sum and difference signals, gain imbalances after the combiner (where the amplifiers and filters are located) do not result in bias errors. The price for achieving this relative insensitivity to gain imbalance is that phase delays become significant. With a sum–difference loop, the error term is given by

$$E(t) = P_s \frac{g_E{}^2 + g_L{}^2}{2} g_\Sigma g_\Delta \cos(\phi_\Sigma - \phi_\Delta)$$

$$\cdot \left\{ \frac{\epsilon}{\Delta} + \frac{\rho^2 - 1}{\rho^2 + 1} \left[ \frac{1}{4} + \frac{\epsilon^2}{\Delta^2} + \frac{N_0 B_{IF}}{P_s} \right] \right.$$

$$\left. + \frac{2\rho}{\rho^2 + 1} \left[ \frac{1}{4} - \frac{\epsilon^2}{\Delta^2} \right] \beta + N_E \right\} \qquad (32)$$

where $\beta \triangleq \sin(\phi_E - \phi_L) \tan(\phi_\Sigma - \phi_\Delta)$. The variance of the tracking jitter is

$$\sigma_{\epsilon/\Delta}{}^2 = \frac{B_L}{B_{IF}} \frac{2(\rho^4 + 1)}{(\rho^2 + 1)^2} \left\{ \frac{1}{2} \left[ 1 + \frac{4(\rho^4 - 1)}{\rho^4 + 1} \frac{\bar{\epsilon}}{\Delta} \right. \right.$$

$$+ \frac{4\bar{\epsilon}^2}{\Delta^2} + \frac{2\rho^2}{\rho^4 + 1} \left( 1 + \frac{4\epsilon^2}{\Delta^2} \right) \tan(\phi_\Sigma - \phi_\Delta)$$

$$+ \frac{\rho(\rho^2 - 1)}{\rho^4 + 1} \left( 1 - \frac{4\bar{\epsilon}^2}{\Delta^2} \right) \beta \right] \frac{N_0 B_{IF}}{P_s}$$

$$+ \left[ 1 + \frac{2\rho^2}{\rho^4 + 1} \tan^2(\phi_\Sigma - \phi_\Delta) \right] \left( \frac{N_0 B_{IF}}{P_s} \right)^2 \right\}. \qquad (33)$$

The tracking bias error is found again by realizing that the expected value of the tracking error will be zero, which gives

$$\frac{\bar{\epsilon}}{\Delta} + \frac{\rho^2 - 1}{\rho^2 + 1} \left[ \frac{1}{4} + \frac{\bar{\epsilon}^2}{\Delta^2} + \frac{N_0 B_{IF}}{P_s} \right] + \frac{2\rho}{\rho^2 + 1} \beta \left[ \frac{1}{4} - \frac{\bar{\epsilon}^2}{\Delta^2} \right]$$

$$= 0. \qquad (34)$$

Some significant points in (34) are that the postcombination gains ($g_\Sigma$ and $g_\Delta$) have no effect on the bias error, but a bias error will exist if $\beta \neq 0$ (i.e., both $\phi_E \neq \phi_L$ and $\phi_\Sigma \neq \phi_\Delta$) even when there is a perfect gain balance. The bias error shown in Fig. 8 also applies to the sum–difference loop for $\beta = 0$; however, it should be remembered that only the mixers and the combiner influence $\rho$ with this loop. Fig. 10 illustrates the bias from phase differences for perfectly balanced gains and for a gain imbalance of 1 dB. For signal-to-noise ratios greater than 5 dB, the biasing effects of gain imbalance and phase imbalance may be considered to be separable. We note that $\beta$ is small even with relatively large phase differences (e.g., for $\phi_\Sigma - \phi_\Delta = \phi_E - \phi_L = 10°$, $\beta = 0.03$); thus, the biasing effect of phase differences in a sum–difference loop should not be overemphasized.

Having results for the TCTL we proceed to examine the sensitivity of the MCTL to gain and phase imbalance. Through similar analysis one obtains the result for the MCTL tracking

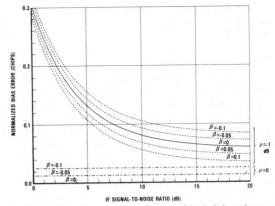

Fig. 10. Tracking bias resulting from phase delay imbalance in reconfigured square-law loop.

error as

$$E(t) = P_s \frac{g_E + g_L}{2} g_0 \left[ 1 - \frac{|\epsilon|}{\Delta} \right] \cos\left( \frac{\phi_E - \phi_L}{2} \right)$$

$$\cdot \cos\left( \phi_0 - \frac{\phi_E + \phi_L}{2} \right) \cdot \left\{ \frac{\epsilon}{\Delta} + \left[ \frac{\rho - 1}{\rho + 1} + \hat{\beta} \right] \right.$$

$$\cdot \left[ \frac{1}{2} + \frac{1}{2\left[ 1 - \frac{|\epsilon|}{\Delta} \right]} \right] \frac{N_0 B_{IF}}{P_s} + \hat{\beta} \frac{\rho - 1}{\rho + 1} \frac{\epsilon}{\Delta} + N_E \right\}$$

$$(35)$$

where

$$\hat{\beta} = \tan\left( \frac{\phi_E - \phi_L}{2} \right) \tan\left( \phi_0 - \frac{\phi_E + \phi_L}{2} \right).$$

The complete expression for $\sigma_{\epsilon/\Delta}{}^2$ is quite involved and has been omitted since it conveys little information pertinent to the present discussion. Solving for the tracking bias error gives

$$\frac{\bar{\epsilon}}{\Delta} + \frac{\rho - (1 - \hat{\beta})/(1 + \hat{\beta})}{\rho + (1 - \hat{\beta})/(1 + \hat{\beta})} \left[ \frac{1}{2} + \frac{1}{2\left[ \frac{1 - |\bar{\epsilon}|}{\Delta} \right]} \right] \frac{N_0 B_{IF}}{P_s} = 0.$$

$$(36)$$

A tracking bias will be present any time $\rho \neq (1 - \hat{\beta})/(1 + \hat{\beta})$, which can occur even with perfect gain balance ($\rho = 1$). As with the sum–difference TCTL, the gain of the on-time or difference channel does not influence the tracking bias, allowing the use of less expensive arm filters and amplifiers.

While it is difficult to present results from (36) in a form that is generally useful, it is possible to make comparisons with the TCTL. If one accepts that bias errors due to gain imbalance are separable from bias errors due to phase imbalance in the region of interest, it becomes possible to make direct comparisons. The dashed curves in Fig. 8 present the bias error for an MCTL if the phase delays were perfectly balanced. At high signal-to-noise ratios this bias is equal to that experienced by the TCTL, but the bias increases more slowly with decreasing SNR for the MCTL. For an SNR of 5 dB, the MCTL experiences only 60 percent of the biasing that a TCTL would experience for a given gain imbalance (given balanced phase delays). It should be recalled that only the mixers and combiner contribute to this imbalance as with the sum–difference loop. When the effects of phase delays are considered (assuming balanced gains) one obtains the result

$$\bar{\epsilon} \approx \frac{(\phi_E - \phi_L)\psi}{4} \tag{37}$$

for both the MCTL and the sum–difference loop in the region where $\sin(x) \approx \tan(x) \approx x$. In (37), $\psi = \phi_\Sigma - \phi_\Delta$ for the sum–difference loop while $\psi = \phi_0 - (\phi_E + \phi_L)/2$ for an MCTL. The conclusion to be drawn is that the MCTL is no more sensitive to gain and phase imbalance than the sum–difference loop (which was configured to minimize these effects) and is considerably less sensitive to gain imbalance at low predetection signal-to-noise ratios.

## IV. SUMMARY

In order to reduce the hardware complexity of a noncoherent delay lock loop in multiple data rate applications, a modified loop has been presented that utilizes the on-time or data channel as the reference. This concept eliminates the need for the traditional loop's sum channel (early signal plus late signal) and, hence, the hardware associated with that channel. This saving may be substantial if the channel were to be optimized for a number of different data rates. With the elimination of an entire IF channel, the MCTL complexity is nearly equivalent to the well-known dithering loop [4], whose primary advantage is its hardware simplicity. However, it was shown in [3] that the MCTL does not suffer the loss in tracking performance (with respect to the TCTL) that the dithering loop experiences. In fact, the MCTL achieves a smaller tracking jitter than the TCTL for equal $S/N$. Upon taking into account the band-limiting effects of the bandpass arm filters, this paper has shown that the MCTL experiences an even greater improvement in tracking performance over the TCTL than predicted in [3]. Furthermore, [3] sighted deficiencies in the acquisition performance of the MCTL; these comparative deficiencies have been lessened when the band-limiting effects are included.

Finally, the noise-free analysis in [3] indicated that the sensitivity of the MCTL to hardware gain and phase imbalances was nearly equivalent to that of a TCTL which had been configured to minimize the resulting tracking bias error.

When noise is included in the analysis, the conclusion of [3] is seen to be true for phase imbalances at all signal-to-noise ratios but is valid only at high SNR's for gain imbalances. At lower IF SNR's, the MCTL has been shown to be significantly less sensitive to gain imbalances resulting in smaller tracking bias errors.

## APPENDIX

The methods used to evaluate the tracking error for the classical square law loop with gain and/or phase errors present are detailed here. Evaluations for the other loops are obtained through more involved but completely similar arguments. The interference is assumed to be Gaussian after despreading.[2]

Using the definitions in (1) but including the gain and phase terms of Fig. 1(a) one obtains the result

$$
\begin{aligned}
E(t) = &\frac{g_E^2}{2} \left[ P_s R_{PN}^2 \left( \epsilon - \frac{\Delta}{2} \right) + 2\sqrt{P_s}\, d(t - \tau) \right. \\
&\left. \cdot R_{PN}\left( \epsilon - \frac{\Delta}{2} \right) N_{CE}(t) + N_{CE}^2(t) + N_{SE}^2(t) \right] \\
&\cdot \frac{g_L^2}{2} \left[ P_s R_{PN}^2 \left( \epsilon + \frac{\Delta}{2} \right) + 2\sqrt{P_s}\, d(t - \tau) \right. \\
&\left. \cdot R_{PN}\left( \epsilon + \frac{\Delta}{2} \right) N_{CL}(t) + N_{CL}^2(t) + N_{SL}^2(T) \right]
\end{aligned} \tag{A1}
$$

where $\epsilon \triangleq \tau - \hat{\tau}$ and the noise components are defined as

$$
N_{CE} \triangleq N_C(t) S_{PN}\left( t - \hat{\tau} + \frac{\Delta}{2} \right) * h_{BPF}(t)
$$

$$
N_{SE} \triangleq N_S(t) S_{PN}\left( t + \hat{\tau} + \frac{\Delta}{2} \right) * h_{BPF}(t) \tag{A2}
$$

and similarly for the late sequence $S_{PN}(t - \hat{\tau} - (\Delta/2))$. The only approximation used in (A1) is that the arm filter's bandwidth is $\ll$ the spread bandwidth so that the PN self-noise may be neglected. Using (3) for the autocorrelation terms results in

$$
E(t) = \frac{g_E^2 + g_L^2}{2} P_s \left\{ \frac{\epsilon}{\Delta} + \alpha \left( \frac{1}{4} + \frac{\epsilon^2}{\Delta^2} \right) + N_A + N_B \right\} \tag{A3}
$$

where

$$
\begin{aligned}
N_A \triangleq &\frac{d(t)}{P_s} \left\{ \left( \frac{1}{2} + \frac{\epsilon}{\Delta} \right) N_{CE} - \left( \frac{1}{2} - \frac{\epsilon}{\Delta} \right) N_{CL} \right. \\
&\left. + \alpha \left[ \left( \frac{1}{2} + \frac{\epsilon}{\Delta} \right) N_{CE} + \left( \frac{1}{2} - \frac{\epsilon}{\Delta} \right) N_{CL} \right] \right\}
\end{aligned}
$$

[2] This is not intended to imply that all interference appears Gaussian after despreading.

$$N_B \triangleq \frac{1}{2P_s} \{ N_{CE}{}^2 - N_{CL}{}^2 + N_{SE}{}^2 - N_{SL}{}^2$$

$$+ \alpha [N_{CE}{}^2 + N_{CL}{}^2 + N_{SE}{}^2 + N_{SL}{}^2] \} \qquad \text{(A4)}$$

$$\alpha \triangleq (g_E{}^2 - g_L{}^2)/(g_E{}^2 + g_L{}^2) = (\rho^2 - 1)/(\rho^2 + 1).$$

The noise term $N_A$ is seen to be zero-mean and Gaussian; however, $N_B$ will be zero-mean only if $g_E = g_L$. Computing power spectral densities for each of these terms results in

$$S_{N_A}(f) = \frac{N_0}{P_s} \left[ (1 + \alpha^2) \left( \frac{1}{4} + \frac{\bar{\epsilon}^2}{\Delta^2} \right) + 2\alpha \frac{\bar{\epsilon}}{\Delta} \right] |H_{\text{BPF}}(f)|^2$$

$$S_{N_B}(f) = \frac{N_0{}^2}{2P_s{}^2} \left[ (1 + \alpha^2) |H_{\text{BPF}}(f)|^2 * |H_{\text{BPF}}(f)|^2 \right.$$

$$\left. + (2\alpha^2 B_{\text{IF}}{}^2) \delta(f) \right] \qquad \text{(A5)}$$

where it has been assumed that the spectrum of the data $d(t)$ is passed through the arm filters without significant attenuation. The term $B_{\text{IF}}$ denotes the noise equivalent bandwidth of the arm filters. Noting that the expected value of $N_B$ is $\alpha N_0 B_{\text{IF}}/P_s$ allows writing (A3) as

$$E(t) = \frac{g_E{}^2 + g_L{}^2}{2} P_s \left\{ \frac{\epsilon}{\Delta} + \alpha \left[ \frac{1}{4} + \frac{\epsilon^2}{\Delta^2} + \frac{N_0 B_{\text{IF}}}{P_s} \right] + N_E \right\}$$

$$\qquad \text{(A6)}$$

where $N_E$ is a zero-mean random variable having

$$S_{N_E}(f) = S_{N_A}(f) + S_{N_B}(f) - \left[ \alpha \frac{N_0 B_{\text{IF}}}{P_s} \right]^2 \delta(f). \qquad \text{(A7)}$$

Assuming that the closed-loop bandwidth is much less than $B_{\text{IF}}$ implies that $S_{N_E}(f)$ is essentially flat over the loop's passband. Thus, for a stationary input, $\epsilon$ is a Gaussian random variable with

$$\sigma_{\epsilon/\Delta}{}^2 = \int_{-\infty}^{\infty} S_{N_E}(f) |H_L(f)|^2 \, df$$

$$= (1 + \alpha^2) \frac{2B_L}{B_{\text{IF}}} \left\{ \left[ \frac{1}{4} + \frac{2\alpha}{1 + \alpha^2} \frac{\bar{\epsilon}}{\Delta} + \frac{\bar{\epsilon}^2}{\Delta^2} \right] \right.$$

$$\left. \cdot \frac{N_0 B_{\text{IF}}}{P_s} + \frac{1}{2} \frac{N_0 B_{\text{IF}}}{P_s} \right\}. \qquad \text{(A8)}$$

In (A8) an ideal bandpass filter was assumed to evaluate the term $|H_{\text{BPF}}(f)|^2 * |H_{\text{BPF}}(f)|^2$ and $H_L(f)$ denotes the closed-loop transfer function.

## REFERENCES

[1] W. J. Gill, "A comparison of binary delay-lock loop implementations," *IEEE Trans. Aerosp. Electron. Syst.*, vol. AES-2, pp. 415–424, July 1966.

[2] M. K. Simon, "Noncoherent pseudonoise code tracking performance of spread spectrum receivers," *IEEE Trans. Commun.*, vol. COM-25, pp. 327–345, Mar. 1977.

[3] R. A. Yost and R. W. Boyd, "A modified PN code tracking loop: Its performance and implementation sensitivities," in *Proc. Nat. Telecommun. Conf.*, Houston, TX, Dec. 1980, pp. 61.5-1–61.5-5.

[4] H. P. Hartmann, "Analysis of a dithering loop for PN code tracking," *IEEE Trans. Aerosp. Electron. Syst.*, vol. AES-10, pp. 2–9, Jan. 1974.

[5] J. J. Spilker, "Delay-lock tracking of binary signals," *IEEE Trans. Space Electron. Telem.*, vol. SET-9, pp. 1–8, Mar. 1963.

[6] M. K. Simon and W. C. Lindsey, "Optimum performance of suppressed carrier receivers with Costas loop tracking," *IEEE Trans. Commun.*, vol. COM-25, pp. 215–227, Feb. 1977.

[7] M. K. Simon, "On the calculation of squaring loss in Costas loops with arbitrary arm filters," *IEEE Trans. Commun.*, vol. COM-26, pp. 181–184, Jan. 1978.

# A Novel Spread-Spectrum Receiver Synchronization Scheme Using a SAW-Tapped Delay Line

PAUL W. BAIER, KLAUS DOSTERT, AND MADHUKAR PANDIT

*Abstract*—In spread-spectrum communication systems, a basic task which has to be performed is the synchronization of the receiver pseudonoise (PN) signal with the pseudonoise signal contained in the input. In this paper, a synchronization system employing a surface acoustic wave tapped delay line (SAW TDL) matched filter for both initial synchronization (acquisition) and tracking is presented. The periodic correlation peaks at the SAW TDL output repeatedly correct the epoch of the local PN signal clock phase and the code generator initial condition to their correct values. As the correlation impulses are distorted and attenuated due to the effects of message modulation, Doppler frequency shifts and unavoidable interferences, a modulation canceller and a differentiating device are employed to improve synchronization performance. Formulas for estimating the performance of a system incorporating these ideas are given. A hardware implementation of the suggested system has been built and tested. Experimental results obtained with the system are presented.

## I. INTRODUCTION

$\mathbf{S}$PREAD-SPECTRUM systems employing a binary pseudo-noise (PN) signal as a subcarrier, find widespread application in the fields of jam-resistant communication and ranging [1]. One of the main tasks to be accomplished at the receiving end of a spread-spectrum system is the synchronization of the PN signal generated locally at the receiver with the PN signal contained in the received signal. Classically, the synchronization is performed in two stages, viz. acquisition or coarse synchronization and tracking or fine synchronization. The advent of surface acoustic wave (SAW) and charge coupled device technology has led to synchronization schemes with fast acquisition characteristics. Such a synchronization system has been described by Hunsinger [2].

In contradistinction to synchronization schemes incorporating separate coarse and fine synchronization subsystems, Baier *et al.* [3] have suggested a synchronization scheme consisting of a circuit for the control of the local PN signal epoch, which performs both the acquisition and tracking. In their scheme the input to the control loop consists of the periodic epoch signals, which are obtained by passing the received (spread) signal through a SAW-tapped delay line with an appropriate impulse response.

The purpose of this paper is to give an alternative synchronization scheme using a SAW-tapped delay line for the combined tasks of acquisition and tracking. In the scheme presented, periodic epoch signals derived from the incoming signal are used to set the epoch of the local PN signal generation.

Manuscript received July 21, 1981; revised January 28, 1982.
The authors are with Fachbereich Elektrotechnik, Universität Kaiserslautern, D6750 Kaiserslautern, West Germany.

tor clock frequency. Implementation of this simple idea, however, involves providing appropriate measures for overcoming phenomena which would otherwise hinder the proper functioning of the synchronization system. Thus, e.g., message modulation could distort the input signal to the matched filter to such a degree that the matched filter impulse response would be inappropriate. Such phenomena and the corresponding measures are also discussed. The synchronization scheme suggested is primarily described for a direct sequencing spread-spectrum system (sometimes also referred to as a PN spread-spectrum system). However, the basic ideas can also be used for setting up a synchronization scheme for frequency hopping systems.

## II. SYNCHRONIZATION SYSTEM BASED ON THE SET-AND-CLOCK PRINCIPLE

### A. Principle of the Synchronization System

Fig. 1 shows the principle of the synchronization system for synchronizing the local PN signal $p_1(t)$ with the PN signal $p(t)$ contained in the received signal

$$s_r(t) = A \cdot p(t) \cdot \cos\left[2\pi f_0 t + \varphi(t)\right] + n(t) \qquad (2.1)$$

with

$p(t)$ = binary PN signal with clock frequency $f_c = \dfrac{1}{T_c}$, period $LT_c$, and with autocorrelation function $R_p(\tau)$, $|R_p(\tau)| \ll R_p(0)$ for $|\tau| > T_c$; typically, $p(t)$ is an $m$-sequence;

$f_0$ = carrier frequency;

$\varphi(t)$ = term containing message modulation generally in the form of angle modulation;

$n(t)$ = noise at receiver input, assumed white Gaussian.

The signal $s_r(t)$ is passed through a surface acoustic wave matched filter (SAW TDL) whose impulse response is $g_1(t)$, with

$$g_1(t) = \begin{cases} p(kT_c - t) \cdot \cos\left(2\pi f_0 t\right), & 0 \leqslant t \leqslant kT_c \\ 0, & \text{otherwise.} \end{cases} \qquad (2.2)$$

The value of $k$ has the same order of magnitude as $L$ and is limited mainly by technological considerations. The output of the matched filter SAW TDL is passed through an envelope

Reprinted from *IEEE Trans. Commun.*, vol. COM-30, pp. 1037–1047, May 1982.

248

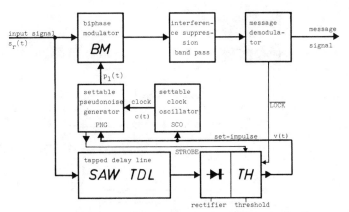

Fig. 1.   Direct sequencing spread-spectrum receiver incorporating the
set-and-clock principle for synchronization.

detector ED to obtain the envelope $u(t)$:

$$u(t) = \text{envelope } \{s_r(t) * g_1(t)\}. \qquad (2.3)$$

The instant $t_s$ at which $u(t)$ crosses the threshold $b$ from below
is detected by TH. The output $v(t)$ of TH is a digital signal
which sets the phase of the clock signal $c(t)$ generated by the
stable clock SCO and the initial condition of the PN code
generator PNG driven by the clock SCO. The clock frequency
of SCO is chosen to be equal to $f_c$ and the connections of PNG
are so chosen that the PN signal at the output of the PNG has
the same code as the code contained in $p(t)$. The value of the
clock phase to which SCO has to be set is given by the clock
phase of $p(t)$ at the instant $t = t_s$. The initial condition to
which the PNG has to be set is given by the epoch of $p(t)$ at
the instant $t = t_s$.

Consider the synchronization system operating under the
idealized conditions $n(t) = 0$ and $\varphi(t) = \text{const.}$ The output
$u(t)$ exhibits main correlation peaks at the instants at which
the epoch of $p(t)$ matches with the impulse response of the
SAW TDL and further partial correlation peaks. The main
correlation peaks are periodic with period $= L \cdot T_c$. By choosing the threshold level $b$ appropriately, it is ensured that only
the main correlation peaks affect the setting of SCO and PNG.
After each setting the PN signal $p_1(t)$ has the same epoch as
$p(t)$ and, furthermore, as, due to clock frequency of SCO $= f_c$,
the relative epoch of $p_1(t)$ and $p(t)$ is constant, synchronism
prevails.

### B. Factors Causing Degradation of the Synchronization Performance and Corresponding Measures

The synchronization system based on Section II-A will function satisfactorily in practice only if the following conditions
are ensured:

1) The clock frequency of $p_1(t)$ must be equal to that of
$p(t)$, namely, $f_c$. This can be achieved by providing stable
clocks of identical frequency at the transmitter and receiver.
However, a Doppler frequency shift of the clock frequency of
$p(t)$ cannot easily be compensated. The difference $\Delta f_c$ between the clock frequencies of $p(t)$ and $p_1(t)$ limits the period

$L \cdot T_c$, which is also equal to the time interval during which
the clocks run free.

2) The setting of the clock phase must be fast and precise.

In practice $n(t)$ is not 0. Furthermore, $\varphi(t)$ varies with time,
as, generally, message modulation is present. These factors in
conjunction with the fact that $\Delta f_c$ is not equal to zero lead to
a degradation of the synchronization performance because

1) not all main correlation peaks are detected, i.e., the
detection probability $P_d < 1$;

2) false set impulses occur, as false alarms are generated
at instants at which $u(t)$ crosses $b$ away from the main peaks;

3) the instants $t_s$ at which $u(t)$ crosses $b$ exhibit random
deviations (jitter) $t_j$ from the instants at which $p(t)$ takes on
the prearranged epoch corresponding to the impulse response
$g_1(t)$ programmed into the SAW TDL; and

4) the information modulation $\varphi(t)$ impairs the satisfactory
operation of the matched filter SAW TDL. The effect of variations of $\varphi(t)$ can be interpreted in terms of the shifting of
the instantaneous carrier frequency [see (2.1)] of the input to
the SAW TDL—as deviations of the center frequency of the
input from $f_0$ lead to a degradation of the matched filter
output, message modulation leads to an attenuation of the
main correlation impulse.

The measures adopted to suppress the deleterious effects
enumerated above are described below with reference to
Fig. 2.

*1) Increasing the Detection Probability $P_d$ While Main-
taining the Probability of False Alarms Low:* If synchronism between $p_1(t)$ and $p(t)$ exists, the approximate time of
arrival of a main correlation peak is known at the receiver.
This knowledge is exploited to generate a strobe impulse
which inhibits the generation of set impulses outside time
intervals beginning at $t_s - T_c$ and ending at $t_s + T_c$, whereby
$t_s$ is the expected instant at which the main correlation peak
crosses $b$. The section of $p(t)$ programmed into SAW TDL is
so chosen that no partial correlation peak of appreciable magnitude occurs in the neighborhood of the main correlation
peak. By employing the strobe signal, the false alarm probability, which is the probability of a false alarm between two
main correlation peaks, is greatly diminished. Now the thresh-

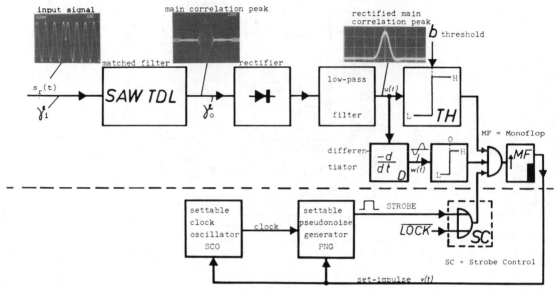

Fig. 2. Generation of the set impulse from the input signal $s_r(t)$.

old can be set lower to obtain a satisfactorily high $P_d$. The strobe signal technique is not without disadvantages—a coarse synchronization has to be performed to approximately align $p_1(t)$ with $p(t)$. This can require complicated acquisition mechanisms.

*2) Reducing the Jitter $\{t_j\}$ of the Set Impulses $v(t)$:* The jitter $\{t_j\}$ can be characterized by its mean deviation $t_{j\text{rms}}$. If the system in Fig. 1 is considered, some thought reveals that a jitter is caused by variations of the input signal amplitude $A$ even without the noise term $n(t)$, and by slowly varying $n(t)$ even with a constant amplitude $A$. Jitter arising under these conditions can be completely eliminated by using the differentiating device $D$ in Fig. 2. The output $w(t)$ of $D$ is the negative time derivative of the input $u(t)$. The instant at which $w(t)$ changes its sign under the condition that at this instant $u(t) > b$ holds is the instant at which the set impulse is generated. The performance of a timing circuit employing differentiation is better, even if the input signal contains additive white noise. Variations of signal amplitude cause practically imperceptible jitter [4].

To fully exploit the possibilities offered by the precise timing signal obtained by the differentiating device and the strobe impulse, it is necessary to employ a highly stable clock with settable phase. One method of realizing a clock fulfilling these requirements is to use a quartz oscillator with frequency $mf_c$ to drive a programmable frequency divider with the dividing ratio $m$ [5]. To set the phase of the clock, the divider is brought to a prescribed state by the set impulse. There is a time uncertainty $(T_c/m)$ in the phase of the clock signal generated by this method. Satisfactory results are obtained for $m > 5$.

*3) Elimination of the Effect of Angle Modulation on the Amplitude of the Main Correlation Peak:* The suppression of the angle modulation is achieved by a modulation canceller which operates by causing an interaction between the message

signal and the modulated carrier $s_r(t)$ [6]. The performance of a SAW TDL for modulated input signals and the working principle of the modulation canceller are dealt with in detail in the next section. At this stage it suffices to note that modulation cancellation is achieved by feedback of the message signal to the modulation canceller MC in Fig. 3.

### III. THE MODULATION CANCELLER

#### A. Necessity of the Modulation Canceller

From the point of view of exact and frequent setting of the clock generator by the correlation impulses, it is necessary to have as large a signal-to-noise ratio (SNR) $\gamma_o$ as possible at the output of the SAW TDL. The SNR improvement $\gamma_o/\gamma_i$ offered by the SAW TDL depends on the length (i.e., number of taps $k$) of the SAW TDL and the ratio of the clock frequency $f_c$ to the bandwidth $B_{RF}'$ of the signal $\cos [2\pi f_0 t + \varphi(t)]$ [see (2.1)] at the receiver.

Fig. 4 shows the SNR improvement as a function of the length $k$ with the normalized frequency deviation $\Delta f/f_c$ as a parameter. $\Delta f$ is the deviation of the center frequency of the received signal $s_r(t)$ from $f_0$. Setting $B_{RF}' \approx 2\Delta f$, the maximum SNR improvement is obtained for a length $k_{\text{opt}}$ approximately equal to $0.7 \cdot f_c/B_{RF}'$ [7].

There is no SNR improvement if $k$ is increased beyond this value. An improvement of SNR can then be achieved only if $B_{RF}'$ is reduced. A reduction of $B_{RF}'$ can be effected by cancelling the modulation contained in the signal $\cos [2\pi f_0 t + \varphi(t)]$.

A device performing this, a modulation canceller, would also eliminate the effects of variations of the center frequency caused by oscillator drift and Doppler shift. The center frequency $f_L$ of the signal input to the SAW TDL would then be determined at the receiver itself.

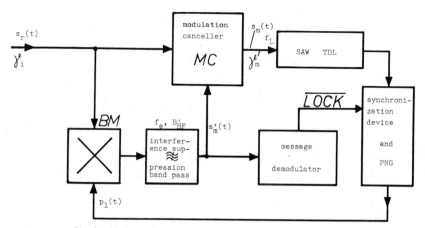

Fig. 3. Modulation canceller and set-and-clock system for a direct sequencing spread-spectrum receiver.

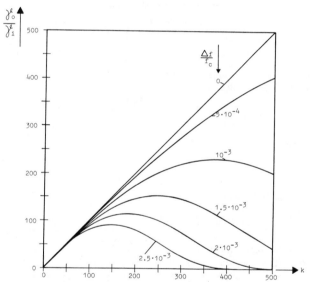

Fig. 4. The SNR gain improvement $\gamma_o/\gamma_i$ of a SAW TDL as a function of the length $k$ with $\Delta f/f_c$ as a parameter.

## B. Principle of the Modulation Canceller

Fig. 5 shows the principle structure of the modulation canceller in the context of a spread-spectrum receiver. Let the input signal to the receiver be

$$s_r(t) = A \cdot p(t) \cdot \cos\left[2\pi f_e t + \varphi(t)\right]. \qquad (3.1)$$

The mixers $M1$ and $M2$ are assumed to be ideal mixers with multiplicative factors 1. The constant delay time $\tau_1$ equalizes the delay times of $s_r(t)$ and $s_m{}''(t)$ and need not be further considered here.

The output of the interference suppression bandpass is the spectrally compressed signal

$$s_m{}'(t) = A_2 \cdot \cos\left[2\pi f_e t + \varphi(t)\right]. \qquad (3.2)$$

In the mixer $M1$, the signals $s_m{}'(t)$ and

$$s_L(t) = A_L \cdot \cos\left(2\pi f_L t\right) \qquad (3.3)$$

are mixed to obtain the output

$$s_m{}''(t) = A_2 A_L \tfrac{1}{2} \{\cos\left[2\pi(f_e + f_L)t + \varphi(t)\right]$$
$$+ \cos\left[2\pi(f_e - f_L)t + \varphi(t)\right]\}. \qquad (3.4)$$

By mixing $s_m{}''(t)$ with $s_r(t)$ in the mixer $M2$, the signal

$$s_m{}'''(t) = A \cdot A_2 \cdot A_L \cdot \tfrac{1}{4} p(t)$$
$$\cdot \{\cos\left[2\pi(2f_e + f_L)t + 2\varphi(t)\right]$$
$$+ \cos\left[2\pi(2f_e - f_L)t + 2\varphi(t)\right]$$
$$+ \cos\left(2\pi f_L t\right) + \cos\left(-2\pi f_L t\right)\}. \qquad (3.5)$$

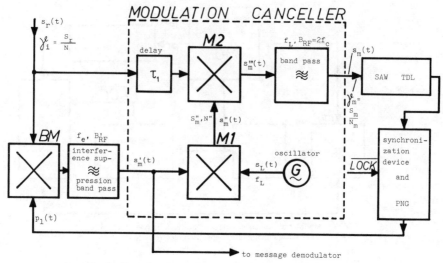

Fig. 5. Example of a modulation canceller for a direct sequencing spread-spectrum receiver; $M1$ and $M2$ may both be balanced mixers as well as *BM*.

is obtained. Filtering $s_m'''(t)$ by means of a bandpass with the center frequency $f_L$ and bandwidth $2 \cdot f_c$, one obtains

$$s_m(t) = A \cdot A_2 \cdot A_L \cdot \tfrac{1}{2} \cdot p_b(t) \cos(2\pi f_L t). \qquad (3.6)$$

$p_b(t)$ is a band-limited version of $p(t)$.

It may be noted that

1) $s_m(t)$ contains the incoming spreading function modulation but not the message modulation $\varphi(t)$;

2) the center frequency $f_L$ of $s_m(t)$ can be precisely set at the receiver corresponding to the needs of the center frequency of the impulse response programmed into the SAW TDL; and

3) the permissible deviation of the center frequency of the received signal $s_r(t)$ is dictated by the tolerance on the interference suppression bandpass and not that on the SAW TDL—this is a great advantage, as the tolerance on the bandpass is not as stringent as that on the SAW TDL.

The performance of the modulation canceller for the case in which the input signal $s_r(t)$ contains a noise component and when synchronism is not achieved is studied in the next section.

## IV. SYSTEM PERFORMANCE

The synchronization system performance can be specified under the following aspects:

1) synchronizing error in the synchronized state, and

2) mean time of acquisition of synchronization.

The synchronizing error, which is the relative shift between the PN signals $p_1(t)$ and $p(t)$, is a random process $\tau(t)$, $-T_c < \tau < T_c$.

A measure for the synchronizing error, which will be used in the following, is given by the degradation of the signal power at the output of the interference suppression bandpass. The relationship between the normalized signal power $K(\tau)$

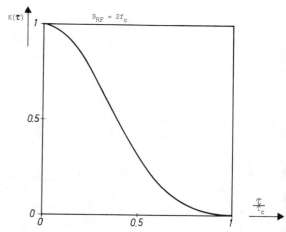

Fig. 6. The normalized signal power $K(\tau)$ at the output of the interference suppression bandpass as a function of the synchronization error $\tau$.

and the epoch error $\tau$ is given by [7] (see Appendix A)

$$K(\tau) = \frac{\left[ \displaystyle\int_{-h}^{h} \text{sinc}^2(\pi x) \cdot \cos\left(2\pi \frac{\tau}{T_c} x\right) dx \right]^2}{\left[ \displaystyle\int_{-h}^{h} \text{sinc}^2(\pi x)\, dx \right]^2}$$

with

$$h = B_{RF}/(2f_c). \qquad (4.1)$$

The run of this function is shown for $h = 1$ in Fig. 6. The mean signal power $\overline{S}$ is then

$$\overline{S} = \int_{-T_c/2}^{T_c/2} K(\tau) \cdot p_\tau(\tau)\, d\tau \qquad (4.2)$$

with

$p_\tau(\tau)$ = probability density function (pdf) of $\{\tau\}$.

In the following $\bar{S}$ is approximated by

$$\bar{S} = K\,(\xi_\tau) \qquad (4.3)$$

where $\xi_\tau$ represents the square root of the second moment $E\{\tau^2\}$ of $\{\tau\}$.

The mean acquisition time $T_{MA}$ is the expected value of the time taken by the system to achieve lock.

In the following, the effects of synchronization error and the mean acquisition time $T_{MA}$ are dealt with for a system without a modulation canceller and one with a modulation canceller separately.

### A. Synchronizing System Without Modulation Canceller

For narrow-band FM systems, the length of the SAW TDL being limited by technological considerations, a modulation canceller is not called for. Thus for a 3 kHz message bandwidth and $f_c = 10$ MHz, a 255 tap SAW TDL can be employed with the main correlation peak being reduced only by less than 10 percent (see Fig. 4). The SNR $\gamma_o$ at the output of the SAW TDL is now constant for a constant input SNR $\gamma_i$ at the input. This fact simplifies the calculation of $\bar{S}$ for a system without modulation canceller.

*1) Synchronizing Error:* In the synchronized state, the strobe impulses black out false alarms in the intervals away from the main correlation peaks and consequently the relative epoch which is also equal to the synchronizing error of $p_1(t)$ with respect to $p(t)$ varies over such an interval by $\Delta f_c \cdot L \cdot T_c$ chips, whereby $\Delta f_c$ is the difference of the clock frequencies of $p(t)$ and $p_1(t)$. Assuming that the relative epoch is approximated by its value midway between the peaks, the synchronizing error $\{\tau\}$ in any interval between two neighboring peaks is given by

$$\{\tau\} = \{\tau_s\} + \{\tau_j\} + \{\tau_f\}, \qquad (4.4)$$

with

$\{\tau_s\}$ = error caused by the clock phase setting;

$\{\tau_j\}$ = error caused by variations of the instant at which the set impulse is generated; and

$\{\tau_f\}$ = error caused by the difference of the clock frequencies of $p(t)$ and $p_1(t)$.

The quantities $\{\tau_s\}$, $\{\tau_j\}$, and $\{\tau_f\}$ may be looked upon to be statistically independent of one another and their variances $\sigma_s^2$, $\sigma_j^2$, and $\sigma_f^2$ can be added to obtain the variance $\sigma_\tau^2$. Furthermore, $E\{\tau\} = E\{\tau_s\} + E\{\tau_j\} + E\{\tau_f\}$, and $\xi_\tau$ in (4.3) is given by $\sqrt{\sigma_\tau^2 + (E\{\tau\})^2}$.

The probability density function (pdf) $p_s(\tau)$ due to setting error is represented by

$$p_s(\tau) = \begin{cases} \dfrac{1}{(T_c/m)}, & -T_c/(2m) \leqslant \tau \leqslant T_c/(2m) \\[2mm] 0, & \text{otherwise.} \end{cases} \qquad (4.5)$$

$T_c/m$ is the maximum time interval which elapses between the occurrence of the set impulse and the setting of the clock phase. The variance $\sigma_s^2$ is $(T_c/m)^2/12$ and $E\{\tau_s\} = 0$.

The pdf of $\{\tau_j\}$ is assumed to be Gaussian with zero mean and variance $\sigma_j^2$, i.e.,

$$p_j(\tau) = \exp\,[-\tau^2/(2\sigma_j^2)]/(\sqrt{2\pi}\sigma_j). \qquad (4.6)$$

For a given synchronization system, $\sigma_j^2$ is determined by the detection circuit and the SNR $\gamma_o$ at the output of the SAW TDL. A formula for calculating $\sigma_j^2$ is given in [4].

The synchronizing error due to the frequency difference $\Delta f_c$ of the clock frequencies increases by $\Delta f_c \cdot L \cdot T_c$ between two neighboring intervals separated by a correlation peak if the correlation peak in question is not detected. The probability of detection $P_d$ of a correlation peak is given by

$$P_d = Q\left(\sqrt{2 \cdot k}\,\gamma,\ b \cdot \frac{\sqrt{2 \cdot \gamma}}{A \cdot \sqrt{k}}\right). \qquad (4.7)$$

$Q(\alpha, \beta)$ is the Marcum $Q$-function, and $\gamma$ is the SNR at the input of the SAW TDL.

Assumed that $L \cdot T_c$ is not too large, the pdf of $\{\tau_f\}$ can be shown to be

$$p_f(\tau) = \sum_{\nu=0}^{\infty} P_d(1 - P_d)^\nu$$

$$\cdot\, \delta\!\left[\tau - (\nu + 0.5) \cdot \frac{\Delta f_c}{f_c} \cdot L \cdot T_c\right]. \qquad (4.8)$$

Corresponding to this pdf, the second moment of $\{\tau_f\}$ is

$$\xi_f^2 = \left[\frac{\Delta f_c}{f_c}\right]^2 \cdot (L \cdot T_c)^2 \cdot \left[\frac{2}{P_d^2} - \frac{2}{P_d} + 0.25\right]. $$

As $\{\tau_s\}$ and $\{\tau_j\}$ are zero mean variables, the second moment of $\{\tau\}$ is

$$\xi_\tau^2 = \sigma_s^2 + \sigma_j^2 + \xi_f^2, \qquad (4.9)$$

using which $\bar{S}$ can be calculated from (4.3).

*2) Acquisition Performance:* During acquisition, it is advantageous to put the strobe mechanism out of action and let all the peaks crossing the threshold $b$ trigger set impulses. The strobe mechanism is switched on as soon as the power at the output of the interference suppression bandpass rises above a predetermined value. If the rise time of the interference bandpass is denoted by $T_r$, false alarms in the time intervals beginning after a delay of $T_r$ after the main correlation peak and ending just before the succeeding correlation peak have no effect on the acquisition time. Only if a false alarm occurs in the interval $(t_\nu, t_\nu + T_r)$, whereby $t_\nu$ is the instant at which the $\nu$th correlation peak occurs, is the acquisition prolonged by a period $L \cdot T_c$, even if the $\nu$th peak is detected. Using a procedure similar to the one adopted in [8], it can be shown

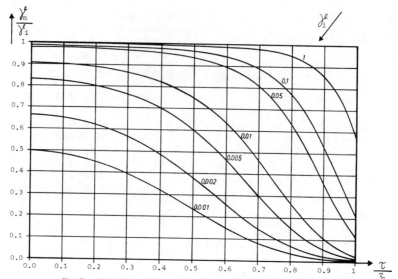

Fig. 7. The ratio $\gamma_m/\gamma_i$ caused by the modulation canceller as a function of the synchronism error $\tau$ with $\gamma_i$ as parameter.

that the mean time for acquisition is (see Appendix B)

$$T_{MA} = \frac{L}{f_c} \left[ \frac{1}{P_d(1-P_f)} - \frac{1}{2} \right], \qquad (4.10)$$

with

$P_f$ = probability of at least one false alarm in an interval of length $T_r$.

The probability $P_f$ can be expressed in terms of the false alarm rate $n_{fa}$, which is the number of upward crossings per second executed by $u(t)$ away from the main correlation peaks by means of the equation

$$P_f = 1 - \exp(-n_{fa}T_r). \qquad (4.11)$$

Generally, the product $n_{fa} \cdot T_r$ is so small that $P_f$ can be approximately set to zero. Then the mean acquisition time is given by

$$T_{MA} = \frac{L}{f_c} \left[ \frac{1}{P_d} - \frac{1}{2} \right]. \qquad (4.12)$$

*B. Synchronization System With Modulation Canceller*

*1) System in the Synchronized State:* For transmission of wide-band FM or frequency shift keying (FSK) at high data rates the use of a modulation canceller is imperative. The modulation canceller efficiency which can be expressed by the ratio $\gamma_m/\gamma_i$ depends on the synchronizing error $\tau$. In Appendix D it is shown that this relationship is given by

$$\frac{\gamma_m}{\gamma_i} = \frac{1}{1 + \dfrac{1}{p \cdot K(\tau)} \cdot \left(1 + \dfrac{1}{\gamma_i}\right)} \qquad (4.13)$$

with $p = B_{RF}/B_{RF}'$, i.e., processing gain.

Fig. 7 shows the run of $\gamma_m/\gamma_i$ for the case of $p = 1000$ and $B_{RF} = 2 \cdot f_c$. In this figure, $\gamma_i$ is the parameter. For an input SNR $\gamma_i > 0.05$ and a synchronizing error $\tau < T_c/2$, the modulation canceller efficiency is practically 100 percent and $\gamma_m \approx \gamma_i$. This means that the receiver operates satisfactorily. For low SNR ratios at the receiver input, say $\gamma_i < 0.01$, the SNR $\gamma_m$ at the input of the SAW TDL is worsened, i.e., $\gamma_m < \gamma_i$, even if the synchronizing error is zero. This lowers the detection probability $P_d$ of the main correlation peaks and increases the jitter of the set impulses. As a result, $\gamma_m$ is further worsened and so on until the chain process ends in a loss of synchronization after $n$ periods of the PN signal. The probability of synchronization loss $P_{s1}$ can be calculated approximately by assuming the initial synchronization error at $t = 0$ to be $\sigma_\tau$. Consider now the $n$th main correlation peak at the instant $n \cdot L \cdot T_c$. The probability that all the $n$ foregoing peaks have not been detected is

$$P_{s1} = \prod_{\nu=1}^{n} (1 - P_{d\nu}), \qquad (4.14)$$

whereby $P_{d\nu}$, the detection probability of the $\nu$th correlation peak, is a function of $\gamma_m(\tau_\nu)$. The synchronization error in the neighborhood of the $\nu$th main correlation peak is

$$\tau_\nu = \sigma_\tau + \frac{\Delta f_c}{f_c} \cdot \nu \cdot L \cdot T_c \qquad (4.15)$$

because the foregoing $\nu$ correlation peaks have not been detected. From (4.15), $\tau_1, \tau_2, \cdots, \tau_n$ are calculated, whereby $n$ is the smallest integer for which $\tau_n > T_c$ holds. Using these values, the SNR $\gamma_m(\tau_\nu)$ and the detection probability $P_{d\nu}$ of the $\nu$th correlation peak are calculated from (4.13) and (4.7), respectively. Finally, the probability of synchronization loss is obtained from (4.14).

*2) Acquisition Performance:* The modulation canceller presents certain disadvantages under the aspect of initial synchronization. Eventually the possibility of providing for message modulation suppression at the transmitter has to be envisaged. It can be shown that the modulation canceller increases the mean acquisition time by a factor $L$. For the case of a circuit with strobe impulses one has then (see Appendix C)

$$T_{MA} = \frac{L^2}{f_c}\left(\frac{1}{P_d} - \frac{1}{2}\right). \tag{4.16}$$

## V. EXPERIMENTAL RESULTS

The principles described in Sections II and III have been implemented in a prototype spread-spectrum transmission system. The details of the prototype system are as follows.

*Spread-Spectrum Modulation*

| | |
|---|---|
| type of modulation: | direct sequencing |
| subcarrier: | binary PN signal with the clock frequency $f_c = 10$ MHz and the period $L = 1278$. |

*Message Modulation*

| | |
|---|---|
| type of modulation: | FM/FSK |
| bandwidth/bit rate: | 3 kHz/200 kbits/s |
| center frequency: | 75 MHz/53, 6 MHz. |

*Receiver Synchronization*

| | |
|---|---|
| type: | set-and-clock |
| SAW TDL: | 255 taps, center frequency 75 MHz |
| modulation canceller: | biphase modulators and adjustable oscillator (PLL) as in Fig. 10. |

Fig. 8 (full line) shows the output SNR $a_b$ in dB at the output of the interference suppression bandpass as a function of the input SNR $a_i$ in dB, with no message modulation present. As the interference suppression bandpass employed has a bandwidth of 80 kHz, the processing gain $p$ is ideally 24 dB (dotted line in Fig. 8). Down to about $a_i = -10$ dB, the processing gain is practically 24 dB. If $a_i$ is reduced further, the full line deviates from the dotted line because of synchronization errors.

The bit error rate $P_e$ for binary FSK modulation, as a function of the input SNR $a_i$ in dB, is shown in Fig. 9. In this case the interference suppression bandpass has a bandwidth of 1 MHz. Now the modulation canceller has to be employed. In Fig. 9 $P_e$ for the two cases 1) frequency deviation $\Delta F_d = 40$ kHz and 2) frequency deviation $\Delta F_d = 45$ kHz is shown as a function of $a_i$. The dotted lines have been determined by calculation [7] using the considerations in Section IV.

Fig. 10 shows the experimental setup including the spread-spectrum receiver for digital message transmission. The module $E0$ contains the modulation canceller in which the center frequency $f_L$ of the SAW TDL can be exactly set. $E1$ contains a bandpass with the center frequency $f_L$ and the bandwidth $B_{RF} = 2f_c$ and a regulated amplifier. The modules $E2$, $E3$, and $E4$ incorporate the functional blocks which are shown in

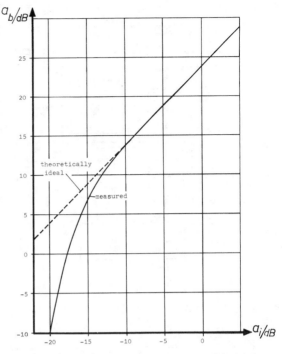

Fig. 8. The signal-to-noise ratio $a_b$ at the output of the interference suppression bandpass as a function of the signal-to-noise ratio $a_i$ at the spread-spectrum receiver input.

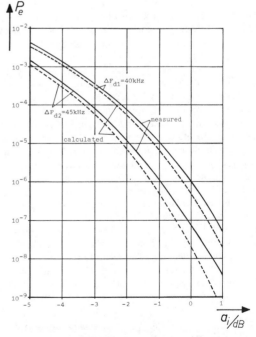

Fig. 9. The bit error probability $P_e$ as a function of the signal-to-noise ratio $a_i$ at a bit rate of 200 kbits/s.

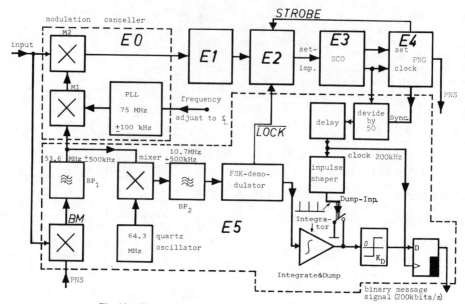

Fig. 10. The functional blocks of the experimental spread-spectrum
receiver for digital signals.

Fig. 2, i.e., the synchronization device and the PNG. The message channel in Fig. 10 is self-explanatory.

## VI. CONCLUSIONS

A synchronization system has been presented for synchronizing the PN signal generator of a spread-spectrum receiver. The presented system uses a surface acoustic wave matched filter for marking the epoch of the received PN signal and a stable settable clock oscillator. A modulation canceller to extend the range of application of the SAW TDL, a strobe impulse mechanism for suppressing the effects of false alarms and a differentiating circuit to improve the timing accuracy of the SAW TDL are necessary for implementing a synchronization system on this principle. The synchronization improvement which these devices offer has been studied. The experimental results indicate that it is indeed feasible to obtain a good synchronizing system based on these ideas. If acquisition is performed using a modulation canceller and a difference frequency procedure, the mean acquisition time is proportional to $L^2$. This result was confirmed qualitatively by experiment. In practical implementation it is necessary that $L$ should not be too large. However, the modulation canceller should be put out of action and the message modulation suppressed at the transmitter during acquisition. Then the mean acquisition time is proportional to $L$. This seems to be a concept which is practicable. Furthermore, the use of the modulation canceller calls for a sufficiently large SNR $\gamma_m$ at the input of the SAW TDL. At low values of the SNR $\gamma_i$, the SNR $\gamma_m$ is worsened over proportionally, as can be inferred from (4.13). This can lead to a chain reaction ending in sync loss. By the setup described in Section V, the critical value of $a_i = 10 \log \gamma_i$ was found to be approximately $-8$ dB.

## APPENDIX A

### CALCULATION OF THE DEGRADATION $K(\tau)$ IN (4.1)

The amplitude of the desired signal at the output of the interference suppression bandpass is determined by the autocorrelation function $R_p(\tau)$ of the PN signal $p(t)$ contained in $s_r(t)$. A PN signal which is not band limited is made up of individual rectangular impulses of the form rect $(t/T_c)$. For a good PN signal with a large period $L \cdot T_c$, the autocorrelation function normalized with respect to its maximum value is

$$R_p'(\tau) = \Lambda\left(\frac{\tau}{T_c}\right)$$

$$= \frac{1}{T_c}\left[\text{rect}\left(\frac{\tau}{T_c}\right) * \text{rect}\left(\frac{\tau}{T_c}\right)\right]. \tag{A.1}$$

If the PN signal is ideally band limited with passband $= B_{RF}$, a single impulse is given by

$$\text{rect}\left(\frac{\tau}{T_c}\right) * B_{RF} \text{ sinc}(\pi B_{RF}\tau)$$

with

$$\text{sinc}(x) = \frac{\sin(x)}{x}.$$

The normalized autocorrelation function $R_{pb}(\tau)$ of the band-limited PN signal is then

$$R_{pb}(\tau) = \frac{1}{T_c}\left[\text{rect}\left(\frac{\tau}{T_c}\right) * B_{RF} \cdot \text{sinc}\left(\pi B_{RF}\tau\right)\right]$$

$$* \left[\text{rect}\left(\frac{\tau}{T_c}\right) * B_{RF} \cdot \text{sinc}\left(\pi B_{RF}\tau\right)\right] \qquad \text{(A.2)}$$

and the Fourier transform of $R_{pb}(\tau)$ is

$$\Phi_{ppb}(f) = T_c \cdot \text{sinc}^2\left(\pi f T_c\right) \cdot \text{rect}\left(\frac{f}{B_{RF}}\right). \qquad \text{(A.3)}$$

It is simpler to calculate $R_{pb}(\tau)$ by first determining $\Phi_{ppb}(f)$ from (A.3) and finding its inverse Fourier transform. Thus,

$$R_{pb}(\tau) = \int_{-\frac{B_{RF}}{2}}^{\frac{B_{RF}}{2}} T_c \cdot \text{sinc}^2\left(\pi f T_c\right) \cdot \cos\left(2\pi f \tau\right) df$$

$$\text{(A.4)}$$

or, with $x = f \cdot T_c$ and $h = B_{RF} \cdot T_c/2$,

$$R_{pb}(\tau) = \int_{-h}^{h} \text{sinc}^2\left(\pi x\right) \cdot \cos\left(2\pi \frac{\tau}{T_c} x\right) dx. \qquad \text{(A.5)}$$

The function $R_{pb}(\tau)$ also represents the normalized amplitude at the output of the interference suppression bandpass. Thereby, $\tau$ denotes the epoch error. The power of the desired signal is then proportional to $R_{pb}{}^2(\tau)$.

By normalizing the power of the desired signal with respect to the power at epoch error $\tau = 0$, one has (4.1)

$$\frac{R_{pb}{}^2(\tau)}{R_{pb}{}^2(0)} = K(\tau)$$

$$= \frac{\left[\int_{-h}^{h} \text{sinc}^2\left(\pi x\right) \cdot \cos\left(2\pi \tau x/T_c\right) dx\right]^2}{\left[\int_{-h}^{h} \text{sinc}^2\left(\pi x\right) dx\right]^2}.$$

$$\text{(A.6)}$$

## APPENDIX B

### DERIVATION OF AN EXPRESSION FOR THE MEAN ACQUISITION TIME $T_{MA}$ FOR ACQUISITION WITHOUT STROBE IMPULSES

Let $P_d$ denote the detection probability of a correlation peak and $P_f$ denote the probability for a false alarm in an interval $(t_\nu, t_\nu + T_r)$. The probability $P_{d1}$ that the first correlation peak leads to lock is $P_d \cdot (1 - P_f)$. The corresponding mean acquisition time $T_{MA1}$ is half the interval length between two correlation peaks, i.e., $T_{MA1} = L/(2 \cdot f_c)$. The

probability $P_{d2}$ that the second correlation peak leads to lock is $(1 - P_{d1}) \cdot P_d (1 - P_f) = P_{d1} \cdot (1 - P_{d1})$ and the corresponding acquisition time is $L/(2f_c) + L/f_c = 3 \cdot L/(2f_c)$. Similarly, the probability that the $\nu$th peak achieves lock is given by $(1 - P_{d1})^{\nu-1} \cdot P_{d1}$ and the corresponding acquisition time is $(2\nu - 1) \cdot L/(2f_c)$. Then, the mean acquisition time $T_{MA}$ is given by

$$T_{MA} = \sum_{\nu=1}^{\infty} P_{d1} \cdot (1 - P_{d1})^{\nu-1} \cdot (2\nu - 1) \cdot L/(2f_c).$$

$$\text{(B.1)}$$

Simplification of the above equation yields

$$T_{MA} = \frac{L}{f_c}\left[\frac{1}{P_d(1 - P_f)} - \frac{1}{2}\right]. \qquad \text{(4.10)}$$

## APPENDIX C

### AN APPROXIMATE VALUE OF THE MEAN ACQUISITION TIME FOR ACQUISITION WITH STROBE IMPULSES

When the strobe mechanism is active, false alarms are suppressed. However, acquisition is achieved only if the correlation peak falls within the slot of interval $(t_s - T_c), (t_s + T_c)$ generated by the local PN signal clock. A frequency offset $\Delta f_c$ between the incoming PN signal clock and the local PN signal clock causes the epoch of the incoming PN signal to slide at a rate of $\Delta f_c$ chips per second with respect to the slot.

As on an average the epoch has to slide through $L/2$ chips before the epoch corresponding to the SAW TDL section occurs, the mean acquisition time would be $L/(2 \cdot \Delta f_c)$ if the first correlation peak achieves lock.

The frequency offset $\Delta f_c$ has to be so chosen that on the one hand $\Delta f_c$ is as large as possible, and on the other, the epoch slide over a period $L \cdot T_c$ of the PN signal is not larger than half the slot width $2 \cdot T_c$. Then, $\Delta f_{c\max} = 1/(L \cdot T_c)$ and the mean acquisition time if the first correlation peak achieves lock is

$$T_{MA1} = \frac{L}{2 \cdot \Delta f_{c\max}} = \frac{L^2}{2 \cdot f_c}. \qquad \text{(C.1)}$$

With $P_d$ = detection probability of a peak, the mean acquisition time is

$$T_{MA} = P_d \cdot \frac{L^2}{2 \cdot f_c} + (1 - P_d)P_d \cdot \frac{3}{2} \cdot \frac{L^2}{f_c} + \cdots, \qquad \text{(C.2)}$$

i.e.,

$$T_{MA} = \frac{L^2}{f_c} \cdot \left(\frac{1}{P_d} - \frac{1}{2}\right). \qquad \text{(4.16)}$$

257

## APPENDIX D

### AN ESTIMATE OF THE EFFICIENCY $\gamma_m/\gamma_i$ OF THE MODULATION CANCELLER

Setting out from the block diagram in Fig. 5, the degradation of the SNR $\gamma_m$ at the input of the SAW TDL with respect to the SNR $\gamma_i$ at the receiver input will be estimated in the following.

The received signal $s_r(t)$ in (2.1) has the signal power $S_r$ and is superimposed with noise $n(t)$ of power $N$. Assume that the noise power spectrum is flat over $(f_0 - B_{RF}/2, f_0 + B_{RF}/2)$. The power $S_m''$ of the spectrally compressed signal $s_m''(t)$ decreases with increasing epoch error $\tau$ by a factor $K(\tau)$ as given by (4.1). The power of $s_m''(t)$ is approximately

$$S_m'' = k_a \cdot K(\tau) \cdot S_r \tag{D.1}$$

whereby $k_a$ represents a constant.

The received noise power $N$ is reduced to the value

$$N'' = \frac{k_a \cdot N}{B_{RF}} \cdot B_{RF}' = k_a \cdot \frac{N}{p} \tag{D.2}$$

due to the action of the interference suppression bandpass (see, e.g., [1]). The signal power $S_m$ and the noise power $N_m$ at the output of the modulation canceller are formed by considering the products of signal and noise components at the input of the multiplier $M2$ and are given by

$$S_m = k_b \cdot S_r \cdot S_m'' = k_a \cdot k_b \cdot K(\tau) \cdot S_r^2 \tag{D.3}$$

whereby $k_b$ represents a constant, and

$$N_m = k_b \cdot (N \cdot S_m'' + N'' \cdot S_r + N \cdot N''). \tag{D.4}$$

From (D.1), (D.2), and (D.4)

$$N_m = k_a k_b \cdot \left[ K(\tau) \cdot S_r \cdot N + \frac{S_r \cdot N}{p} + \frac{N^2}{p} \right] \tag{D.5}$$

follows. The SNR at the SAW TDL input is obtained from (D.3) and (D.5) as

$$\gamma_m = \frac{S_m}{N_m} = \frac{K(\tau) \cdot S_r^2}{K(\tau) \cdot S_r \cdot N + \frac{S_r \cdot N}{p} + \frac{N^2}{p}}. \tag{D.6}$$

Finally, substituting $S_r/N = \gamma_i$, the modulation canceller efficiency turns out to be

$$\frac{\gamma_m}{\gamma_i} = \frac{1}{1 + \frac{1}{p \cdot K(\tau)} \cdot (1 + 1/\gamma_i)}. \tag{4.13}$$

## ACKNOWLEDGMENT

The authors are indebted to Dipl.-Ing. H. Grammüller and his colleagues at Siemens AG, Munich, Germany, for their helpful discussions and for providing SAW-tapped delay lines.

## REFERENCES

[1] C. R. Cahn *et al.*, *AGARD Lecture Series No. 58—Spread Spectrum Communications*, 1973.

[2] B. J. Hunsinger, "Surface acoustic wave devices and applications—3. Spread spectrum processors," *Ultrasonics*, pp. 254–262, Nov. 1973.

[3] W. P. Baier, H. Grammüller, and M. Pandit, "Combined acquisition and fine synchronization system for spread spectrum receivers using a tapped delay line correlator," in *Proc. AGARD Conf.*, 1977, no. 230.

[4] K. Dostert and M. Pandit, "Performance of a SAW tapped delay line in a synchronizing circuit," *IEEE Trans. Commun.*, vol. COM-30, pp. 219–222, Jan. 1982.

[5] P. W. Baier, K. Dostert, and M. Pandit, West German Patent Applications 3020 481.0 and 3020 463.8, 1980.

[6] P. W. Baier, K. Dostert, M. Pandit, and R. Simons, West German Patent 28 54832, 1978.

[7] K. Dostert, "Ein neues Spread-Spectrum Empfängerkonzept auf der Basis angezapfter Verzögerungsleitungen für akustische Oberflächenwellen," Ph.D. dissertation, Kaiserslautern Univ., Kaiserslautern, Germany, 1980.

[8] M. Pandit, "The mean acquisition time of active and passive correlation acquisition systems for spread spectrum communication systems," *Proc. Inst. Elec. Eng.*, Part F, 1981.

# A Simple Modulator for Sinusoidal Frequency Shift Keying

RICHARD LaROSA, SENIOR MEMBER, IEEE, THOMAS J. MARYNOWSKI, MEMBER, IEEE, AND KENNETH J. HENRICH

*Abstract*—**This paper describes a method of generating a constant-envelope sinusoidal-frequency-shift-keyed spread-spectrum signal from a sequence of phase coded rectangular pulses. In-phase and quadrature pulse streams are combined and passed through a SAW filter. The constant time delay property of the SAW filter is used to transform the rectangular pulses into the proper shape. Time and frequency domain outputs are shown for a modulator operating at 70 MHz and a chip rate of 5 Mchips/s.**

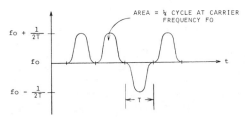

Fig. 1.   SFSK instantaneous frequency versus time.

## INTRODUCTION

IN a spread-spectrum communication system employing sinusoidal frequency shift keying (SFSK), one information sample (digital or analog) is transmitted by a coded sequence of overlapping pulses called "chips." These chips occur at a rate $1/T$ and they all have the same carrier frequency $f_0$. This chip envelope is also characterized by having a continuous rate of change which makes the use of SFSK advantageous in some communication systems since this produces a sharp spectral sidelobe rolloff [1].

The SFSK waveform can be described by its instantaneous frequency versus time function which is shown in Fig. 1. The frequency function consists of a sequence of positive or negative excursions about $f_0$ of duration $T$ and area equal to 1/4 cycle. The polarity of successive excursions is determined by the coding of the sequence. The name "sinusoidal frequency shift keying" is derived from the fact that the rate of change of frequency during each one of these excursions is sinusoidal. The quarter cycle area of each excursion means that successive chips in the sequence are in quadrature. The peak of each chip occurs when $f = f_0$ at which time only one chip exists. At all other times, there are two overlapping chips whose $f_0$ carrier phases are in quadrature.

One possible method of generating an SFSK sequence is to apply a keyed sine wave to a linear integrator whose output is fed into a voltage-controlled oscillator [2]. This method, however, has tolerance problems which makes implementing it a relatively difficult and costly task.

Another possible method of producing an SFSK sequence is shown in Fig. 2 where a carrier source is split into two quadrature voltages, and then the two carriers are amplitude modulated with the exact SFSK envelope shape, which is drawn to scale in Fig. 3. Odd pulses would be applied to one modulator and even pulses would be applied to the other modulator. When the two modulator outputs are combined, the output is the desired constant envelope waveform frequency modulated as in Fig. 1.

Manuscript received July 1, 1981; revised October 29, 1981.
The authors are with the Hazeltine Corporation, Greenlawn, NY 11740.

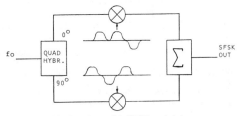

Fig. 2.   A possible SFSK modulator.

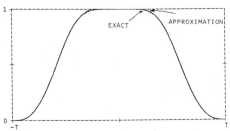

Fig. 3.   SFSK chip envelope.

The problem with the Fig. 2 scheme is that most modulators have a nonlinear output amplitude versus modulation control signal characteristic. The exact modulation transfer function must be known and the modulation waveform must be shaped to compensate for the nonlinearity. This difficult design problem is compounded by temperature drift, aging, and component tolerances.

A third approach is shown in Fig. 4, which differs from Fig. 2 in that bipolar rectangular pulses are applied to the modulator and a SAW filter is used to shape the leading and trailing edges of each rectangular RF pulse to produce the required envelope shape. The advantage of this scheme is that it is much easier to modulate with rectangular pulses because they have only two properties: duration and amplitude. Their duration can be accurately controlled, and the modulator output amplitude is not a strong function of modulator control voltage for sufficiently large control voltages:

Reprinted from *IEEE Trans. Commun.*, vol. COM-30, pp. 1052–1056, May 1982.

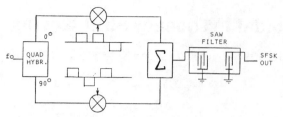

Fig. 4.   Gated carrier SFSK modulator.

It would at first seem that the rounding of the rectangular pulses could be accomplished with a bandpass filter using coils and capacitors rather than a SAW filter. However, the constant time delay (linear phase) property of the SAW filter is essential. This ensures that the resulting pulse envelope is symmetrical, i.e., the leading edge is the mirror image of the trailing edge. A filter made from circuit elements would require phase equalization to achieve this property.

We have found that for the SFSK and similar waveforms, the circuit of Fig. 4 gives satisfactory results, and in the following, its design and experimental results are discussed.

## DESIGN PARAMETERS FOR SFSK MODULATOR

The instantaneous frequency $f(t)$ during a phase transition excursion of Fig. 1 is

$$f(t) = f_0 \pm \frac{1}{4T} \left(1 - \cos\frac{2\pi t}{T}\right) \qquad 0 \leqslant t \leqslant T. \tag{1}$$

The time origin is taken at the start of a phase transition. The instantaneous phase $\phi(t)$ is obtained by integrating (1):

$$\phi(t) = \phi(0) + 2\pi f_0 t \pm \left(\frac{\pi}{2}\frac{t}{T} - \frac{1}{4}\sin\frac{2\pi t}{T}\right). \tag{2}$$

The complex vector whose horizontal projection is the constant envelope output function is

$$\exp j(\phi(0) + 2\pi f_0 t) \left\{\cos\left(\frac{\pi}{2}\frac{t}{T} - \frac{1}{4}\sin\frac{2\pi t}{T}\right)\right.$$

$$\left. \pm j \sin\left(\frac{\pi}{2}\frac{t}{T} - \frac{1}{4}\sin\frac{2\pi t}{T}\right)\right\}. \tag{3}$$

The cosine term in (3) is the amplitude of the chip which is centered at $t = 0$, while the sine term represents the amplitude of the following chip, which starts at $t = 0$. Fig. 3 shows the cosine term.

Examination of Fig. 3 reveals that the leading and trailing edges closely resemble segments of parabolas. This suggests that the pulse can be closely approximated by convolving three rectangular envelope pulses which all have the same carrier frequency $f_0$. This is illustrated in Fig. 5. One of these is the rectangular envelope out of the modulator, while the other two are the respective impulse responses of the two SAW transducers.

The longest pulse has a duration $T_1$ equal to the desired envelope duration at the half-amplitude points. The inter-

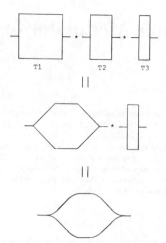

Fig. 5.   Convolution of three rectangular pulses.

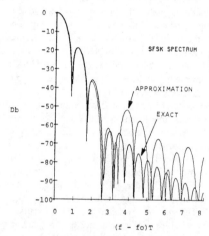

Fig. 6.   SFSK pulse spectra.

mediate length pulse of duration $T_2$ controls the slope at the half-amplitude points, while the shortest pulse duration $T_3$ controls the rounding of the corners. The pulse duration $T_1$, $T_2$, $T_3$ can therefore be chosen by examination of the SFSK envelope.

A more elegant method of choosing parameters is to examine the spectrum of an individual SFSK pulse which is reproduced in Fig. 6 using an FFT algorithm. The first three zeros of this spectrum are approximately equally spaced in frequency by $0.87/T$. This suggest a sinc $(X)$ spectrum corresponding to a rectangular pulse of duration $T_1 = 1.15T$. The third zero of Fig. 6 is wider than the others, suggesting a double zero. This would correspond to a sinc $(X)$ spectrum whose first null is at $2.61/T$. This indicates that $T_2 = 0.38T$. The fourth null in Fig. 6 is located at $3.22/T$, which would correspond to first null of the spectrum of a rectangular pulse of duration $T_3 = 0.31T$.

After convolving these three rectangles together and then comparing the obtained SFSK approximation envelope to the actual envelope, it was found that the rectangles have to be

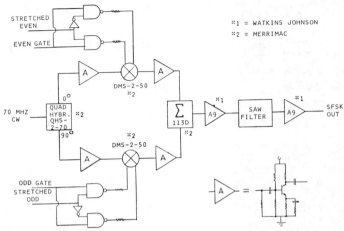

Fig. 7. SFSK prototype circuit.

lengthened slightly to obtain the best approximation. The duration of the rectangular pulses for the best envelope approximation turns out to be $1.17T$, $0.39T$, and $0.32T$, respectively. The approximation to the SFSK envelope is shown in Fig. 3 where it is seen to be in excellent agreement with actual SFSK pulse.

The spectrum of the convolution of the three rectangles is shown in Fig. 6. It can be seen that the approximation is good out to the fourth null. The actual SFSK spectrum is 70 dB below the peak at $f - f_0 = 4/T$, whereas the approximation is 55 dB below the peak at this point. The finite bandwidth of amplifiers and coupling circuits in the modulator system can be expected to modify this response. Furthermore, system engineers will want to limit the spectrum to a width much less than $8/T$ in order to avoid interference with adjacent channels. This means that the actual transmitted waveform will depart slightly from the SFSK frequency pattern of Fig. 1, and also will not be exactly a constant envelope waveform.

In the foregoing narrow-band approximation, the spectrum multiplication or time convolution result is independent of which rectangular pulse is used as the input to the SAW filter. In systems where the subcarrier $f_0$ is accurately generated, it seems preferable to use the longest pulse as the input to the SAW filter. This uses the greatest number of cycles of the accurately generated $f_0$ and ensures the greatest accuracy of the output frequency. This would also seem to ensure the greatest accuracy of the quadrature relationship between successive chips. Effects of turn-on and turn-off transients and dissimilarities between the two mixers are minimized by this choice. This was the configuration implemented in the experimental program.

## EXPERIMENTAL RESULTS

An SFSK modulator was constructed and tested using standard off-the-shelf devices. The prototype was implemented at a 70 MHz carrier frequency and a chip rate of 5 Mbits/s. The actual SAW filter used was fabricated on a lithium niobate substrate and had two unapodized transducers; one had 11 split fingers and the other had 9 split fingers. This configuration resulted in $T_2$ and $T_3$ being equal to $0.375T$ and $0.300T$, respectively, based on measured null locations of the filter frequency transfer characteristic. Due to $T_2$ and $T_3$ being shorter than the required durations of $0.39T$ and $0.32T$, the pulse duration $T_1$ had to be shortened to $1.09T$ to compensate for a nonconstant envelope output which resulted from the decrease in the durations of $T_2$ and $T_3$. The shortening of the three pulses causes the resulting waveform to be no longer an exact SFSK approximation, but rather a similar member of the family of constant envelope waveforms.

The prototype circuit is shown in Fig. 7. The 70 MHz source is split into quadrature signals and applied to both mixers through amplifiers which provide both isolation and gain. The isolation prevents the impedance mismatch of the mixer from unbalancing the quadrature hybrid and the gain offsets circuit losses to provide a net 0 dB gain. The mixers chosen allow differential drive capability, and therefore can be driven from standard TTL. The buffered mixer outputs are summed in phase before being operated on by the SAW filter. SAW drive and detect functions are performed by using standard 50 Ω amplifiers.

Operation of the modulator requires a 70 MHz CW signal and correctly timed modulator signals. The modulator drive signals are derived by pulling the even and odd bits from the PN sequence and stretching them to two bit periods. Next, using an AND logic circuit, the stretched bits are gated with a precise timing pulse. The result is differentially applied to the mixer to provide phased bursts of 70 MHz CW. The even output will either be $0°$ or $180°$ out of phase with respect to the input. The odd output will be either $90°$ or $270°$ out of phase. The sequence of pulses entering the SAW filter contains rectangular bumps in its envelope because the individual pulses overlap. Becuase the carriers of the overlapping pulses are in quadrature, the bumps are $\sqrt{2}$ times the amplitude of an individual pulse. These combined overlapping pulses are then applied to the SAW filter which shapes the pulses into SFSK chips. The output of the SAW filter is a constant amplitude SFSK coded sequence.

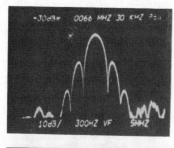

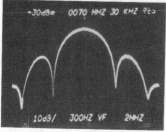

Fig. 8.   Output spectrum of experimental SFSK modulator.

Photographs of experimental results (Fig. 8) indicate close agreement to theoretical predictions (Fig. 6). From these data, Table I is derived.

The results indicate excellent agreement for the first and second sidelobes. Higher order sidelobes have large deviation from the predicted values. This is due to both the approximation used and bulk wave effects in the SAW filter. However, this discrepancy has a minor effect since the higher order lobes are filtered out prior to transmission.

## CONCLUSION

We have shown a simple method of generating members of a class of constant envelope waveforms using a SAW filter to shape the coded output chips. The results of the experimental system constructed showed generally good agreement with theoretical predictions.

In addition to generating the SFSK waveform, it is possible to use this scheme to generate other members of the constant amplitude waveform family. This can be accomplished by simply changing the modulating pulse duration and modifying the impulse response of the shaping filter, with the latter being relatively easy to implement using SAW technology.

### TABLE I
### COMPARISON OF RESULTS

|  | THEORETICAL dB | EXPERIMENT dB |
|---|---|---|
| Main Lobe | 0 | 0 |
| 1st Side Lobe | −19.2 | −17 |
| 2nd Side Lobe | −35.2 | −34 |
| 3rd Side Lobe | −64.2 | −53 |

## REFERENCES

[1]   M. K. Simon, "A generalization of minimum-shift-keying (MSK) type signal based upon input data symbol pulse shaping," *IEEE Trans. Commun.*, vol. COM-24, pp. 845–856, Aug. 1976.
[2]   F. Amoroso, "Pulse and spectrum manipulation in the minimum (frequency) shift keying (MSK) format," *IEEE Trans. Commun.*, vol. COM-24, pp. 381–384, Mar. 1976.

# Randomization Effect of Errors by Means of Frequency-Hopping Techniques in a Fading Channel

MITSUHIKO MIZUNO, MEMBER, IEEE

*Abstract*—This paper describes the effective randomization of errors by means of frequency hopping (FH) in a selective Rayleigh fading channel.

The major advantage of the randomization is the option to use random error-correcting codes in burst error channels, including the fading channel. Bit interleaving, which operates in the time domain, is very useful to randomize the errors, but equivalent operation is possible by FH in the frequency domain. That is based on low correlation between frequency chips, which is realized by hopping over a sufficiently wide bandwidth in selective fading channels.

In this paper, bit error rates of FH with majority decision are analyzed to show the effect of randomization in various spread bandwidths. It is also shown that FH and interleaving are equivalent in the sense that spread bandwidth corresponds to interleaving degree with respect to its randomization effect.

## I. INTRODUCTION

ERROR reduction is important for digital mobile radio in a fading channel to secure high reliability of communications. One useful method is diversity reception, and another is the use of error-correcting codes. In wide-band communication systems, such as spread-spectrum (SS) systems, some specific fading reduction techniques are available if sufficiently wide bandwidth is used. For example, a frequency-hopping (FH) system used in selective fading channels is capable of reducing the correlation between the adjacent frequency chips. Therefore, the burst errors caused by slow fading are randomized by means of the FH, and the application of random error-correcting codes becomes effective. Bit interleaving, which operates in the time domain, is very useful, but the FH will provide equivalent performance in the frequency domain.

Randomization by FH is also effective against CW or partial band interference as well as fading. This effect is described in many papers as an interference rejection capability, so that, in this paper, the effect against fading alone is discussed.

The effect of randomization depends on the spread bandwidth and the frequency correlation characteristics of fading channels, and the burst errors will remain unrandomized if the spread bandwidth is not sufficiently wide. But the error performance of FH in fading channels has been analyzed up to now on the assumption of independent errors [1], and it is not clear how wide a bandwidth is necessary for this assumption, and how much degradation is to be expected if the errors

Manuscript received July 1, 1981; revised February 8, 1982.

The author is with the Communication System and Apparatus Division, Radio Research Laboratories, Ministry of Posts and Telecommunications, Koganei-shi, Tokyo, 184, Japan.

considered are not independent. This paper describes the effect of randomization on the FH performance when majority decision between frequency chips is used, and then shows that the FH spread bandwidth and the degree of interleaving are exchangeable with each other with respect to obtaining the randomization.

## II. PERFORMANCE OF MAJORITY DECISION LOGIC IN A FADING CHANNEL

### A. Modeling Assumptions

For the analysis of burst error characteristics, one useful method is the use of Markov chains [2]. However, in this paper, the bit error performance of majority decision in the correlated Rayleigh fading environment is described without referring to the burst error characteristics, such as burst error length. The following assumptions are applied.

1) Rayleigh fading.
2) Slow fading.
3) No intersymbol interference due to multipath delay spread.
4) Gaussian noise. No pulsed interference from other users.
5) No diversity combining.

### B. Bit Error Rate with Independent Symbol Errors

The error rate of binary noncoherent FSK in additive white Gaussian noise is given by [3]

$$p_s = p_s(\gamma) = \tfrac{1}{2} e^{-\frac{\gamma}{2}} \qquad (2.1)$$

where $\gamma$ is the SNR per symbol. The information bit error rate $P_b$ with two out of three majority decision is given as

$$P_b = 3p_s^2(1-p_s) + p_s^3 = 3p_s^2 - 2p_s^3. \qquad (2.2)$$

Furthermore, in a Rayleigh fading environment, if the symbol errors are considered to be independent by means of wide spreading or sufficient interleaving, the symbol (chip) error rate is written as

$$\overline{p_s} = \int_0^\infty p_s(\gamma)p(\gamma)\,d\gamma = \frac{1}{2+\gamma_0} \qquad (2.3)$$

where

$$p(\gamma) = \frac{1}{\gamma_0} e^{-\frac{\gamma}{\gamma_0}} \qquad (2.4)$$

Reprinted from *IEEE Trans. Commun.*, vol. COM-30, pp. 1048–1051, May 1982.

is the probability density function of $\gamma$ and $\gamma_0$ is the mean SNR in the fading. In this case, the bit error rate is obtained by replacing the $p_s$ in (2.2) by the $\overline{p}_s$ in (2.3), so that $P_b$ is given by

$$P_b = 3\overline{p}_s{}^2 - 2\overline{p}_s{}^3 = 3\left(\frac{1}{2+\gamma_0}\right)^2 - 2\left(\frac{1}{2+\gamma_0}\right)^3. \quad (2.5)$$

### C. Bit Error Rate in a Very Slow Fading Channel

Another case is when the signal level is kept constant during three symbol durations. In this case, the information bit error rate with majority decision (2.2) should be averaged over the fading, so that

$$P_b = \int_0^\infty \{3p_s{}^2(\gamma) - 2p_s{}^3(\gamma)\}p(\gamma)\,d\gamma$$

$$= \frac{1}{4}\left(\frac{3}{1+\gamma_0} - \frac{1}{1+3/2\gamma_0}\right). \quad (2.6)$$

The error rate of (2.6) is much greater than the random error case of (2.5).

### D. Bit Error Rate in General Case

In general, neither of the preceding cases may be applicable to the exact analysis of error performance. In such a case, $P_b$ is obtained by the following calculation:

$$P_b = \int_0^\infty \int_0^\infty \int_0^\infty \{p_s(\gamma_1)p_s(\gamma_2) + p_s(\gamma_2)p_s(\gamma_3)$$

$$+ p_s(\gamma_3)p_s(\gamma_1) - 2p_s(\gamma_1)p_s(\gamma_2)p_s(\gamma_3)\}$$

$$\cdot p(\gamma_1, \gamma_2, \gamma_3)\,d\gamma_1\,d\gamma_2\,d\gamma_3 \quad (2.7)$$

where $\gamma_1$, $\gamma_2$, $\gamma_3$ are the SNR for each symbol and $p(\gamma_1, \gamma_2, \gamma_3)$ is the joint pdf in the fading channel. In (2.7), the integral contains the terms of the form $p_s(\gamma_i)p_s(\gamma_j)$ and $p_s(\gamma_1)p_s(\gamma_2)p_s(\gamma_3)$.

The integral of the term of $p_s(\gamma_1)p_s(\gamma_2)p_s(\gamma_3)$ is calculated as follows. Let $\Gamma$ be the sum of the SNR of each chip. That is,

$$\Gamma = \gamma_1 + \gamma_2 + \gamma_3. \quad (2.8)$$

Using (2.1) and (2.8), the integral becomes

$$\overline{p_s(\gamma_1)p_s(\gamma_2)p_s(\gamma_3)}$$

$$= \int_0^\infty \int_0^\infty \int_0^\infty p_s(\gamma_1)p_s(\gamma_2)p_s(\gamma_3)$$

$$\cdot p(\gamma_1, \gamma_2, \gamma_3)\,d\gamma_1\,d\gamma_2\,d\gamma_3$$

$$= \frac{1}{8}\int_0^\infty \int_0^\infty \int_0^\infty \exp\left(-\frac{\gamma_1 + \gamma_2 + \gamma_3}{2}\right)$$

$$\cdot p(\gamma_1, \gamma_2, \gamma_3)\,d\gamma_1\,d\gamma_2\,d\gamma_3$$

$$= \frac{1}{8}\int_0^\infty \exp\left(-\frac{\Gamma}{2}\right)p(\Gamma)\,d\Gamma \quad (2.9)$$

where $p(\Gamma)$ is the probability density function of $\Gamma$.

Equation (2.9) has the same form as the error rate with maximal ratio diversity combining, in which the combined SNR is the sum of the branch SNR's as in (2.8). Therefore, the analytical method of diversity can be applied to evaluate (2.9). Furthermore, the integral in (2.9) is the Laplace transform of $p(\Gamma)$,

$$G_\Gamma(s) = \int_0^\infty e^{-s\Gamma}p(\Gamma)\,d\Gamma \quad (2.10)$$

in which $s = \frac{1}{2}$. Since $\Gamma$ is expressed as a Hermitian quadratic form shown in (2.8), the Laplace transform $G_\Gamma(s)$ is represented by [3, eq. (10-10-21)] as

$$G_\Gamma(s) = \frac{1}{\det(I + sR)} \quad (2.11)$$

where $I$ is the identity matrix, $R$ is the covariance matrix, which is in this case,

$$R = R_3 = \gamma_0\begin{pmatrix} 1 & \rho_{12} & \rho_{13} \\ \rho_{12}{}^* & 1 & \rho_{23} \\ \rho_{13}{}^* & \rho_{23}{}^* & 1 \end{pmatrix}, \quad (2.12)$$

and where $R_n$ stands for the $n$th order of covariance matrix. Thus, using (2.10) and (2.11), (2.9) becomes

$$\overline{p_s(\gamma_1)p_s(\gamma_2)p_s(\gamma_3)}$$

$$= \frac{1}{8}G_\Gamma(\tfrac{1}{2}) = \frac{1}{8\det(I_3 + R_3/2)}. \quad (2.13)$$

Next, the integral of $p_s(\gamma_i)p_s(\gamma_j)$, which is

$$\overline{p_s(\gamma_i)p_s(\gamma_j)} = \int_0^\infty \int_0^\infty p_s(\gamma_i)p_s(\gamma_j)p(\gamma_i, \gamma_j)\,d\gamma_i\,d\gamma_j,$$

$$(2.14)$$

is obtained in a similar way to (2.13). In this case, the covariance matrix is given as

$$R = R_2 = \gamma_0\begin{pmatrix} 1 & \rho_{ij} \\ \rho_{ij}{}^* & 1 \end{pmatrix}. \quad (2.15)$$

Thus,

$$\overline{p_s(\gamma_i)p_s(\gamma_j)} = \frac{1}{4\det(I_2 + R_2/2)}$$

$$= \frac{1}{\gamma_0{}^2(1 - |\rho_{ij}|^2) + 4\gamma_0 + 4}. \quad (2.16)$$

Consequently, using (2.12), (2.13), (2.15), and (2.16), the bit

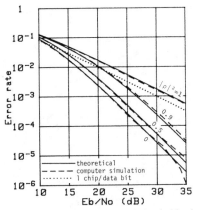

Fig. 1.   Bit error rates of two out of three majority decision in Rayleigh fading channels.

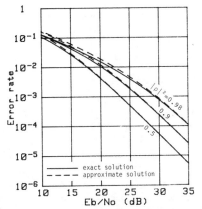

Fig. 2.   Bit error rates of two out of three majority decision, exact and approximate solution.

error rate of (2.7) becomes

with the curve of majority decision at $|\rho|^2 = 1$, the latter is worse than the former. This means that because the division of signal energy of an information bit into many chips causes a reduction of chip SNR and an increase in chip errors, the improvement of bit error rate by the majority decision in the burst error fading channel cannot compensate for the performance degradation due to the division of signal energy.

Fig. 2 shows the comparison of the approximate solution obtained from (2.16), which is applied to the first three terms of (2.17), with the exact solution which is evaluated from all the terms in (2.17). The approximate solution becomes larger than the exact solution when the error rate is large. But the range where $P_b$ is nearly $10^{-3}$ and $|\rho|^2$ is small is of practical interest, and the approximate solution is applicable to that case.

## III. BIT ERROR RATE OF FH AND INTERLEAVING WITH MAJORITY DECISION

### A. Interleaving

The correlation of signal power between the symbols is obtained from the autocorrelation of the fading. In an urban multipath environment, where many waves arrive whose incident directions are uniformly distributed, the autocorrelation is described as [4]

$$\rho(\tau) = J_0(2\pi f_m \tau) \tag{3.1}$$

where $f_m$ is the maximum Doppler frequency, and $J_0(\ )$ is the zero-order Bessel function.

Now, we define the correlation time $T_c$, which is the time separation when the correlation between two samples of the envelope of the received signal becomes 0.5. The correlation given by (3.1) is that of complex Gaussian variates, and the correlation of envelopes of the received signal is described as [5]

$$\rho_e(\tau) \simeq J_0^2(2\pi f_m \tau) = \rho^2(\tau). \tag{3.2}$$

$$P_b = \frac{1}{\gamma_0^2(1 - |\rho_{12}|^2) + 4\gamma_0 + 4} + \frac{1}{\gamma_0^2(1 - |\rho_{23}|^2) + 4\gamma_0 + 4} + \frac{1}{\gamma_0^2(1 - |\rho_{31}|^2) + 4\gamma_0 + 4} - \frac{1}{4}$$

$$\cdot \frac{1}{(1 + \gamma_0/2)^3 - (1 + \gamma_0/2)(\gamma_0/2)^2(|\rho_{12}|^2 + |\rho_{23}|^2 + |\rho_{31}|^2) + 2(\gamma_0/2)^3 \operatorname{Re}(\rho_{12}\rho_{23}\rho_{31})}. \tag{2.17}$$

Setting $\rho_{12} = \rho_{23} = \rho_{31} = 0$, (2.17) reduces to (2.5), which is the random error case, and setting $\rho_{12} = \rho_{23} = \rho_{31} = 1$, (2.17) reduces to (2.6). Hence, (2.17) is the general expression of bit error rate, including (2.5) and (2.6) as special cases.

Fig. 1 shows the bit error rate as a function of $E_b/N_0 = 3\gamma_0$ with various values of $|\rho|^2$, and the dashed curves show the results of a computer simulation, which is carried out by a linear transformation of complex Gaussian variates. The comparison between the theoretical and simulation results indicates good agreement. The dotted curve shows the bit error rate without majority decision, that is, the system of one chip per data bit. In the comparison of this dotted curve

Thus, $T_c$ is given by

$$J_0(2\pi f_m T_c) = 1/\sqrt{2} \tag{3.3}$$

or

$$T_c = \frac{1.1264}{2\pi f_m}. \tag{3.4}$$

Then the correlation between the interleaved symbols becomes

$$\rho_{12} = \rho_{23} = J_0(2\pi f_m i/r_s) = J_0\{1.1264i/(r_s T_c)\} \tag{3.5}$$

$$\rho_{31} = J_0(4\pi f_m i/r_s) = J_0\{2 \cdot 1.1264i/(r_s T_c)\} \tag{3.6}$$

where $r_s$ is the symbol rate and $i$ is the interleaving degree. Thus, $i/r_s$ indicates the time separation between the interleaved chips. Substituting (3.5) and (3.6) into (2.17), the bit error rate of majority decision with interleaved symbols is obtained.

### B. FH of One Data Bit per Frame

Now we choose a simple FH model whose hopping pattern consists of three frequency chips and whose data bit is given by majority decision of those three chips. Furthermore, we assume that the frequencies of the chips are equally separated. In this case, the bit error rate is obtained from (2.18) in which the correlation coefficient is given as the frequency correlation of the fading instead of the time correlation as in the interleaved case. In fact, since these three chips are not transmitted simultaneously, some time differences exist among these chips. Therefore, in some cases, to evaluate the chip correlation, it may be necessary to consider the time correlation as well as the frequency correlation. But if the system is used in a slow fading channel and its hopping rate is sufficiently rapid, i.e., the maximum Doppler frequency is 40 Hz and the hopping rate is more than several khops/s, the time correlation in (3.1) can be approximated by unity, and it is necessary to consider only the frequency correlation.

The frequency correlation function is obtained by [5]

$$\rho(\delta) = \frac{1}{1 + j\delta/B_c} \qquad (3.7)$$

where $\delta$ is the frequency separation and $B_c$ is the coherence bandwidth, which is the frequency separation corresponding to the envelope correlation of 0.5. Let $\Delta$ be the frequency distance between chips. Then the correlation is described as

$$\rho_{12} = \rho_{23} = |\rho(\Delta)| = \frac{1}{\sqrt{1 + (\Delta/B_c)^2}} \qquad (3.8)$$

$$\rho_{31} = |\rho(-2\Delta)| = \frac{1}{\sqrt{1 + (2\Delta/B_c)^2}} \qquad (3.9)$$

where almost the same value of error rate is obtained when the correlation coefficient is calculated directly from (3.7) as a complex value and from (3.8) and (3.9) as its absolute value.

Fig. 3 shows the error rate of FH and interleaving. $\Delta_0 = \Delta/B_c$ is the normalized frequency separation and $i_0 = i/(r_s T_c)$ is the normalized time separation between the symbols, which is the interleaving degree normalized by the chip rate and correlation time. Fig. 4 represents the degradation of gain of majority decision from the case of independent errors. For the case of interleaving, the Bessel functions of (3.5) and (3.6) reduce to zero when $i_0 = 2.4$, and the chip error characteristics are considered to be independent. As $i_0$ further increases, the degradation reduces to zero with a slight oscillation. The degradation becomes less than 1 dB when $\Delta_0 = 1.10$ or $i_0 = 1.02$. For example, let $B_c = 250$ kHz and $T_c = 4.48$ ms ($f_m = 40$ Hz), and let $r_s = 60$ kbits/s (data rate is 20 kbits/s). Then the required frequency separation $\Delta$ of FH is 275 kHz, and the equivalent required interleaving degree $i$ is 275.

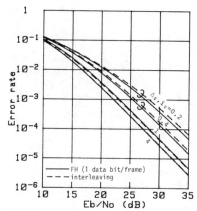

Fig. 3.  Bit error rates of FH and interleaving with majority decision.

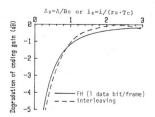

Fig. 4.  Degradation of the gain of majority decision from the case of independent errors.

### IV. PERFORMANCE OF THREE OUT OF FIVE MAJORITY DECISION

A bit error rate of a three out of five majority decision is easily obtained as an extension of the case of a two out of three decision. To simplify the discussion, the correlations between the chips are assumed to equal one another. When errors occur independently, the bit error rate without fading is given by

$$P_b = p_s^5 + 5p_s^4(1 - p_s) + 10p_s^3(1 - p_s)^2$$
$$= 6p_s^5 - 15p_s^4 + 10p_s^3 \qquad (4.1)$$

where $p_s$ is the chip error rate described in (2.1).

In a Rayleigh fading channel, considering the chip correlations, the bit error rate is obtained by the following calculation:

$$P_b = \int_0^\infty \cdots \int_0^\infty \left[ 6 \prod_{i=1}^5 p_s(\gamma_i) - 3 \sum_{i=1}^5 \left\{ \prod_{\substack{j=1 \\ j \neq i}}^5 p_s(\gamma_j) \right\} \right.$$

$$\left. + \sum_{\substack{i,j,k \\ i \neq j \neq k}}^5 p_s(\gamma_i) p_s(\gamma_j) p_s(\gamma_k) \right]$$

$$\cdot p(\gamma_1, \gamma_2, \cdots, \gamma_5) \, d\gamma_1 \, d\gamma_2, \cdots, d\gamma_5 \qquad (4.2)$$

where $\gamma_i (i = 1, 2, \cdots, 5)$ is SNR per each symbol and $p(\gamma_1, \gamma_2, \cdots, \gamma_5)$ is the joint pdf of the fading channel. For simplic-

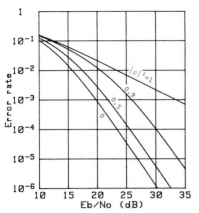

Fig. 5. Bit error rates of three out of five majority decision in Rayleigh fading channels.

ity, let the correlations between the chips be equal. With this assumption, $\gamma_1, \gamma_2, \cdots, \gamma_5$ are interchangeable with each other, and hence, (4.2) reduces to

$$P_b = 6 \int_0^\infty \cdots \int_0^\infty \prod_{i=1}^5 p(\gamma_i) p(\gamma_1, \gamma_2, \cdots, \gamma_5) d\gamma_1 d\gamma_2 \cdots d\gamma_5$$

$$-15 \int_0^\infty \cdots \int_0^\infty \prod_{i=1}^4 p(\gamma_i)$$

$$\cdot p(\gamma_1, \gamma_2, \cdots, \gamma_4) d\gamma_1 d\gamma_2 \cdots d\gamma_4$$

$$+ 10 \int_0^\infty \cdots \int_0^\infty \prod_{i=1}^3 p(\gamma_i)$$

$$\cdot p(\gamma_1, \gamma_2, \gamma_3) d\gamma_1 d\gamma_2 d\gamma_3. \tag{4.3}$$

The integral in (4.3) is calculated in a similar way to obtain (2.13). Thus, (4.3) becomes

$$P_b = \frac{6}{2^5 \det (I_5 + R_5/2)} - \frac{15}{2^4 \det (I_4 + R_4/2)}$$

$$+ \frac{10}{2^3 \det (I_3 + R_3/2)} \tag{4.4}$$

where $R_n$ is the $n$th-order covariance matrix.

The bit error rate calculated from (4.4) is shown in Fig. 5. The figure indicates that if chip correlation is nearly zero, the three out of five decision improves the gain by 3 dB as compared to the two out of three decision at an error rate of $10^{-3}$. But if the chip correlation is nearly unity, the performance of the three out of five decision becomes inferior to the case of the two out of three decision.

## V. CONCLUSION

The improvement caused by a majority decision in FH is described based on the chip correlation due to fading. This result can be expanded to the analysis of block error-correcting codes.

In slow selective fading channels, the effect of randomization of errors is expected by FH as well as by interleaving. Typically, in the majority decision case, several hundred kilohertz of spread bandwidth used with FH give performance equivalent to several hundreds of bits of interleaving. In the comparison of FH with interleaving, however, the following differences are indicated.

1) A large memory is required when bit interleaving is used in slow fading channels, and hence the delay of data bits may not be negligible.

2) When the vehicle stops, the interleaving cannot cause randomization of errors, but the randomization by FH is available if sufficiently wide bandwidth is used.

3) Wide bandwidth is necessary for randomization by FH. Therefore, in practice, the efficient use of the spectrum by the multiple access users is required.

4) In FH, phase correlation as well as amplitude correlation between the chips becomes small if sufficiently randomized. Thus, PSK or DPSK between adjacent chips is not possible. Instead, either DPSK over a frame or noncoherent FSK are the two feasible schemes.

Experiments are now in progress using an FH transmitter and receiver pair and a wide-band fading simulator which is capable of generating the selective Rayleigh fading. Results from these experiments, including comparisons with the analyses described above, will be reported in forthcoming papers.

## ACKNOWLEDGMENT

The author would like to acknowledge the encouragement of S. Miyajima, Director of the Communication System and Apparatus Division, and the valuable discussions with Dr. Y. Kadokawa and other members of the System Performance Research Section, Radio Research Laboratories. He also wishes to thank Dr. F. Ikegami and Dr. S. Yoshida of Kyoto University for their helpful suggestions.

## REFERENCES

[1] D. J. Goodman, P. S. Henry, and V. K. Prabhu, "Frequency-hopped multilevel FSK for mobile radio," *Bell Syst. Tech. J.*, vol. 59, pp. 1257–1275, Sept. 1980.

[2] B. D. Fritchman, "A binary channel characterization using partitioned Markov chains," *IEEE Trans. Inform. Theory*, vol. IT-13, pp. 221–227, Apr. 1967.

[3] M. Schwartz, W. R. Bennett, and S. Stein, *Communication Systems and Techniques*. New York: McGraw-Hill, 1966.

[4] M. J. Gans, "A power-spectral theory of propagation in the mobile-radio environment," *IEEE Trans Vehic. Technol.*, vol. VT-21, pp. 27–38, Feb. 1972.

[5] W. C. Jakes, Jr., *Microwave Mobile Communications*. New York: Wiley, 1974.

# A Spread-Spectrum Simulcast MF Radio Network

ARTHUR A. GIORDANO, MEMBER, IEEE, HENRY A. SUNKENBERG, MEMBER, IEEE, HUGO E. DE PEDRO, MEMBER, IEEE, PETER STYNES, MEMBER, IEEE, DAVID W. BROWN, MEMBER, IEEE, AND SANG C. LEE, MEMBER, IEEE

*Abstract*—This paper describes the design and performance of a spread-spectrum medium frequency (MF) radio network. The network utilizes a simulcast time slotted transmission scheme in which radios that receive a data message in one time slot repeat the message in the next slot. Spread-spectrum radio designs that operate efficiently with simulcast under a variety of channel conditions are investigated. The performance of the radio network is characterized in terms of radio connectivity and message reaction time.

## I. INTRODUCTION

THIS paper describes a spread-spectrum simulcast MF radio network used to provide survivable communications for an MX command, control, and communication ($C^3$) system[1] based in a region of Nevada and Utah. The requirement of *survivability* imposes unusual constraints on the radio network design, i.e.:

1) Use of buried antennas spaced at intervals which permit ground wave communications. These antennas at MF (0.3–3.0 MHz) offer a reasonable trade between performance and complexity.

2) Use of a network architecture that dynamically and rapidly interconnects operating radios when an unknown subset of radios in the network malfunction or are destroyed. Simulcast accomplishes this objective efficiently by automatically reconfiguring a network of the operating radios with simple protocols.

3) Use of a spread-spectrum MF radio design that operates efficiently with the simulcast network and provides reliable communication in the presence of interference, jamming, and atmospheric noise.

The simulcast MF radio network is used to deliver messages with a desired reliability to a specified average fraction of the nodes within a specified interval of time. Several interrelated parameters, system features, and environmental conditions determine whether or not a particular radio network can achieve this required performance. A virtually limitless set of radio parameters, waveform/receiver designs, network architectures, propagation effects, and channel conditions are possible. Analytic determination of radio network perform-

Manuscript received July 6, 1981; revised December 21, 1981.
A. A. Giordano, H. A. Sunkenberg, H. E. dePedro, and P. Stynes are with the Strategic Systems Division, GTE, Westborough, MA 01581.
D. W. Brown and S. C. Lee are with the Computer Sciences Corporation, Falls Church, VA 22046.

[1] The analyses and results provided herein are directed at the multiple protective shelter basing for MX, whereby each MX missile and its associated communications equipment are randomly and deceptively moved between hardened shelters. This basing approach for MX is no longer being considered.

ance is therefore intractable. As a result, a sophisticated radio network simulation was developed to estimate radio network performance by modeling simulcast network configurations in the Nevada/Utah basing. Thus, the results presented apply for a specific radio network configuration. However, a methodology and performance criterion for designing survivable spread-spectrum radio networks is established. Key features of the simulcast network and its performance for several radio waveform designs are subsequently described.

The network utilizes a time slotted simulcast discipline to disseminate messages. This system is a form of time division multiple access (TDMA) within which a radio that receives a message in one time interval repeats the message in the next time interval. Each message then crosses the network in a series of simulcast relay hops. The network is dynamically formed in that communication links are not preestablished but are set up as the transmissions are made. For this reason the network is insensitive to the loss of individual links which may occur due to terminal destruction, malfunction, or unusually poor channel conditions. Thus, for a sufficiently dense original network, simulcast dynamically reconfigures a network of the operating radios and provides great reliability of message throughput. An unfortunate aspect of the simulcast architecture is that multiple radios transmit with unknown phase from unpredictable locations, thereby causing the received signal to fade. An important radio design objective is then to identify a waveform that possesses sufficient diversity to counter the fading imposed by simulcast.

Several MF radio waveform designs have been investigated. They all employ noncoherent frequency-hopped $M$-ary FSK modulation in combination with forward error detection and correction. This class of waveforms provides the required diversity to combat the signal fading caused by simulcast transmission. Compatible coding techniques that have been investigated include dual-$k$, rate $1/v$ convolutional codes concatenated with either BCH codes or nonbinary Reed–Solomon (RS) codes. These waveforms have been investigated in the presence of tone jamming and fullband white Gaussian noise (WGN) channel conditions.

The measures of performance for the simulcast MF radio network are reaction time and communication connectivity. Reaction time is the time duration required to accomplish a message transmission. Communication connectivity is the statistical average fraction of the network that receives a given transmitted message. Connectivity is a function of the number of simulcast hops and its value depends on several radio network parameters. Some of these parameters include

Reprinted from *IEEE Trans. Commun.*, vol. COM-30, pp. 1057–1069, May 1982.

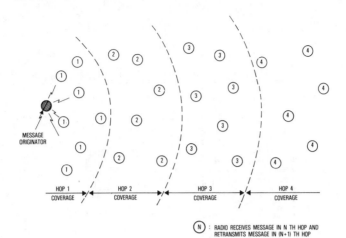

Fig. 1. Simulcast relay.

radio bit and message error rate performance, propagation and channel conditions, number and location of operating network nodes, transmitter power, transmit antenna gains, and data transmission rate. Connectivity performance is estimated by averaging many simulcast transmissions in a Monte Carlo simulation of the network. For these simulations the maximum number of simulcast hops required to span the network is determined. The maximum number of hops can then be utilized in conjunction with a selected network protocol to estimate the reaction time.

Section II of this paper presents an explicit description of simulcast and an overview of the network operation and protocols. Section III provides a system block diagram, candidate waveforms, and a description of the channel model. Section IV presents procedures for estimating network connectivity and reaction time for various channel conditions, system parameters, and radio designs. Detailed descriptions of both the radio and network simulations needed to estimate the connectivity performance are included. Section V presents new results obtained by simulation for bit and message error rates and concludes with examples of connectivity performance. These latter results represent the principal design data incorporated in the conceptual development of the MX MF simulcast radio network.

## II. NETWORK DESCRIPTION

### A. Description of Simulcast

In its simplest form, simulcast propagates messages through the network in a broadcasting mode with radio terminals situated farther and farther away from the originator sequentially participating in the message transmission until they have all received the message. The terminal initially having a message to transmit must first access the network; this is necessary so that other terminals will be receiving the initiator's message when it transmits and no other terminals will be transmitting a different message at the same time. The message may be generated at any time. This results in a variable time delay before the message can be transmitted by the terminal depending on the particular multiple access scheme. Upon transmission, some other terminals in the network may receive the message correctly. The number of terminals in the network receiving the message will depend on several factors: modulation and coding technique used, particular noise conditions during transmission time, the presence of jamming and interference, and the geographical placement of the transmitter and other terminals. All of these factors will determine how wide an area is covered by the first transmission. A terminal that has correctly received the message and validated its authenticity will then immediately retransmit it in the next time interval. Geographically separated transmitters will then send the same message, extending the coverage area and acting as relays for message dissemination.

The simulcast system is a time slotted discipline, within which a radio that receives a message in one time interval repeats it in the next time interval. This time interval is called a simulcast hop and a message that is transmitted across the network in $N$ time intervals is said to require $N$ hops. A pictorial representation of the process is illustrated in Fig. 1. The originator initiates the transmission and subsequent radios repeat the message resulting in a wave-like transmission across the network. Typically, an actual TDMA protocol will limit the number of hops available to cross the network. This limit is chosen so that the selected system will distribute messages to all receivers with high probability.

The main idea of simulcast is that the communication links are not preestablished, but are set up when the transmission is made. For a dense enough network where many terminals exist and for a wide enough transmission range, simulcast provides great reliability of message throughput. The network is insensitive to the loss of individual links which may occur due to terminal destruction or malfunction, or particularly bad channel conditions. This flexibility provides the network with high communications connectivity and low vulnerability to jamming.

When more than one terminal transmits on the same channel, there is mutual signal interference or fading at the receiver due to different propagation distances and phase

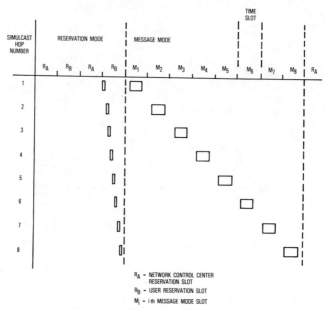

Fig. 2.  Reservation and message mode descriptions.

variations between the transmitters. However, the average power of the received signal is increased over that of an individual steady signal and computer simulations of network performance have illustrated that the simulcast power gain more than offsets the degradation due to fading. Another factor affecting signal propagation is the presence of a jammer or interference. The effect of a jammer will ultimately be to reduce the propagation distance by essentially increasing the channel noise level. Signal processing at the receiver combined with the use of spread-spectrum techniques can significantly reduce the jammer's deleterious effects.

The simulcast network connectivity is very flexible, since links are established during transmission. A disadvantage of this technique is that only one message at a time can exist in the channel being monitored by the receiving nodes. When a node transmits a message, all nodes within its propagation radius must be monitoring that channel in order to receive and repeat the message. Thus, the nodes must operate in a half-duplex mode and the protocol must be such that no messages from other nodes are expected when a node is transmitting.

### B. Network Operation

The radio network utilizes a simulcast transmission scheme in which all radios are assumed to be synchronized. Network operation comprises a reservation mode and a message mode in a time slotted format. It is assumed that there is one control center for the network and that the remaining network nodes can access the network on a reduced availability basis. Specifically, the reservation slots are assigned to the network control center and to the network nodes on an alternating basis as illustrated in Fig. 2. Each reservation slot accommodates a sufficient number of simulcast hops to ensure that the reserva-

tion message is propagated across the network in the duration of a reservation slot. The message mode is initiated at the completion of the reservation mode. For the message mode, a fixed number of simulcast hops (assumed to be eight in Fig. 2) is assigned for a message transmission through the network. At the completion of the message mode, the network reverts to the reservation mode.

The actual number of hops needed to establish connectivity will be determined by the user's position in the network. Those users located in the center of the network will require less hops to reach the entire network and therefore will be assigned a smaller number of hops for message transmission than those users located in the edges of the network. The number of hops selected for simulcast message operation depends on the connectivity exhibited by a specific radio network. This connectivity is a function of the network configuration after an attack, terrain, channel conditions, waveform design, and radio system parameters.

### C. Network Protocols

The network protocols are organized into levels or layers by using a layered architecture approach. This structuring technique separates the network functions into levels and simplifies the protocol design. In order to interconnect the layers, protocols must be defined, which are the set of rules followed within each layer for the interaction of the radio terminals. The layering procedure makes the structure of each layer independent of the structure of higher layers. Each layer serves the layers above it and interacts with other layers by a set of rules called an interface.

Four layers, defined in the radio network, include the application, transport, transmission, and physical layers. The functions performed by each of the layers may be viewed

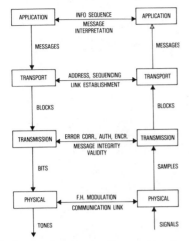

Fig. 3. Overview of protocol layers and interfaces.

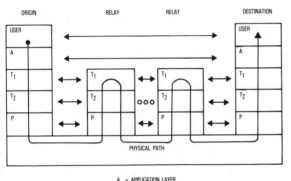

A = APPLICATION LAYER
$T_1$ = TRANSPORT LAYER
$T_2$ = TRANSMISSION LAYER
P = PHYSICAL LAYER

Fig. 4. Simulcast propagation.

as functions performed at different levels of data organization, i.e., messages, blocks, bits, etc. Fig. 3 provides an overview of the protocol layers and interfaces. Each radio terminal will transfer information at each layer. This information will be inserted at the originator and stripped at the corresponding layer of the destination or relay terminals. The application layer will transfer information bits and provide for message interpretation. The transport layer will transfer address information in order to establish a link between the originator and the destination. In addition, sequence numbers for multislot messages will be included. The transmission layer provides for message integrity and validity by the use of error-correcting coding, message authentication, and encryption/decryption. The physical layer will exchange information by using FSK modulation and frequency hopping to provide protection against the jammer.

The implementation of the simulcast scheme in the protocol layer architecture occurs in the transport layer. Fig. 4 shows a message propagating through the network. The message is initiated by the user at the originating node. Each of the protocol layers will process the message and add and remove

control data associated with that layer. The lowest layer, the physical layer, will send the message to other radios through the physical path. The message will be received by other radios which in turn will act as relays. The figure illustrates that a simulcast transmission that is to be relayed by a user and that is not destined to that user will only reach the transport layer where simulcast operations are performed. The application layer at a node is not concerned with the content of messages whose destination is not that node.

## III. RADIO COMMUNICATION SYSTEM

### A. System Block Diagram

In the previous section a description of the MF radio network was provided. This section focuses on the radio communication system utilized in the simulcast network. A simplified block diagram of the transmitter and receiver are illustrated in Fig. 5. Several details of the system are omitted. These details include functions relating to the transmitter and receiver amplifier, front end receiver processing and automatic gain control, synchronization, and message processing. The simplified model is used for communications performance analysis assuming perfect receiver synchronization. Impulsive atmospheric noise is assumed to be converted by nonlinear front end receiver processing to white Gaussian noise with an equivalent received noise power. Even with these simplifying assumptions, receiver error rate performance with various waveforms and channel environments is difficult to obtain analytically. Available analytical results are compared with simulation results in Section V.

The transmitter and receiver functions depicted in Fig. 5 are digital operations associated with a baseband low-pass equivalent model. A data source is assumed to deliver information bits to the transmitter at $R$ bits/s. The information bits are formatted into messages and are message encoded by a rate $R_B$ encoder. Message encoding is required to satisfy requirements for detected, undetected, and missed message probabilities. The message encoded data are next channel encoded by a rate $R_C$ encoder to provide improved performance for the simulcast signal in a variety of channel environments. The encoded data are then modulated by using $M$-ary FSK and frequency hopped over a bandwidth $B$. In $M$-ary FSK the $M = 2^k$ symbols are represented by $M$ different frequencies. In the MF radio design the $M$ different frequencies set aside for each successive $k$-bit symbol are selected pseudorandomly from among the set of tones available for frequency hopping in the assigned bandwidth. These tones are orthogonally spaced at the hopping rate. Pseudorandom spacing of the $M = 2^k$ symbol tones rather than the conceptually simpler block hopping was chosen to maximize the effectiveness of the AJ processing gain against an intelligent jammer. The transmission signaling rate over the channel is $R_0 = R/R_B R_c k$ symbols/s.

The received signal is a sum of up to $P$ transmitted signals, noise, and jamming. For the parameters of interest, limited processing gain $B/R$ is available. Improved performance can then be obtained with several jamming threats by use of frequency domain excision techniques for jammer mitigation.

271

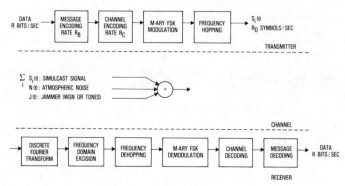

Fig. 5.   System block diagram.

The remaining receiver functions are simply the inverse of the transmitter functions, i.e., frequency dehopping, FSK demodulation, channel decoding, and message decoding. Specific waveforms and appropriate receiver processing will be described next.

### B. Waveform Candidates

The constraints established in the previous section allow selection of a large set of waveforms and an even larger set of waveform parameters. In this section waveforms are presented that have widely varying system performance and implementation complexity. The waveforms are summarized in Table I. The first case uses a rate 0.432, 10 bit error-correcting binary BCH code, no channel encoding, and a binary FSK modulation. Case 2 uses a rate 0.5, five-symbol correcting Reed–Solomon (RS) code with a 64-ary code alphabet, no channel encoding, and 32-ary FSK modulation. Case 3 utilizes the ten-error correcting binary BCH code for message coding, a dual-5 rate 1/5 convolutional code for channel encoding followed by 32-ary FSK modulation. Case 4 uses the five-symbol error-correcting RS code for message encoding, the dual-5 rate 1/5 convolutional code for channel encoding, and 32-ary FSK modulation. For subsequent performance data, a message is assumed to be a single codeword output from the message encoder.

Case 1 was selected since it represents a waveform employing a powerful block code with a simple modulation. Case 2 utilizes an $M$-ary modulation with a nonbinary block code. This waveform type has been successfully implemented in spread-spectrum applications (e.g., TATS [1]). Cases 3 and 4 are concatenated coding schemes where the message encoding is the outer code and the channel coding is the inner code. These waveforms are examples of integrated modulation and coding designs that have exhibited superior performance on fading channels [2]. In all four cases a high-rate message code is selected that provides significant message error protection.

In all cases, the modulation alphabet size is selected to be compatible with the encoder. In cases 2–4, a 32-ary alphabet is chosen since this alphabet provides good performance and permits reasonable implementation complexity. Cases 3 and 4 use dual-$k$ codes that operate at a rate $R_C = 1/v$, where

TABLE I
WAVEFORM CANDIDATES

| Case | Message Encoding | Channel Encoding | FSK Modulation Alphabet Size, $M = 2^k$ |
|---|---|---|---|
| 1 | BCH (111, 48, 10) | None | Binary |
| 2 | RS (20, 10, 5) over GF ($2^6$) | None | 32-ary |
| 3 | BCH (111, 48, 10) | Dual-5 $R_C$=1/5 | 32-ary |
| 4 | RS (20, 10, 5) | Dual-5 $R_C$=1/5 | 32-ary |

$k = v = 5$. For this class of codes, $k$ bits at a time are fed into the encoder and $v$ $k$-bit output symbols are produced. Various combinations of $k$ and $v$ were analyzed relative to the symbol error rate (assuming soft decision Viterbi decoding) under various noise and jamming conditions. Analytical results indicated that performance generally improved with increasing values of $k = 2, 3, 4, 5$—with $v = k$ for each case providing a relatively broad optimization. Implementation complexities limit the practical value of $k$ to 5, resulting in a 32-ary FSK system, i.e., a dual-5 rate 1/5 convolutional code with soft decision Viterbi decoding. In this paper only the above heuristic arguments are invoked to motivate the waveform selection. Further justification is not only beyond the scope of this paper, but also detracts from the central theme provided in Section IV.

### C. Channel Model

The channel is depicted schematically in Fig. 5. In the general case, multiple transmissions are received in the presence of atmospheric noise and jamming. If the received signal arrives from a single dominant transmitter, the received signal is assumed to have been transmitted over a nonfading channel. If multiple transmissions are received, the received signal is assumed to exhibit a fading envelope distribution. The receiver is designed to mitigate the impulsive atmospheric noise by use of an adaptive nonlinear front end clipper adjusted to avoid jammer capture. As a result, the additive atmospheric noise following clipping is modeled as white Gaussian noise (WGN) with an average power corresponding to the worst case summer noise level for the Southwestern U.S. The jammer is assumed to be one of two generic types, i.e., WGN or optimum fraction of the band jamming (OFOBJ). Subsequent connectivity performance then postulates that the additive interference is either WGN consisting of the sum of fullband noise jamming and externally generated atmospheric noise or OFOBJ. In the connectivity performance estimates, the jammer is assumed to be equally effective at each radio in the network. Although this jamming scenario is impossible to attain for realistic threats, a worst case performance estimate results.

In spread-spectrum systems, OFOBJ causes the worst performance that a nonintelligent jammer, such as a repeat back jammer, can achieve. In OFOBJ the jammer's average power is spread over the fraction of the available bandwidth (determined by the received signal-to-jamming ratio) to force the worst error rate performance. Use of this jamming threat in effect converts a WGN channel into a fading channel with a corresponding deleterious effect on error rate performance [7]. To combat this jamming strategy, waveform designs with sufficient diversity (e.g., cases 3 and 4 in Section III-B) are required. For a frequency-hopped FSK system OFOBJ takes the form of tone jamming. Ignoring additive noise, this strategy assumes that the jammer knows the level setting of the frequency domain clipper and spreads his signal power across as many tones as necessary to ensure that the tones are not clipped. That is, the jammer attempts to maintain a signal-to-jammer level per tone of 0 dB. Since the optimum spreading factor is a function of the received signal-to-jammer level which will be unknown to the jammer in the simulcast scheme, the results with jamming represent the best performance that the jammer can expect.

In simulcast operation it was previously pointed out that when multiple transmissions are received, the received signal undergoes fading. A description of the fading process will now be provided. Assume that a given radio receives the same frequency transmission from $P$ transmitters. Because of different transmission losses, the contributions from each source will not be of equal power. Moreover, due to transmitter phase and path length differences, the contributions will not be in phase. The total received signal will be

$$Z = \sum_{k=1}^{P} a_k e^{j\phi_k} \tag{1}$$

where $a_k$ and $\phi_k$ are the envelope and phase, respectively, of the $k$th received signal component. The phases $\phi_k$ are uniformly distributed and independent. By averaging over the phases, we can obtain an estimate of the received signal power, $P_R$, i.e.,

$$P_R = \langle |Z|^2 \rangle = \sum_{k=1}^{P} |a_k|^2. \tag{2}$$

Thus, the fading results from the vector addition of noncoherent signals arriving from diverse transmission sources. Given the immensity of the simulcast network and the variety of

potential geographic configurations, an analytical fading model is desirable. To model the message reception process in the fading channel, two alternate techniques were extensively investigated. These techniques are as follows:

1) a vector addition of the signals at the receiver input and a symbol-by-symbol simulation of the message reception using a nonfading receiver performance curve;

2) simple power addition of the input signals to characterize the mean of the fading signal at the receiver input and use of a fading performance curve for the receiver to determine an average probability of symbol error.

The analytic model selected was a Rayleigh-distributed envelope with $p(r)$, the envelope distribution of the received signal, given by

$$p(r) = \frac{r}{P_R} \exp\left(-r^2/2P_R\right). \tag{3}$$

This model was chosen for two reasons. First, for a sufficiently large number of sources the resultant fading is Rayleigh distributed as a consequence of the central limit theorem. Second, receiver simulation results using multiple tones showed that error rate performance in most instances was upper bounded by Rayleigh fading performance over a wide range of symbol error rates.

Simulcast network simulation runs were then performed using the vector and power addition methods discussed above. The connectivity results for the two methods were found after extensive simulation to be either identical or were more pessimistic using the Rayleigh fading model, particularly in sparsely configured networks. Since the vector addition method was computationally burdensome, the more pessimistic Rayleigh fading model was adopted. Therefore, all connectivity performance results can be interpreted as worst case estimates.

## IV. RADIO NETWORK MODEL

### A. Network Performance Methodology

In this system performance is only partially assessed in terms of reliable bit and/or message error rates. Network connectivity and reaction time are more important performance criteria. Since these criteria are an atypical measure of network performance [3], it is important to discuss the methodology associated with estimating performance.

Fig. 6 and Table II present the performance estimation methodology and performance parameters, respectively. In Fig. 6 the left side illustrates conventional communication performance estimation. Bit and message error rate performance are computed for a selected waveform, e.g., cases 1–4 above and specified channel conditions. On the right-hand side of the figure, radio system parameters such as transmit power, antenna gain, data rate, etc., are selected. The radio network configuration is then specified by fixing the percentage of radios that are to be active during a particular transmission. This configuration can be chosen either by a random geographic selection of $L$ radios from a set of $U$ possible users representing a random destruction of nodes in the network, or

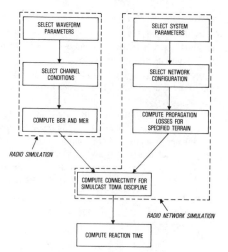

Fig. 6.  Methodology of performance estimation.

TABLE II
PERFORMANCE PARAMETERS

WAVEFORM PARAMETERS

    Modulation type and alphabet size - 32-ary FSK, BFSK.

    Message Encoding - BCH (111, 48, 10), RS (20, 10, 5)

    Channel Encoding - Dual 5 Rate 1/5, none

CHANNEL PARAMETERS

    WGN Jamming

    OFOBJ

    Non-Fading

    Rayleigh Fading

RADIO SYSTEM PARAMETERS

    Transmit Power

    Transmit Antenna Gain

    Data Rate

    Receiver Bandwidth

RADIO NETWORK PARAMETERS

    Terrain Elevation and Roughness Data

    Soil Conductivity for Mountains and Valleys

    Number and Location of Active Radios

    Type of Network Distribution - Random, Swath

by selection of $L$ radios from $U$ in a manner that forces a large geographic swath of a few or no users. Once the network configuration is specified, propagation path losses are computed for assumed terrain roughness and conductivity parameters. For a particular type of network configuration, i.e., either random or swath, the connectivity or average fraction of radios that receive a message is computed. This average

is obtained by estimating the expected value of the fraction of radios that received the message for a given type of network configuration over the ensemble of configurations with a fixed percentage of active users. From these data the number of simulcast hops required to achieve 100 percent connectivity (or some other criteria) is determined from a radio network simulation using a selected set of parameters from Table II. Reaction time for a given message can then be estimated from the number of hops, the message duration, and the average time for a user to access the network.

*B. Radio Simulation Description*

Bit and message error rate estimates derived though simulation are based on the functional model of the demodulator, channel decoder, and message decoder shown in Fig. 5. An ideal receiver is assumed with time and frequency synchronization which dehops the received signal to produce the $M$-ary baseband signal. The basic demodulator uses $M$ matched filters, each matched to one of the $M$-ary FSK tones at the hop rate. The output of each matched filter is envelope detected, sampled at the end of each frequency hop interval, and quantized into $Q$ bits. The output of an envelope detector is actually simulated by drawing a sample from the known distribution at that point. For example, if the input to the matched filter is a tone plus Gaussian noise, the output of its associated envelope detector has a Rician distribution.

The quantized output level is used as the metric in a soft decision Viterbi decoder and represents the likelihood that the associated $M$-ary alphabet symbol was transmitted during the frequency hop period. The Viterbi decoder works with a path history of ten $M$-ary symbols and the output symbols are converted to binary bits. For a 32-ary alphabet and a code rate of $R_C = 1/5$, there are five chips (symbols) transmitted each code frame, and the set of 160 quantized samples or metrics per frame are used to form the 1024 possible trellis branch metrics for the dual-5 code simulated.

To simplify the simulation model, the "all zeros" message is assumed to be transmitted and bit errors are determined by counting "ones" in the binary output stream from the Viterbi decoder. To eliminate any bias in the decoder when using this technique, a random choice is made among nodes which have equal metric values. Message errors are assumed to occur when the number of symbol errors in the codeword exceeds the correction capability of the message code.

*C. Network Simulation Description*

The performance of a simulcast radio network, in terms of message connectivity, depends on many factors: ground conductivity, propagation characteristics, terrain topography, distance between nodes, and distribution of nodes within the network area. The estimation of network performance, then, requires complex modeling of the regional topography and local environmental characteristics within the area as well as the evaluation of propagation losses and accurate determination of radio performance.

A simulcast network achieves high communications survivability by providing the maximum number of links possible to route messages. When a node transmits, every other node in the network that has not already received the message monitors the channel and waits for the transmission. In many cases during a transmission, the signal is too weak relative to noise or jamming and there will be too many errors for valid reception. The message propagation through the network, then, is highly probabilistic and dependent upon the particular network configuration and the location of the originating node.

It is difficult to predict communications connectivity performance from propagation considerations alone, or from radio receiver probability of error curves alone. A network simulation is required that models both terrain effects and receiver performance in order to obtain valid estimates of network performance. This section describes the basic elements of the network model: the propagation characteristics, terrain effects, noise and jamming characteristics, and program operation.

In the network model $L$ radios are distributed over $U$ nodes where $U$ is approximately 200, $L$ is 90 or less, and at most one radio is resident at a particular node. The $U$ nodes are geographically located in the valleys of a Nevada and Utah region that is typical of mountainous Western U.S. terrain. (See Fig. 7.) The size of the region is approximately 170 km by 390 km.

Communications in the radio network results from MF groundwave propagation including losses due to mountainous terrain. The received signal power $P_R$ from a single transmitter is given by

$$P_R = P_T + G_T - L_P \qquad (4)$$

where

$P_T$ = transmitter power at the antenna terminals including line losses in dBW

$G_T$ = transmitter antenna gain in dB referenced to a short lossless vertical radiator (SLVR)

$L_P$ = propagation path loss in dB between two SLVR's.

The sum $P_T + G_T$ is the effective radiated power. The propagation path loss can be modeled in two parts: 1) the ground wave propagation loss over a smooth homogeneous earth, and 2) a loss due to terrain roughness. For a smooth homogeneous earth

$$L_P = 32.44 + 20 \log f_{MHz} + 20 \log R + CR - 2K \qquad (5)$$

where

$R$ = transmitter to receiver distance in km

$C$ = constant dependent on soil conductivity (see Table III)

$f_{MHz}$ = frequency in MHz

$K$ = gain of a short vertical dipole = 1.76 dB.

This equation is consistent with CCIR [4] for the ranges of interest.

When a mountain is located between the transmitter and receiver, an additional loss is introduced that is approximately constant beyond the shadow zone of the mountain. This loss,

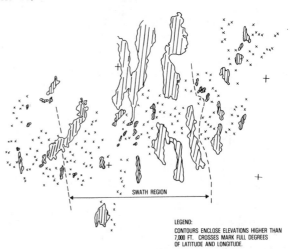

Fig. 7. User locations in geographic region.

LEGEND:
CONTOURS ENCLOSE ELEVATIONS HIGHER THAN 7,000 FT. CROSSES MARK FULL DEGREES OF LATITUDE AND LONGITUDE.

TABLE III
GROUND ABSORPTION AT 450 kHz

| Soil Conductivity in s/m | Path Constant, C $\frac{dB}{km}$ |
|---|---|
| .01 | .05 |
| .0065 | .08 |
| .003 | .14 |

referred to as mountain loss $L_m$, is a function of mountain height, mountain conductivity, valley conductivity, and the ratio of mountain height to mountain width. The computed mountain loss is the difference between the loss over rough terrain and the loss over a smooth homogeneous path. For rough terrain, path loss $L_P$ is computed from

$$L_P = 138.45 + 20 \log f_{MHz} - E \qquad (6)$$

where $E$ = field strength between two SLVR's on a perfectly conducting ground referenced to 1 kW and is computed from the Hufford integral wave solution [4], [5], [8]. A typical mountain loss is on the order of 2-3 dB for a single mountain 1 km in height. In the network simulation, smooth homogeneous earth propagation path loss is computed from (5) using the locations of the $L$ active radios selected from the $U$ possible locations. Estimating mountain losses between the $L$ radio sites would involve a vast computational effort for this network. Instead this computation is simplified by utilizing a mountain loss matrix which specifies mountain losses between the active radios. This matrix is obtained by an off-line computation utilizing (6) for a set of mountains with various heights, widths, and conductivities. This set is selected to characterize the topographic region shown in Fig. 7.

The received interference results from atmospheric noise and jamming. Atmospheric noise power $N$ is computed from

$$N = kTBF_a \tag{7}$$

where

$kT$ = the receiver thermal noise at room temperature

$$\left(-204 \frac{\text{dBW}}{\text{Hz}}\right)$$

$B$ = receiver bandwidth in Hz

$F_a$ = noise factor obtained from CCIR 322 [6].

For a 90 percent time availability and a 90 percent service probability under worst case summer noise conditions in the Southwestern U.S., the noise factor is approximately 104 dB and is the level of atmospheric noise assumed. The jamming level $J$ is computed from assumed values for jammer transmitter power, antenna gain, and location. To compute a bound on the worst case performance, the largest average interference power experienced by any radio in the network is assumed to be the average power received by every radio in the network.

To decide whether or not a specific receiver successfully receives a message in the simulation, several steps are needed. These include the following.

1) The received signal power is computed from each transmitting site that has previously received the message according to the propagation prediction formulas described above.

2) The received interference level is specified, and a signal-to-interference ratio (SIR) is formed. From the ratio of the bandwidth $B$ and data rate $R$, an $E_b/I_0$ is computed according to

$$E_b/I_0 = \text{SIR } B/R \tag{8}$$

where $E_b/I_0$ is the energy contrast ratio.

3) A random variable distributed from 0 to 1 is selected. This number is used to extract a test value of $E_b/I_0$ from a message error rate performance curve for the appropriate channel, i.e., fading or nonfading. If the $E_b/I_0$ received exceeds the test value of $E_b/I_0$, the message is declared to be successfully received.

From the above equations the $E_b/I_0$ available at the receiver from a single transmitter can be expressed as

$$E_b/I_0 = K_0 - 20 \log R - CR - L_m \tag{9}$$

where $K_0 = P_T + G_T - 20 \log f_{\text{MHz}} + 2K - I + 10 \log B/R$ and $I$ denotes the total interference power. Simulcast gain from multiple transmitters is determined within the network simulation. For simulation purposes all parameters except for range are fixed with nominal values.

A selection of $L$ radios from $U$ is made in either a random or swath configuration. A message originator is randomly selected to initiate the simulcast operation. The fraction of radios in the network that receive the message in each hop of

the simulcast is then computed in accordance with the above procedure for determining successful message reception. The simulation is then repeated for another selection of $L$ radios from $U$ in the same configuration and another random message originator. These operations are repeated and the results are averaged to produce the average conductivity performance for the network. The number of repetitions is selected to be sufficiently large to produce a specified level of confidence in the result.

## V. PERFORMANCE RESULTS

### A. Bit and Message Error Rates

In this section bit error rate (BER) and message error rate (MER) performance are presented for the waveforms indentified in Section III-B. BER is measured at the output of the channel decoder using a Monte Carlo simulation of receiver functions. For dual-$k$ channel encoding BER upper bounding techniques are used to check the simulation results. MER performance is estimated by simulation alone in order to include the effects of burst error patterns exhibited by the channel decoder. If independent errors occurred at the channel decoder output, theoretical MER performance estimates could be obtained. However, messages are assumed to be short and noncontinuous, rendering interleaving techniques as an ineffective means of dispersing error bursts. Before presenting the simulation results, the dual-$k$ BER upper bounding techniques will be described.

In the most general case, a dual-$k$ encoder with $v$ code symbols producing a $1/v$ rate code can be repeated $D$ times to introduce diversity. The bit error rate performance for dual-$k$ convolutional codes with soft decision decoding is then upper bounded using a union bound approach to obtain a result of the form

$$\text{BER} \leqslant \frac{2^{k-1}}{2^k - 1} \sum_{i=2v}^{\infty} b_i P_2(iD). \tag{10}$$

$P_2(iD)$ above is the appropriate expression for the probability of error when pairwise comparing the correct path and an incorrect path over a distance $i$ dual-$k$ branch symbols in the convolutional code trellis. In general, the $P_2(iD)$ expression is a function of the channel (fading or nonfading), interference type, rate of the convolutional code, and signal-to-interference ratio. The $b_i$'s are related to the derivative of the code transfer function. The tightness of the bound is a function of the number of terms used in the evaluation and the bit error rate values being considered. For the present case, the bounds can be expected to be fairly tight for bit error rates less than 0.01.

To present expressions for $P_2(iD)$ it is convenient to define a chip as the basic signaling element occurring over the transmission interval. In general the dual-$k$ encoder produces $Dv$ chips in the $M$-ary alphabet at the encoder output for every $k$ bits at the input. Defining bit and chip energies as $E_b$ and $E_c$, respectively, the chip energy contrast ratio is related to the bit energy contrast ratio by

$$\frac{E_c}{I_0} = \frac{k}{vD} \frac{E_b}{I_0}. \tag{11}$$

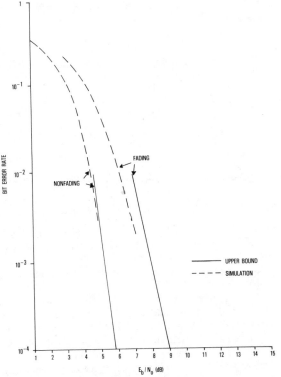

Fig. 8. Bit error rate performance for dual-5 rate 1/5 32-ary FSK waveform in WGN.

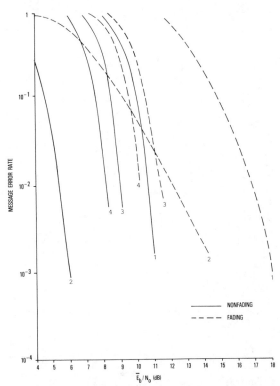

Fig. 9. Simulated message error rate performance for waveform candidates in WGN.

For a nonfading additive WGN channel [9] with $d = iD$

$$P_2(d) = \frac{1}{2^{2d-1}} \exp\left(\frac{-d}{2}\frac{E_c}{I_0}\right) \sum_{n=0}^{d-1}\left(\frac{(dE_c/I_0)^n}{2^n n!}\right)$$

$$\cdot \sum_{k=0}^{d-1-n}\binom{2d-1}{k}. \tag{12}$$

For a Rayleigh fading channel with additive WGN [2]

$$P_2(d) = p^d\left(\sum_{j=0}^{d-1}\binom{d-1+j}{j}(1-p)^j\right) \tag{13}$$

where

$$p = \frac{1}{2 + \bar{E}_c/I_0}$$

and $\bar{E}_c/I_0$ denotes the mean value of $E_c/I_0$. In the latter expression the fading is assumed to be constant for any chip duration and the frequency hopping produces fading that is independent from chip to chip. It should be noted that the expressions for $P_2(d)$ are exact. Performance for the waveforms identified in Section III-B will now be presented for nonfading and Rayleigh fading channels assuming either additive WGN or OFOBJ.

Fig. 8 compares simulated BER performance with the theoretical upper bounds for fading and nonfading WGN channels. The waveform is a dual-5 rate 1/5 32-ary FSK that is soft decision Viterbi decoded without frequency domain excision. The results indicate good agreement between simulated and theoretical performance at high energy contrast ratios.

MER performance for the four waveform candidates identified in Section III-B will now be presented for nonfading and Rayleigh fading channels. Frequency domain excision is applied in all cases. The two waveforms employing convolutional channel encoding utilize soft decision Viterbi decoding with eight-level quantization. MER performance with WGN and OFOBJ is depicted in Figs. 9 and 10, respectively.

Comparing the nonfading results in Fig. 9 it can be seen that the nonbinary waveforms exhibit improved performance over the binary waveform and that the convolutional channel code introduces a performance penalty in comparison with the nonbinary RS code used in waveform 2. This penalty can be attributed to noncoherent combining losses. Comparing the fading results in Fig. 9 two distinct features can be observed. First, binary FSK performance with the BCH (111, 48, 10) code alone is poor and second, performance for the nonbinary waveforms depends on the MER desired due to crossing of MER performance curves. For message error rates below $10^{-2}$ the concatenated coding schemes used in waveforms 3 and 4 yield improved performance over the RS (20, 10, 5) 32-ary FSK case as a result of the code diversity.

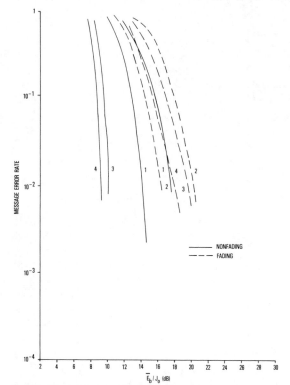

Fig. 10.  Simulated message error rate performance for waveform candidates in OFOBJ.

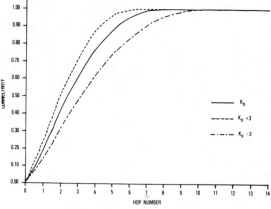

Fig. 11.  Connectivity performance.

Comparing the OFOBJ nonfading results in Fig. 10 it can be seen that the concatenated coding cases exhibit improved performance over the cases without channel encoding. Comparing waveforms 1 and 2 it can be seen that the BCH (111, 48, 10) binary FSK waveform is preferred in comparison with the RS (20, 10, 5) 32-ary FSK waveform. This result can be attributed to the increased probability of jamming 32 tones rather than two tones. Fading performance in the OFOBJ channel is very similar for all four waveforms. In this case the best performance is achieved by the BCH (111, 48, 10) binary FSK case and is closely followed by the RS (20, 10, 5) dual-5 rate 1/5 32-ary FSK case.

Examining Figs. 9 and 10 for WGN and OFOBJ channels in fading and nonfading cases, it can be concluded that no single waveform performs best in all cases. It can also be observed that the concatenated waveforms exhibit robust performance; these waveforms either offer the best performance or exhibit only a slight degradation in performance from the waveform that is best. In all channels the RS (20, 10, 5) dual-5 rate 1/5 32-ary FSK performs better than the BCH (111, 48, 10) dual-5 rate 1/5 32-ary FSK. This is a result of the fact that the nonbinary RS code is better suited for dealing with symbol errors at the Viterbi decoder output which tend to occur in bursts.

*B. Connectivity Performance*

Connectivity performance is presented in this section for the two network configurations corresponding to the two types of attack considered, i.e., random and swath. Fig. 7 depicts the locations of the 200 possible users from which a specific network configuration can be selected. For a swath configuration the users resident within the boundaries of the graph are substantially reduced.

Fig. 11 illustrates connectivity performance for BCH (111, 48, 10) dual-5 rate 1/5 32-ary FSK waveform with a random distribution in OFOBJ. The curve is generated from an average of 200 simulation runs. At each hop the connectivity is in fact a random variable with an estimated mean that is plotted on the curve. Invoking the central limit theorem, a confidence interval on the connectivity at a specific hop number can be obtained.

If the connectivity has an empirical mean and variance given by $c$ and $s^2$, respectively, a 95 percent confidence interval for connectivity is given by

$$c \pm \frac{1.97s}{\sqrt{M}}$$

where $M$ is the number of simulation runs. Simulation estimates of the empirical variance at a connectivity $c = 0.95$ yield $s = \sqrt{0.005}$ for $M = 200$. Thus, a 95 percent connectivity confidence interval at 95 percent mean connectivity is (94 percent, 96 percent).

Fig. 11 also illustrates connectivity results for $K_0 + 3$ dB and $K_0 - 3$ dB [see (9)]. A 3 dB variation in $K_0$ is equivalent to a 3 dB variation in any one (or a combination) of several performance parameters including transmit power, antenna gain, data rate, etc. The curves illustrate that a 3 dB increase in $K_0$ results in about a 1 hop improvement at the 95 percent connectivity level whereas a 3 dB decrease yields about a 2 hop degradation.

Figs. 12–15 illustrate connectivity performance for the four waveforms identified in Section III-B for various combinations of network dsitributions and channel conditions. Figs. 12 and 13 illustrate connectivity performance for a random network distribution with WGN jamming and OFOBJ, respectively; Figs. 14 and 15 show connectivity performance for a swath configuration with WGN jamming and OFOBJ, respectively. Comparison of Figs. 14 and 15 with Figs. 12

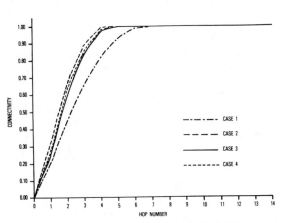

Fig. 12.   Random distribution in WGN jamming.

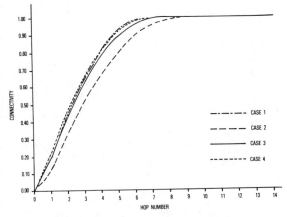

Fig. 13.   Random distribution in OFOBJ.

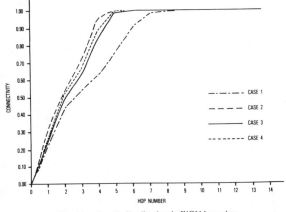

Fig. 14.   Swath distribution in WGN jamming.

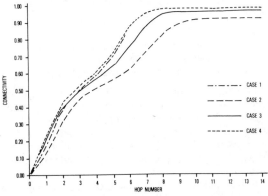

Fig. 15.   Swath distribution in OFOBJ.

and 13 reveals that the connectivity curves in the swath configuration exhibit reduced slope in the vicinity of the swath (at hop 4 or 5). Another general feature that can be observed is that only a low degree of connectivity is attained on the first hop and that any moderate level of connectivity requires several hops, typically 3 to 7 depending on channel conditions. Since on the second and subsequent hops multiple radios are participating in each simulcast transmission, fading channel conditions will control connectivity performance.

One unexpected result can be observed by examining Figs. 9 and 14. In Fig. 9 waveform 2 performs better than waveforms 3 and 4 for MER values higher than $10^{-1}$ and poorer than waveforms 3 or 4 for MER values lower than $10^{-2}$. However, in Fig. 14 connectivity performance for waveform 2 is better than the connectivity performance for waveforms 3 and 4. This indicates that unlike typical data communications networks, poor message error rate conditions are important in simulcast operation. At extended ranges near the boundary of a hop, where signal-to-noise ratio conditions are poor, there is still a high probability of correct message reception due to a large number of available receivers. In general, receiver message error rates vary widely within a hop depending upon the location of the receiver within the hop. As a result, it is difficult to predict which waveforms offer the best connectivity performance from knowledge of MER performance.

Another feature that can be observed from examination of Figs. 12–15 is that the robust MER performance of the concatenated codes (waveforms 3 and 4) under various channel conditions leads to robust connectivity performance for various network configurations. For example, waveform 2 performs poorer than the concatenated schemes for OFOBJ in random and swath network distributions. (See Figs. 13 and 15.) This is a direct result of MER performance in a fading channel (see Fig. 9). Similarly, the degraded connectivity performance of waveform 1 relative to the concatenated codes is directly attributable to the poor MER performance for waveform 1 in a fading channel (see Fig. 9).

### C. Reaction Time

The message reaction time $T_R$ can be computed from

$$T_R = N_{\max} N_s T_s + T_a \tag{14}$$

where

$T_s$ = slot durations in seconds
$N_s$ = integer number of slots to transmit a message
$T_a$ = average network access time
$N_{max}$ = maximum number of simulcast hops.

Connectivity performance for a specified connectivity criterion determines the maximum number of simulcast hops required to span the network. Network protocols and message formats determine the number of slots to transmit a message and the average network access time. The radio design and message processing delays determine the slot duration.

In the computation of reaction time, an interesting trade-off exists between simulcast connectivity and data rate. If the data rate is increased, reaction time to transmit a message through the network decreases. However, an excessively high data rate can reduce or sever connectivity. This optimization is critically dependent upon the radio network design as well as the region of operation and the threats against the network.

## VI. CONCLUSIONS

In this paper a spread-spectrum simulcast MF radio network applicable to the MX command, control, and communications system has been described. In this application connectivity and reaction time have been shown to be important criteria for measuring performance of a survivable radio network. To obtain estimates of these parameters, a sophisticated network simulation was developed that utilizes bit and message error rate performance as an input. Good radio network performance is attained by use of a robust waveform, such as a concatenated code design. This design provides acceptable performance under a variety of channel conditions and possesses sufficient diversity to operate efficiently with simulcast induced fading.

The design of a geographically constrained radio network that is survivable and operates satisfactorily in a hostile interference environment requires a successful union of a spread-spectrum waveform and a dynamically reconfigurable network. Selection of the frequency-hopped concatenated code waveform with a simulcast network accomplishes this objective. The connectivity results demonstrate that simulcast is an effective means of communicating among radios whose locations are unknown prior to transmission. As a result, the networks are dynamically reconfigured and radio connectivity is established under a variety of adverse conditions, e.g., radio destruction/malfunction, jamming, and atmospheric noise. In addition, the data indicate that reductions in several system parameters, e.g., transmitter power, antenna gain, interference power, etc., cause a graceful degradation in radio network connectivity performance.

## REFERENCES

[1] P. Drouilhet and S. Bernstein, "TATS—A bandspread modem-demodulation system for multiple access tactical satellite communications," in *IEEE EASCON '69 Rec.*, pp. 126–132.
[2] J. G. Proakis and I. Rahman, "Performance of concatenated dual-*k* codes on a Rayleigh fading channel with a bandwidth constraint," *IEEE Trans. Commun.*, vol. COM-27, pp. 801–806, May 1979.
[3] R. H. West and T. L. Taylor, "A simulcast radio relay system model," in *Proc. Int. Conf. Commun.*, June 1980, pp. 12.5.1–12.5.5.
[4] "Transmission loss in studies of radio systems," in *Proc. CCIR 10th Plenary Assembly*, vol. III, *Fixed and Mobile Services, Standard Frequencies and Time Signals, Monitoring of Emissions*. Geneva, Switzerland: Int. Telecommun. Union, 1963, Rep. 112.
[5] J. R. Johler and S. Horowitz, "Propagation of LORAN-C ground and ionospheric wave pulses," Off. Telecommun., Rep. 73-20, 1973.
[6] "World distribution and characteristics of atmospheric radio noise," in *Proc. CCIR 10th Plenary Assembly*. Geneva, Switzerland: Int. Telecommun. Union, 1963, Rep. 322.
[7] A. Viterbi and I. Jacobs, "Advances in coding and modulation for noncoherent channels affected by fading, partial band and multiple-access interference," in *Advances in Communication Systems*, vol. 4, A. J. Viterbi, Ed. New York: Academic, 1975.
[8] R. H. Ott, "A new method for prediction of groundwave attenuation over inhomogeneous and irregular terrain," Inst. Telecommun. Sci., Res. Rep. 7, Jan. 1971.
[9] J. Proakis, "On the probability of error for multichannel reception of binary signals," *IEEE Trans. Commun.*, vol. COM-16, pp. 68–71, Feb. 1968.

# Author Index

# Subject Index

# Editors' Biographies

**Charles E. Cook** (S'49–A'51–M'54–SM'56–F'72) received the S.B. degree from Harvard College, Cambridge, MA, in 1949, and the M.E.E. degree from the Polytechnic Institute of Brooklyn, Brooklyn, NY, in 1954.

After two years with Melpar, Inc., he joined the Sperry Corporation in 1951, where he was responsible for original research on the development and application of large time–bandwidth signal processing to high-power radar systems. Since 1971 he has been with the MITRE Corporation, Bedford, MA, where he is on the Division Staff of the Communications Division and is a principal investigator on programs concerning the vulnerability and survivability of command, control, and communications systems. He holds several basic patents on pulse compression and spread-spectrum radar and communications signal processing techniques, and is coauthor (with M. Bernfeld) of *Radar Signals—An Introduction to Theory and Application* (New York: Academic). He has been a Guest Lecturer at the University of Pennsylvania, the Polytechnic Institute of Brooklyn, and for the IEEE Boston Section Radar Series. He has authored or coauthored a number of journal papers on the design and application of large time–bandwidth and spread-spectrum signals, air traffic control beacon interference, and the antijam effectiveness of netted communications links. He was elected a Fellow of the IEEE in 1972 for contributions to signal processing theory and radar design. He is listed in *American Men and Women of Science* and *Who's Who in Engineering*, and is an Editor of the IEEE TRANSACTIONS ON COMMUNICATIONS for Communication Systems Disciplines.

Mr. Cook is a member of Sigma Xi.

**Fred W. Ellersick** (M'57–SM'62) was born in Jersey City, NJ, on May 12, 1933. He received the B.E.E. degree from Rensselaer Polytechnic Institute, Troy, NY, in 1954, the M.E.E. degree from Syracuse University, Syracuse, NY, in 1961, and the Ph.D. degree in electrical engineering from the University of Maryland, College Park, in 1967.

From 1954 to 1969 he was with the IBM Corporation, where he was engaged in a variety of projects—from exploratory research through product engineering—concerning communications/computer systems; most of his work at IBM was concerned with system engineering or planning for military or space systems. Since 1969 he has been with the MITRE Corporation, Bedford, MA. His work at MITRE has emphasized planning and system engineering for Air Force communications systems and other military command, control, and communications systems. He is presently on the Division Staff of the Communications Division.

Dr. Ellersick has published numerous technical papers and reports, and holds several patents for his work in space communications. He is a member of Tau Beta Pi, Sigma Xi, Eta Kappa Nu, and Phi Kappa Phi. Long active in the IEEE, he is currently the Editor-in-Chief of IEEE COMMUNICATIONS MAGAZINE and Vice Chairman of the Technical Committee on Communication Systems Disciplines; former IEEE posts he held include Chairman of the Washington Chapter of the Information Theory Group and an Editor for the IEEE TRANSACTIONS ON COMMUNICATIONS.

**Laurence B. Milstein** (S'66–M'68–SM'77) was born in Brooklyn, NY, on October 28, 1942. He received the B.E.E. degree from the City College of New York, New York, NY, in 1964 and the M.S. and Ph.D. degrees, both in electrical engineering, from the Polytechnic Institute of Brooklyn, Brooklyn, NY, in 1966 and 1968, respectively.

From 1968 to 1974 he was employed by the Space and Communications Group of Hughes Aircraft Company, and from 1974 to 1976 he was with the Department of Electrical and Systems Engineering, Rensselaer Polytechnic Institute, Troy, NY. Since 1976 he has been with the Department of Electrical Engineering and Computer Sciences, University of California, San Diego, where he is a Professor, working in the area of digital communication theory with special emphasis on spread-spectrum communication systems. He has been a Consultant to government and industry in the areas of radar and communications.

Dr. Milstein is an Associate Editor for Communication Theory for the IEEE TRANS-ACTIONS ON COMMUNICATIONS and an Associate Editor for IEEE COMMUNICATIONS MAGAZINE. He is a member of the Board of Governors of Com Soc, the Communication Theory Technical Committee of the IEEE Communications Society, Eta Kappa Nu, Tau Beta Pi, and Sigma Xi.

**Donald L. Schilling** (S'56–M'58–SM'69–F'75) was born in Brooklyn, NY, on June 11, 1935. He received the B.E.E. degree from the City College of New York, New York, NY, the M.S.E.E. degree from Columbia University, New York, and the Ph.D. degree in electrical engineering from the Polytechnic Institute of Brooklyn, Brooklyn, NY, in 1956, 1958, and 1962, respectively.

From 1956 to 1962 he was a Lecturer in the Department of Electrical Engineering, City College of New York, a Lecturer in the Department of Physics, Brooklyn College, Brooklyn, and a member of the Technical Staff of the Electronic Research Laboratories, Columbia University. In 1962 he was appointed Assistant Professor in the Department of Electrical Engineering, Polytechnic Institute of Brooklyn, and in 1966 he became an Associate Professor. He is currently Professor of Electrical Engineering in the Department of Electrical Engineering, City College of New York. He is coauthor of *Electronic Circuits: Discrete and Integrated, Principles of Communication Systems, Digital and Analog Systems, Circuits and Devices,* and *Digital Integrated Electronics.* He is also a Consultant to several companies in the communications and radar theory areas.

Dr. Schilling was Director of Publications and a member of the Board of Governors of Com Soc, and served as Editor of the IEEE TRANSACTIONS ON COMMUNICATIONS. He is immediate Past President of the IEEE Communications Society. He is a member of Sigma Xi and Eta Kappa Nu.